Further Pure Mathematics

BRIAN AND MARK GAULTER

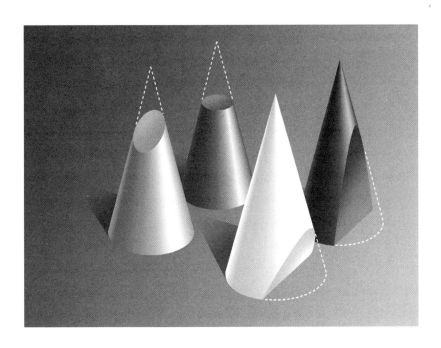

OXFORD
UNIVERSITY PRESS

OXFORD
UNIVERSITY PRESS

Great Clarendon Street, Oxford OX2 6DP

Oxford University Press is a department of the University of Oxford.
It furthers the University's objective of excellence in research, scholarship,
and education by publishing worldwide in

Oxford New York

Auckland Cape Town Dar es Salaam Hong Kong Karachi
Kuala Lumpur Madrid Melbourne Mexico City Nairobi
New Delhi Shanghai Taipei Toronto

With offices in

Argentina Austria Brazil Chile Czech Republic France Greece
Guatemala Hungary Italy Japan Poland Portugal Singapore
South Korea Switzerland Thailand Turkey Ukraine Vietnam

Oxford is a registered trade mark of Oxford University Press
in the UK and in certain other countries

British Library Cataloguing in Publication Data

Data available

ISBN: 978-0-19-914735-9

10 9

Typeset and illustrated by Tech-Set Ltd, Gateshead, Tyne and Wear
Printed in Thailand by Imago

Contents

Preface

Our aim has been to produce a single, comprehensive text which contains all the pure mathematics in A-level Further Mathematics, all the further part of A-level Pure Mathematics, and all the content of AS Further Pure Mathematics.

We have endeavoured to develop the subject in such a way as to be accessible to all students who are engaged with the syllabuses, both linear and modular, which came into force in September 2000. Indeed, we have even anticipated possible future change by including, in the chapter on groups, sections on dihedral groups and vector spaces.

It is anticipated that most students will use the book with the guidance of a teacher, but we have taken into account the needs of those students who are studying on their own. Thus we have designed the contents to be suitable for self-study and for revision. In each chapter, every section introducing a topic is followed by several detailed worked examples on that topic, which are typical of, and lead on to, the questions in the exercises which end each section. By following through the worked examples, students will both deepen their understanding of each topic and acquire the techniques needed to make significant progress with the exercises. The exercises, most of which contain a comprehensive selection of recent examination questions, will provide ample practice in applying these techniques and will help to develop those important, fundamental skills which

every student of pure mathematics at this level should have.

The order and style in which we have covered the topics are broadly in accord with our approach to teaching the subject. Naturally, this will not suit everyone, and so the text is generously cross-referenced to help those who want to proceed differently.

While the text is focused on the needs of A-level students, we expect that it will prove to be of great value to students in higher education who are following courses in pure mathematics, physics and the various engineerings and who therefore need an introductory text on the advanced mathematical techniques which underpin these subjects.

We are grateful to AQA, EDEXCEL, MEI, NICCEA, OCR, WJEC and SQA for permission to use their questions. The answers provided for these questions are the sole responsibility of the authors.

Thanks are also due to Rob Fielding and James Nicholson for checking the answers to the exercises and to the examination questions.

Finally, we wish to express our thanks to John Day for his painstaking work in editing the book.

Brian Gaulter
Mark Gaulter
October 2000

1 Complex numbers

In all our previous mathematics work, we have assumed that it is not possible to have a square root of a negative number. For example, on page 26 of *Introducing Pure Mathematics* where we considered the solution of quadratic equations, $ax^2 + bx + c = 0$, we noted that when $b^2 - 4ac$ is less than zero, the equation is said to have no real roots.

In fact, such an equation has **two complex roots**.

Take, for example, the solution of $x^2 + 2x + 3 = 0$. Using the quadratic formula, we obtain

$$x = \frac{-2 \pm \sqrt{4 - 12}}{2}$$

$$= \frac{-2 \pm \sqrt{-8}}{2}$$

$$= \frac{-2 \pm \sqrt{8}\sqrt{-1}}{2}$$

$$= \frac{-2 \pm 2\sqrt{2}\sqrt{-1}}{2}$$

$$= -1 \pm \sqrt{2}\sqrt{-1}$$

There is no real number which is $\sqrt{-1}$, as the square of any real number is always positive.

Therefore, we say that $\sqrt{-1}$ is an **imaginary number**. We denote $\sqrt{-1}$ by i.

So, using i, we can express the roots of the equation above in the form

$$-1 \pm \sqrt{2}i$$

or $\quad -1 - \sqrt{2}i \quad$ and $\quad -1 - \sqrt{2}i$

Note j is also used to represent $\sqrt{-1}$.

What is a complex number?

A **complex number** is a number of the form

$$a + ib$$

where a and b are real numbers and $i^2 = -1$.

For example, $3 + 5i$ is a complex number.

If $a = 0$, the number is said to be **wholly imaginary**. If $b = 0$, the number is **real**. If a complex number is 0, both a and b are 0.

We usually use $x + iy$ to represent an unknown complex number, and z to represent $x + iy$. So, when the unknown in an equation is a complex number, we denote it by z: for example, $z^2 - 40z + 40 = 0$, whose roots are $2 \pm 6i$.

In a similar way, we use w to represent a second unknown complex number, where $w = u + iv$.

The complex conjugate

The complex number $x - iy$ is called the **complex conjugate** (or often just the **conjugate**) of $x + iy$, and is denoted by z^* or $\bar{z}$.

For example, $2 - 3i$ is the complex conjugate of $2 + 3i$, and the complex conjugate of $-8 - 9i$ is $-8 + 9i$.

Calculating with complex numbers

When we work with complex numbers, we use ordinary algebraic methods. That means that we **cannot** combine a real number with an i-term. For example, $2 + 3i$ cannot be simplified.

For two complex numbers to be equal, **their real parts must be equal and their imaginary parts must be equal.**

This is a **necessary condition** for the equality of two complex numbers.

Hence, if $a + ib = c + id$, then $a = c$ and $b = d$.

For example, if $2 + 3i = x + iy$, then $x = 2$ and $y = 3$.

Addition and subtraction

When adding two complex numbers, we add the real terms and **separately** add the i-terms. For example,

$$(3 + 7i) + (4 - 6i) = (3 + 4) + (7i - 6i)$$
$$= 7 + i$$

Generally, for addition we have

$$(x + iy) + (u + iv) = (x + u) + i(y + v)$$

and for subtraction

$$(x + iy) - (u + iv) = (x - u) + i(y - v)$$

Example 1 Subtract $8 - 4i$ from $7 + 2i$.

SOLUTION

$$7 + 2i - (8 - 4i) = 7 - 8 + (2i + 4i)$$
$$= -1 + 6i$$

Example 2 Find x **and** y **if** $x + 2\mathrm{i} + 2(3 - 5\mathrm{i}y) = 8 - 13\mathrm{i}.$

SOLUTION

Equating real terms, we get

$$x + 6 = 8$$

$$\Rightarrow \quad x = 2$$

Equating imaginary terms, we get

$$2 - 10y = -13$$

$$\Rightarrow \quad 15 = 10y$$

$$\Rightarrow \quad y = 1\tfrac{1}{2}$$

Multiplication

We apply the general algebraic method for multiplication. For example,

$$(2 + 3\mathrm{i})(4 - 5\mathrm{i}) = 2(4 - 5\mathrm{i}) + 3\mathrm{i}(4 - 5\mathrm{i})$$

$$= 8 - 10\mathrm{i} + 12\mathrm{i} - 15\mathrm{i}^2$$

Since $\mathrm{i}^2 = -1$, this simplifies to

$$8 - 10\mathrm{i} + 12\mathrm{i} - 15 \times -1 = 8 - 10\mathrm{i} + 12\mathrm{i} + 15$$

$$= 23 + 2\mathrm{i}$$

Generally, we have

$$(a + \mathrm{i}b)(c + \mathrm{i}d) = ac - bd + \mathrm{i}(ad + bc) \quad \text{since } \mathrm{i}^2 = -1$$

Note It is simpler to multiply out the numbers every time than to memorise this formula.

Division

To be able to divide by a complex number, we have to change it to a real number. Take, for example, the fraction

$$\frac{2 + 3\mathrm{i}}{4 + 5\mathrm{i}}$$

In the simplification of surds on page 408 of *Introducing Pure Mathematics*, we noted that $\dfrac{1}{1 + \sqrt{3}}$ could be simplified by multiplying the numerator and the denominator of this fraction by $1 - \sqrt{3}$.

Similarly, to simplify $\dfrac{2 + 3\mathrm{i}}{4 + 5\mathrm{i}}$ we multiply its numerator and its denominator by

$4 - 5\mathrm{i}$, which is the **complex conjugate** of the denominator. Thus, we have

$$\frac{2 + 3\mathrm{i}}{4 + 5\mathrm{i}} = \frac{(2 + 3\mathrm{i})(4 - 5\mathrm{i})}{(4 + 5\mathrm{i})(4 - 5\mathrm{i})}$$

$$= \frac{8 + 12\mathrm{i} - 10\mathrm{i} - 15\mathrm{i}^2}{4^2 - (5\mathrm{i})^2}$$

$$= \frac{23 + 2i}{16 + 25} \quad \text{[Note: } -(5i)^2 = -(-25) = +25]$$

$$= \frac{23}{41} + \frac{2}{41}i$$

Example 3 Simplify $\dfrac{3 + i}{7 - 3i}$.

SOLUTION

Multiplying the numerator and the denominator by the complex conjugate of $7 - 3i$, which is $7 + 3i$, we obtain

$$\frac{3 + i}{7 - 3i} = \frac{(3 + i)(7 + 3i)}{(7 - 3i)(7 + 3i)}$$

$$= \frac{21 + 7i + 9i + 3i^2}{7^2 - (3i)^2}$$

$$= \frac{21 + 16i - 3}{49 + 9} \quad \text{[Note: } -(3i)^2 = -(-9) = +9]$$

$$= \frac{18}{58} + \frac{16i}{58}$$

$$= \frac{9}{29} + \frac{8}{29}i \quad \text{or} \quad \frac{1}{29}(9 + 8i)$$

Example 4 Simplify $\dfrac{(5 - 3i)(7 + i)}{2 - i}$.

SOLUTION

First, we simplify the numerator:

$$\frac{(5 - 3i)(7 + i)}{2 - i} = \frac{35 + 5i - 21i - 3i^2}{2 - i}$$

$$= \frac{35 - 16i + 3}{2 - i}$$

$$= \frac{38 - 16i}{2 - i}$$

We then multiply the numerator and the denominator of this fraction by the complex conjugate of $2 - i$, which is $2 + i$:

$$\frac{(38 - 16i)(2 + i)}{(2 - i)(2 + i)} = \frac{76 + 16 + 38i - 32i}{4 + 1}$$

$$= \frac{92 + 6i}{5} \quad \text{or} \quad 18\tfrac{2}{5} + 1\tfrac{1}{5}i$$

Exercise 1A

1 Simplify each of the following.

a) i^3 **b)** i^4 **c)** i^6 **d)** i^9

2 Express each of the following complex numbers in the form $a + ib$.

a) $3 + 2\sqrt{-1}$ **b)** $6 - 3\sqrt{-1}$ **c)** $-4 + \sqrt{-9}$

d) $-2 + \sqrt{-8}$ **e)** $\sqrt{-100} - \sqrt{-64}$

3 Write down the complex conjugate of z when z is:

a) $3 + 4i$ **b)** $2 - 6i$ **c)** $-4 - 3i$ **d)** $-8 + 5i$

4 Solve each of the following equations.

a) $z^2 + 2z + 4 = 0$ **b)** $z^2 - 3z + 6 = 0$ **c)** $2z^2 + z + 1 = 0$ **d)** $4z - 3 - 2z^2 = 0$

5 Simplify each of the following.

a) $(8 + 4i) + (2 - 6i)$ **b)** $(-7 + 3i) + (8 - 4i)$ **c)** $2 - 4i + 3(-1 + 2i)$

d) $4(-2 + 5i) + 5(2 + 7i)$ **e)** $(8 + 3i) - (7 + 2i)$ **f)** $(7 + 6i) - (4 - 2i)$

g) $2(9 - 3i) - 4(2 - 6i)$ **h)** $3(8 + i) - 2(3 - 5i)$

6 Evaluate each of these expressions.

a) $(3 + i)(2 + 3i)$ **b)** $(4 - 2i)(5 + 3i)$ **c)** $(8 - i)(9 + 2i)$

d) $(9 - 3i)(5 - i)$ **e)** $i(2 - 3i)(i + 4)$ **f)** $(3 - 2i)(7 - 5i)$

7 Express each of these fractions in the form $a + ib$, where $a, b \in \mathbb{R}$.

a) $\dfrac{2 + 3i}{4 - i}$ **b)** $\dfrac{4 + 3i}{5 + i}$ **c)** $\dfrac{8 - i}{2 + 3i}$ **d)** $\dfrac{2 + 5i}{-3 + 2i}$

8 Solve each of the following equations in x and y.

a) $x + iy = 4 - 2i$ **b)** $x + iy + 3 - 2i = 4(-2 + 5i)$

c) $x + iy = (2 + i)(3 - 2i)$ **d)** $x + iy = (3 - 5i)(4 + i)$

e) $x + iy = \dfrac{7 + i}{2 - i}$ **f)** $x + iy = (2 - 3i)^2$

9 If $z = 3 + i$, find the value of $z + \dfrac{1}{z}$.

10 Find the solution of each of the following equations.

a) $x^2 + 4x + 7 = 0$ **b)** $x^2 + 2x + 6 = 0$ **c)** $2x^2 + 6x + 9 = 0$ **d)** $x^2 - 5x + 25 = 0$

Argand diagram

The French mathematician Jean Robert Argand (1768–1822) is credited with the invention and development of the graphical representation of complex numbers and the operations upon them, although others had anticipated his work. So, this graphical representation has become known as the **Argand diagram**.

In the Argand diagram, the complex number $a + ib$ is represented by the point (a, b), as shown on the right.

Real numbers are represented on the x-axis and imaginary numbers on the y-axis. Thus, the general complex number $(x + iy)$ is represented by the point (x, y).

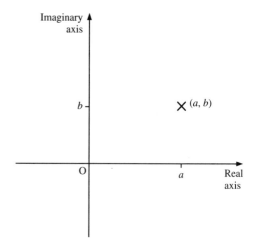

Example 5 Represent the complex number $2 + 3i$ on an Argand diagram. Show its complex conjugate.

SOLUTION

The number $2 + 3i$ is represented by the point $P(2, 3)$.

The complex conjugate is $2 - 3i$, which is represented by the point $P'(2, -3)$.

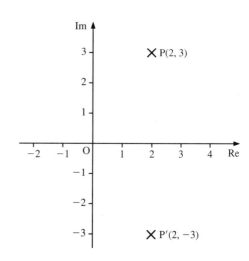

Note The position of the complex conjugate z^* can always be obtained by reflecting the position of z in the real axis.

Modulus–argument or polar form of complex numbers

The position of point $P(x, y)$ on the Argand diagram can be given in terms of OP, the distance of P from the origin, and θ, the angle in the **anticlockwise** sense which OP makes with the positive real axis.

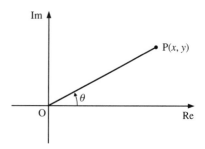

The length OP is the **modulus** of z, denoted by $|z|$, and this length $|z|$ is **always** taken to be **positive**.

The angle θ (normally in radians) is the **argument** of z, denoted by $\arg z$. The **principal value** of θ is taken to be between $-\pi$ and π.

Connection between the $x + iy$ form and the modulus–argument form

From the diagram on the right, we have

$$r = |z| = \sqrt{x^2 + y^2}$$

$$x = r\cos\theta \quad \text{and} \quad y = r\sin\theta$$

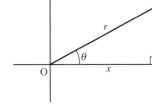

which give

$$z \equiv x + iy = r\cos\theta + ir\sin\theta$$

$$= r(\cos\theta + i\sin\theta)$$

To find θ, we use

$$\tan\theta = \frac{y}{x}$$

but we need to take care when either x or y is **negative**. (See part **b** in Example 6.)

Example 6 Find the modulus and argument of each of these complex numbers.

a) $2 + 2\sqrt{3}i$ **b)** $-1 - i$

SOLUTION

a)

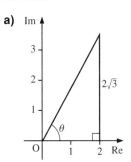

The modulus of $2 + 2\sqrt{3}i$ is given by

$$\sqrt{2^2 + (2\sqrt{3})^2} = 4$$

Its argument, θ, is given by

$$\tan^{-1}\sqrt{3} = \frac{\pi}{3}$$

b)

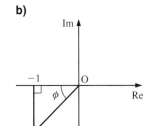

The modulus of $-1 - i$ is given by

$$\sqrt{1^2 + 1^2} = \sqrt{2}$$

Angle ϕ is $\frac{\pi}{4}$. Therefore, the argument (the angle from the positive real axis) is

$$-\frac{\pi}{2} - \frac{\pi}{4} = -\frac{3\pi}{4}$$

Note If the angle in Example 6 is measured **anticlockwise** from the positive real axis, its value is $\frac{5\pi}{4}$, but this is not between π and $-\pi$. Thus, we take the clockwise angle, which is $-\frac{3\pi}{4}$. The minus sign denotes that the angle is measured in the clockwise sense.

Multiplication of two complex numbers in modulus–argument form

Consider the complex numbers z_1, and z_2 given by

$$z_1 \equiv r_1(\cos\theta_1 + i\sin\theta_1) \quad \text{and} \quad z_2 \equiv r_2(\cos\theta_2 + i\sin\theta_2)$$

Multiplying z_1 by z_2, we get

$$z_1 z_2 = r_1(\cos\theta_1 + i\sin\theta_1)\, r_2(\cos\theta_2 + i\sin\theta_2)$$

$$= r_1 r_2[(\cos\theta_1\cos\theta_2 - \sin\theta_1\sin\theta_2) + i(\sin\theta_1\cos\theta_2 + \cos\theta_1\sin\theta_2)]$$

$$= r_1 r_2[\cos(\theta_1 + \theta_2) + i\sin(\theta_1 + \theta_2)]$$

We can state this result as follows:

To find the product of two complex numbers, **multiply their moduli** and **add their arguments**.

Division of two complex numbers in modulus–argument form

Dividing z_1 by z_2, we get

$$\frac{z_1}{z_2} = \frac{r_1(\cos\theta_1 + i\sin\theta_1)}{r_2(\cos\theta_2 + i\sin\theta_2)} = \frac{r_1}{r_2}\frac{\cos\theta_1 + i\sin\theta_1}{\cos\theta_2 + i\sin\theta_2}$$

Multiplying the numerator and the denominator by the complex conjugate of $\cos\theta_2 + i\sin\theta_2$, we have

$$\frac{z_1}{z_2} = \frac{r_1}{r_2}\frac{(\cos\theta_1 + i\sin\theta_1)(\cos\theta_2 - i\sin\theta_2)}{(\cos\theta_2 + i\sin\theta_2)(\cos\theta_2 - i\sin\theta_2)}$$

$$= \frac{r_1}{r_2}\frac{\cos\theta_1\cos\theta_2 + \sin\theta_1\sin\theta_2 + i(\sin\theta_1\cos\theta_2 - \cos\theta_1\sin\theta_2)}{(\cos^2\theta_2 + \sin^2\theta_2)}$$

$$= \frac{r_1}{r_2}[\cos(\theta_1 - \theta_2) + i\sin(\theta_1 - \theta_2)] \quad \text{since } \cos^2\theta_2 + \sin^2\theta_2 \equiv 1$$

We can state this result as follows:

To find the quotient of two complex numbers, **divide their moduli** and **subtract their arguments**.

Example 7 Find the modulus and argument of each of the following.

a) $z = 1 + i$ **b)** $w = -1 + \sqrt{3}i$ **c)** zw **d)** z^2 **e)** $\dfrac{w}{z}$

SOLUTION

a) From the diagram, we have

$$\text{Modulus of } z = \sqrt{2}$$

$$\text{Argument of } z = \frac{\pi}{4}$$

b) Modulus of $w = \sqrt{1^2 + (\sqrt{3})^2} = 2$

$$\text{Argument of } w = \pi - \frac{\pi}{3} = \frac{2\pi}{3}$$

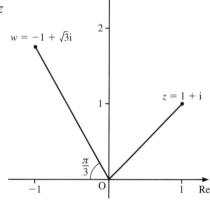

c) Modulus of $zw = |z| \times |w| = 2\sqrt{2}$

Argument of zw is

$$\arg z + \arg w = \frac{\pi}{4} + \frac{2\pi}{3} = \frac{11\pi}{12}$$

d) Using $z^2 = z \times z$, we have

Modulus of $z^2 = |z| \times |z| = \sqrt{2} \times \sqrt{2} = 2$

Argument of z^2 is

$$\arg z + \arg z = \frac{\pi}{4} + \frac{\pi}{4} = \frac{\pi}{2}$$

e) Modulus of $\dfrac{w}{z} = \dfrac{|w|}{|z|} = \dfrac{2}{\sqrt{2}} = \sqrt{2}$

Argument of $\dfrac{w}{z}$ is

$$\arg w - \arg z = \frac{2\pi}{3} - \frac{\pi}{4} = \frac{5\pi}{12}$$

Exercise 1B

1 Represent each of the following on an Argand diagram.

a) $2 + 2i$ **b)** $-3 + 3i$ **c)** $-2 + 2\sqrt{3}i$

d) $-1 - i$ **e)** $4i$ **f)** $5 + 12i$

g) -4 **h)** $6 + \sqrt{13}i$

2 Find the modulus and argument of each of the complex numbers in Question **1**.

3 Given that $z = 3 + 4i$,

a) calculate   **i)** z^2 **ii)** z^3

b) find   **i)** $|z|$ **ii)** $|z^2|$ **iii)** $|z^3|$

c) evaluate   **i)** $\arg z$ **ii)** $\arg z^2$ **iii)** $\arg z^3$

4 Express the complex number z in its $a + ib$ form when:

a) $|z| = 2$ and $\arg z = \dfrac{\pi}{3}$ **b)** $|z| = 4$ and $\arg z = \dfrac{\pi}{4}$ **c)** $|z| = 1$ and $\arg z = -\dfrac{\pi}{2}$

d) $|z| = 4$ and $\arg z = \dfrac{3\pi}{4}$ **e)** $|z| = 2$ and $\arg z = \dfrac{5\pi}{6}$ **f)** $|z| = 6$ and $\arg z = \dfrac{7\pi}{6}$

5 a) Simplify $\dfrac{1 - i}{-3 - i}$.

b) Find the modulus and argument of the complex number $-5 + 12i$ (WJEC)

6 Given that $z = \dfrac{3 + 4i}{5 - 12i}$, find the modulus and argument of z.   (WJEC)

7 Given that $z = \dfrac{1+i}{1-2i}$, find

    a) z in the form $a + ib$

    b) the modulus and argument of z.   (WJEC)

8 **i)** Given that $z_1 = 5 + i$ and $z_2 = -2 + 3i$,

    a) show that $|z_1|^2 = 2|z_2|^2$

    b) find $\arg(z_1 z_2)$.

    ii) Calculate, in the form $a + ib$, where $a, b \in \mathbb{R}$, the square roots of $16 - 30i$.   (EDEXCEL)

9 Given that

$$z = \tan\alpha + i, \text{ where } 0 < \alpha < \tfrac{1}{2}\pi$$
$$w = 4[\cos(\tfrac{1}{10}\pi) + i\sin(\tfrac{1}{10}\pi)]$$

  find in their simplest forms

    i) $|z|$ **ii)** $|zw|$ **iii)** $\arg z$ **iv)** $\arg\left(\dfrac{z}{w}\right)$ (OCR)

10 The complex number z is given by $z = \sin^2\alpha + i\sin\alpha\cos\alpha$, where $0 < \alpha < \tfrac{1}{2}\pi$. Simplifying your answers as far as possible, find

    i) $|z|$ **ii)** $\arg z$ (OCR)

11 The complex numbers z and w are such that

$$z = -2 + 5i \qquad zw = 14 + 23i$$

    a) Find w in the form $p + qi$, where p and q are real.

    b) Display z and w on the same Argand diagram.

    c) Find $\arg z$, in radians, giving your answer to two decimal places.

    d) Write down the complex number that represents the mid-point M of the line joining the points z and zw.   (EDEXCEL)

12 **a)** Find the roots of the equation $z^2 + 4z + 7 = 0$, giving your answers in the form $p \pm i\sqrt{q}$, where p and q are integers.

    b) Show these roots on an Argand diagram.

    c) Find for each root

       i) the modulus

       ii) the argument, in radians

    giving your answers to three significant figures.   (EDEXCEL)

13 By putting $z = z + iy$, find the complex number z which satisfies the equation

$$z + 2z^* = \frac{15}{2-i}$$

  where z^* denotes the complex conjugate of z.   (NEAB)

14 Given that $z_1 = 1 + 2i$ and $z_2 = \tfrac{3}{5} + \tfrac{4}{5}i$, write $z_1 z_2$ and $\dfrac{z_1}{z_2}$ in the form $p + iq$, where p and $q \in \mathbb{R}$.

  In an Argand diagram, the origin O and the points representing $z_1 z_2$, $\dfrac{z_1}{z_2}$, z_3 are the vertices of a rhombus. Find z_3 and sketch the rhombus on this Argand diagram.

  Show that $|z_3| = \dfrac{6\sqrt{5}}{5}$.   (EDEXCEL)

15 The complex numbers z_1 and z_2 are given by

$$z_1 = 5 + i \qquad z_2 = 2 - 3i$$

a) Show the points representing z_1 and z_2 on an Argand diagram.

b) Find the modulus of $z_1 - z_2$.

c) Find the complex number $\dfrac{z_1}{z_2}$ in the form $a + ib$, where a and b are rational numbers.

d) Hence find the argument of $\dfrac{z_1}{z_2}$, giving your answer in radians to three significant figures.

e) Determine the values of the real constants p and q such that

$$\frac{p + iq + 3z_1}{p - iq + 3z_2} = 2i \qquad \text{(EDEXCEL)}$$

16 $\qquad z_1 = -3 + 4i \qquad z_2 = 1 + 2i$

a) Express $z_1 z_2$ and $\dfrac{z_1}{z_2}$ each in the form $a + ib$ where $a, b \in \mathbb{R}$.

b) Display z_1 and z_2 on the same Argand diagram.

c) Find $\arg z_1$, giving your answer in radians to one decimal place.

Given that $z_1 + (p + iq)z_2 = 0$, where $p, q \in \mathbb{R}$,

d) obtain the value of p and the value of q. (EDEXCEL)

17 The complex number z is given by $z = -2 + 2i$.

a) Find the modulus and argument of z.

b) Write down the modulus and argument of $\dfrac{1}{z}$.

c) Show on an Argand diagram the points A, B and C representing the complex numbers z, $\dfrac{1}{z}$ and $z + \dfrac{1}{z}$ respectively.

d) State the value of $\angle ACB$. (EDEXCEL)

18 $\qquad z_1 = -30 + 15i$

a) Find $\arg z_1$, giving your answer in radians to two decimal places.

The complex numbers z_2 and z_3 are given by $z_2 = -3 + pi$ and $z_3 = q + 3i$, where p and q are real constants and $p > q$.

b) Given that $z_2 z_3 = z_1$, find the value of p and the value of q.

c) Using your values of p and q, plot the points corresponding to z_1, z_2 and z_3 on an Argand diagram.

d) Verify that $2z_2 + z_3 - z_1$ is real and find its value. (EDEXCEL)

19 i) Evaluate the square roots of the complex number $5 + 12i$ in the form $a + bi$, where a and b are real.

ii) If θ is the argument of either of these square roots, obtain the value of $\cos 4\theta$ as an **exact** fraction.   (NICCEA)

20 a) The complex numbers z and w are such that $z = (4 + 2i)(3 - i)$ and $w = \dfrac{4 + 2i}{3 - i}$. Express each of z and w in the form $a + ib$, where a and b are real.

b) i) Write down the modulus and argument of each of the complex numbers $4 + 2i$ and $3 - i$. Give each modulus in an exact surd form and each argument in radians between $-\pi$ and π.

ii) The points O, P and Q in the complex plane represent the complex numbers $0 + 0i$, $4 + 2i$ and $3 - i$ respectively. Find the exact length of PQ and hence, or otherwise, show that triangle OPQ is right-angled.   (AEB 97)

Loci in the complex plane

We know from our previous work on vector geometry that the vector $\mathbf{a} - \mathbf{b}$ connects the point with position vector $\mathbf{b}$ to the point with position vector $\mathbf{a}$. (See *Introducing Pure Mathematics*, page 498.) Similarly, in the complex plane, $z - z_1$ joins the point z_1 to the point z.

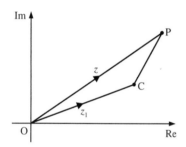

From the diagram, we have
$$\overrightarrow{OC} = z_1 \quad \text{and} \quad \overrightarrow{OP} = z$$
Therefore, we obtain
$$\overrightarrow{CP} = \overrightarrow{CO} + \overrightarrow{OP}$$
$$= -z_1 + z$$
$$= z - z_1$$

Using this fact, we can identify a number of loci.

Loci which should be recognised

- $|z - z_1| = r$

$|z - z_1|$ is the modulus or length of $z - z_1$. That is, the length of the line joining z_1 to a variable point z.

Thus, $|z - z_1| = r$ is the locus of a point, z, moving so that the length of the line joining a fixed point z_1 to z is always r. Hence, the locus of z is a circle, centre z_1 and radius r.

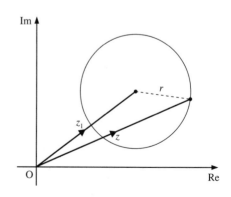

Example 8 State and sketch the locus of $|z - 2 - 3i| = 3$.

SOLUTION

This locus is $|z - (2 + 3i)| = 3$, which is a circle, centre $(2, 3)$ and radius 3.

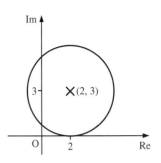

Note When sketching this locus, show clearly that the circle **touches** the x-axis and **cuts** the y-axis twice.

- **$\arg(z - z_1) = \theta$**

The point z satisfies this locus when the line joining z_1 to z has argument θ.

This is the **half-line**, starting at z_1, inclined at θ to the real axis. (It is called a half-line because we want only that part of the line which starts at z_1.)

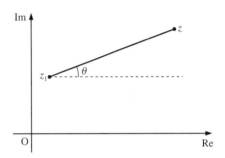

Example 9 State and sketch the locus of $\arg(z - 2) = \dfrac{\pi}{3}$.

SOLUTION

This locus is the half-line starting at $(2, 0)$, inclined at an angle of $\dfrac{\pi}{3}$ to the real axis.

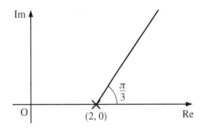

- **$|z - z_1| = |z - z_2|$**

The line joining z to z_1 is equal in length to the line joining z to z_2. Therefore, the locus of z is the perpendicular bisector of the line joining z_1 to z_2.

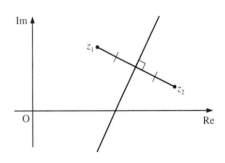

Example 10 State the locus of $|z - 3| = |z - 2i|$.

SOLUTION

This locus is the perpendicular bisector of the line joining $+3$ to $+2i$.

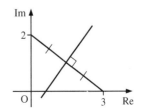

- $|z - z_1| = k|z - z_2|$, where $k \neq 1$

The locus of $P(z)$ is drawn so that the length of the line joining P to z_1 is k times the length of the line joining P to z_2.

Assuming $z \equiv x + iy$, $z_1 \equiv x_1 + iy_1$ and $z_2 \equiv x_2 + iy_2$, Pythagoras' theorem gives

$$|z - z_1| = \sqrt{(x - x_1)^2 + (y - y_1)^2}$$

and $\quad |z - z_2| = \sqrt{(x - x_2)^2 + (y - y_2)^2}$

Therefore, $|z - z_1| = k|z - z_2|$ can be expressed as

$$\sqrt{(x - x_1)^2 + (y - y_1)^2} = k\sqrt{(x - x_2)^2 + (y - y_2)^2}$$

Squaring both sides, we get

$$(x - x_1)^2 + (y - y_1)^2 = k^2[(x - x_2)^2 + (y - y_2)^2]$$
$$\Rightarrow \quad x^2 - 2xx_1 + x_1^2 + y^2 - 2yy_1 + y_1^2 = k^2x^2 - 2k^2xx_2 + k^2x_2^2 + k^2y^2 - 2k^2yy_2 + k^2y_2^2$$
$$\Rightarrow \quad (1 - k^2)x^2 + (1 - k^2)y^2 - x(2x_1 - 2k^2x_2) - y(2y_1 - 2k^2y_2) + x_1^2 + y_1^2 - k^2x_2^2 - k^2y_2^2 = 0$$

In this equation, the coefficients of x and y are the same, and there is no term in xy. Therefore, the locus of z is a circle.

By symmetry, a diameter of this circle lies on the line joining z_1 to z_2.

Note We recall from earlier work (*Introducing Pure Mathematics*, page 220) that the equation of a circle, centre (a, b) and radius r, is

$$(x - a)^2 + (y - b)^2 = r^2$$

This equation may also be written as

$$x^2 + y^2 + 2gx + 2fy + c = 0$$

To find the centre and the radius of a circle when its equation is written in this form, we use the method of completing the square:

$$x^2 + y^2 + 2gx + 2fy + c = 0$$
$$(x + g)^2 + (y + f)^2 = g^2 + f^2 - c$$

Therefore, the centre of the circle is $(-g, -f)$, and its radius is $\sqrt{g^2 + f^2 - c}$.

Example 11 Find the locus of $|z - 2| = 3|z + 2|$.

SOLUTION

Let A be $(-2, 0)$ and B be $(2, 0)$.

The locus required is the locus of P when $BP = 3AP$.

To find this circle, we determine the two points at which it intersects the line joining A to B.

The point $(-1, 0)$ satisfies this condition.

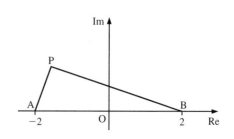

The other point on the line AB which satisfies this condition is never between A and B, but on the line AB produced.

The point $(-4, 0)$ is the other point which satisfies the locus.

The points $(-1, 0)$ and $(-4, 0)$ identify the diameter of the locus's circle. Therefore, the circle has centre $(-2\frac{1}{2}, 0)$ and radius $1\frac{1}{2}$.

Its equation is $|z + 2\frac{1}{2}| = \frac{3}{2}$.

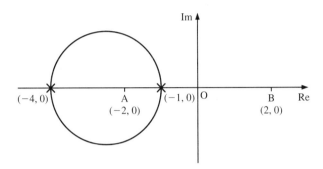

Example 12 Find the locus of $|z - 18| = 2|z + 18i|$.

SOLUTION

To find the circle, we determine the two points at which it intersects the line joining z_1 to z_2, where $z_1 = 18$ and $z_2 = -18i$.

The two points satisfying the locus are $6 - 12i$ and $-18 - 36i$.

These two points identify the diameter of the locus's circle. Therefore, the circle has its centre at $-6 - 24i$ and has a radius of $12\sqrt{2}$.

Hence, its equation is $|z + 6 + 24i| = 12\sqrt{2}$.

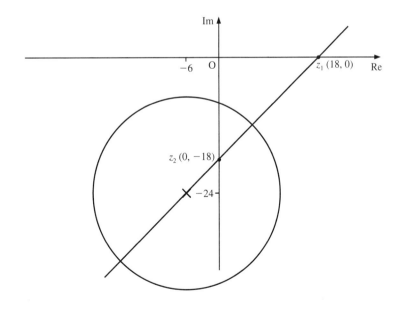

- $\text{arg} \dfrac{(z - z_1)}{(z - z_2)} = \theta$

To find this locus, we use the relationship

$$\text{arg} \frac{u}{v} = \text{arg}\, u - \text{arg}\, v$$

Putting $u = z - z_1$ and $v = z - z_2$, we get

$$\text{arg} \frac{z - z_1}{z - z_2} = \text{arg}\,(z - z_1) - \text{arg}\,(z - z_2)$$

$$\Rightarrow \quad \text{arg}\,(z - z_1) - \text{arg}\,(z - z_2) = \theta$$

Angles in the same segment are equal. Therefore, the locus of z is part of the circle through z_1 and z_2 (shown dashed).

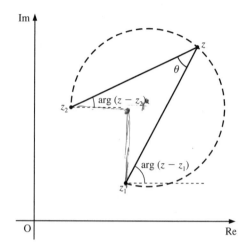

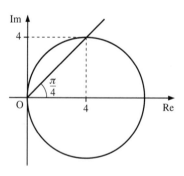

Example 13 Show the locus of z when

a) $|z - 4| = 4$ **b)** $\text{arg}\, z = \dfrac{\pi}{4}$

Find the point which satisfies both loci.

SOLUTION

The two loci required are shown in the diagram on the right.

The point which satisfies both loci is $(4, 4)$ or $(4 + 4i)$.

Note   Usually, it is possible to find a common point on two separate loci by using simple geometry and common sense. In Example 12, the point $(4, 4)$ can readily be seen to be on both loci. To calculate a common point may involve complicated algebra.

Example 14 Find the locus of $\dfrac{\pi}{4} < \arg(z - 2) < \dfrac{\pi}{3}$.

SOLUTION

We draw the two separate loci

$$\frac{\pi}{4} = \arg(z - 2) \quad \text{and} \quad \arg(z - 2) = \frac{\pi}{3}$$

ensuring that we select the correct sector.

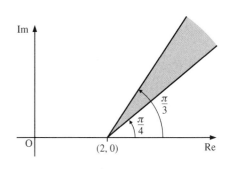

- $|z - z_1| + |z - z_2| = c$

This locus is an ellipse, with z_1 and z_2 as foci (see section on ellipses, pages 222–6). To find the position of the ellipse, we have to find four points which satisfy the locus:

- two points on the line joining z_1 to z_2 produced, and
- two points on the perpendicular bisector of the line joining z_1 to z_2.

Example 15 Find the locus of z when $|z - 4| + |z + 2| = 10$.

SOLUTION

First, we identify on the diagram the points A and B representing z_1 and z_2. These are $(4, 0)$ and $(-2, 0)$.

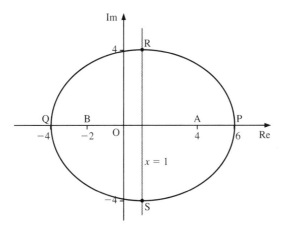

We then extend AB in both directions, where AB is of length 6.

Therefore, the points satisfying the locus are P(6, 0) and Q(−4, 0), so that PA = 2 and PB = 8, which gives PA + PB = 10.

Also, we have QA = 8 and QB = 2, which gives QA + QB = 10.

The perpendicular bisector of PQ is the line $x = 1$.

The points satisfying the locus on this line are R(1, 4) and S(1, −4), so that RA = 5, RB = 5 and hence RA + RB = 10.

These four points, P, Q, R and S, identify the major and minor axes of the ellipse.

Cube roots of unity

If z is a cube root of 1, we have

$$z^3 = 1$$
$$\Rightarrow \quad z^3 - 1 = 0$$
$$\Rightarrow \quad (z - 1)(z^2 + z + 1) = 0$$

Therefore, either: $z = 1$, which is the real root, or

$$z^2 + z + 1 = 0$$

If w is a **complex** cube root of 1, $w \neq 1$ and satisfies the equation $z^2 + z + 1 = 0$. Hence, we have

$$w^2 + w + 1 = 0$$
$$\Rightarrow \quad w = \frac{-1 \pm \sqrt{1 - 4}}{2}$$
$$= -\frac{1}{2} \pm \frac{\sqrt{3}}{2}i$$

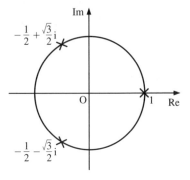

If we plot these three roots of 1 on an Argand diagram, we find them to be symmetrically positioned on the circumference of a circle of radius 1, as shown in the diagram on the right.

Square of a complex cube root of unity

If w is a complex cube root of 1, w^2 is also a complex cube root of 1.

Proof

If w is a complex cube root of 1, then $w^3 = 1$. Therefore, we have

$$(w^2)^3 = w^6 = (w^3)^2 = 1$$

That is, w^2 is also a complex cube root of 1.

Note   We found earlier that $w = -\dfrac{1}{2} \pm \dfrac{\sqrt{3}}{2}i$. Hence, we have

$$w^2 = \left(-\frac{1}{2} + \frac{\sqrt{3}}{2}i\right)^2 \quad \text{or} \quad -\frac{1}{2} - \frac{\sqrt{3}}{2}i$$

Or we have

$$w^2 = \left(-\frac{1}{2} - \frac{\sqrt{3}}{2}i\right)^2 \quad \text{or} \quad -\frac{1}{2} + \frac{\sqrt{3}}{2}i$$

Thus, we obtain

$$1 + w + w^2 = \left[1 + \left(-\frac{1}{2} + \frac{\sqrt{3}}{2}i\right) + \left(-\frac{1}{2} - \frac{\sqrt{3}}{2}i\right)\right] = 0$$

which agrees with the equation found above.

Example 16 If w is the complex root of 1, find the value of $w^4 + w^8$.

SOLUTION

$$w^4 + w^8 = w \times w^3 + w^2(w^3)^2$$

Since $w^3 = 1$, we get

$$w^4 + w^8 = w + w^2$$

Since $1 + w + w^2 = 0$, we find

$$w^4 + w^8 = -1$$

Example 17 If p is a cube root of 1, find the possible values of $p^2 + p^4$.

SOLUTION

$$p^2 + p^4 = p^2 + p \times p^3$$
$$= p^2 + p \quad \text{since } p^3 = 1$$

If p is real, $p = 1$, and thus $p^2 + p = 2$.

If p is a **complex** cube root, we have

$$p^2 + p = -1$$

Therefore, the possible values of $p^2 + p^4$ are 2 and -1.

Exercise 1C

1 Sketch the locus of z when:

a) $|z| = 5$ b $|z| = 3$ c) $|z - 2| = 3$ d) $|z - 2i| = 4$

e) $|z + 2 + 2i| = 2\sqrt{2}$ f) $|z + 3 - \sqrt{3}i| = 2\sqrt{3}$ g) $2|z - i| = 3$

2 Sketch the locus of z when:

a) $\arg z = \dfrac{\pi}{3}$ b) $\arg z = -\dfrac{3\pi}{4}$ c) $\arg(z + 2) = \dfrac{\pi}{2}$

d) $\arg(z - 3i) = \dfrac{\pi}{3}$ e) $\arg(z + 1 + i) = \dfrac{\pi}{4}$ f) $\arg(z - 2 - \sqrt{3}i) = -\dfrac{2\pi}{3}$

3 Sketch the locus of z when:

a) $|z - 2| = |z - 4|$ b) $|z - 6| = |z + 3|$ c) $|z - i| = |z - 2i|$

d) $|z + 2i| = |z - 2|$ e) $\left|\dfrac{z - 1 - i}{z + 2 + 2i}\right| = 1$ f) $\left|\dfrac{z - 4i}{z + 4}\right| = 1$

4 Sketch the locus of z when:

a) $|z - 1| = 3|z + 2|$ b) $|z + i| = 2|z - 2i|$ c) $|z - i| = 4|z + 3i|$

d) $|z - 2 - i| = 3|z + 6 + 3i|$ e) $\left|\dfrac{z - 2}{z + 2i}\right| = 3$

5 Sketch each of the following.

a) $\arg\left(\dfrac{z}{z-2}\right) = \dfrac{\pi}{4}$

b) $\arg\left(\dfrac{z-1}{z-3}\right) = \dfrac{\pi}{3}$

c) $\arg\left(\dfrac{z+2i}{z-2i}\right) = \dfrac{\pi}{4}$

d) $\arg\left(\dfrac{z}{z+4i}\right) = \dfrac{\pi}{6}$

6 If w is a complex root of 1, simplify each of these.

a) $w^4 + w^8$

b) $w^9 + w^{18}$

c) $w^3 + w^7 + w^{11}$

7 If w is a cube root of 1, find the possible values of each of the following.

a) $1 + w^4 + w^8$

b) $w^3 + w^6$

c) $\dfrac{w + w^4}{w^2 + w^5}$

d) $w^8 + w^{10}$

8 Find the solutions of $(z-2)^3 = 1$.

9 With the aid of a sketch, explain why there is no complex number which satisfies both

$$\arg z = \frac{\pi}{3} \quad \text{and} \quad |z-2-i| = |z-4+i|$$

10 The complex number $z = x + iy$ satisfies the equation

$$|z - 9 + 4i| = 3|z - 1 - 4i|$$

The complex number z is represented by the point P in the Argand diagram.

a) Show that the locus of P is a circle.
b) State the centre and radius of this circle.
c) Sketch the circle on an Argand diagram.   (EDEXCEL)

11 A complex number z satisfies the inequality

$$|z + 2 - (2\sqrt{3})i| \leq 2$$

Describe in geometrical terms, with the aid of a sketch, the corresponding region in an Argand diagram. Find

i) the least possible value of $|z|$
ii) the greatest possible value of $\arg z$.   (OCR)

12 The region R in an Argand diagram is defined by the inequalities

$$|z| \leq 4 \quad \text{and} \quad |z| \geq |z-2|$$

Draw a clearly labelled diagram to illustrate R.   (OCR)

13 The region R of an Argand diagram is defined by the inequalities

$$0 \leq \arg(z + 4i) \leq \tfrac{1}{4}\pi \quad \text{and} \quad |z| \leq 4$$

Draw a clearly labelled diagram to illustrate R.   (OCR)

14 Two complex numbers, z and w, satisfy the inequalities

$$|z - 3 - 2i| \leq 2 \quad \text{and} \quad |w - 7 - 5i| \leq 1$$

By drawing an Argand diagram, find the least possible value of $|z - w|$.   (OCR)

15 The point P in the Argand diagram represents the complex number z and the point Q
represents the complex number w, where $w = \dfrac{1}{z+1}$.

  i) Find w when

  a) $z = -i$ **b)** $z = i$

  expressing your answers in the form $u + iv$.

  ii) Find z in terms of w.

  iii) Given that P lies on the circle with centre the origin and radius 1, prove that $|w| = |w - 1|$.

  iv) Sketch the locus represented by $|w| = |w - 1|$.   (OCR)

16 a) The point P in the complex plane represents the complex number z. Describe the locus of P
in each of the following cases:

  i) $|z - 2| = 1$ **ii)** $\arg(z - 2) = \dfrac{2\pi}{3}$

On the same diagram of the complex plane, draw each of the loci defined in parts **i)** and **ii)**
above.

  b) **i)** The point A in the complex plane represents the complex number $w = a + ib$ (where a
  and b are real), and is such that $|w - 2| = 1$ **and** $\arg(w - 2) = \dfrac{2\pi}{3}$. Determine the value
  of a and the value of b, giving each answer in an exact form.

  ii) Write down the value of $\arg(w)$, and hence find the least positive integer n for which
  $\arg(w^n) > 2.5$.   (AEB 98)

17 The complex numbers z_1 and z_2 are such that $z_1 = 1 + ai$ and $z_2 = a + i$, for some integer
$a \geqslant 0$.

  a) Given that $w = z_1 + z_2$, show that $|w| = (1 + a)\sqrt{2}$ and write down $\arg(w)$, the argument
  of w.

  Hence find, in terms of a, the value of the complex number w^4.

  b) In the case when $a = 2$, the complex numbers z_1 and z_2 are represented in the complex plane
  by the points P_1 and P_2 respectively.

  Determine a cartesian equation of the locus of the point P, which represents the complex
  number z, given that $|z - z_1| = |z - z_2|$.

  c) In the case when $a = 0$, the complex numbers z_1 and z_2 are represented in the complex plane
  by the points Q_1 and Q_2 respectively.

  Describe fully, and sketch, the locus of the point Q, which represents the complex number z,
  given that $\arg\left(\dfrac{z - z_1}{z - z_2}\right) = \dfrac{\pi}{2}$.   (AEB 97)

2 Further trigonometry with calculus

If the triangles were to make a God they would give him three sides.
MONTESQUIEU

General solutions of trigonometric equations

In *Introducing Pure Mathematics* (page 341), we solved the trigonometric
equation $\cos\theta = \frac{1}{2}$ by obtaining the solution of $60°$ from a calculator and using
the graphs of $y = \cos\theta$ and $y = \frac{1}{2}$ to obtain the other solutions. When we have
several solutions to find, this method is very time-consuming and tends to
induce errors.

The usual method of finding more than one solution of such trigonometric
equations is to use the **general solution**.

General solutions for cosine curves

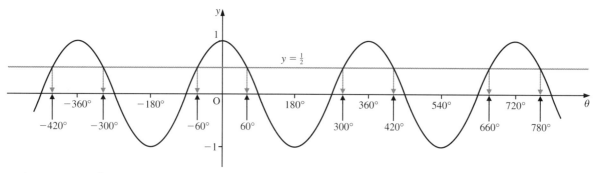

When $\cos\theta = \frac{1}{2}$, we will find from the graph of $y = \cos\theta$ (above) that the
solutions for θ are

$$\ldots, -300°, -60°, 60°, 300°, 420°, 660°, 780°, 1020°, 1140°, \ldots$$

or $360n° \pm 60°$ for any integer, n.

Hence, the general solution of $\cos\theta = \cos\alpha$ is given by

$$\theta = 360\,n° \pm \alpha \quad \text{for any integer, } n$$

where θ and α are measured in degrees.

If θ and α are measured in radians, this general solution would be

$$\theta = 2n\pi \pm \alpha$$

Example 1 Find the values of θ from $0°$ to $720°$ for which $\cos\theta = \dfrac{1}{\sqrt{2}}$.

SOLUTION

The calculator gives $\cos^{-1}\left(\dfrac{1}{\sqrt{2}}\right)$ as $45°$. Hence, the first solution, or α, is $45°$.

Putting $\alpha = 45°$ into the general solution, $\theta = 360n° \pm \alpha$, we get the following solutions:

$$\text{When } n = 0 \quad \theta = 45°$$
$$\text{When } n = 1 \quad \theta = 315° \text{ or } 405°$$
$$\text{When } n = 2 \quad \theta = 675°$$

Example 2 Find the values of θ from $0°$ to $360°$ for which $\cos 5\theta = \dfrac{\sqrt{3}}{2}$.

SOLUTION

After removing the cos term, we apply the general solution, using different values of n until we have a full range of solutions.

The calculator gives $\cos^{-1}\left(\dfrac{\sqrt{3}}{2}\right)$ as $30°$. Hence, the first solution, or α, is $30°$.

In this case, the general solution is an equation in **5θ**. So, with $\alpha = 30°$, we have

$$5\theta = 360n° \pm 30°$$
$$\Rightarrow \quad \theta = 72n° \pm 6°$$

Therefore, the solutions are as follows:

When $n = 0$	$\theta = 6°$	When $n = 3$	$\theta = 210°$ or $222°$
When $n = 1$	$\theta = 66°$ or $78°$	When $n = 4$	$\theta = 282°$ or $294°$
When $n = 2$	$\theta = 138°$ or $150°$	When $n = 5$	$\theta = 354°$

Note We can always check these values on a graphics calculator, after having selected the correct **range** or **view window**.

General solutions for sine curves

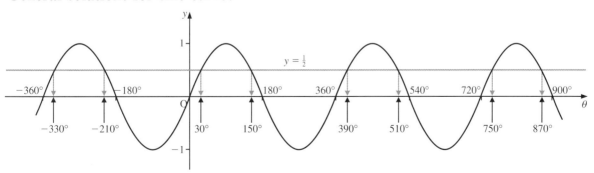

When $\sin\theta = \frac{1}{2}$, we will find from the graph of $y = \sin\theta$ (above) that the solutions for θ are

$$\ldots, -330°, -210°, 30°, 150°, 390°, 510°, 750°, \ldots$$

which can be written as

$$\ldots, -360° + 30°, -180° - 30°, 30°, 180° - 30°, 360° + 30°, 540° - 30°, 720° + 30°, \ldots$$

Hence, the general solution of $\sin\theta = \sin\alpha$ is given by

$$\theta = 180n° + (-1)^n\alpha \quad \text{for any integer, } n$$

where θ and α are measured in degrees.

If θ and α are measured in radians, this general solution would be

$$\theta = n\pi + (-1)^n\alpha$$

Example 3 Find the values of θ between $0°$ and $720°$ for which $\sin\theta = \dfrac{\sqrt{3}}{2}$.

SOLUTION

The calculator gives $\sin^{-1}\left(\dfrac{\sqrt{3}}{2}\right)$ as $60°$. Hence, the first solution, or α, is $60°$.

From the general solution, $\theta = 180n° + (-1)^n\alpha$, we have

$$\theta = 180n° + (-1)^n 60°$$

Therefore, the solutions are as follows:

When $n = 0$ $\theta = 60°$
When $n = 1$ $\theta = 180° - 60° = 120°$
When $n = 2$ $\theta = 360° + 60° = 420°$
When $n = 3$ $\theta = 540° - 60° = 480°$
When $n = 4$ $\theta = 720° + 60° = 780°$

But $\theta = 780°$ is out of the required range. Therefore, there are four solutions: $\theta = 60°, 120°, 420°$ and $480°$.

Example 4 Find the values of θ between $0°$ and $360°$ for which $\sin 3\theta = \dfrac{1}{\sqrt{2}}$.

SOLUTION

The calculator gives $\sin^{-1}\left(\dfrac{1}{\sqrt{2}}\right)$ as $45°$. Hence, the first solution, or α, is $45°$.

In this case, the general solution is an equation in 3θ. So, with $\alpha = 45°$, we have

$$3\theta = 180n° + (-1)^n 45°$$
$$\Rightarrow \quad \theta = 60n° + (-1)^n 15°$$

Therefore, the solutions are as follows:

When $n = 0$ $\theta = 15°$ When $n = 3$ $\theta = 165°$
When $n = 1$ $\theta = 45°$ When $n = 4$ $\theta = 255°$
When $n = 2$ $\theta = 135°$ When $n = 5$ $\theta = 285°$

That is, there are six solutions: $\theta = 15°, 45°, 135°, 165°, 255°$ and $285°$.

General solutions for tangent curves

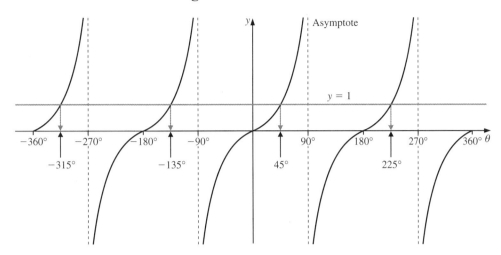

When $\tan \theta = 1$, we will find from the graph of $y = \tan \theta$ (above) that the solutions for θ are

$$\ldots, -135°, 45°, 225°, 405°, \ldots$$

or $180n° + 45°$ for any integer, n.

Hence, the general solution of $\tan \theta = \tan \alpha$ is given by

$$\theta = 180n° + \alpha \quad \text{for any integer, } n$$

where θ and α are measured in degrees.

If θ and α are measured in radians, this general solution would be

$$\theta = n\pi + \alpha$$

Example 5 Find the values between $0°$ and $360°$ for which $\tan 4\theta = -\sqrt{3}$.

SOLUTION

The calculator gives $\tan^{-1}(-\sqrt{3})$ as $-60°$. Hence, the first solution, or α, is $-60°$.

In this case, the general solution is an equation in **4θ**. So, with $\alpha = -60°$, we have

$$4\theta = 180n° - 60°$$
$$\Rightarrow \quad \theta = 45n° - 15°$$

Therefore, the solutions are $30°, 75°, 120°, 165°, 210°, 255°, 300°, 345°$.

Exercise 2A

In Questions **1** to **4** and **8** to **15**, find the general solution of each equation in **a)** radians, and **b)** degrees. In Questions **5** to **7**, find the general solution of each equation in radians only.

1 $\sin \theta = \dfrac{1}{\sqrt{2}}$ **2** $\cos \theta = -\frac{1}{2}$ **3** $\sin 2\theta = \frac{1}{2}$

4 $\tan 3\theta = 1$

5 $\sin\left(2x + \dfrac{\pi}{4}\right) = 1$

6 $\cos\left(3x - \dfrac{\pi}{3}\right) = \dfrac{1}{2}$

7 $\sin\left(2x + \dfrac{\pi}{3}\right) = \cos\left(2x + \dfrac{\pi}{3}\right)$

8 $\sin^2 4\theta = \frac{1}{2}$

9 $\sin^2 3\theta + \cos 3\theta + 1 = 0$

10 $\cos 2\theta = \cos\theta - 1$

11 $\tan\theta = 2\operatorname{cosec} 2\theta$

12 $\sin 5\theta - \sin\theta = \sin 2\theta$

13 $3\cos^2\theta + \cos^2 2\theta = 4$

14 $2\cos 2\theta = \sin\theta - 1$

15 $\sin 7\theta + \cos 3\theta = 0$

16 Show that $\sin 3x \equiv 3\sin x - 4\sin^3 x$. Find, in radians, the general solution of the equation $\sin 3x = 2\sin x$. (EDEXCEL)

17 Find, in radians in terms of π, the general solution of the equation $\cos\theta = \sin 2\theta$.

(EDEXCEL)

18 Find the general solution of the equation $\cos 2x = \cos\left(x + \dfrac{\pi}{3}\right)$, giving your answer in terms of π. (EDEXCEL)

19 Given that $t = \tan x$, write down an expression for $\tan 2x$ in terms of t. Hence, or otherwise, find the general solution, in radians, of the equation $\tan x + \tan 2x = 0$. (AEB 97)

20 Show that the general solution of the equation $\tan\left(3x - \dfrac{\pi}{4}\right) = \tan x$ is $x = \dfrac{(4n + 1)\pi}{8}$, where n is an integer. (NICCEA)

Harmonic form

As explained on page 374 of *Introducing Pure Mathematics*, the harmonic form is

$$R\cos(\theta \pm \alpha) \quad \text{or} \quad R\sin(\theta \pm \alpha)$$

where $R > 0$ is a constant.

Turning $a\cos\theta + b\sin\theta$ into $R\cos(\theta \pm \alpha)$ or $R\sin(\theta \pm \alpha)$

This is used when solving trigonometric equations, when finding the maximum and minimum of trigonometric expressions, and sometimes when solving problems in simple harmonic motion, where R is the amplitude of the motion.

- $a\cos\theta + b\sin\theta = R\cos(\theta - \alpha)$

Expanding $R\cos(\theta - \alpha)$, we obtain

$$a\cos\theta + b\sin\theta = R\cos\theta\cos\alpha + R\sin\theta\sin\alpha$$

Equating the coefficients of $\cos\theta$ gives: $a = R\cos\alpha$

Equating the coefficients of $\sin\theta$ gives: $b = R\sin\alpha$

Therefore, we have

$$a^2 + b^2 = R^2\cos^2\alpha + R^2\sin^2\alpha$$
$$= R^2(\cos^2\alpha + \sin^2\alpha)$$
$$= R^2$$
$$\Rightarrow \quad R = \sqrt{a^2 + b^2}$$

Note that R is **always** taken to be **positive**.

To find α, we use

$$\cos\alpha = \frac{a}{R} \quad \text{and} \quad \sin\alpha = \frac{b}{R}$$

which give

$$\tan\alpha = \frac{b}{a}$$

but we need to take care when either a or b is **negative**.

When using $a\cos\theta$ or $b\sin\theta$, if either a or b is negative, always use

$$\cos\alpha = \frac{a}{R} \quad \textbf{and} \quad \sin\alpha = \frac{b}{R},$$

and ensure that both give the same value for α. If they do not, use the value of α which is **not** between $0°$ and $90°$.

Example 6 Turn $3\cos\theta - 4\sin\theta$ into $R\cos(\theta - \alpha)$.

SOLUTION

We have

$$R = \sqrt{a^2 + b^2} = \sqrt{3^2 + 4^2} = 5$$

which gives

$$\cos\alpha = \frac{3}{5} \qquad \sin\alpha = -\frac{4}{5}$$
$$\alpha = 53.1° \qquad \alpha = -53.1° \quad \text{(from the calculator)}$$

Note $-53.1°$ is also a solution of $\cos\alpha = \frac{3}{5}$, but a calculator always gives the angle between $0°$ and $90°$ wherever possible.

Therefore, we use $\alpha = -53.1°$, as this is the value found from both $\cos\alpha = \frac{3}{5}$ and $\sin\alpha = -\frac{4}{5}$ which is **not** between $0°$ and $90°$.

Hence, we get

$$3\cos\theta - 4\sin\theta = 5\cos(\theta + 53.1°)$$

Example 7 Find the general solution of $5\cos\theta - 12\sin\theta = 6.5$. Hence find the solutions which lie between $0°$ and $360°$.

SOLUTION

Using $5\cos\theta - 12\sin\theta = R\cos(\theta + \alpha)$, we find

$$R = \sqrt{5^2 + 12^2} = 13$$

which gives

$$\cos\alpha = \frac{5}{13} \quad \Rightarrow \quad \alpha = 67.4°$$

Therefore, we have

$$5\cos\theta - 12\sin\theta = 13\cos(\theta + 67.4°)$$

So, $5\cos\theta - 12\sin\theta = 6.5$ becomes

$$13\cos(\theta + 67.4°) = 6.5$$

$$\Rightarrow \quad \cos(\theta + 67.4°) = \frac{6.5}{13} = 0.5$$

which gives

$$\theta + 67.4° = 360n° \pm 60°$$

Therefore, the general solution is

$$\theta = 360n° \pm 60° - 67.4°$$

When $n = 0$, both solutions are negative and are outside the required range. Therefore, the solutions required are 232.6° and 352.6° (when $n = 1$).

- **$a\sin\theta + b\cos\theta = R\sin(\theta + \alpha)$**

Expanding $R\sin(\theta + \alpha)$, we obtain

$$a\sin\theta + b\cos\theta = R\sin\theta\cos\alpha + R\cos\theta\sin\alpha$$

Equating the coefficients of $\sin\theta$ gives: $a = R\cos\alpha$

Equating the coefficients of $\cos\theta$ gives: $b = R\sin\alpha$

Therefore, we again have

$$R = \sqrt{a^2 + b^2}$$

$$\cos\alpha = \frac{a}{R} \quad \text{and} \quad \sin\alpha = \frac{b}{R}$$

Example 8 Turn $24\sin\theta + 7\cos\theta$ into $R\sin(\theta + \alpha)$.

SOLUTION

We have

$$R = \sqrt{24^2 + 7^2} = 25$$

which gives

$$\cos\alpha = \frac{24}{25} \quad \Rightarrow \quad \alpha = 16.3°$$

Hence, we get

$$24\sin\theta + 7\cos\theta = 25\sin(\theta + 16.3°)$$

Note To avoid the problem of possibly obtaining two different values for α, we select whichever one of $R\cos(\theta - \alpha)$, $R\cos(\theta + \alpha)$, $R\sin(\theta - \alpha)$ or $R\sin(\theta + \alpha)$ contains the same sign as the expression being simplified.

Thus, we would convert $3\sin\theta - 4\cos\theta$ into the form $R\sin(\theta - \alpha)$, which is the only trigonometric formula giving $a\sin\theta - b\cos\theta$. In this case, we have

$$R = \sqrt{3^2 + 4^2} = 5$$

$$\cos\alpha = \frac{3}{5} \quad \Rightarrow \quad \alpha = 53.1°$$

which give

$$3\sin\theta - 4\cos\theta = 5\sin(\theta - 53.1°)$$

Example 9 For each of

a) $f(x) = 24\cos\theta + 7\sin\theta$ **b)** $f(x) = \dfrac{1}{2 + 4\sin\theta - 3\cos\theta}$

find

i) the range of values for $f(x)$ **ii)** a maximum point   **iii)** a minimum point

SOLUTION

a) i) Using $24\cos\theta + 7\sin\theta = R\cos(\theta - \alpha)$, we have

$$R = \sqrt{24^2 + 7^2} = 25$$

$$\cos\alpha = \frac{24}{25} \quad \Rightarrow \quad \alpha = 16.3°$$

which give

$$24\cos\theta + 7\sin\theta = 25\cos(\theta - 16.3°)$$

Now, $\cos\theta$ has a maximum of $+1$ and a minimum of -1. Therefore, the range of values of $\cos(\theta - 16.3°)$ is -1 to $+1$, which gives the range of values of $25\cos(\theta - 16.3°)$ as -25 to $+25$. That is,

$$-25 \leqslant f(x) \leqslant 25$$

ii) For the maximum point, we have $\cos(\theta - 16.3°) = 1$. Therefore,

$$\theta - 16.3° = 0 \quad \Rightarrow \quad \theta = 16.3°$$

Hence, the maximum point is $(16.3°, 25)$.

iii) For the minimum point, we have $\cos(\theta - 16.3°) = -1$. Therefore,

$$\theta - 16.3° = 180° \quad \Rightarrow \quad \theta = 196.3°$$

Hence, the minimum point is $(196.3°, -25)$.

b) i)
$$f(x) = \frac{1}{2 + 4\sin\theta - 3\cos\theta}$$

We found in Example 6 that $3\cos\theta - 4\sin\theta = 5\cos(\theta + 53.1°)$. Therefore, we have

$$f(x) = \frac{1}{2 - 5\cos(\theta + 53.1°)}$$

The **denominator** has a range from -3 to 7. (Remember that $1 \div 0 = \infty$.) Therefore, $f(x)$ has a range

$$f(x) \geqslant \tfrac{1}{7} \quad \text{and} \quad f(x) \leqslant -\tfrac{1}{3}$$

ii) The maximum point is found where $\cos(\theta + 53.1°)$ is $+1$. That is,

$$\theta + 53.1° = 360° \quad \Rightarrow \quad \theta = 306.9°$$

Therefore, the maximum point is $(306.9°, -\tfrac{1}{3})$.

iii) The minimum point is found where $\cos(\theta + 53.1°)$ is -1. That is,

$$\theta + 53.1° = 180° \quad \Rightarrow \quad \theta = 126.9°$$

Therefore, the minimum point is $(126.9°, \tfrac{1}{7})$.

Exercise 2B

1 Find the value of R and of α in each of the following identities.

a) $5\cos\theta + 12\sin\theta \equiv R\cos(\theta - \alpha)$ **b)** $3\cos\theta - 4\sin\theta \equiv R\cos(\theta + \alpha)$

c) $3\sin\theta - 4\cos\theta \equiv R\sin(\theta - \alpha)$ **d)** $\cos 2\theta + \sin 2\theta \equiv R\cos(2\theta - \alpha)$

e) $6\sin 3\theta + 8\cos 3\theta \equiv R\sin(3\theta + \alpha)$

2 For each of the following expressions, find

i) the maximum and minimum values
ii) the smallest non-negative value of x for which this occurs.

Where necessary, give your answer correct to one decimal place.

a) $12\cos\theta - 9\sin\theta$ **b)** $8\cos 2\theta + 6\sin 2\theta$ **c)** $\dfrac{4}{8 - 3\cos\theta - 4\sin\theta}$

d) $\dfrac{6}{8 + 4\sin\theta - 2\cos\theta}$ **e)** $\dfrac{2}{1 + 3\cos\theta + 4\sin\theta}$ **f)** $\dfrac{3}{8 + 8\sin\theta + 6\cos\theta}$

3 Find the general solution, in degrees, of each of these equations.

a) $3\cos\theta + 4\sin\theta = 2.5$ **b)** $12\cos\theta - 5\sin\theta = 6.5$ **c)** $\cos 2\theta + \sin 2\theta = \dfrac{1}{\sqrt{2}}$

d) $\sin 3\theta - \cos 3\theta = \dfrac{1}{\sqrt{2}}$ **e)** $2\sin 6\theta + 3\sin^2 3\theta = 0$

Inverse trigonometric functions

The inverse function $\sin^{-1}x$, or $\arcsin x$, is defined as the angle whose sine is x. For example,

$$\sin\left(\frac{\pi}{6}\right) = \frac{1}{2} \quad \Rightarrow \quad \sin^{-1}\left(\frac{1}{2}\right) = \frac{\pi}{6}$$

Hence, if $\theta = \sin^{-1}x$, then $\sin\theta = x$.

Sketching inverse trigonometric functions

Inverse sine graph

The graph of $y = \sin^{-1} x$ is obtained by reflecting the graph of $y = \sin x$ in the line $y = x$.

To enable the sketch to be drawn to an acceptable degree of accuracy, we need to find the gradient of the sine curve at the origin. So, we differentiate $y = \sin x$, which gives

$$\frac{dy}{dx} = \cos x$$

At the origin, where $x = 0$, we have

$$\frac{dy}{dx} = \cos 0 = 1$$

So, the gradient of $y = \sin x$ at the origin is 1.

We then proceed as follows:

- First, draw the line $y = x$. Show this as a dashed line.
- Next, carefully sketch the graph of $y = \sin x$, remembering that $y = x$ is a tangent to $y = \sin x$ at the origin.
- Finally, carefully sketch the reflection of $y = \sin x$ in the line $y = x$, to give the graph shown below.

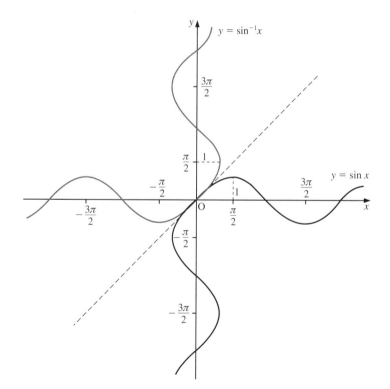

The graphs of other inverse trigonometric functions are found similarly: that is, by reflecting the graph of the relevant trigonometric function in the line $y = x$. If the curve of the function passes through the origin, start by finding its gradient at that point.

Inverse tan graph

Differentiating $y = \tan x$ to find the gradient, we get

$$\frac{dy}{dx} = \sec^2 x$$

At the origin, where $x = 0$, we have

$$\frac{dy}{dx} = \sec^2 0 = \frac{1}{\cos^2 0} = 1$$

Thus, the gradient of $y = \tan x$ at the origin is 1.

The graphs of $y = \tan x$ and $y = \tan^{-1} x$ (or arctan x) are shown below.

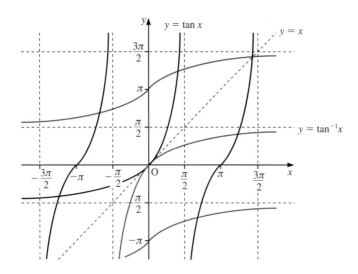

Inverse cosine graph

The graphs of $y = \cos x$ and $y = \cos^{-1} x$ (or $y = \arccos x$) are shown below.

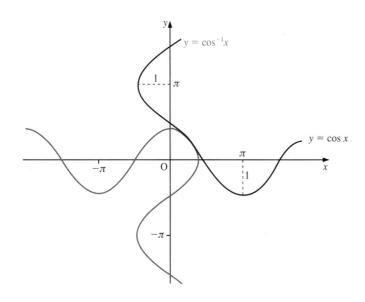

Exercise 2C

1 Find the value of each of these inverse functions.

a) $\sin^{-1}0.5$ **b)** $\sin^{-1}\left(-\dfrac{1}{2}\right)$ **c)** $\cos^{-1}\left(-\dfrac{\sqrt{3}}{2}\right)$

d) $\tan^{-1}1$ **e)** $\sec^{-1}\sqrt{2}$ **f)** $\cot^{-1}3$

2 Sketch the graph of each of these inverse functions.

a) $\sec^{-1}x$ **b)** $\operatorname{cosec}^{-1}x$ **c)** $\cot^{-1}x$

3 If $\cos^{-1}x = \dfrac{2\pi}{5}$, find $\sin^{-1}x$.

4 Prove that $\tan^{-1}\left(\dfrac{1+x}{1-x}\right) = \dfrac{\pi}{4} + \tan^{-1}x$

5 Find the general solution of the equation

$$3\cos\theta - 7\sin\theta = -6$$

Give your answer in degrees correct to two decimal places.   (NICCEA)

6 $5\cos x - 12\sin x \equiv R\cos(x+\alpha)$

where $R > 0$ and α is acute and measured in degrees.

a) Find the value of R.
b) Find the value of α to one decimal place.
c) Hence, or otherwise, find the general solution of the equation

$$5\cos x - 12\sin x = 4 \quad \text{(EDEXCEL)}$$

7 i) Write $f(\theta) = 7\cos\theta - 3\sin\theta$ in the form $R\cos(\theta+\alpha)$, where R is positive and α is acute.
ii) Find the maximum and minimum values of $f(\theta)$.
iii) Solve $7\cos\theta - 3\sin\theta = 1$, giving the general solution in degrees.   (NICCEA)

8 a) Find all values of x between $0°$ and $360°$ satisfying

$$3\cos x + \sin x = -1$$

b) Find the general solution of the equation

$$\sin 2x + \sin 4x = \cos 2x + \cos 4x \quad \text{(WJEC)}$$

9 Given that

$$7\cos\theta + 24\sin\theta \equiv R\cos(\theta - \alpha)$$

where $R > 0$, $0 \leqslant \alpha \leqslant 90°$,

a) find the values of the constants R and α.

Hence find

b) the general solution of the equation $7\cos\theta + 24\sin\theta = 15$
c) the range of the function $f(\theta)$ where

$$f(\theta) \equiv \frac{1}{5 + (7\cos\theta + 24\sin\theta)^2} \qquad 0 \leqslant \theta < 360° \qquad \text{(EDEXCEL)}$$

10 Find, in degrees, the value of the acute angle α for which

$$\cos \theta - (\sqrt{3}) \sin \theta \equiv 2 \cos (\theta + \alpha)$$

for all values of θ.

Solve the equation

$$\cos x - (\sqrt{3}) \sin x = \sqrt{2} \qquad 0° \leqslant x \leqslant 360° \qquad \text{(EDEXCEL)}$$

11 Express $\cos \theta + \sqrt{3} \sin \theta$ in the form $R \cos (\theta - \alpha)$, where $R > 0$ and $0° < \alpha < 90°$.

Hence find the general solution of the equation

$$\cos \theta + \sqrt{3} \sin \theta = 2 \cos 40°$$

giving your answers in degrees. (AEB 96)

12 The angle α is such that $0 < \alpha < \dfrac{\pi}{2}$ and $R \cos (\theta + \alpha) \equiv 84 \cos \theta - 13 \sin \theta$, where R is some positive real number.

a) State the value of R and find α, in radians, correct to three decimal places.
b) Hence determine the general solution, in radians, of the equation

$$84 \cos \theta - 13 \sin \theta = 17 \qquad \text{(AEB 98)}$$

13 $$f(x) \equiv 7 \cos x - 24 \sin x$$

Given that $f(x) \equiv R \cos (x + \alpha)$, where $R \geqslant 0$, $0 \leqslant \alpha \leqslant \dfrac{\pi}{2}$, and x and α are measured in radians,

a) find R and show that $\alpha = 1.29$ to two decimal places.

Hence write down

b) the minimum value of $f(x)$
c) the value of x in the interval $0 \leqslant x \leqslant 2\pi$ which gives this minimum value.
d) Find the smallest two positive values of x for which

$$7 \cos x - 24 \sin x = 10 \qquad \text{(EDEXCEL)}$$

14 i) $$f(\theta) \equiv 9 \sin \theta + 12 \cos \theta$$

Given that $f(\theta) \equiv R \sin (\theta + \alpha)$ where $R > 0$, $0 \leqslant \alpha \leqslant 90°$,

a) find the values of the constants R and α.
b) Hence find the values of θ, $0 \leqslant \theta < 360°$, for which

$$9 \sin \theta + 12 \cos \theta = -7.5$$

giving your answers to the nearest tenth of a degree.

ii) Find, in radians in terms of π, the general solution to the equation

$$\sqrt{3} \sin (\theta - \tfrac{1}{6}\pi) = \sin \theta \qquad \text{(EDEXCEL)}$$

Differentiation of inverse trigonometric functions

$\sin^{-1}x$ or $\arcsin x$

If $y = \sin^{-1}x$, then $\sin y = x$.

Differentiating $\sin y = x$, we obtain

$$\cos y \frac{dy}{dx} = 1$$

$$\Rightarrow \quad \frac{dy}{dx} = \frac{1}{\cos y} = \frac{1}{\sqrt{1 - \sin^2 y}} = \frac{1}{\sqrt{1 - x^2}}$$

Therefore, we have

$$\int \frac{dx}{\sqrt{1 - x^2}} = \sin^{-1}x + c$$

Similarly, if $y = \sin^{-1}\left(\dfrac{x}{a}\right)$, then $\sin y = \dfrac{x}{a}$.

Differentiating, we get

$$\cos y \frac{dy}{dx} = \frac{1}{a}$$

$$\Rightarrow \quad \frac{dy}{dx} = \frac{1}{a\cos y} = \frac{1}{a\sqrt{1 - \sin^2 y}}$$

$$\Rightarrow \quad \frac{dy}{dx} = \frac{1}{a\sqrt{1 - \left(\dfrac{x}{a}\right)^2}} = \frac{1}{\sqrt{a^2 - x^2}}$$

Therefore, we have

$$\frac{d}{dx}\sin^{-1}\left(\frac{x}{a}\right) = \frac{1}{\sqrt{a^2 - x^2}}$$

which gives

$$\int \frac{dx}{\sqrt{a^2 - x^2}} = \sin^{-1}\left(\frac{x}{a}\right) + c$$

If $y = \cos^{-1}x$, we can show that

$$\frac{d}{dx}\cos^{-1}x = \frac{-1}{\sqrt{1 - x^2}}$$

which gives

$$\int \frac{dx}{\sqrt{1 - x^2}} = -\cos^{-1}x + c$$

In the diagram on the right, $\sin\theta = x$, $\cos\phi = x$ and $\phi = \dfrac{\pi}{2} - \theta$.

Therefore, we have $\theta = \sin^{-1}x$ and $\phi = \cos^{-1}x$, giving

$$\sin^{-1}x = \frac{\pi}{2} - \cos^{-1}x$$

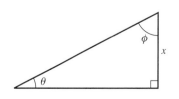

So, we get

$$\int \frac{dx}{\sqrt{1-x^2}} = \sin^{-1}x + c$$

$$= -\cos^{-1}x + c'$$

where $c' = \frac{\pi}{2} + c$.

Hence, it is unusual to use a function in $\cos^{-1}x$ in differentiation or in integration, as it is simply an alternative to $\sin^{-1}x$.

$\tan^{-1}x$ or $\arctan x$

If $y = \tan^{-1}\left(\frac{x}{a}\right)$, then $\tan y = \frac{x}{a}$.

Differentiating $\tan y = \frac{x}{a}$, we obtain

$$\sec^2 y \frac{dy}{dx} = \frac{1}{a}$$

$$\Rightarrow \quad \frac{dy}{dx} = \frac{1}{a\sec^2 y} = \frac{1}{a(1 + \tan^2 y)} = \frac{1}{a\left[1 + \left(\frac{x}{a}\right)^2\right]}$$

$$\Rightarrow \quad \frac{dy}{dx} = \frac{a}{a^2 + x^2}$$

Therefore, we have

$$\frac{d}{dx} \tan^{-1}\left(\frac{x}{a}\right) = \frac{a}{a^2 + x^2}$$

which gives

$$\int \frac{dx}{a^2 + x^2} = \frac{1}{a} \tan^{-1}\left(\frac{x}{a}\right) + c$$

Note $\quad \dfrac{d}{dx} \tan^{-1}x = \dfrac{1}{1 + x^2}$ $\quad$ and $\quad \displaystyle\int \dfrac{dx}{1 + x^2} = \tan^{-1}x + c$

Example 10 Differentiate each of the following inverse functions.

a) i) $\sin^{-1}\left(\dfrac{x}{3}\right)$ $\quad$ **ii)** $\sin^{-1}4x$ $\quad$ **b)** $\tan^{-1}\left(\dfrac{x}{5}\right)$

SOLUTION

a) Using $\dfrac{d}{dx} \sin^{-1}\left(\dfrac{x}{a}\right) = \dfrac{1}{\sqrt{a^2 - x^2}}$, we have

i) $\quad \dfrac{d}{dx} \sin^{-1}\left(\dfrac{x}{3}\right) = \dfrac{1}{\sqrt{9 - x^2}}$

ii) $$\frac{d}{dx} \sin^{-1}4x = \frac{d}{dx} \sin^{-1}\left(\frac{x}{\frac{1}{4}}\right) = \frac{1}{\sqrt{\frac{1}{16} - x^2}}$$

$$\Rightarrow \quad \frac{d}{dx} \sin^{-1}4x = \frac{4}{\sqrt{1 - 16x^2}}$$

b) Using $\dfrac{d}{dx} \tan^{-1}\left(\dfrac{x}{a}\right) = \dfrac{a}{a^2 + x^2}$, we have

$$\frac{d}{dx} \tan^{-1}\left(\frac{x}{5}\right) = \frac{5}{25 + x^2}$$

Example 11 Evaluate **a)** $\displaystyle\int_0^2 \frac{1}{\sqrt{4 - x^2}}\, dx$ **b)** $\displaystyle\int_0^1 \frac{1}{\sqrt{4 - 3x^2}}\, dx$

SOLUTION

a) $$\int_0^2 \frac{1}{\sqrt{4 - x^2}}\, dx = \left[\sin^{-1}\left(\frac{x}{2}\right)\right]_0^2$$

$$= \sin^{-1}1 - \sin^{-1}0 = \frac{\pi}{2} - 0$$

Therefore, we have

$$\int_0^2 \frac{1}{\sqrt{4 - x^2}}\, dx = \frac{\pi}{2}$$

b) For integrals in this form, we **always** reduce the coefficient of x^2 to unity before integrating. Hence, in this case, we have

$$\int_0^1 \frac{1}{\sqrt{4 - 3x^2}}\, dx = \frac{1}{\sqrt{3}}\int_0^1 \frac{1}{\sqrt{\frac{4}{3} - x^2}}\, dx$$

$$= \frac{1}{\sqrt{3}}\int_0^1 \frac{1}{\sqrt{\left(\frac{2}{\sqrt{3}}\right)^2 - x^2}}\, dx$$

$$= \frac{1}{\sqrt{3}}\left[\sin^{-1}\left(\frac{x}{\frac{2}{\sqrt{3}}}\right)\right]_0^1$$

$$= \frac{1}{\sqrt{3}}\left[\sin^{-1}\left(\frac{\sqrt{3}x}{2}\right)\right]_0^1$$

$$= \frac{1}{\sqrt{3}}\left(\sin^{-1}\left(\frac{\sqrt{3}}{2}\right) - \sin^{-1}0\right) = \frac{1}{\sqrt{3}}\frac{\pi}{3}$$

Hence, we obtain

$$\int_0^1 \frac{1}{\sqrt{4 - 3x^2}}\, dx = \frac{\pi}{3\sqrt{3}} \quad \text{or} \quad \frac{\pi\sqrt{3}}{9}$$

Example 12 Evaluate $\displaystyle\int_0^3 \frac{1}{9 + x^2}\, dx$.

SOLUTION

$$\int_0^3 \frac{1}{9 + x^2}\, dx = \left[\frac{1}{3} \tan^{-1}\left(\frac{x}{3}\right)\right]_0^3$$

$$= \frac{1}{3} \tan^{-1} 1 - \frac{1}{3} \tan^{-1} 0 = \frac{1}{3}\frac{\pi}{4}$$

Therefore, we have

$$\int_0^3 \frac{1}{9 + x^2}\, dx = \frac{\pi}{12}$$

Example 13 Find $\displaystyle\int \frac{1}{16 + 25x^2}\, dx$.

SOLUTION

Remember   Reduce the coefficient of x^2 to unity before integrating. (See Example 11.)

Hence, we have

$$\int \frac{1}{16 + 25x^2}\, dx = \frac{1}{25} \int \frac{1}{\dfrac{16}{25} + x^2}\, dx$$

$$= \frac{1}{25} \int \frac{1}{\left(\dfrac{4}{5}\right)^2 + x^2}\, dx$$

$$= \frac{1}{25}\frac{1}{\dfrac{4}{5}} \tan^{-1}\left(\frac{x}{\dfrac{4}{5}}\right) + c$$

Therefore, we have

$$\int \frac{1}{16 + 25x^2}\, dx = \frac{1}{20} \tan^{-1}\left(\frac{5x}{4}\right) + c$$

Example 14 Find $\displaystyle\int \frac{dx}{x^2 + 6x + 25}$.

SOLUTION

When it is anticipated that the integral will be an inverse trigonometric function, we start by using the method of completing the square to turn the quadratic denominator into the form $a(x + b)^2 + c$. Then we reduce the coefficient of $(x + b)^2$ to unity so that we can use the standard integration formula with $(x + b)$ replacing x.

Hence, we have

$$x^2 + 6x + 25 = (x + 3)^2 + 16$$

which gives

$$\int \frac{dx}{x^2 + 6x + 25} = \int \frac{dx}{(x + 3)^2 + 16}$$

The integral we have obtained is now in the same form as $\int \dfrac{dx}{x^2 + a^2}$, with $(x+3)$ replacing x and 4 replacing a. Thus, we have

$$\int \frac{dx}{(x+3)^2 + 16} = \frac{1}{4}\tan^{-1}\left(\frac{x+3}{4}\right) + c$$

Example 15 Find $\displaystyle\int \frac{dx}{\sqrt{11 - 8x - 4x^2}}$.

SOLUTION

To convert $11 - 8x - 4x^2$ into the form $a(x+b)^2 + c$, it is easier first to factorise out the minus sign, and then take the sign back inside when the square is completed.

Note   The minus sign must be kept **within the square root**.

So, factorising out the minus sign, we have

$$\sqrt{11 - 8x - 4x^2} = \sqrt{-(4x^2 + 8x - 11)}$$

$$= \sqrt{-4\left(x^2 + 2x - \frac{11}{4}\right)}$$

Then, completing the square, we get

$$\sqrt{11 - 8x - 4x^2} = \sqrt{-4\left[(x+1)^2 - \frac{15}{4}\right]}$$

$$= \sqrt{15 - 4(x+1)^2}$$

$$= 2\sqrt{\frac{15}{4} - (x+1)^2}$$

Substituting this into the given integral, we have

$$\int \frac{dx}{\sqrt{11 - 8x - 4x^2}} = \frac{1}{2}\int \frac{dx}{\sqrt{\dfrac{15}{4} - (x+1)^2}}$$

$$= \frac{1}{2}\sin^{-1}\left(\frac{x+1}{\sqrt{\dfrac{15}{4}}}\right) + c$$

which gives

$$\int \frac{dx}{\sqrt{11 - 8x - 4x^2}} = \frac{1}{2}\sin^{-1}\left(\frac{2(x+1)}{\sqrt{15}}\right) + c$$

Exercise 2D

1 Differentiate each of the following with respect to x.

    a) $\sin^{-1}5x$ **b)** $\tan^{-1}3x$ **c)** $\sin^{-1}\sqrt{2}x$ **d)** $\tan^{-1}\frac{3}{4}x$

    e) $\sin^{-1}x^2$ **f)** $\tan^{-1}\left(\dfrac{x}{1+x^2}\right)$ **g)** $(\sin^{-1}2x)^3$ **h)** $(3\tan^{-1}5x)^4$

    i) $\sec^{-1}x$ **j)** $\cot^{-1}x$

2 Find each of the following integrals.

    a) $\displaystyle\int\dfrac{dx}{\sqrt{4-x^2}}$ **b)** $\displaystyle\int\dfrac{dx}{\sqrt{9-x^2}}$ **c)** $\displaystyle\int\dfrac{dx}{\sqrt{25-4x^2}}$

    d) $\displaystyle\int\dfrac{dx}{\sqrt{16-9x^2}}$ **e)** $\displaystyle\int\dfrac{dx}{9+x^2}$ **f)** $\displaystyle\int\dfrac{dx}{16+x^2}$

    g) $\displaystyle\int\dfrac{dx}{25+16x^2}$ **h)** $\displaystyle\int\dfrac{dx}{9+25x^2}$

3 Evaluate each of the following definite integrals, giving the exact value of your answer.

    a) $\displaystyle\int_0^1\dfrac{dx}{\sqrt{1-x^2}}$ **b)** $\displaystyle\int_0^2\dfrac{dx}{4+x^2}$ **c)** $\displaystyle\int_0^3\dfrac{dx}{\sqrt{9-x^2}}$ **d)** $\displaystyle\int_0^2\dfrac{dx}{4+3x^2}$

    e) $\displaystyle\int_{-\frac{1}{5}}^{\frac{1}{5}}\dfrac{dx}{\sqrt{1-25x^2}}$

4 Evaluate each of the following definite integrals, giving your answer correct to three significant figures.

    a) $\displaystyle\int_0^{0.1}\dfrac{dx}{\sqrt{4-25x^2}}$ **b)** $\displaystyle\int_1^2\dfrac{dx}{4+9x^2}$ **c)** $\displaystyle\int_1^2\dfrac{dx}{\sqrt{3-(x-1)^2}}$

    d) $\displaystyle\int_0^1\dfrac{dx}{4(x+1)^2+5}$ **e)** $\displaystyle\int_0^2\dfrac{dx}{\sqrt{20-8x-x^2}}$ **f)** $\displaystyle\int_0^1\dfrac{dx}{16x^2+20x+35}$

5 Find the exact value of $\displaystyle\int_0^{\frac{5}{8}}\dfrac{1}{\sqrt{(25-16x^2)}}\,dx$.   (OCR)

6 Express $5+4x-x^2$ in the form $a-(x-b)^2$, where a and b are positive constants. Hence find the exact value of

$$\int_{\frac{7}{2}}^5\dfrac{1}{\sqrt{(5+4x-x^2)}}\,dx \qquad \text{(OCR)}$$

7 Express $\dfrac{2x^3+5x^2+11x+13}{(x+1)(x^2+4)}$ in partial fractions.

    Show that

$$\int_0^1\dfrac{2x^3+5x^2+11x+13}{(x+1)(x^2+4)}\,dx = 2+\ln\left(\frac{5}{2}\right)+\frac{1}{2}\tan^{-1}\left(\frac{1}{2}\right) \qquad \text{(OCR)}$$

8 Given that $y = x - \sqrt{(1 - x^2)}\sin^{-1}x$, show that

$$\frac{dy}{dx} = \frac{x\sin^{-1}x}{\sqrt{(1 - x^2)}}$$

Hence, or otherwise, evaluate

$$\int_0^{\frac{1}{4}\sqrt{3}} \frac{2x\sin^{-1}(2x)}{\sqrt{(1 - 4x^2)}}\, dx$$

giving your answer in terms of π and $\sqrt{3}$.   (OCR)

9 Given that $z = \tan^{-1}x$, derive the result $\dfrac{dz}{dx} = \dfrac{1}{1 + x^2}$.

[No credit will be given for merely quoting the result from the *List of Formulae*.]

Hence express $\dfrac{d}{dx}(\tan^{-1}(xy))$ in terms of x, y and $\dfrac{dy}{dx}$.

Given that x and y satisfy the equation

$$\tan^{-1}x + \tan^{-1}y + \tan^{-1}(xy) = \tfrac{11}{12}\pi$$

prove that, when $x = 1$, $\dfrac{dy}{dx} = -1 - \tfrac{1}{2}\sqrt{3}$.   (OCR)

10 i) Given that $y = \sin^{-1}x$, derive the result $\dfrac{dy}{dx} = \dfrac{1}{\sqrt{(1 - x^2)}}$.

[No credit will be given for merely quoting the result from the *List of Formulae*.]

ii) Find $\dfrac{d}{dx}\sqrt{(1 - x^2)}$.

iii) Using the above results, find $\displaystyle\int_0^1 \sin^{-1}x\, dx$.   (OCR)

11 Given that $x = \dfrac{1}{y}$, show that

$$\int \frac{1}{x\sqrt{(x^2 - 1)}}\, dx = -\int \frac{1}{\sqrt{(1 - y^2)}}\, dy$$

Find $\displaystyle\int \frac{1}{x\sqrt{(x^2 - 1)}}\, dx$.   (OCR)

12 OAC is a quadrant of the circle whose equation is $x^2 + y^2 = 1$. From B, a point on the circumference, a perpendicular is dropped to D, a point on the radius OC, so that the x-coordinate of D is t. This is shown in the figure on the right.

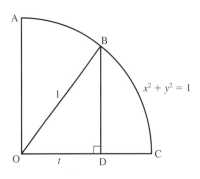

i) Show that the area of the sector AOB is $\tfrac{1}{2}\sin^{-1}t$.

ii) Find the area of the triangle OBD in terms of t.

iii) Hence show that

$$\sin^{-1}t = 2\int_0^t \sqrt{1 - x^2}\, dx - t\sqrt{1 - t^2}$$

iv) By using integration by parts, show that

$$\int_0^t \sqrt{1 - x^2}\, \mathrm{d}x = t\sqrt{1 - t^2} + \int_0^t \frac{x^2}{\sqrt{1 - x^2}}\, \mathrm{d}x$$

v) By using parts **iii** and **iv**, prove that

$$\sin^{-1} t = \int_0^t \frac{\mathrm{d}x}{\sqrt{1 - x^2}} \qquad \text{(NICCEA)}$$

3 Polar coordinates

All places are distant from Heaven alike.
ROBERT BURTON

Position of a point

The position of a point, P, in a plane may be given in terms of its distance from a fixed point, O, called the **pole**, and the angle which OP makes with a fixed line, called the **initial line**. When the position of a point is given in this way, we have the **polar coordinates** of the point.

In the diagram on the right, the cartesian coordinates of point P would be given as (x, y).

Its position in polar coordinates would be given as (r, θ), where $r (\geqslant 0)$ is the distance of P from the origin, O, and θ is the **anticlockwise angle** which OP makes with the x-axis, which is normally taken as the initial line.

θ is normally measured in radians and its **principal value** is taken to be between $-\pi$ and π.

Example 1 Plot the point P with coordinates $\left(4, \dfrac{\pi}{6}\right)$ and the point Q with coordinates $\left(2, -\dfrac{\pi}{3}\right)$.

SOLUTION

a) Draw the line OP at $\dfrac{\pi}{6}$ radians to the x-axis.

Make OP = 4 units.

Then P is the point identified.

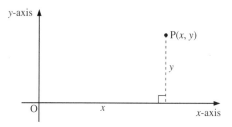

b) Draw the line OQ at $-\dfrac{\pi}{3}$ radians to the x-axis. The

negative value of the angle means that $\dfrac{\pi}{3}$ is measured

in a **clockwise** direction from the x-axis.
Make OQ = 2 units.

Then Q is the point identified.

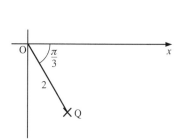

Exercise 3A

Plot the points with the following polar coordinates.

1 $\left(3, \dfrac{\pi}{4}\right)$

2 $\left(2, \dfrac{2\pi}{3}\right)$

3 $\left(3, -\dfrac{\pi}{3}\right)$

4 $\left(2, \dfrac{3\pi}{2}\right)$

5 $\left(4, -\dfrac{\pi}{4}\right)$

Connection between polar and cartesian coordinates

In the diagram on the right, the point P is (x, y) in cartesian coordinates and (r, θ) in polar coordinates.

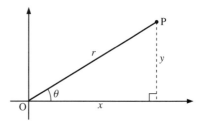

We see that

$$x = r\cos\theta \qquad y = r\sin\theta$$

$$r = \sqrt{x^2 + y^2} \qquad \tan\theta = \frac{y}{x}$$

If either x or y is negative, we should refer to the position of the point to determine the value of θ.

We can use the above equations to convert the equation of a curve from its cartesian form to its polar form, or vice versa.

Example 2 Find the polar equation of the curve $x^2 + y^2 = 2x$.

SOLUTION

Substituting $x = r\cos\theta$, $y = r\sin\theta$ into $x^2 + y^2 = 2x$ (shown on the right), we have

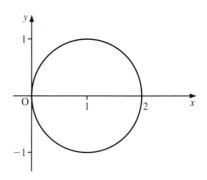

$$r^2\cos^2\theta + r^2\sin^2\theta = 2r\cos\theta$$

$$\Rightarrow \quad r^2(\cos^2\theta + \sin^2\theta) = 2r\cos\theta$$

$$\Rightarrow \quad r^2 = 2r\cos\theta$$

$$\Rightarrow \quad r = 2\cos\theta \quad \text{(since } r \neq 0\text{)}$$

Hence, the polar equation of the given curve is $r = 2\cos\theta$.

Exercise 3B

1 Find the cartesian equation of each of these curves.

a) $r = 4$

b) $r\cos\theta = 3$

c) $r\sin\theta = 7$

d) $r = a(1 + \cos\theta)$

e) $r = a(1 - \cos\theta)$

f) $\dfrac{2}{r} = 1 + \cos\theta$

2 Find the polar equation of each of these curves.

a) $x^2 + y^2 = 9$ b) $xy = 16$ c) $\dfrac{x^2}{9} + \dfrac{y^2}{16} = 1$ d) $x^2 + y^2 = 6x$

e) $x^2 + y^2 + 8y = 16$ f) $(x^2 + y^2)^2 = x^2 - y^2$

Sketching curves given in polar coordinates

The normal way to sketch a curve expressed in polar coordinates is to plot points roughly using simple values of θ.

Example 3 Sketch $r = a\cos 3\theta$.

SOLUTION

Part of a table giving values for r is shown below.

$\boldsymbol{\theta}$	0	$\dfrac{\pi}{18}$	$\dfrac{\pi}{9}$	$\dfrac{\pi}{6}$	$\dfrac{\pi}{2}$	$\dfrac{5\pi}{9}$	$\dfrac{11\pi}{18}$	$\dfrac{2\pi}{3}$	$\dfrac{13\pi}{18}$	$\dfrac{7\pi}{9}$	$\dfrac{5\pi}{6}$
$\boldsymbol{r}$	a	$\dfrac{\sqrt{3}}{2}a$	$\dfrac{1}{2}a$	0	0	$\dfrac{1}{2}a$	$\dfrac{\sqrt{3}}{2}a$	a	$\dfrac{\sqrt{3}}{2}a$	$\dfrac{1}{2}a$	0

Note   When $\theta = \dfrac{2\pi}{9}$, $r = a\cos\left(\dfrac{2\pi}{3}\right) = -\dfrac{1}{2}a$. Therefore, since r must

always be positive, the curve does not exist when $\theta = \dfrac{2\pi}{9}$.

Similarly, the curve does not exist for any value of θ between

$$\dfrac{\pi}{6} \text{ and } \dfrac{\pi}{2}, \dfrac{5\pi}{6} \text{ and } \dfrac{7\pi}{6}, \dfrac{9\pi}{6} \text{ and } \dfrac{11\pi}{6}, \dfrac{13\pi}{6} \text{ and } \dfrac{15\pi}{6}$$

Plotting the values given in the table and joining the points gives a curve with three loops or lobes.

Notice that the lines $\theta = \dfrac{\pi}{6}$, $\theta = \dfrac{\pi}{2}$ and $\theta = \dfrac{5\pi}{6}$

are all **tangents** to the loops. The tangents meet at the origin or pole. All three loops are **congruent**.

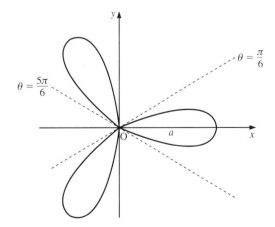

Example 4 Sketch $r = 1 + 2\cos\theta$.

SOLUTION

Part of a table giving values for r is shown below.

θ	0	$\dfrac{\pi}{6}$	$\dfrac{\pi}{3}$	$\dfrac{\pi}{2}$	$\dfrac{2\pi}{3}$	$-\dfrac{2\pi}{3}$	$-\dfrac{\pi}{2}$	$-\dfrac{\pi}{3}$	$-\dfrac{\pi}{6}$	0
r	3	$1 + \sqrt{3}$	2	1	0	0	1	2	$1 + \sqrt{3}$	3

To find when r is negative, we solve $r = 0$:

$$1 + 2\cos\theta = 0$$

$$\Rightarrow \quad \cos\theta = -\tfrac{1}{2}$$

$$\Rightarrow \quad \theta = \frac{2\pi}{3}, \frac{4\pi}{3} \left(\text{or} -\frac{2\pi}{3}\right)$$

We note that $1 + 2\cos\theta$ is negative for $\dfrac{2\pi}{3} < \theta < \dfrac{4\pi}{3}$. However, in A-level examinations **only positive values** of r are required, which means that, as far as you are concerned, the curve does not exist for values of θ between $\dfrac{2\pi}{3}$ and $\dfrac{4\pi}{3}$, and should not be shown.

When most models of the graphics calculator display this curve, they include negative values of r, which you should ignore.

The sketch of $r = 1 + 2\cos\theta$ is shown below.

The dashed part represents the negative values of r which are commonly displayed by graphics calculators.

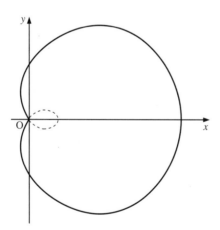

Most polar curves are sketched in the same way.

Here are three tips which will help you to sketch curves given in polar coordinates

- Look for any **symmetry**. If r is a function of $\cos\theta$ only, there is symmetry about the initial line. If r is a function of $\sin\theta$ only, there is symmetry about the line $\theta = \dfrac{\pi}{2}$.
- The equations $r = a\sin\theta$ and $r = a\cos\theta$ are **circles**.

Example 5 Find the cartesian equation of the curve $r = a\cos\theta$.

SOLUTION

Multiplying $r = a\cos\theta$ by r, we get

$$r^2 = ar\cos\theta$$

Substituting $r^2 = x^2 + y^2$ and $x = r\cos\theta$, we have

$$x^2 + y^2 = ax$$

which gives

$$\left(x - \frac{a}{2}\right)^2 + y^2 = \left(\frac{a}{2}\right)^2$$

This is a circle with centre $\left(\dfrac{a}{2}, 0\right)$ and radius a.

- When a polar equation contains $\sec\theta$ or $\operatorname{cosec}\theta$, it is often easier to **use its cartesian equation**.

Example 6 Sketch $r = a\sec\theta$.

SOLUTION

$$r = a\sec\theta = \frac{a}{\cos\theta}$$
$$\Rightarrow \quad r\cos\theta = a$$
$$\Rightarrow \quad x = a$$

This is the straight line shown on the right.

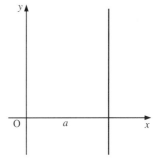

Example 7 Sketch $r = a\sec(\alpha - \theta)$.

SOLUTION

Transposing terms, we have

$$r\cos(\alpha - \theta) = a$$

which gives

$$r\cos\theta\cos\alpha + r\sin\theta\sin\alpha = a$$

Replacing $r\cos\theta$ with x and $r\sin\theta$ with y, we get

$$x\cos\alpha + y\sin\alpha = a.$$

which is the straight line shown on the right.

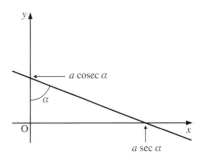

Exercise 3C

1 Sketch each of the curves given in Question **1** of Exercise 3B.

2 Sketch each of the following curves.

a) $r = a\sin 2\theta$, $\quad 0 < \theta < 2\pi$ **b)** $r = a\cos 4\theta$, $\quad 0 < \theta < 2\pi$

c) $r = 2 + 3\cos\theta$, $\quad -\pi < \theta < \pi$ **d)** $r = a\theta$, $\quad 0 < \theta < 2\pi$

e) $r = 4\sec\theta$, $\quad -\dfrac{\pi}{2} < \theta < \dfrac{\pi}{2}$

Area of a sector of a curve

Let A be the area bounded by the curve $r = f(\theta)$ and the two radii at α and at θ.

As θ increases by $\delta\theta$, the increase in area, δA, shown shaded, is given by

$$\tfrac{1}{2}r^2\delta\theta \leqslant \delta A \leqslant \tfrac{1}{2}(r + \delta r)^2\delta\theta \quad \text{(using areas of sectors)}$$

Dividing throughout by $\delta\theta$, we obtain

$$\tfrac{1}{2}r^2 \leqslant \frac{\delta A}{\delta\theta} \leqslant \tfrac{1}{2}(r + \delta r)^2$$

As $\delta\theta \to 0$, $\dfrac{\delta A}{\delta\theta} \to \dfrac{\mathrm{d}A}{\mathrm{d}\theta}$ and $\delta r \to 0$. Therefore, we have

$$\frac{\mathrm{d}A}{\mathrm{d}\theta} = \frac{1}{2}r^2$$

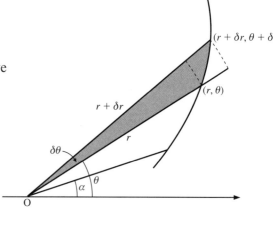

Integrating both sides with respect to θ, we obtain

$$\int \frac{\mathrm{d}A}{\mathrm{d}\theta}\,\mathrm{d}\theta = \frac{1}{2}\int r^2\,\mathrm{d}\theta$$

$$\Rightarrow \quad A = \frac{1}{2}\int r^2\,\mathrm{d}\theta$$

Therefore, the general equation for the area of a sector of a curve is

$$A = \frac{1}{2}\int_\alpha^\beta r^2\,\mathrm{d}\theta$$

when the area is bounded by the radii $\theta = \alpha$ and $\theta = \beta$.

Example 8 Find the area of one loop of the curve $r = a\cos 3\theta$.

SOLUTION

One loop is bounded by the tangent lines $\theta = \dfrac{\pi}{6}$ and $\theta = -\dfrac{\pi}{6}$ (see page 46).

Therefore, its area, A, is given by

$$A = \frac{1}{2}\int_{-\frac{\pi}{6}}^{\frac{\pi}{6}} r^2\,\mathrm{d}\theta \quad \Rightarrow \quad A = \frac{1}{2}\int_{-\frac{\pi}{6}}^{\frac{\pi}{6}} a^2\cos^2 3\theta\,\mathrm{d}\theta$$

Using the double-angle formula to integrate, we have

$$A = \frac{1}{2}a^2 \int_{-\frac{\pi}{6}}^{\frac{\pi}{6}} \frac{1}{2}(\cos 6\theta + 1)\,d\theta$$

$$= \frac{a^2}{4}\left[\frac{\sin 6\theta}{6} + \theta\right]_{-\frac{\pi}{6}}^{\frac{\pi}{6}}$$

$$= \frac{a^2}{4}\left(\frac{\pi}{6} + \frac{\pi}{6}\right) = \frac{a^2\pi}{12}$$

So, the area of one loop of $r = a\cos 3\theta$ is $\dfrac{a^2\pi}{12}$.

Note   It is often preferable to use only the area in the first quadrant when a curve is symmetrical in other quadrants. Thus, in Example 8, instead of using $\dfrac{1}{2}\displaystyle\int_{-\frac{\pi}{6}}^{\frac{\pi}{6}} a^2\cos^2 3\theta\,d\theta$, we could have used $2 \times \dfrac{1}{2}\displaystyle\int_{0}^{\frac{\pi}{6}} a^2\cos^2 3\theta\,d\theta$.

Example 9 Find the area bounded by the curve $r = k\theta$ and the lines $\theta = \dfrac{\pi}{2}$ and $\theta = \pi$.

SOLUTION

The curve $r = k\theta$ is shown for $\dfrac{\pi}{2} \leqslant \theta \leqslant \pi$.

The area, A, required is given by

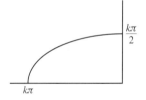

$$A = \frac{1}{2}\int_{\frac{\pi}{2}}^{\pi} k^2\theta^2\,d\theta$$

$$= \frac{k^2}{2}\left[\frac{\theta^3}{3}\right]_{\frac{\pi}{2}}^{\pi} = \frac{k^2}{2}\left(\frac{\pi^3}{3} - \frac{\pi^3}{24}\right) = \frac{7k^2\pi^3}{48}$$

Hence, the area required is $\dfrac{7k^2\pi^3}{48}$.

Example 10 Sketch the curves $r = 1 + \cos\theta$ and $r = \sqrt{3}\sin\theta$. Find

a) the points where the curves meet

b) the area contained between the curves.

SOLUTION

Before sketching the two curves, we note that

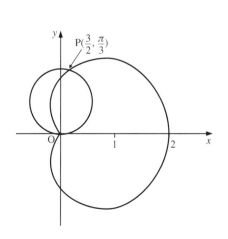

- $r = 1 + \cos\theta$ is similar to $1 + 2\cos\theta$ (page 46)
- $r = \sqrt{3}\sin\theta$ is similar to $r = a\cos\theta$ (page 47).

a) The two curves meet when

$$1 + \cos\theta = \sqrt{3}\sin\theta$$

Using

$$\sin\theta \equiv \sin\left(\frac{\theta}{2} + \frac{\theta}{2}\right) \equiv 2\sin\left(\frac{\theta}{2}\right)\cos\left(\frac{\theta}{2}\right)$$

and

$$\cos\theta \equiv 2\cos^2\left(\frac{\theta}{2}\right) - 1$$

we express $1 + \cos\theta = \sqrt{3}\sin\theta$ as

$$1 + 2\cos^2\left(\frac{\theta}{2}\right) - 1 = 2\sqrt{3}\sin\left(\frac{\theta}{2}\right)\cos\left(\frac{\theta}{2}\right)$$

$$\Rightarrow \quad 2\cos^2\left(\frac{\theta}{2}\right) = 2\sqrt{3}\sin\left(\frac{\theta}{2}\right)\cos\left(\frac{\theta}{2}\right)$$

which gives

$$\cos\left(\frac{\theta}{2}\right) = \sqrt{3}\sin\left(\frac{\theta}{2}\right) \quad \text{or} \quad \cos\left(\frac{\theta}{2}\right) = 0$$

$$\Rightarrow \quad \tan\left(\frac{\theta}{2}\right) = \frac{1}{\sqrt{3}} \qquad\qquad \Rightarrow \quad \theta = \pi$$

$$\Rightarrow \quad \frac{\theta}{2} = \frac{\pi}{6} \quad \Rightarrow \quad \theta = \frac{\pi}{3}$$

Therefore, the curves meet at $\left(\frac{3}{2}, \frac{\pi}{3}\right)$ and $(0, \pi)$

Remember These are polar coordinates (r, θ).

b) To find the area contained between the curves, we draw the line OP and consider separately the two areas so formed.

The area shaded in the diagram immediately right is bounded by the curve $r = \sqrt{3}\sin\theta$ and the two radii $\theta = 0$ and $\theta = \frac{\pi}{3}$.

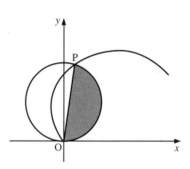

Hence, this area is $\dfrac{1}{2}\displaystyle\int_0^{\frac{\pi}{3}} (\sqrt{3}\sin\theta)^2 \, d\theta$.

The area shaded in the lower diagram on the right is bounded by the curve $r = 1 + \cos\theta$ and the two radii $\theta = \frac{\pi}{3}$ and π.

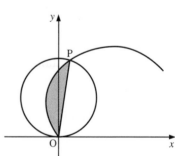

Hence, this area is $\dfrac{1}{2}\displaystyle\int_{\frac{\pi}{3}}^{\pi} (1 + \cos\theta)^2 \, d\theta$.

Therefore, the area contained between the two curves is given by

$$\frac{1}{2}\int_0^{\frac{\pi}{3}} (\sqrt{3}\sin\theta)^2 d\theta + \frac{1}{2}\int_{\frac{\pi}{3}}^{\pi} (1 + \cos\theta)^2 \, d\theta$$

$$= \frac{1}{2}\int_0^{\frac{\pi}{3}} 3\sin^2\theta \, d\theta + \frac{1}{2}\int_{\frac{\pi}{3}}^{\pi} (1 + 2\cos\theta + \cos^2\theta) \, d\theta$$

$$= \frac{3}{2} \int_0^{\frac{\pi}{3}} \frac{1}{2}(1 - \cos 2\theta)\, \mathrm{d}\theta + \frac{1}{2} \int_{\frac{\pi}{3}}^{\pi} \left[1 + 2\cos\theta + \frac{1}{2}(\cos 2\theta + 1) \right] \mathrm{d}\theta$$

$$= \frac{3}{4} \left[\theta - \frac{1}{2}\sin 2\theta \right]_0^{\frac{\pi}{3}} + \frac{1}{2} \left[\frac{3}{2}\theta + 2\sin\theta + \frac{1}{4}\sin 2\theta \right]_{\frac{\pi}{3}}^{\pi}$$

$$= \frac{3}{4} \left(\frac{\pi}{3} - \frac{\sqrt{3}}{4} \right) + \frac{1}{2} \left(\frac{3\pi}{2} - \frac{3\pi}{6} - \sqrt{3} - \frac{\sqrt{3}}{8} \right)$$

$$= \frac{3\pi}{4} - \frac{3\sqrt{3}}{4}$$

Therefore, the area contained within the curves is $\dfrac{3\pi}{4} - \dfrac{3\sqrt{3}}{4}$.

Exercise 3D

1 Find the area bounded by the curve $r = a\theta$ and the radii $\theta = \dfrac{\pi}{2}, \theta = \pi$.

2 For each of the following curves, find the area enclosed by one loop.

   a) $r = a\cos 2\theta$ **b)** $r = a\sin 2\theta$ **c)** $r = a\cos 4\theta$

3 Find the area enclosed by the curve $r = a\cos\theta$.

4 Find the area enclosed by the curve $r = 2 + 3\cos\theta$.

5 **a)** Find the polar equation of the curve $(x^2 + y^2)^3 = y^4$.
   b) Hence, **i)** sketch the curve, and **ii)** find the area enclosed by the curve.

6 Find where the following two curves intersect.

$$r = 2\sin\theta \qquad 0 \leqslant \theta < \pi$$
$$\text{and} \qquad r = 2(1 - \sin\theta) \qquad -\pi < \theta < \pi$$

Hence, find the area which is between the two curves.

7 In this question you may use the identity $\sin 3\theta \equiv 3\sin\theta - 4\sin^3\theta$. The cartesian equation of a curve C is

$$(x^2 + y^2)(x^2 + y^2 - 3ay) + 4ay^3 = 0$$

where $a > 0$.

   a) Show that, in terms of polar coordinates (r, θ), the equation of C is $r = a\sin 3\theta$.
   b) The curve consists of three equal loops, as shown in the diagram. The point O is the pole, and OL is the initial line.

Find, in terms of a, the exact value of the area of one of these loops.   (NEAB)

8 **a)** Sketch the curve with polar equation

$$r = a(2 + \cos\theta) \qquad 0 \leqslant \theta < 2\pi$$

and a is a positive constant.

Mark on your sketch the polar coordinates of the points where the curve meets the half-lines $\theta = 0$, $\theta = \pi$, $\theta = \dfrac{\pi}{2}$ and $\theta = \dfrac{3\pi}{2}$.

b) Find the area of the region enclosed by this curve, giving your answer in terms of π and a.

(EDEXCEL)

9 The diagram shows a sketch of the loop whose polar equation is

$$r = 2(1 - \sin\theta)\sqrt{(\cos\theta)} \qquad -\tfrac{1}{2}\pi \leqslant \theta \leqslant \tfrac{1}{2}\pi$$

where O is the pole.

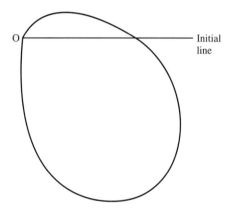

a) Show that the area enclosed by the loop is $\dfrac{16}{3}$.

b) Show that the initial line divides the area enclosed by the loop in the ratio $1:7$.   (NEAB)

Equations of the tangents to a curve

The tangents to $r = a\cos 3\theta$ **perpendicular** to the initial line are shown on the right.

These are at points A, where x is at a maximum, B and C where x is at a minimum, and D, where x has a point of inflexion.

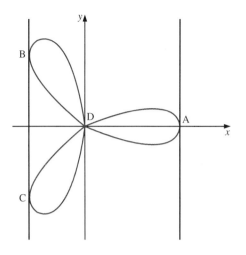

We note that $x = r\cos\theta$. Therefore, to find the maximum and minimum values of x, we find the maximum and the minimum values of $r\cos\theta$.

Since $r = a\cos 3\theta$, we have

$$x = a\cos 3\theta \cos\theta$$

which gives

$$\frac{\mathrm{d}x}{\mathrm{d}\theta} = -3a\sin 3\theta \cos\theta - a\cos 3\theta \sin\theta$$

The maximum and the minimum values occur when $\dfrac{\mathrm{d}x}{\mathrm{d}\theta} = 0$. That is, when

$$-3a\sin 3\theta \cos\theta - a\cos 3\theta \sin\theta = 0$$

We simplify this expression using the factor formulae:

$$\sin A\cos B = \frac{1}{2}[\sin(A+B) + \sin(A-B)]$$

and

$$\cos A\sin B = \frac{1}{2}[\sin(A+B) - \sin(A-B)]$$

which give $\dfrac{\mathrm{d}x}{\mathrm{d}\theta} = 0$ when

$$\frac{3}{2}[\sin 4\theta + \sin 2\theta] + \frac{1}{2}[\sin 4\theta - \sin 2\theta] = 0$$

$$\Rightarrow \quad 2\sin 4\theta + \sin 2\theta = 0$$

Applying the double-angle formula, we get

$$4\sin 2\theta \cos 2\theta + \sin 2\theta = 0$$
$$\sin 2\theta(4\cos 2\theta + 1) = 0$$

which gives

$$\sin 2\theta = 0 \quad \text{or} \quad \cos 2\theta = -\tfrac{1}{4}$$

$$\sin 2\theta = 0 \quad \Rightarrow \quad \theta = 0, \frac{\pi}{2}, \pi, \ldots$$

$$\cos 2\theta = -\tfrac{1}{4} \quad \Rightarrow \quad \theta = n\pi \pm 0.912$$

We have to ensure that the curve exists at these points. For example, when $\theta = 0.912$, $\cos 3\theta$ is negative, thus r is negative and the curve does not exist.

Hence, the values of θ at the points where the tangent is perpendicular to the initial line are

$$\theta = 0 \text{ at A} \qquad\qquad \theta = \pi - 0.912 \text{ at B}$$

$$\theta = -\pi + 0.912 \text{ at C} \qquad \theta = \frac{\pi}{2} \text{ at D}$$

So, the equations of the tangents **perpendicular** to the initial line are

$$x = a$$
$$x = 0$$

and

$$x = -0.9858a \quad \text{or} \quad a\cos\frac{3}{2}\left[\cos^{-1}\left(-\frac{1}{4}\right)\right]$$

The tangents to $r = a\cos 3\theta$ **parallel** to the initial line are shown below. These are at the points P, Q, R and S. To find these points, we find the maximum and minimum values of $r\sin\theta$ in a similar way to that shown above.

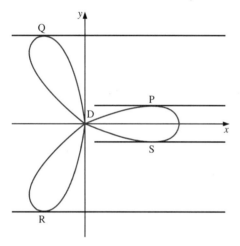

Exercise 3E

1 Find the equation of each tangent to the curve $r = a\cos 3\theta$ which is parallel to the initial line.

2 Find the equation of the tangent to the curve $r = e^\theta$ which is

a) parallel to the initial line
b) perpendicular to the initial line.

3 Give in polar coordinates the points on the curve $r = a\cos 2\theta$ where the tangents are

a) parallel to the initial line
b) perpendicular to the initial line.

4 The diagram (top of page 55) shows a square PQRS with sides parallel to the axes Ox and Oy. The square circumscribes a curve C whose cartesian equation is $(x^2 + y^2)^{\frac{3}{2}} = xy$.

a) Show that, in terms of polar coordinates (r, θ), the equation of C is $r = \frac{1}{2}\sin 2\theta$.

b) Find the area bounded by C.

c) The coordinates of a variable point on C are (x, y).

  i) Show that $x = \sin\theta - \sin^3\theta$.

  ii) Show that, as θ varies, the maximum value of x occurs when $\sin\theta = \dfrac{1}{\sqrt{3}}$.

  iii) Calculate the area of the square PQRS.

(NEAB)

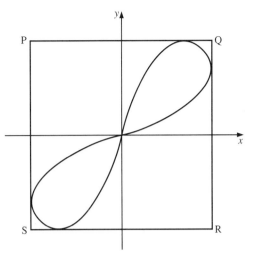

5 The diagram shows a sketch of the curve C whose polar equation is

$$r = \sqrt{3} - \cos\theta \quad (-\pi < \theta \leqslant \pi)$$

The line L touches the curve at A and B. Express in terms of θ the x-coordinate of a general point, P, on C and determine the values of θ for which this coordinate has a stationary value.

Deduce that at A, $\theta = \dfrac{\pi}{6}$.

Show that the area of the region bounded by C and L, shown shaded in the diagram, is

$$\frac{17\sqrt{3}}{16} - \frac{7\pi}{12} \quad \text{(NEAB)}$$

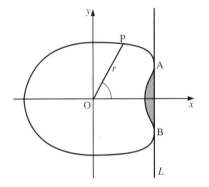

6 a) Sketch the curve with polar equation

$$r = \cos 2\theta \quad -\frac{\pi}{4} \leqslant \theta \leqslant \frac{\pi}{4}$$

At the distinct points A and B on this curve, the tangents to the curve are parallel to the initial line, $\theta = 0$.

b) Determine the polar coordinates of A and B, giving your answers to three significant figures. (EDEXCEL)

7 The figure on the right shows a sketch of the circle with polar equation $r = a$ and the cardioid with polar equation $r = a(1 - \cos\theta)$, where a is a positive constant.

a) Verify that the curves intersect where $\theta = \pm\dfrac{\pi}{2}$.

b) Find the area of the shaded region, giving your answer in terms of a and π. (EDEXCEL)

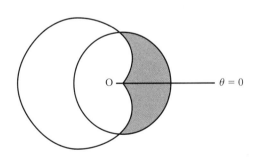

8 The curves C_1 and C_2 have polar equations

$$C_1: \quad r = 4\sin^2\theta \quad 0 \leqslant \theta < 2\pi$$
$$C_2: \quad r = (2\sqrt{3})\sin 2\theta \quad 0 \leqslant \theta < \tfrac{1}{2}\pi$$

a) Sketch C_1 and C_2 on the same diagram.

b) Find the polar coordinates of all points of intersection of C_1 and C_2.

c) Find, to two decimal places, the area of the region R which is inside both C_1 and C_2.

(EDEXCEL)

9 Relative to the origin O as pole and initial line $\theta = 0$, find an equation in polar coordinate form for

a) a circle, centre O and radius 2

b) a line perpendicular to the initial line and passing through the point with polar coordinates $(3, 0)$

c) a straight line through the points with polar coordinates $(4, 0)$ and $\left(4, \dfrac{\pi}{3}\right)$.   (EDEXCEL)

4 Differential equations

Change and decay in all around I see.
H. F. LYTE

We have already solved first-order differential equations in which the variables are separable (see pages 457–60 in *Introducing Pure Mathematics*.)

We will now consider three other main types of differential equation.

First-order equations requiring an integrating factor

This is the other main type of first-order differential equation.

Equations of this type are of the form

$$\frac{dy}{dx} + Py = Q$$

where P and Q are functions of x.

Such an equation can be solved by first multiplying both sides by the **integrating factor** $e^{\int P\,dx}$.

Multiplying $\frac{dy}{dx} + Py = Q$ by $e^{\int P\,dx}$, we get

$$e^{\int P\,dx}\frac{dy}{dx} + Pe^{\int P\,dx}y = Qe^{\int P\,dx}$$

Since the left-hand side is the differential of $ye^{\int P\,dx}$, we therefore have

$$\frac{d}{dx}\left(ye^{\int P\,dx}\right) = Qe^{\int P\,dx}$$

which gives

$$ye^{\int P\,dx} = \int Qe^{\int P\,dx}\,dx$$

The right-hand side is often integrated by parts.

Example 1 If $\dfrac{dy}{dx} + 3y = x$, find y.

SOLUTION

The integrating factor is $e^{\int 3\,dx}$, which is e^{3x}.

Multiplying both sides by e^{3x}, we obtain

$$e^{3x}\frac{dy}{dx} + e^{3x}3y = xe^{3x}$$

$$\Rightarrow \quad \frac{d}{dx}(ye^{3x}) = xe^{3x}$$

Integrating by parts, we have

$$ye^{3x} = \int xe^{3x}\,dx$$

$$= \frac{1}{3}e^{3x} \times x - \int \frac{1}{3}e^{3x}\,dx$$

which gives

$$ye^{3x} = \frac{1}{3}xe^{3x} - \frac{1}{9}e^{3x} + c$$

Multiplying both sides by e^{-3x}, **including** c, we obtain

$$y = \frac{1}{3}x - \frac{1}{9} + ce^{-3x}$$

Note   The constant term, c, has now become a function of x.

Example 2 Solve the differential equation $x\dfrac{dy}{dx} - 2y = x^4$.

SOLUTION

Dividing both sides by x to make the first term $\dfrac{dy}{dx}$, we obtain

$$\frac{dy}{dx} - \frac{2y}{x} = x^3$$

The integrating factor is

$$e^{\int -(2/x)\,dx} = e^{-2\ln x} = e^{\ln x^{-2}}.$$

Applying the result $e^{\ln u} = u$, we have $e^{\ln x^{-2}} = \dfrac{1}{x^2}$.

We now multiply the differential equation by the integrating factor, $\dfrac{1}{x^2}$, to obtain

$$\frac{1}{x^2}\frac{dy}{dx} - \frac{2}{x^3}y = x$$

which we express as

$$\frac{d}{dx}\left(\frac{1}{x^2}y\right) = x$$

$$\Rightarrow \quad \frac{1}{x^2}y = \int x\,dx$$

$$\Rightarrow \quad \frac{1}{x^2}y = \frac{x^2}{2} + c$$

Multiplying both sides by x^2, we obtain the **general solution**

$$y = \tfrac{1}{2}x^4 + cx^2$$

Note To obtain a **particular solution**, we need to be given a specific point which lies on the curve. Hence, we can find the value of c. This extra fact is called a **boundary condition**. Example 3 illustrates such a situation.

Example 3 Solve the differential equation $\dfrac{dy}{dx} + \dfrac{1}{x}y = x^2$, given that $y = 3$ when $x = 2$.

SOLUTION

The integrating factor is $e^{\int (1/x)\,dx} = e^{\ln x} = x$.

Multiplying the differential equation by the integrating factor, x, we have

$$x\frac{dy}{dx} + y = x^3$$

which we express as

$$\frac{d}{dx}(xy) = x^3$$

$$\Rightarrow \quad xy = \frac{1}{4}x^4 + c$$

When $x = 2$, $y = 3$, which gives

$$6 = 4 + c \quad \Rightarrow \quad c = 2$$

Therefore, the solution is

$$xy = \frac{1}{4}x^4 + 2 \quad \text{or} \quad y = \frac{1}{4}x^3 + \frac{2}{x}$$

Exercise 4A

1 Simplify each of the following.

a) $e^{\ln x^2}$ **b)** $e^{\frac{1}{2}\ln(x^2+1)}$ **c)** $e^{-3\ln x}$ **d)** $e^{\int \tan x\,dx}$

e) $e^{\int x/(x^2-1)\,dx}$ **f)** $e^{3x\ln 2}$

In each of Questions **2** to **7**, find the general solution.

2 $\dfrac{dy}{dx} + 3y = x$ **3** $\dfrac{dy}{dx} - 5y = e^{2x}$ **4** $x\dfrac{dy}{dx} + y = x^2$

5 $x\dfrac{dy}{dx} - 2y = x^3$ **6** $\dfrac{dy}{dx} - \dfrac{4y}{x-1} = 5(x-1)^3$ **7** $\tan x\dfrac{dy}{dx} + y = e^{2x}\tan x$

8 A curve C in the x–y plane passes through the point $(1,0)$. At any point (x,y) on C,

$$\frac{dy}{dx} + y = e^{-x}$$

a) Find the general solution of this differential equation.
b) **i)** Hence find the equation of C, giving your answer in the form $y = f(x)$.
     ii) Write down the equation of the asymptote of C.   (NEAB)

9 Find the general solution of the differential equation

$$\frac{dy}{dx} - 3x^2 y = x e^{x^3}$$

giving y explicitly in terms of x in your answer.

Find also the particular solution for which $y = 1$ when $x = 0$.   (OCR)

10 Find the general solution of the differential equation

$$(\cos x)\frac{dy}{dx} + (\sin x)y = \cos^2 x$$

expressing y in terms of x.   (OCR)

11 Find the general solution of the differential equation

$$x\frac{dy}{dx} + 4y = x$$

giving y explicitly in terms of x in your answer.

Find also the particular solution for which $y = 1$ when $x = 1$.   (OCR)

12 Find, in the form $y = f(x)$, the general solution of the differential equation

$$\frac{dy}{dx} + \frac{4}{x}y = 6x - 5 \quad x > 0 \quad \text{(EDEXCEL)}$$

13 A car moves from rest along a straight road. After t seconds the velocity is v metres per second. The motion is modelled by

$$\frac{dv}{dt} + \alpha v = e^{\beta t}$$

where α and β are positive constants.

  i) Find v in terms of α, β and t.
  ii) Show that, as long as the above model applies, the car does not come to rest.   (OCR)

14 The variables v and t are related by the differential equation

$$\frac{dv}{dt} = 20 + \tfrac{1}{10}v\tan\left(\tfrac{1}{10}t\right)$$

Given that $v = 1$ when $t = 0$, find v when $t = 2$.   (OCR)

15 i) Find the general solution of the differential equation

$$\frac{dy}{dx} + y\tan x = \cos x$$

  ii) If $y = 2$ when $x = 0$, find the particular solution.   (NICCEA)

16 Given that

$$\frac{dy}{dx} + (2x + 1)y = 12x^3 e^{-x^2 - x}$$

and that $y = 5$ when $x = 0$, find y in terms of x.   (OCR)

17 The number, N, of animals of a certain species at time t years increases at a rate of λN per year by births, but decreases at a rate of μt per year by deaths, where λ and μ are positive constants.

Modelled as continuous variables, N and t are related by the differential equation

$$\frac{dN}{dt} = \lambda N - \mu t$$

Given that $N = N_0$ when $t = 0$, find N in terms of t, λ, μ and N_0.   (OCR)

18 i) Find the general solution of the differential equation

$$\frac{dy}{dx} = k(x + y)$$

where k is a constant, giving your answer in the form $y = f(x)$.

ii) The gradient at any point $P(x, y)$ of a curve is proportional to the sum of the coordinates of P. The curve passes through the point $(1, -2)$ and its gradient at $(1, -2)$ is -4.

   a) Find the equation of the curve.

   b) Show that the line $y = -x - \frac{1}{4}$ is an asymptote to the curve.   (OCR)

19 i) Show that the appropriate integrating factor for

$$\frac{dy}{dx} + (2 \cot x)y = f(x)$$

is $\sin^2 x$.

ii) Hence find the general solution of the differential equation

$$\sin x \frac{dy}{dx} + 2y \cos x = \cos x \qquad \text{(NICCEA)}$$

20 Find the general solution of the differential equation

$$(4 + t^2) \frac{ds}{dt} = 1$$

Given that $s = 0$ when $t = 2$, express s in terms of t.   (EDEXCEL)

21 a) Find the general solution of the differential equation

$$x \frac{dy}{dx} - y = x^2 e^{-x}$$

giving your answer in the form $y = f(x)$.

b) i) Verify that the graphs of all solutions of the differential equation pass through the origin O, and find the particular solution which is such that $\frac{dy}{dx} = -1$ at O.

ii) For this particular solution, state the limiting value of y as $x \to \infty$.   (NEAB)

Second-order differential equations

An equation is termed **second order** when it contains the second derivative, $\frac{d^2 y}{dx^2}$.

Initially, we will consider equations of the form

$$a \frac{d^2 y}{dx^2} + b \frac{dy}{dx} + cy = 0$$

where a, b and c are constants.

To solve the equation $a\dfrac{d^2y}{dx^2} + b\dfrac{dy}{dx} + cy = 0$, we make the substitution $y = Ae^{nx}$. Hence, we have

$$\frac{dy}{dx} = nAe^{nx} \quad \text{and} \quad \frac{d^2y}{dx^2} = n^2Ae^{nx}$$

which give

$$an^2Ae^{nx} + bnAe^{nx} + cAe^{nx} = 0$$

That is,

$$an^2 + bn + c = 0$$

This quadratic equation is called the **auxiliary equation**.

The solution of a second-order differential equation depends on the type of solution which satisfies its auxiliary equation. There are three types of solution of a quadratic equation:

1 Two real and different roots
2 Two real and equal roots.
3 Two complex roots.

Type 1 solution.

The auxiliary equation has two **real**, **different roots**, n_1 and n_2. So, the solution of $a\dfrac{d^2y}{dx^2} + b\dfrac{dy}{dx} + cy = 0$ is

$$y = Ae^{n_1 x} + Be^{n_2 x}$$

where A and B are arbitrary constants.

To verify that this is the full solution, we need to confirm that the following two conditions obtain:

- There are two arbitrary constants, as it is a second-order differential equation.
- The solution does satisfy the equation

$$a\frac{d^2y}{dx^2} + b\frac{dy}{dx} + cy = 0$$

We notice that there are indeed the two required arbitrary constants.

To prove that the solution, $y = Ae^{n_1 x} + Be^{n_2 x}$, satisfies the differential equation, we substitute it and its derivatives in the LHS of

$$a\frac{d^2y}{dx^2} + b\frac{dy}{dx} + cy = 0$$

which gives

$$a\frac{d^2y}{dx^2} + b\frac{dy}{dx} + cy = a(n_1^2 Ae^{n_1 x} + n_2^2 Be^{n_2 x}) + b(n_1 Ae^{n_1 x} + n_2 Be^{n_2 x}) + c(Ae^{n_1 x} + Be^{n_2 x})$$

$$= Ae^{n_1 x}(an_1^2 + bn_1 + c) + Be^{n_2 x}(an_2^2 + bn_2 + c)$$

$$= 0$$

since n_1 and n_2 are roots of the equation $an^2 + bn + c = 0$.

To find the values of A and B, we need **two boundary conditions**. Usually, these are either

- the values of y at two different values of x, or
- the value of y and that of $\dfrac{dy}{dx}$ for one value of x.

Example 4 Find y when $2\dfrac{d^2y}{dx^2} - \dfrac{dy}{dx} - 3y = 0$, given that $x = 0$ when $y = 2$ and y is finite as x tends to infinity.

SOLUTION

Substituting $y = Ae^{nx}$ and its derivatives in $2\dfrac{d^2y}{dx^2} - \dfrac{dy}{dx} - 3y = 0$, we get

$$2n^2 - n - 3 = 0$$
$$\Rightarrow \quad (2n - 3)(n + 1) = 0$$
$$\Rightarrow \quad n = \frac{3}{2} \quad \text{and} \quad -1$$

Therefore, we have

$$y = Ae^{\frac{3}{2}x} + Be^{-x}$$

When $x = 0$, $y = 2$, which gives

$$2 = A + B$$

We know that as x tends to infinity, y is finite. Therefore, $A = 0$ because the limit of $e^{\frac{3}{2}x}$ as x tends to infinity is not finite.

Hence, $B = 2$, which gives $y = 2e^{-x}$.

Type 2 solution

The auxiliary equation has two **real, equal roots**, n. In this case, we cannot, as in Type 1, use just $y = Ae^{nx} + Be^{nx}$, since this simplifies to $y = (A + B)e^{nx}$ or $y = Ce^{nx}$, which has only **one** arbitrary constant. The solution is, therefore,

$$y = (A + Bx)e^{nx}$$

To prove this is the solution, we must show that it satisfies the equation

$$a\frac{d^2y}{dx^2} + b\frac{dy}{dx} + cy = 0$$

Differentiating $y = (A + Bx)e^{nx}$ twice, we get

$$\frac{dy}{dx} = Be^{nx} + ne^{nx}(A + Bx)$$

$$\frac{d^2y}{dx^2} = Bne^{nx} + n^2e^{nx}(A + Bx) + ne^{nx}B$$

$$= n^2(A + Bx)e^{nx} + 2nBe^{nx}$$

Substituting these in the LHS of $a\dfrac{d^2y}{dx^2} + b\dfrac{dy}{dx} + cy = 0$, we have

$$a\frac{d^2y}{dx^2} + b\frac{dy}{dx} + cy = a[n^2(A + Bx)e^{nx} + 2nBe^{nx}] + b[Be^{nx} + ne^{nx}(A + Bx)] + c(A + Bx)e^{nx}$$

$$= (A + Bx)e^{nx}(an^2 + bn + c) + (2na + b)Be^{nx}$$

Since n is a root of $an^2 + bn + c = 0$, the first term is zero.

Consider now the quadratic formula, $n = \dfrac{-b \pm \sqrt{b^2 - 4ac}}{2a}$. When its roots are coincident, $b^2 - 4ac = 0$. Therefore, we have

$$n = -\frac{b}{2a} \quad \Rightarrow \quad 2na + b = 0$$

So, the second term is also zero.

Hence, $a\dfrac{d^2y}{dx^2} + b\dfrac{dy}{dx} + cy$ does equal zero, and $y = (A + Bx)e^{nx}$ is indeed the required solution.

Example 5 Solve $\dfrac{d^2y}{dx^2} + 6\dfrac{dy}{dx} + 9y = 0$.

SOLUTION

Substituting $y = Ae^{nx}$ and its derivatives in $\dfrac{d^2y}{dx^2} + 6\dfrac{dy}{dx} + 9y = 0$, we get

$$n^2 + 6n + 9 = 0$$
$$\Rightarrow \quad (n + 3)(n + 3) = 0$$
$$\Rightarrow \quad n = -3$$

Therefore, the general solution is

$$y = (A + Bx)e^{-3x}$$

Type 3 solution

The auxiliary equation has two **complex roots**, $n_1 \pm in_2$.

Therefore, the solution of $a\dfrac{d^2y}{dx^2} + b\dfrac{dy}{dx} + cy = 0$ is

$$y = Ae^{(n_1 + in_2)x} + Be^{(n_1 - in_2)x}$$
$$= e^{n_1 x}(Ae^{in_2 x} + Be^{-in_2 x})$$
$$= e^{n_1 x}[A\cos n_2 x + iA\sin n_2 x + B\cos(-n_2 x) + iB\sin(-n_2 x)]$$
$$= e^{n_1 x}(A\cos n_2 x + iA\sin n_2 x + B\cos n_2 x - iB\sin n_2 x)$$
$$= e^{n_1 x}[(A + B)\cos n_2 x + i(A - B)\sin n_2 x]$$

Since A and B are arbitrary constants, we can combine $(A + B)$ to give an arbitrary constant C, and we can combine $i(A - B)$ to give an arbitrary constant D. So, we have

$$y = e^{n_1 x}(C\cos n_2 x + D\sin n_2 x)$$

Example 6 Solve $\dfrac{d^2y}{dx^2} - 2\dfrac{dy}{dx} + 3y = 0$, given that $y = 0$ and $\dfrac{dy}{dx} = 6$, when $x = 0$.

SOLUTION

Substituting $y = Ae^{nx}$ and its derivatives in $\dfrac{d^2y}{dx^2} - 2\dfrac{dy}{dx} + 3y = 0$, we get

$$n^2 - 2n + 3 = 0$$

$$\Rightarrow \quad n = \dfrac{2 \pm \sqrt{4 - 12}}{2} = 1 \pm \sqrt{2}i$$

Therefore, the general solution is

$$y = e^x(C\cos\sqrt{2}x + D\sin\sqrt{2}x)$$

To find C and D, we use the boundary conditions.

When $x = 0$, $y = 0$, which gives

$$0 = C\cos 0 + D\sin 0 \quad \Rightarrow \quad C = 0$$

Hence, we have

$$y = De^x \sin\sqrt{2}x$$

As one boundary condition is given in terms of $\dfrac{dy}{dx}$, we differentiate the above:

$$\dfrac{dy}{dx} = De^x \sin\sqrt{2}x + \sqrt{2}De^x \cos\sqrt{2}x$$

When $x = 0$, $\dfrac{dy}{dx} = 6$, which gives

$$6 = D\sin 0 + \sqrt{2}D\cos 0$$

$$\Rightarrow \quad 6 = \sqrt{2}D \quad \Rightarrow \quad D = 3\sqrt{2}$$

Therefore, the solution is $y = 3\sqrt{2}e^x \sin\sqrt{2}x$.

Alternative notation for derivatives

Sometimes it is more convenient to denote $\dfrac{dy}{dx}$ by y' or f', and $\dfrac{d^2y}{dx^2}$ by y'' or f'', where $y = f(x)$.

Exercise 4B

In Questions **1** to **12**, find the general solution of each differential equation.

1 $\dfrac{d^2y}{dx^2} - 6\dfrac{dy}{dx} - 8y = 0$ **2** $\dfrac{d^2y}{dx^2} + 3\dfrac{dy}{dx} + 2y = 0$ **3** $2\dfrac{d^2y}{dx^2} - \dfrac{dy}{dx} - 6y = 0$

4 $3\dfrac{d^2y}{dx^2} + 4\dfrac{dy}{dx} - 7y = 0$ **5** $\dfrac{d^2x}{dt^2} - 7\dfrac{dx}{dt} - 8x = 0$ **6** $\dfrac{d^2x}{dt^2} - 11\dfrac{dx}{dt} + 28x = 0$

7 $\dfrac{d^2y}{dx^2} + 4\dfrac{dy}{dx} + 4y = 0$ **8** $\dfrac{d^2y}{dx^2} - 6\dfrac{dy}{dx} + 9y = 0$ **9** $\dfrac{d^2y}{dx^2} + \dfrac{dy}{dx} + y = 0$

10 $\dfrac{d^2y}{dx^2} + 4\dfrac{dy}{dx} + 8y = 0$ **11** $\dfrac{d^2x}{dt^2} - 6\dfrac{dx}{dt} + 7x = 0$ **12** $\dfrac{d^2x}{dt^2} + 2\dfrac{dx}{dt} + 13x = 0$

Second-order differential equations of the type

$$a\frac{d^2y}{dx^2} + b\frac{dy}{dx} + cy = f(x)$$

If $y = g(x)$ is the solution of

$$a\frac{d^2y}{dx^2} + b\frac{dy}{dx} + cy = 0$$

and $y = h(x)$ is the solution of

$$a\frac{d^2y}{dx^2} + b\frac{dy}{dx} + cy = f(x)$$

then we have

$$y = h(x) + \lambda g(x)$$

as the **general solution** of

$$a\frac{d^2y}{dx^2} + b\frac{dy}{dx} + cy = f(x)$$

Proof

Substituting $y = h + \lambda g$ and its derivatives in the LHS of

$$a\frac{d^2y}{dx^2} + b\frac{dy}{dx} + cy = f(x)$$

we have

$$ay'' + by' + cy = a(h'' + \lambda g'') + b(h' + \lambda g') + c(h + \lambda g)$$
$$= ah'' + bh' + ch + \lambda(ag'' + by' + cy)$$
$$= f(x)$$

since h is a solution of $ah'' + bh' + ch = f(x)$, and g is a solution of $ag'' + bg' + cg = 0$.

Therefore,

$$y = h(x) + \lambda g(x)$$

is the general solution of $ay'' + by' + cy = f(x)$

$g(x)$ is called the **complementary function (CF)**, and $h(x)$ is called the **particular integral (PI)**.

The particular solution is obtained by inserting boundary conditions into the general solution.

Types of particular integral

The particular integral depends on the function $f(x)$.

We will consider three types of function $f(x)$:

- polynomial
- exponential
- trigonometric

• f(x) is a polynomial of degree n

In this case, the particular integral will also be a polynomial of degree n.

Example 7 By finding **a)** the complementary function and **b)** the particular integral, solve the equation

$$\frac{d^2x}{dt^2} + 3\frac{dx}{dt} - 4x = 8$$

SOLUTION

a) For the complementary function, we use

$$\frac{d^2x}{dt^2} + 3\frac{dx}{dt} - 4x = 0$$

Substituting $x = Ae^{nt}$ and its derivatives in the above equation, we get

$$n^2 + 3n - 4 = 0$$
$$\Rightarrow \quad (n+4)(n-1) = 0$$
$$\Rightarrow \quad n = 1 \quad \text{or} \quad -4$$

So, the CF is $x = Ae^t + Be^{-4t}$.

b) For the particular integral, f(x) is a polynomial of degree 0. Hence, we need consider only $x = c$ for the particular integral.

Substituting $x = c$ in $\dfrac{d^2x}{dt^2} + 3\dfrac{dx}{dt} - 4x = 8$, we get

$$-4c = 8 \quad \Rightarrow \quad c = -2$$

So, the PI is $x = -2$.

Therefore, the general solution is $x = Ae^t + Be^{-4t} - 2$.

Example 8 Find the solution of $\dfrac{d^2y}{dx^2} + 3\dfrac{dy}{dx} - 4y = 2 + 8x^2$, given that, when $x = 0$, $y = 0$ and $\dfrac{dy}{dx} = 1$.

SOLUTION

To find the CF, we use

$$\frac{d^2y}{dx^2} + 3\frac{dy}{dx} - 4y = 0$$

Substituting $y = Ae^{nx}$ and its derivatives in the above equation, we get

$$n^2 + 3n - 4 = 0$$
$$\Rightarrow \quad (n+4)(n-1) = 0$$
$$\Rightarrow \quad n = 1 \quad \text{or} \quad -4$$

So, the CF is $y = Ae^x + Be^{-4x}$.

To find the PI, we substitute $y = a + bx + cx^2$ and its derivatives in

$$\frac{d^2y}{dx^2} + 3\frac{dy}{dx} - 4y = 3 + 8x^2$$

which gives

$$2c + 3(b + 2cx) - 4(a + bx + cx^2) = 3 + 8x^2$$

Equating coefficients of x^2: $-4c = 8$ $\Rightarrow$ $c = -2$

Equating coefficients of x: $6c - 4b = 0$ $\Rightarrow$ $b = -3$

Letting $x = 0$ in the above equation, we get

$$2c + 3b - 4a = 3$$
$$\Rightarrow \quad a = -4$$

So, the PI is $y = -4 - 3x - 2x^2$.

Therefore, the general solution is

$$y = Ae^x + Be^{-4x} - 4 - 3x - 2x^2$$

We now need to find values for A and B.

When $x = 0$, $y = 0$, which gives

$$0 = A + B - 4$$
$$\Rightarrow \quad A + B = 4 \qquad\qquad [1]$$

Differentiating $y = Ae^x + Be^{-4x} - 4 - 3x - 2x^2$, we have

$$\frac{dy}{dx} = Ae^x - 4Be^{-4x} - 3 - 4x$$

When $x = 0$, $\dfrac{dy}{dx} = 1$, which gives

$$1 = A - 4B - 3$$
$$\Rightarrow \quad A - 4B = 4 \qquad\qquad [2]$$

From [1] and [2], we get $A = 4$ and $B = 0$.

Therefore, the general solution is $y = 4e^x - 4 - 3x - 2x^2$.

• f(x) is an exponential function

Take, for example, the equation

$$\frac{d^2y}{dx^2} + 3\frac{dy}{dx} - 4y = 3e^{7x}$$

In this case, $f(x) = 3e^{7x}$. The particular integral will be of the same form: Ce^{7x}.

Therefore, the CF is $y = Ae^x + Be^{-4x}$ (see Example 8).

To find the PI, we substitute $y = Ce^{7x}$ and its derivatives in

$$\frac{d^2y}{dx^2} + 3\frac{dy}{dx} - 4y = 3e^{7x}$$

which gives

$$49Ce^{7x} + 21Ce^{7x} - 4Ce^{7x} = 3e^{7x}$$

$$\Rightarrow \quad 66C = 3 \quad \Rightarrow \quad C = \tfrac{1}{22}$$

So, the PI is $y = \tfrac{1}{22}e^{7x}$.

Therefore, the general solution is $y = Ae^x + Be^{-4x} + \dfrac{e^{7x}}{22}$.

● **f(x) is a trigonometric function of the form $a \sin nx$**

Take, for example, $f(x) = 4 \sin 2x$. The particular integral will be of the form
$C \sin 2x + D \cos 2x$

Example 9 Solve $\dfrac{d^2y}{dx^2} + 3 \dfrac{dy}{dx} - 4y = 4 \sin 2x$.

SOLUTION

The CF is $y = Ae^x + Be^{-4x}$ (see Example 8).

Caution Suppose we were simply to consider $y = C \sin 2x$ as the PI.
Because there is only a $\sin 2x$ term on the right-hand side, we would
obtain

$$\frac{dy}{dx} = 2C \cos 2x \quad \text{and} \quad \frac{d^2y}{dx^2} = -4C \sin 2x$$

Substituting these in $\dfrac{d^2y}{dx^2} + 3 \dfrac{dy}{dx} - 4y = 4 \sin 2x$, we would obtain

$$-4C \sin 2x + 3 \times 2C \cos 2x - 4C \sin 2x = 4 \sin 2x$$

which includes only one term in $\cos 2x \left(\text{from } \dfrac{dy}{dx} \right)$.

This means that this equation **cannot be solved**.

Hence, the PI used **must** contain **both** $\sin 2x$ **and** $\cos 2x$ terms. That is,

$$y = C \sin 2x + D \cos 2x$$

Differentiating this, we have

$$y' = 2C \cos 2x - 2D \sin 2x$$
$$y'' = -4C \sin 2x - 4D \cos 2x$$

Substituting y' and y'' in $\dfrac{d^2y}{dx^2} + 3 \dfrac{dy}{dx} - 4y = 4 \sin 2x$, we get

$$-4C \sin 2x - 4D \cos 2x + 6C \cos 2x - 6D \sin 2x - 4C \sin 2x - 4D \cos 2x = 4 \sin 2x$$

Equating coefficients of $\sin 2x$: $\quad -8C - 6D = 4$

$$\Rightarrow \quad -4C - 3D = 2 \qquad [1]$$

Equating coefficients of $\cos 2x$: $\quad -8D + 6C = 0$

$$\Rightarrow \quad -4D + 3C = 0 \qquad [2]$$

Solving the simultaneous equations [1] and [2], we get

$$C = -\frac{8}{25} \quad \text{and} \quad D = -\frac{6}{25}$$

Therefore, the PI is

$$y = -\frac{8}{25}\sin 2x - \frac{6}{25}\cos 2x$$

Hence, the general solution is

$$y = Ae^x + Be^{-4x} - \frac{8}{25}\sin 2x - \frac{6}{25}\cos 2x$$

Example 10 Solve $\dfrac{d^2y}{dx^2} - \dfrac{dy}{dx} - 2y = 3e^{2x}$, given that $y = 0$ and $\dfrac{dy}{dx} = 11$ when $x = 0$.

SOLUTION

To find the CF, we use

$$\frac{d^2y}{dx^2} - \frac{dy}{dx} - 2y = 0$$

Substituting $y = Ae^{nx}$ and its derivatives in the above equation, we get

$$n^2 - n - 2 = 0$$
$$\Rightarrow \quad (n-2)(n+1) = 0$$
$$\Rightarrow \quad n = 2 \quad \text{or} \quad -1$$

So, the CF is $y = Ae^{2x} + Be^{-x}$.

To find the PI, we let $y = Cxe^{2x}$.

(**Note** xe^{2x} is used here because e^{2x} already forms part of the CF.)

Differentiating $y = Cxe^{2x}$, we have

$$\frac{dy}{dx} = Ce^{2x} + 2Cxe^{2x}$$

$$\frac{d^2y}{dx^2} = 2Ce^{2x} + 2Ce^{2x} + 4Cxe^{2x}$$

Substituting these in $\dfrac{d^2y}{dx^2} - \dfrac{dy}{dx} - 2y = 3e^{2x}$, we get

$$4Ce^{2x} + 4Cxe^{2x} - Ce^{2x} - 2Cxe^{2x} - 2Cxe^{2x} = 3e^{2x}$$

(**Note** The x-terms should cancel at this stage.)

$$3Ce^{2x} = 3e^{2x} \quad \Rightarrow \quad C = 1$$

Therefore, the PI is $y = xe^{2x}$.

Hence, the general solution is $y = Ae^{2x} + Be^{-x} + xe^{2x}$.

At this stage, after adding the CF and the PI, we insert the boundary conditions:

$$y = 0 \text{ when } x = 0 \quad \Rightarrow \quad 0 = A + B$$

$$\frac{dy}{dx} = 2Ae^{2x} - Be^{-x} + e^{2x} + 2xe^{2x}$$

$$\frac{dy}{dx} = 11 \text{ when } x = 0 \quad \Rightarrow \quad 11 = 2A - B + 1 \quad \Rightarrow \quad 10 = 2A - B$$

Since $0 = A + B$, we have

$$A = \frac{10}{3} \quad \text{and} \quad B = -\frac{10}{3}$$

The solution is, therefore,

$$y = \left(\frac{10}{3} + x\right)e^{2x} - \frac{10}{3}e^{-x}$$

Example 11 Solve $y'' - 4y' + 4y = 3e^{2x}$.

SOLUTION

To find the CF, we substitute $y = Ae^{nx}$ and its derivatives in $y'' - 4y' + 4y = 0$, which gives

$$n^2 - 4n + 4 = 0$$
$$\Rightarrow \quad (n - 2)(n - 2) = 0$$
$$\Rightarrow \quad n = 2 \quad \text{(repeated root)}$$

Therefore, the CF is $y = (A + Bx)e^{2x}$.

To find the PI, we need to use a term in x^2e^{2x}, since both e^{2x} **and** xe^{2x} already form terms in the CF. Therefore, we let $y = Cx^2e^{2x}$, which gives

$$y' = 2Cx^2e^{2x} + 2Cxe^{2x}$$
$$y'' = 4Cx^2e^{2x} + 4Cxe^{2x} + 2Ce^{2x} + 4Cxe^{2x}$$
$$= 4Cx^2e^{2x} + 8Cxe^{2x} + 2Ce^{2x}$$

Substituting these in $y'' - 4y' + 4y = 3e^{2x}$, we have

$$4Cx^2e^{2x} + 8Cxe^{2x} + 2Ce^{2x} - 4(2Cx^2e^{2x} + 2Cxe^{2x}) + 4Cx^2e^{2x} = 3e^{2x}$$

(**Note** The terms in x^2 and x should cancel at this stage.)

$$2Ce^{2x} = 3e^{2x} \quad \Rightarrow \quad C = \tfrac{3}{2}$$

Therefore, the PI is $y = \tfrac{3}{2}x^2e^{2x}$.

Hence, the general solution is $y = (A + Bx + \tfrac{3}{2}x^2)e^{2x}$.

CHAPTER 4 DIFFERENTIAL EQUATIONS

Example 12 Solve $y'' + 16y = 2\cos 4x$.

SOLUTION

To find the CF, we substitute $y = Ae^{nx}$ and its second derivative in $y'' + 16y = 0$, which gives

$$n^2 + 16 = 0 \quad \Rightarrow \quad n = \pm 4i$$

The CF is, therefore, $y = A\cos 4x + B\sin 4x$.

Note that for the PI we need to use terms in $x\cos 4x$ and $x\sin 4x$, since the CF already contains the terms $\cos 4x$ and $\sin 4x$. Therefore, the PI is given by

$$y = Cx\cos 4x + Dx\sin 4x$$

So, we have

$$y' = C\cos 4x - 4Cx\sin 4x + D\sin 4x + 4Dx\cos 4x$$
$$y'' = -4C\sin 4x - 4C\sin 4x - 16Cx\cos 4x + 4D\cos 4x + 4D\cos 4x - 16D\sin 4x$$

Substituting the above in $y'' + 16y = 2\cos 4x$, we get

$$-8C\sin 4x - 16Cx\cos 4x + 8D\cos 4x - 16Dx\sin 4x + 16Cx\cos 4x +$$
$$+16Dx\sin 4x = 2\cos 4x$$

Simplifying, equating sin and cos terms, and remembering that the terms in x should cancel, we find

$$C = 0 \quad \text{and} \quad D = \tfrac{1}{4}$$

Therefore, the PI is $y = \tfrac{1}{4}x\sin 4x$.

Hence, the solution is

$$y = A\sin 4x + B\cos 4x + \frac{x}{4}\sin 4x$$

Exercise 4C

In Questions **1** to **12**, find the general solution of each differential equation.

1 $\dfrac{d^2y}{dx^2} + 7\dfrac{dy}{dx} - 8y = 16x$ 　　**2** $\dfrac{d^2y}{dx^2} + 4\dfrac{dy}{dx} + 3y = 4e^{-2x}$ 　　**3** $2\dfrac{d^2y}{dx^2} - 3\dfrac{dy}{dx} - 5y = 10x^2 + 1$

4 $3\dfrac{d^2y}{dx^2} + 2\dfrac{dy}{dx} - y = 4\sin 5x$ 　　**5** $\dfrac{d^2x}{dt^2} - 4\dfrac{dx}{dt} - 5x = 3e^{3t}$ 　　**6** $\dfrac{d^2s}{dt^2} - 8\dfrac{ds}{dt} + 15s = 5\cos 2t$

7 $\dfrac{d^2y}{dx^2} + 5\dfrac{dy}{dx} + 4y = 2e^{-x}$ 　　**8** $\dfrac{d^2y}{dx^2} - 6\dfrac{dy}{dx} + 9y = 5e^{3x}$ 　　**9** $\dfrac{d^2y}{dx^2} - 2\dfrac{dy}{dx} + 3y = 22e^{4x}$

10 $\dfrac{d^2y}{dx^2} + 6\dfrac{dy}{dx} + 10y = 3e^{-4x}$ 　　**11** $\dfrac{d^2x}{dt^2} - 2\dfrac{dx}{dt} + x = 4e^t$ 　　**12** $\dfrac{d^2x}{dt^2} + 16x = 3\cos 4t$

13 Solve the differential equation

$$\frac{d^2x}{dt^2} - 2\frac{dx}{dt} + 5x = 0$$

if $x = -3$ and $\dfrac{dx}{dt} = 1$ when $t = 0$. 　　(NICCEA)

14 a) Find the general solution of the differential equation

$$\frac{d^2y}{dx^2} - 4y = 10e^{3x}$$

b) Hence find the solution for which $y = -2$ at $x = 0$, and $\frac{dy}{dx} = -6$ at $x = 0$. (EDEXCEL)

15 Find the general solution of the equation $\frac{d^2y}{dx^2} = e^{2x} + \cos \frac{1}{2}x$.

State what extra information would be needed to enable a particular solution to be obtained.

(NEAB/SMP 16–19)

16 i) Find the solution of the differential equation

$$\frac{d^2y}{dx^2} + 4\frac{dy}{dx} + 13y = 0$$

for which $y = 4$ and $\frac{dy}{dx} = 1$ at $x = 0$.

ii) Given that

$$\cos x \frac{dy}{dx} + 2y\sin x = \cos^3 x + \sin x \quad 0 < x < \frac{1}{2}\pi$$

and that $y = 1$ at $x = \frac{1}{3}\pi$, find the value of y at $x = \frac{1}{4}\pi$. (EDEXCEL)

17 Find the general solution of the differential equation $\frac{d^2y}{dx^2} - 4\frac{dy}{dx} + 5y = \sin 2x$. (EDEXCEL)

18 i) Solve the differential equation $\frac{d^2x}{dt^2} + 16x = 0$ to find its general solution.

ii) If $x = 3$ and $\frac{dx}{dt} = -8$ when $t = 0$, show that the particular solution of the differential

equation above is

$$x = 3\cos 4t - 2\sin 4t$$

iii) By writing the particular solution as $R\cos(4t + \alpha)$, find the first positive value of t for which x is maximum. (NICCEA)

19 Obtain the solution of the differential equation

$$20\frac{d^2x}{dt^2} + 4\frac{dx}{dt} + x = 2t + 11$$

given that, when $t = 0$, $x = 3$ and $\frac{dx}{dt} = 2.8$. Show that $x \approx 2t + 3$ for large positive t. (OCR)

20 Find the general solution of the differential equation

$$\frac{d^2x}{dt^2} + 5\frac{dx}{dt} + 4x = 15\cos 3t - 5\sin 3t$$ (OCR)

21 Find the general solution of the differential equation

$$\frac{d^2y}{dx^2} - 3\frac{dy}{dx} - 4y = 50\sin 2x$$

Given that $y = 0$ when $x = 0$ and that y remains finite as $x \to \infty$, find y in terms of x. (OCR)

22 i) Find the general solution of the differential equation

$$\frac{d^2x}{dt^2} - 4\frac{dx}{dt} + 29x = -16\cos 2t + 50\sin 2t$$

ii) If $x = 3$ and $\frac{dx}{dt} = 10$ when $t = 0$, find the particular solution.   (NICCEA)

23 a) Solve the equation $\frac{dy}{dx} = x + xy$. You do not need to make y the subject of your solution.

b) Find the complementary function and a particular integral for the equation

$$\frac{dy}{dx} - 3y = 2x + e^{4x}$$

Hence write down the general solution of the equation.   (NEAB/SMP 16–19)

24 a) Find the general solution of the differential equation

$$\frac{d^2y}{dx^2} + 4\frac{dy}{dx} + 13y = 0$$

b) Given that $y = a\cos 3x + b\sin 3x$ is a particular integral of the differential equation

$$\frac{d^2y}{dx^2} + 4\frac{dy}{dx} + 13y = 6\cos 3x - 8\sin 3x$$

find the values of a and b.

c) Show that this particular integral has maximum and minimum values of $\frac{\sqrt{10}}{4}$ and $-\frac{\sqrt{10}}{4}$ respectively.

d) Find the solution of the differential equation

$$\frac{d^2y}{dx^2} + 4\frac{dy}{dx} + 13y = 6\cos 3x - 8\sin 3x$$

for which $y = 0$ and $\frac{dy}{dx} = 0$ at $x = 0$.   (EDEXCEL)

25 a) Find the general solution of the differential equation

$$2\frac{d^2y}{dx^2} - 7\frac{dy}{dx} - 4y = 8\sin x - 19\cos x$$

b) Hence find the solution for which $y = 0$ at $x = 0$ and $\frac{dy}{dx} = 11$ at $x = 0$.   (EDEXCEL)

26 The value of the stock held by a large business organisation t years after 1st January 1998 is $(10 + x)$ million dollars. The variation of x, which may be regarded as a continuous variable, is modelled by the differential equation

$$4\frac{d^2x}{dt^2} + 8\frac{dx}{dt} + 5x = 2\cos t - 16\sin t$$

i) Find the general solution for x in terms of t.

ii) Given that $x = 1$ and $\frac{dx}{dt} = 3$ when $t = 0$, find, correct to four significant figures, the predicted value of the stock held on 1st January 2000.   (OCR)

27 Find the values of the constants p and q for which $y = px \sin 2x + qx \cos 2x$ is a particular integral of the differential equation

$$\frac{d^2y}{dx^2} + 4y = \sin 2x$$

Find the general solution of this differential equation.

Show that when $x = n\pi$, where n is a large positive integer, $y \approx -\frac{1}{4}n\pi$, whatever the initial conditions, and find a corresponding approximation for y when $x = (n + \frac{1}{2})\pi$. (OCR)

28 Given that $x = At^2 e^{-t}$ satisfies the differential equation

$$\frac{d^2x}{dt^2} + 2\frac{dx}{dt} + x = e^{-t}$$

a) find the value of A.

b) Hence find the solution of the differential equation for which $x = 1$ and $\dfrac{dx}{dt} = 0$ at $t = 0$.

c) Use your solution to prove that for $t \geqslant 0$, $x \leqslant 1$. (EDEXCEL)

Solution of differential equations by substitution

We can now solve the following three types of differential equation:

- First order in which variables are separable.
- First order requiring an integrating factor.
- Second order of the form $ay'' + by' + cy = f(x)$, where a, b and c are constants.

Substitutions can be used to make a differential equation, which is one of these three types, more manageable.

For example, to solve

$$(m + 5kM) + t\frac{dm}{dt} = (m + 5kM)^3$$

we would make the substitution $p = m + 5kM$, which changes this equation into

$$p + t\frac{dp}{dt} = p^3$$

In this form, the equation looks less daunting and is easier to solve.

Substitutions can also be used to convert a more difficult form of differential equation to one of the above three types. (In an A-level examination, these kinds of substitution will normally be given.)

Two such substitutions which you will meet frequently are $y = ux$ and $x = e^u$, where u is a function of x. Their application is shown respectively in Examples 13 and 14.

Example 13 Solve $x^2 \dfrac{dy}{dx} = 4x^2 + xy + y^2$, given that when $x = 1$, $y = 2$.

SOLUTION

Notice that in this equation the power of each term, treating x and y as the same, is 2.

Such equations are called **homogeneous equations**, for which the usual substitution is $y = ux$.

Differentiating $y = ux$ with respect to x, we have

$$\frac{dy}{dx} = \frac{du}{dx} x + u$$

Substituting for $\dfrac{dy}{dx}$ and for y in $x^2 \dfrac{dy}{dx} = 4x^2 + xy + y^2$, we get

$$x^2 \left(x \frac{du}{dx} + u \right) = 4x^2 + ux^2 + u^2 x^2$$

Dividing through by x^2 and rearranging the terms, we have

$$x \frac{du}{dx} = 4 + u^2$$

$$\Rightarrow \quad \int \frac{du}{4 + u^2} = \int \frac{dx}{x}$$

which gives (see page 36)

$$\frac{1}{2} \tan^{-1} \left(\frac{u}{2} \right) = \ln x + c$$

$$\Rightarrow \quad \frac{1}{2} \tan^{-1} \left(\frac{y}{2x} \right) = \ln x + c$$

Now when $x = 1$, $y = 2$. Therefore, $c = \dfrac{\pi}{8}$. Hence, we have

$$\frac{1}{2} \tan^{-1} \left(\frac{y}{2x} \right) = \ln x + \frac{\pi}{8}$$

$$\Rightarrow \quad \frac{y}{2x} = \tan \left(\frac{\pi}{4} + 2 \ln x \right)$$

$$\Rightarrow \quad y = 2x \tan \left(\frac{\pi}{4} + 2 \ln x \right)$$

Example 14 Solve $x^2 \dfrac{d^2 y}{dx^2} - 2x \dfrac{dy}{dx} - 10y = 0$ using the substitution $x = e^u$.

SOLUTION

We need to replace $\dfrac{dy}{dx}$ by a term in $\dfrac{dy}{du}$, and $\dfrac{d^2 y}{dx^2}$ by a term in $\dfrac{d^2 y}{du^2}$.

So, first we differentiate $x = e^u$ with respect to u, which gives $\dfrac{dx}{du} = e^u$.

Using $\dfrac{dy}{dx} = \dfrac{dy}{du}\dfrac{du}{dx}$, we get

$$\frac{dy}{dx} = \frac{1}{e^u}\frac{dy}{du}$$

$$\Rightarrow \quad \frac{dy}{dx} = e^{-u}\frac{dy}{du}$$

We now differentiate this equation with respect to x, noting that the RHS is differentiated as a product and using

$$\frac{d}{dx}\left(\frac{dy}{du}\right) = \frac{d}{du}\left(\frac{dy}{du}\right)\frac{du}{dx} = \frac{d^2y}{du^2}\frac{du}{dx}$$

Hence, we arrive at

$$\frac{d^2y}{dx^2} = -e^{-u}\frac{du}{dx}\frac{dy}{du} + e^{-u}\frac{d^2y}{du^2}\frac{du}{dx}$$

Since $\dfrac{du}{dx} = e^{-u}$, we therefore have

$$\frac{d^2y}{dx^2} = -e^{-2u}\frac{dy}{du} + e^{-2u}\frac{d^2y}{du^2}$$

Substituting for $\dfrac{dy}{dx}$ and $\dfrac{d^2y}{dx^2}$ in $x^2\dfrac{d^2y}{dx^2} - 2x\dfrac{dy}{dx} - 10y = 0$, we get

$$e^{2u}\left(-e^{-2u}\frac{dy}{du} + e^{-2u}\frac{d^2y}{du^2}\right) - 2e^u e^{-u}\frac{dy}{du} - 10y = 0$$

$$\Rightarrow \quad \frac{d^2y}{du^2} - 3\frac{dy}{du} - 10y = 0$$

Substituting $y = Ae^{nu}$ and its derivatives in the above equation, we obtain

$$n^2 - 3n - 10 = 0$$

$$\Rightarrow \quad n = 5 \quad \text{or} \quad -2$$

Therefore, the general solution is

$$y = Ae^{5u} + Be^{-2u}$$

Using $x = e^u$, we have

$$e^{5u} = (e^u)^5 = x^5 \quad \text{and} \quad e^{-2u} = (e^u)^{-2} = x^{-2}$$

which give

$$y = Ax^5 + \frac{B}{x^2}$$

Exercise 4D

1 Using the substitution $y = ux$, find the general solution of each of the following.

a) $\dfrac{dy}{dx} = \dfrac{x - 3y}{x}$

b) $xy\dfrac{dy}{dx} = x^2 + y^2$

c) $x^2y\dfrac{dy}{dx} = x^3 + x^2y - y^3$

d) $3x^3\dfrac{dy}{dx} = y^3 - x^2y$

2 Using the substitution $p = x + y$, find the general solution of

$$\frac{dy}{dx} = \frac{3x + 3y + 4}{x + y + 1}$$

3 Use the substitution $p = 2x + 3y$ to find the general solution of

$$\frac{dy}{dx} = \frac{4x + 6y - 5}{2x + 3y + 1}$$

4 Using the substitution $x = e^u$, find the general solution of

a) $x^2 \dfrac{d^2y}{dx^2} + 2x \dfrac{dy}{dx} - 2y = 0$
 b) $x^2 \dfrac{d^2y}{dx^2} - 5x \dfrac{dy}{dx} - 6y = 0$

c) $x^2 \dfrac{d^2y}{dx^2} - 3x \dfrac{dy}{dx} + 4y = 0$
 d) $x^2 \dfrac{d^2y}{dx^2} + 2x \dfrac{dy}{dx} + y = 0$

5 Given that $x = t^{\frac{1}{2}}$, $x > 0$, $t > 0$, and y is a function of x, find $\dfrac{dy}{dx}$ in terms of $\dfrac{dy}{dt}$ and t.

Assuming that $\dfrac{d^2y}{dx^2} = 4t \dfrac{d^2y}{dt^2} + 2 \dfrac{dy}{dt}$, show that the substitution $x = t^{\frac{1}{2}}$, transforms the differential equation

$$\frac{d^2y}{dx^2} + \left(6x - \frac{1}{x}\right) \frac{dy}{dx} - 16x^2 y = 4x^2 e^{2x^2} \qquad \text{[I]}$$

into the differential equation

$$\frac{d^2y}{dt^2} + 3 \frac{dy}{dt} - 4y = e^{2t}$$

Hence find the general solution of [I], giving y in terms of x. (EDEXCEL)

6 a) Find the general solution of the equation

$$\frac{dz}{dx} + z = e^x$$

b) Make the substitution $y = xz$ in the equation

$$x \frac{dy}{dx} + (x - 1)y = x^2 e^x$$

Hence write down the solution of this equation. (NEAB/SMP 16–19)

7 a) Show that the substitution $v = xy$ transforms the differential equation

$$x \frac{d^2y}{dx^2} + 2(1 + 2x) \frac{dy}{dx} + 4(1 + x)y = 32e^{2x} \quad x \neq 0$$

into the differential equation

$$\frac{d^2v}{dx^2} + 4 \frac{dv}{dx} + 4v = 32e^{2x}$$

b) Given that $v = ae^{2x}$, where a is a constant, is a particular integral of this transformed equation, find a.

c) Find the solution of the differential equation

$$x \frac{d^2y}{dx^2} + 2(1 + 2x) \frac{dy}{dx} + 4(1 + x)y = 32e^{2x}$$

for which $y = 2e^2$ and $\frac{dy}{dx} = 0$ at $x = 1$.

d) Determine whether or not this solution remains finite as $x \to \infty$.   (EDEXCEL)

8 The variables x and y are functions of t, and satisfy the differential equations

$$\frac{dx}{dt} + 2x = y \qquad (*)$$

$$\frac{dy}{dt} + x = 0$$

By eliminating y, show that

$$\frac{d^2x}{dt^2} + 2 \frac{dx}{dt} + x = 0$$

Find the general solution of this differential equation for x and deduce by substitution in (*) the general solution for y.

Hence, or otherwise, find x and y in terms of t, given that $x = 1$ and $y = 0$ when $t = 0$.

(NEAB)

9 a) Find, in the form $y = f(x)$, the general solution of the equation

$$(x^2 - 1) \frac{dy}{dx} + xy = 1 \quad x > 1$$

b) i) Given that $y = \frac{u}{x}$, show that

$$\frac{d^2y}{dx^2} = \frac{1}{x} \frac{d^2u}{dx^2} - \frac{2}{x^2} \frac{du}{dx} + \frac{2u}{x^3}$$

ii) Hence find the general solution of the differential equation

$$\frac{d^2y}{dx^2} + \frac{2}{x} \frac{dy}{dx} + 25y = 0 \quad x > 0 \qquad \text{(EDEXCEL)}$$

5 Determinants

In algebra, to mention only one thing of many, Jacobi cast the theory of determinants into the simple form now familiar to every student.
E. T. BELL

Definition of 2 × 2 and 3 × 3 determinants

The 2 × 2 determinant $\begin{vmatrix} a & b \\ c & d \end{vmatrix}$ represents the expression $ad - bc$.

For example, we have

$$\begin{vmatrix} 3 & 4 \\ 7 & 8 \end{vmatrix} = 3 \times 8 - 4 \times 7 = 24 - 28 = -4$$

The 3 × 3 determinant $\begin{vmatrix} a & b & c \\ d & e & f \\ g & h & i \end{vmatrix}$ represents the expression

$$a\begin{vmatrix} e & f \\ h & i \end{vmatrix} - b\begin{vmatrix} d & f \\ g & i \end{vmatrix} + c\begin{vmatrix} d & e \\ g & h \end{vmatrix}$$

which is

$$a(ei - fh) - b(di - fg) + c(dh - eg)$$

We see that the determinant of a 3 × 3 matrix is found by expanding the matrix along its first row. In turn, we take each **element**, or **entry**, in the first row, cover up its column and the first row, and find the determinant of the 2 × 2 matrix which is left. We then combine the three results. Notice the minus sign for the b-term, which relates to the fact that b is an **odd** number of places from the first element, a.

Note   It is much easier to learn the method for evaluating a determinant than to remember its formula.

Example 1 Evaluate $\begin{vmatrix} 3 & 7 & 8 \\ 4 & 2 & 5 \\ 1 & 9 & 15 \end{vmatrix}$

SOLUTION

$$\begin{vmatrix} 3 & 7 & 8 \\ 4 & 2 & 5 \\ 1 & 9 & 15 \end{vmatrix} = 3\begin{vmatrix} 2 & 5 \\ 9 & 15 \end{vmatrix} - 7\begin{vmatrix} 4 & 5 \\ 1 & 15 \end{vmatrix} + 8\begin{vmatrix} 4 & 2 \\ 1 & 9 \end{vmatrix}$$

$$= 3(30 - 45) - 7(60 - 5) + 8(36 - 2)$$

$$= -45 - 385 + 272$$

$$= -158$$

Determinants, unlike matrices, **always** consist of a **square array** of elements.

The determinant of the square matrix **A** is denoted either by $|\mathbf{A}|$ or by det **A**.

Because determinants are always square, the expansion method just described can be applied to determinants of any size. Thus to evaluate the determinant of a 4×4 matrix, we first expand it along its top row to get an expression involving four 3×3 matrices, remembering to **alternate the plus and minus signs**. For example,

$$\begin{vmatrix} 1 & 3 & 4 & 2 \\ 5 & -1 & -3 & -4 \\ 2 & -3 & 4 & 7 \\ 1 & 8 & 5 & 6 \end{vmatrix} = 1 \begin{vmatrix} -1 & -3 & -4 \\ -3 & 4 & 7 \\ 8 & 5 & 6 \end{vmatrix} - 3 \begin{vmatrix} 5 & -3 & -4 \\ 2 & 4 & 7 \\ 1 & 5 & 6 \end{vmatrix} +$$

$$+ 4 \begin{vmatrix} 5 & -1 & -4 \\ 2 & -3 & 7 \\ 1 & 8 & 6 \end{vmatrix} - 2 \begin{vmatrix} 5 & -1 & -3 \\ 2 & -3 & 4 \\ 1 & 8 & 5 \end{vmatrix}$$

We then proceed to evaluate each 3×3 matrix as before.

Rules for the manipulation of determinants

Changing a determinant without changing its value

We can alter the rows and the columns of a determinant in three ways **without changing its value**. Two are given below.

Adding any row, or column, to any other row, or column

If we add the corresponding elements in two rows (or columns), the value of the determinant is unaltered. For example, we have

$$\begin{vmatrix} a & b & c \\ d & e & f \\ g & h & i \end{vmatrix} = \begin{vmatrix} a+b & b & c \\ d+e & e & f \\ g+h & h & i \end{vmatrix}$$

The rule also applies to the **subtraction** of the corresponding elements in two rows (or columns). So, we have

$$\begin{vmatrix} a & b & c \\ d & e & f \\ g & h & i \end{vmatrix} = \begin{vmatrix} a & b & c \\ d-g & e-h & f-i \\ g & h & i \end{vmatrix}$$

Example 2 Evaluate $\begin{vmatrix} 1 & 1 & 1 \\ 0 & -1 & -1 \\ 4 & 6 & 8 \end{vmatrix}$

SOLUTION

The most efficient way to evaluate this determinant is to add the second row to the first row.

Note If you cannot quickly spot this simplification, it is better to expand using 2×2 determinants, rather than to spend time trying various possible simplifications.

So, we have

$$\begin{vmatrix} 1 & 1 & 1 \\ 0 & -1 & -1 \\ 4 & 6 & 8 \end{vmatrix} = \begin{vmatrix} 1+0 & 1-1 & 1-1 \\ 0 & -1 & -1 \\ 4 & 6 & 8 \end{vmatrix}$$

$$= \begin{vmatrix} 1 & 0 & 0 \\ 0 & -1 & -1 \\ 4 & 6 & 8 \end{vmatrix}$$

Expanding this simplified determinant, we get

$$\begin{vmatrix} 1 & 0 & 0 \\ 0 & -1 & -1 \\ 4 & 6 & 8 \end{vmatrix} = 1 \times \begin{vmatrix} -1 & -1 \\ 6 & 8 \end{vmatrix} = 1 \times (-8+6) = -2$$

Adding any multiple of any row, or column, to any other row, or column

If we add the same multiple of the elements of a column (or row) to the corresponding elements of another column (or row), the value of the determinant is unaltered. For example, we have

$$\begin{vmatrix} a & b & c \\ d & e & f \\ g & h & i \end{vmatrix} = \begin{vmatrix} a+5b & b & c \\ d+5e & e & f \\ g+5h & h & i \end{vmatrix}$$

The rule also applies to negative multiples. So, we have

$$\begin{vmatrix} a & b & c \\ d & e & f \\ g & h & i \end{vmatrix} = \begin{vmatrix} a & b & c \\ d-3a & e-3b & f-3c \\ g & h & i \end{vmatrix}$$

Example 3 Evaluate $\begin{vmatrix} 4 & 6 & 8 \\ 0 & 1 & 4 \\ 1 & 3 & 4 \end{vmatrix}$

SOLUTION

This determinant is best simplified by subtracting $2 \times$ the third row from the first row.

Again, if you cannot quickly spot this, it is better to expand using 2×2 determinants.

So, we have

$$\begin{vmatrix} 4 & 6 & 8 \\ 0 & 1 & 4 \\ 1 & 3 & 4 \end{vmatrix} = \begin{vmatrix} 4-2 & 6-6 & 8-8 \\ 0 & 1 & 4 \\ 1 & 3 & 4 \end{vmatrix} = \begin{vmatrix} 2 & 0 & 0 \\ 0 & 1 & 4 \\ 1 & 3 & 4 \end{vmatrix}$$

Expanding this simplified determinant, we get

$$\begin{vmatrix} 2 & 0 & 0 \\ 0 & 1 & 4 \\ 1 & 3 & 4 \end{vmatrix} = 2 \times \begin{vmatrix} 1 & 4 \\ 3 & 4 \end{vmatrix} = -16$$

Two rows or columns can be interchanged by changing the sign of the determinant

For example, by switching columns 1 and 2 in the left-hand determinants, we get

$$\begin{vmatrix} a & b & c \\ d & e & f \\ g & h & i \end{vmatrix} = - \begin{vmatrix} b & a & c \\ e & d & f \\ h & g & i \end{vmatrix}$$

and

$$\begin{vmatrix} 2 & 0 & 0 \\ 0 & 1 & 4 \\ 1 & 3 & 4 \end{vmatrix} = - \begin{vmatrix} 0 & 2 & 0 \\ 1 & 0 & 4 \\ 3 & 1 & 4 \end{vmatrix}$$

When any two rows or any two columns are equal, the determinant is zero

Say, for example, the corresponding elements in columns 1 and 3 are equal, as in the determinant below. If we subtract column 3 from column 1, column 1 becomes a column of zeros. Hence, the value of the determinant must be zero.

$$\begin{vmatrix} 4 & 1 & 4 \\ 2 & 3 & 2 \\ 3 & -5 & 3 \end{vmatrix} = \begin{vmatrix} 0 & 1 & 4 \\ 0 & 3 & 2 \\ 0 & -5 & 3 \end{vmatrix} = 0$$

Example 4 Evaluate $\begin{vmatrix} 2 & 0 & 0 & 7 & 0 \\ 0 & 1 & 4 & 6 & 4 \\ 1 & 3 & 4 & 3 & 4 \\ 14 & 2 & 3 & 3 & 3 \\ 2 & 0 & 2 & 12 & 2 \end{vmatrix}$

SOLUTION

Evaluating this determinant by normal expansion would be very time consuming. However, we notice that columns 3 and 5 are identical, and so the value of the determinant is 0.

Multiplying any row, or any column, by k, multiplies the value of the determinant by k

If we multiply all the elements of one row (or column) by k, this is the same as multiplying the value of the determinant by k. For example, we have

$$\begin{vmatrix} a & kb & c \\ d & ke & f \\ g & kh & i \end{vmatrix} = k \begin{vmatrix} a & b & c \\ d & e & f \\ g & h & i \end{vmatrix}$$

If we multiply **every element** in the determinant by k, we obtain

$$\begin{vmatrix} ka & kb & kc \\ kd & ke & kf \\ kg & kh & ki \end{vmatrix}$$

We can take the factor k out of each column. Hence, we obtain

$$\begin{vmatrix} ka & kb & kc \\ kd & ke & kf \\ kg & kh & ki \end{vmatrix} = k^3 \begin{vmatrix} a & b & c \\ d & e & f \\ g & h & i \end{vmatrix}$$

Transpose of a determinant

The **transpose** of a determinant is obtained by reflecting the determinant in its **leading diagonal**. (This is the diagonal from the top left corner to the bottom right corner. It is also known as the **principal diagonal**.)

The value of the transpose of a determinant is the **same** as the determinant's **original value**. For example, we have

$$\begin{vmatrix} a & b & c \\ d & e & f \\ g & h & i \end{vmatrix} = \begin{vmatrix} a & d & g \\ b & e & h \\ c & f & i \end{vmatrix}$$

Example 5 Evaluate $\begin{vmatrix} 2 & 8 & 9 \\ 0 & -1 & 3 \\ 0 & 4 & 1 \end{vmatrix}$

SOLUTION

To simplify the calculation, we replace the given determinant by its transpose:

$$\begin{vmatrix} 2 & 8 & 9 \\ 0 & -1 & 3 \\ 0 & 4 & 1 \end{vmatrix} = \begin{vmatrix} 2 & 0 & 0 \\ 8 & -1 & 4 \\ 9 & 3 & 1 \end{vmatrix}$$

which gives

$$\begin{vmatrix} 2 & 0 & 0 \\ 8 & -1 & 4 \\ 9 & 3 & 1 \end{vmatrix} = 2\begin{vmatrix} -1 & 4 \\ 3 & 1 \end{vmatrix} = 2(-1 - 12) = -26$$

Factorisation of determinants

The easier way to find the factors of a determinant is to use the rules for manipulating determinants. Rarely, if ever, do we multiply out the determinant and then factorise the result.

In Example 6, the factors are obtained by subtracting, in turn, one column from another. In Example 7, a factor is obtained by first adding **all** three rows.

Example 6 Factorise $\begin{vmatrix} a & b & c \\ a^2 & b^2 & c^2 \\ a^3 & b^3 & c^3 \end{vmatrix}$

First, we take out the factors a, b and c, which gives

$$\begin{vmatrix} a & b & c \\ a^2 & b^2 & c^2 \\ a^3 & b^3 & c^3 \end{vmatrix} = abc \begin{vmatrix} 1 & 1 & 1 \\ a & b & c \\ a^2 & b^2 & c^2 \end{vmatrix}$$

Next, we subtract column 1 from column 2, and take out a fourth factor:

$$abc \begin{vmatrix} 1 & 1 & 1 \\ a & b & c \\ a^2 & b^2 & c^2 \end{vmatrix} = abc \begin{vmatrix} 1 & 0 & 1 \\ a & b-a & c \\ a^2 & b^2-a^2 & c^2 \end{vmatrix}$$

$$= abc(b-a) \begin{vmatrix} 1 & 0 & 1 \\ a & 1 & c \\ a^2 & b+a & c^2 \end{vmatrix}$$

Then, we subtract column 1 from column 3, and complete the factorisation:

$$abc(b-a) \begin{vmatrix} 1 & 0 & 0 \\ a & 1 & c-a \\ a^2 & b+a & c^2-a^2 \end{vmatrix} = abc(b-a)(c-a) \begin{vmatrix} 1 & 0 & 0 \\ a & 1 & 1 \\ a^2 & b+a & c+a \end{vmatrix}$$

$$= abc(b-a)(c-a)[(c+a)-(b+a)]$$

$$= abc(b-a)(c-a)(c-b)$$

$$= abc(a-b)(b-c)(c-a)$$

Example 7

a) Factorise $\begin{vmatrix} a & b & c \\ b & c & a \\ c & a & b \end{vmatrix}$

b) Hence, find the factors of $a^3 + b^3 + c^3 - 3abc$.

a) First, we add rows 2 and 3 to row 1, which gives

$$\begin{vmatrix} a & b & c \\ b & c & a \\ c & a & b \end{vmatrix} = \begin{vmatrix} a+b+c & a+b+c & a+b+c \\ b & c & a \\ c & a & b \end{vmatrix}$$

Next, we take out the factor $(a+b+c)$, which gives

$$(a+b+c) \begin{vmatrix} 1 & 1 & 1 \\ b & c & a \\ c & a & b \end{vmatrix}$$

Then, we subtract column 1 from columns 2 and 3, and complete the factorisation:

$$(a+b+c) \begin{vmatrix} 1 & 1 & 1 \\ b & c & a \\ c & a & b \end{vmatrix} = (a+b+c) \begin{vmatrix} 1 & 0 & 0 \\ b & c-b & a-b \\ c & a-c & b-c \end{vmatrix}$$

$$= (a+b+c)[(c-b)(b-c) - (a-b)(a-c)]$$

$$= (a+b+c)(bc + ac + ab - a^2 - b^2 - c^2)$$

b) Expanding the determinant, we obtain

$$\begin{vmatrix} a & b & c \\ b & c & a \\ c & a & b \end{vmatrix} = a(cb - a^2) - b(b^2 - ac) + c(ab - c^2)$$

$$= 3abc - a^3 - b^3 - c^3$$
$$= -(a^3 + b^3 + c^3 - 3abc)$$

Hence, we have

$$a^3 + b^3 + c^3 - 3abc = -\begin{vmatrix} a & b & c \\ b & c & a \\ c & a & b \end{vmatrix}$$

That is,

$$a^3 + b^3 + c^3 - 3abc = -(a + b + c)(bc + ca + ab - a^2 - b^2 - c^2)$$
$$= (a + b + c)(a^2 + b^2 + c^2 - bc - ca - ab)$$

Exercise 5A

1 Find the value of each of these determinants.

a) $\begin{vmatrix} 3 & 8 & 5 \\ 9 & 2 & -2 \\ 2 & 5 & 1 \end{vmatrix}$
b) $\begin{vmatrix} 3 & 3 & 3 \\ 1 & -4 & 1 \\ 6 & -7 & 5 \end{vmatrix}$
c) $\begin{vmatrix} 2 & 5 & 1 \\ 6 & 3 & 3 \\ 8 & -2 & 4 \end{vmatrix}$
d) $\begin{vmatrix} 4 & 3 & 1 \\ 1 & -5 & 2 \\ -5 & -1 & 9 \end{vmatrix}$

2 Factorise each of these determinants.

a) $\begin{vmatrix} 1 & a & a^3 \\ 1 & b & b^3 \\ 1 & c & c^3 \end{vmatrix}$
b) $\begin{vmatrix} 3p & 3q & 3r \\ 2p & 2q & r \\ 5p & -3q & 2r \end{vmatrix}$
c) $\begin{vmatrix} 1 & a^2 & a^3 \\ 1 & b^2 & b^3 \\ 1 & c^2 & c3 \end{vmatrix}$

d) $\begin{vmatrix} 0 & x - y & x^2 - y^2 \\ x - y & x & x^2 + 2xy + y^2 \\ y - x & y & 0 \end{vmatrix}$

3 Express the determinant

$$D = \begin{vmatrix} a^3 + a^2 & a & 1 \\ b^3 + b^2 & b & 1 \\ c^3 + c^2 & c & 1 \end{vmatrix}$$

as the product of four linear factors.

Given that no two of a, b and c are equal and that $D = 0$, find the value of $a + b + c$. (NEAB)

4 Show that

$$\det \begin{pmatrix} 2 & 2k & 1 \\ 1 & k - 1 & 1 \\ 2 & 1 & k + 1 \end{pmatrix}$$

has the same value for all values of k. (SQA/CSYS)

Solution of three equations in three unknowns

Consider the three equations

$$a_1x + b_1y + c_1z + d_1 = 0$$
$$a_2x + b_2y + c_2z + d_2 = 0$$
$$a_3x + b_3y + c_3z + d_3 = 0$$

It can be shown by algebraic elimination that their general solution is given by

$$\frac{x}{\begin{vmatrix} b_1 & c_1 & d_1 \\ b_2 & c_2 & d_2 \\ b_3 & c_3 & d_3 \end{vmatrix}} = -\frac{y}{\begin{vmatrix} a_1 & c_1 & d_1 \\ a_2 & c_2 & d_2 \\ a_3 & c_3 & d_3 \end{vmatrix}} = \frac{z}{\begin{vmatrix} a_1 & b_1 & d_1 \\ a_2 & b_2 & d_2 \\ a_3 & b_3 & d_3 \end{vmatrix}} = -\frac{1}{\begin{vmatrix} a_1 & b_1 & c_1 \\ a_2 & b_2 & c_2 \\ a_3 & b_3 & c_3 \end{vmatrix}}$$

Note the following five important facts:

- The determinant under x does not include any of the x-coefficients.

- The determinant under y does not include any of the y-coefficients.

- The determinant under z does not include any of the z-coefficients.

- The y-fraction and the unit fraction carry a minus sign. (The minus sign alternates as in the expansion of a determinant.)

- If one of the determinants is zero, the corresponding unknown is also zero.

  For example,

  $$\text{if } \begin{vmatrix} b_1 & c_1 & d_1 \\ b_2 & c_2 & d_2 \\ b_3 & c_3 & d_3 \end{vmatrix} = 0, \text{ then } x = 0$$

From the equations above, we have

$$x = -\frac{\begin{vmatrix} b_1 & c_1 & d_1 \\ b_2 & c_2 & d_2 \\ b_3 & c_3 & d_3 \end{vmatrix}}{\begin{vmatrix} a_1 & b_1 & c_1 \\ a_2 & b_2 & c_2 \\ a_3 & b_3 & c_3 \end{vmatrix}} \qquad y = \frac{\begin{vmatrix} a_1 & c_1 & d_1 \\ a_2 & c_2 & d_2 \\ a_3 & c_3 & d_3 \end{vmatrix}}{\begin{vmatrix} a_1 & b_1 & c_1 \\ a_2 & b_2 & c_2 \\ a_3 & b_3 & c_3 \end{vmatrix}} \qquad z = \frac{\begin{vmatrix} a_1 & b_1 & d_1 \\ a_2 & b_2 & d_2 \\ a_3 & b_3 & d_3 \end{vmatrix}}{\begin{vmatrix} a_1 & b_1 & c_1 \\ a_2 & b_2 & c_2 \\ a_3 & b_3 & c_3 \end{vmatrix}}$$

Hence, these three equations **have a unique solution unless** $\begin{vmatrix} a_1 & b_1 & c_1 \\ a_2 & b_2 & c_2 \\ a_3 & b_3 & c_3 \end{vmatrix} = 0$

Conversely, they **do not have a unique solution if** $\begin{vmatrix} a_1 & b_1 & c_1 \\ a_2 & b_2 & c_2 \\ a_3 & b_3 & c_3 \end{vmatrix} = 0$

Geometric interpretation of three equations in three unknowns

Each of the equations $a_i x + b_i y + c_i z + d_i = 0$ $(i = 1, 2, 3)$ may be considered as the equation of a plane in three-dimensional space.

With three planes, there are seven possible configurations.

- The three planes intersect in a single point. In this case, the three equations have a **unique solution**.

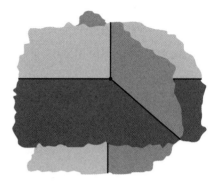

- The three planes form a triangular prism. In this case, there is no point where all three planes intersect. Hence, the equations are said to be **inconsistent**, as they have no solutions.

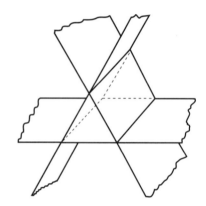

- Two of the planes are parallel and separate, and are intersected by the third plane. Again, there is no point where all three planes intersect, and so the equations are inconsistent in this case, too.

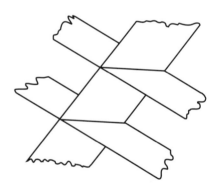

- Two other configurations in which the planes have no common point and therefore their equations are inconsistent are:

  ○ All three planes are parallel and separate.
  ○ Two of the planes are coincident and the third plane is parallel but separate.

The two remaining configurations correspond to the three equations having infinitely many solutions.

- The three planes have a common line, giving an infinite number of points (x, y, z) which satisfy all three equations. In this case, the equations are said to be **linearly dependent**, and the configuration is called a **sheaf of planes** or a **pencil of planes**.

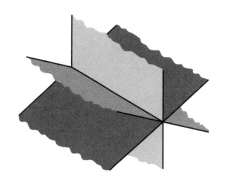

- All three planes coincide, giving an infinite number of points which satisfy all three equations.

Example 8 How many solutions are there to these three equations?

$$4x - \lambda y + 6z = 2$$
$$2y + \lambda z = 1$$
$$x - 2y + 4z = 0$$

SOLUTION

First, we find the determinant

$$\begin{vmatrix} a_1 & b_1 & c_1 \\ a_2 & b_2 & c_2 \\ a_3 & b_3 & c_3 \end{vmatrix} = \begin{vmatrix} 4 & -\lambda & 6 \\ 0 & 2 & \lambda \\ 1 & -2 & 4 \end{vmatrix}$$

$$= 4 \begin{vmatrix} 2 & \lambda \\ -2 & 4 \end{vmatrix} + \lambda \begin{vmatrix} 0 & \lambda \\ 1 & 4 \end{vmatrix} + 6 \begin{vmatrix} 0 & 2 \\ 1 & -2 \end{vmatrix}$$

$$= 4(8 + 2\lambda) + \lambda(-\lambda) + 6(-2)$$

$$= -\lambda^2 + 8\lambda + 20$$

Therefore, there is a unique solution unless

$$-\lambda^2 + 8\lambda + 20 = 0$$
$$\Rightarrow \quad \lambda^2 - 8\lambda - 20 = 0$$
$$\Rightarrow \quad (\lambda - 10)(\lambda + 2) = 0$$

That is, there is a unique solution unless $\lambda = 10$ or $\lambda = -2$.

If $\lambda = 10$, the equations are

$$4x - 10y + 6z = 2 \qquad [1]$$
$$2y + 10z = 1 \qquad [2]$$
$$x - 2y + 4z = 0 \qquad [3]$$

We now use equations [1] and [3] to get the same expression as that on the left-hand side of equation [2]. Subtracting $4 \times$ equation [3] from equation [1], we have

$$-2y - 10z = 2$$
$$\Rightarrow \quad 2y + 10z = -2$$

This contradicts equation [2], and so the equations have no solution. That is, the three equations are **inconsistent**.

If $\lambda = -2$, the equations are

$$4x + 2y + 6z = 2 \qquad [4]$$
$$2y - 2z = 1 \qquad [5]$$
$$x - 2y + 4z = 0 \qquad [6]$$

Proceeding as before, we subtract equation [4] from $4 \times$ equation [6], which gives

$$-10y + 10z = -2$$
$$\Rightarrow \quad 2y - 2z = \frac{1}{5}$$

This contradicts equation [5], and so the equations have no solution. That is, the three equations are inconsistent.

Example 9 Solve the equations

$$2x - 3y + 4z = 1 \qquad [1]$$
$$3x - y = 2 \qquad [2]$$
$$x + 2y - 4z = 1 \qquad [3]$$

SOLUTION

First, we calculate the determinant $\begin{vmatrix} 2 & -3 & 4 \\ 3 & -1 & 0 \\ 1 & 2 & -4 \end{vmatrix}$ and find that its value is zero.

Therefore, there is not a unique solution to the three equations, and so we cannot use the general formula for the solution of three equations.

Adding equations [1] and [3], we obtain

$$3x - y = 2$$

which is equation [2].

Since one equation is a combination of the other two, the equations are said to be **linearly dependent**.

We cannot find a unique solution for two equations in three unknowns.

To solve these equations, we let $x = t$. Hence, x is no longer an unknown. We thereby have only two unknowns in these two equations, and so we can solve them.

Using equation [2], we obtain $y = 3t - 2$. Substituting this in equation [3], we get

$$4z = t + 2(3t - 2) - 1$$
$$\Rightarrow \quad z = \frac{7t - 5}{4}$$

So, the solution is $\left(t, 3t - 2, \dfrac{7t - 5}{4} \right)$.

Each value of the parameter t gives a different point. Since there is only one parameter, this solution represents a line.

Exercise 5B

1 Express the determinant

$$\begin{vmatrix} a & bc & b+c \\ b & ca & c+a \\ c & ab & a+b \end{vmatrix}$$

as the product of four linear factors.

Hence, or otherwise, find the values of a for which the simultaneous equations

$$ax + 2y + 3z = 0$$
$$2x + ay + (1+a)z = 0$$
$$x + 2ay + (2+a)z = 0$$

have a solution other than $x = y = z = 0$.

Solve the equations when $a = -3$.   (NEAB)

2 Consider the following system of simultaneous equations

$$x - y + 2z = 6$$
$$2x + 3y - z = 7$$
$$x + 9y - 8z = -4$$

i) By evaluating an appropriate determinant, show that this system does not have a unique solution.

ii) Solve this system of simultaneous equations.   (NICCEA)

3 Consider the system of simultaneous equations

$$3x + y - 2z = -4$$
$$x + 2y + 3z = 11$$
$$3x - 4y - 13z = -41$$

i) Solve this system of equations.

ii) Hence show in a sketch how the planes defined by the above equations are arranged so that the solution is of the form found in part **i**.   (NICCEA)

4 Show that the equations

$$x + \lambda y + z = 2a$$
$$x + y + \lambda z = 2b$$
$$\lambda x + y + \lambda z = 2c$$

where $a, b, c \in \mathbb{R}$, have a unique solution for x, y, z provided that $\lambda \neq 1$ and $\lambda \neq -1$.

a) In the case when $\lambda = 1$, state the condition to be satisfied by a, b and c for the equations to be consistent.

b) In the case when $\lambda = -1$, show that for the equations to be consistent

$$a + c = 0$$

Solve the equations in this case.

Give a geometrical description of the configuration of the three planes represented by the equations in the cases:

i) $\lambda = -1$ and $a + c = 0$

ii) $\lambda = -1$ and $a + c \neq 0$.   (NEAB)

5 Find the values of k for which the simultaneous equations

$$kx + 2y + z = 0$$
$$3x - 2z = 4$$
$$3x - 6ky - 4z = 14$$

do not have a unique solution for x, y and z.

Show that, when $k = -2$, the equations are inconsistent, and give a geometrical interpretation of the situation in this case.   (OCR)

6 Show that if $a \neq 3$ then the system of equations

$$x + 3y + 4z = -5$$
$$2x + 5y - z = 5a$$
$$3x + 8y + az = b$$

has a unique solution.

Given that $a = 3$, find the value of b for which the equations are consistent.   (OCR)

7 Given that

$$\mathbf{M} = \begin{pmatrix} 1 & 1 & -1 \\ 1 & 2 & -k \\ 1 & -k & -1 \end{pmatrix}$$

find $\det \mathbf{M}$ in terms of k.

Determine the values of k for which the simultaneous equations

$$x + y - z = 1$$
$$x + 2y - kz = 0$$
$$x - ky - z = 1$$

have a unique solution.

i) Solve these equations in the case when $k = 2$.
ii) Show that the equations have no solution when $k = 1$.
iii) Find the general solution when $k = -1$.

Give a geometrical interpretation of the equations in each of the three cases $k = 2$, $k = 1$ and $k = -1$.   (NEAB)

8 a) Express the determinant

$$D = \begin{vmatrix} 1 & 1 & 1 \\ a & b & c \\ a^3 & b^3 & c^3 \end{vmatrix}$$

as the product of four linear factors.

b) Two points, A and B, have coordinates $(1, 2, 8)$ and $(1, 3, 27)$, respectively. A third point, C, which is distinct from A and B, has coordinates $(1, c, c^3)$. Given that the vectors $\overrightarrow{OA}$, $\overrightarrow{OB}$, $\overrightarrow{OC}$ are linearly dependent, find the value of c.   (NEAB)

9 i) Show that the system of equations

$$x + 4y + 12z = 5$$
$$x + ay + 6z = a - 0.5$$
$$3x + 12y + 4az = b - 3$$

has a unique solution provided $a \neq 4$ and $a \neq 9$

ii) Find the solution in the case where $a = 3$ and $b = 42$.

iii) Show that when $a = 9$ the equations do not have a solution unless $b = 18$.

iv) Give a geometrical interpretation of the system in the case where $a = 9$ and $b = 13$.

(OCR)

10 It is given that

$$\mathbf{A} = \begin{pmatrix} 1 & -1 & 2 \\ 1 & p & -3 \\ 1 & -1 & q \end{pmatrix} \quad \text{and} \quad \mathbf{b} = \begin{pmatrix} 4 \\ -5 \\ 13 \end{pmatrix}$$

i) Find the determinant of $\mathbf{A}$ in terms of p and q.

ii) Hence show that if $p \neq -1$ and $q \neq 2$ then the system of equations defined by $\mathbf{Ax} = \mathbf{b}$ has a unique solution.

iii) Show that if $p = -1$ then the system does not have a solution unless q has a particular value, q_1, which is to be found.

iv) Give a geometrical interpretation of the system in the case where $p = -1$ and $q = q_1$.

(OCR)

11 Show that the only real value of λ for which the simultaneous equations

$$(2 + \lambda)x - y + z = 0$$
$$x - 2\lambda y - z = 0$$
$$4x - y - (\lambda - 1)z = 0$$

have a solution other than $x = y = z = 0$ is -1.

Solve the equations in the case when $\lambda = -1$, and interpret your result geometrically. (NEAB)

12 Consider the system of equations x, y and z,

$$2x + 3y - z = p$$
$$x - 2z = -5$$
$$qx + 9y + 5z = 8$$

where p and q are real.

Find the values of p and q for which this system has:

i) a unique solution
ii) an infinite number of solutions
iii) no solution. (NICCEA)

6 Vector geometry

The Great Bear is looking so geometrical.
One would think that something or other could be proved.
CHRISTOPHER FRY

Vector equation of a line

In *Introducing Pure Mathematics* (page 506), we found the vector equation of a line, AB.

From this, it follows that the general equation of a line through the point A and in the direction of **b** is

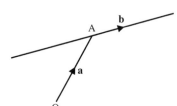

$$\mathbf{r} = \mathbf{a} + t\mathbf{b}$$

where **a** is the position vector of A, and each value of the parameter t corresponds to a point on the line.

Example 1

a) Find the equation of the line through the point (2, 4, 5) in the direction $-2\mathbf{i} + 3\mathbf{j} + 8\mathbf{k}$.

b) Find p and q so that the point $(p, 10, q)$ lies on this line.

SOLUTION

a) The equation of the line is $\mathbf{r} = \begin{pmatrix} 2 \\ 4 \\ 5 \end{pmatrix} + t \begin{pmatrix} -2 \\ 3 \\ 8 \end{pmatrix}$

b) If the point $(p, 10, q)$ lies on the line, then for some t we have

$$\begin{pmatrix} 2 \\ 4 \\ 5 \end{pmatrix} + t \begin{pmatrix} -2 \\ 3 \\ 8 \end{pmatrix} = \begin{pmatrix} p \\ 10 \\ q \end{pmatrix}$$

Considering these coordinates, we have

$$\begin{array}{lll} \text{For } \mathbf{i}: & 2 - 2t = p & [1] \\ \text{For } \mathbf{j}: & 4 + 3t = 10 & [2] \\ \text{For } \mathbf{k}: & 5 + 8t = q & [3] \end{array}$$

From [2], we get $t = 2$.

Substituting $t = 2$ in [1] and [3], we get

$$p = -2 \quad \text{and} \quad q = 21$$

Cartesian equation of a line

To find the three-dimensional cartesian equation of a line which passes through the point (x_1, y_1, z_1) in the direction $\begin{pmatrix} l \\ m \\ n \end{pmatrix}$, we use the vector equation

$$\mathbf{r} = \mathbf{a} + t\mathbf{b}$$

Hence, we obtain the vector equation of this line as

$$\mathbf{r} = \begin{pmatrix} x_1 \\ y_1 \\ z_1 \end{pmatrix} + t \begin{pmatrix} l \\ m \\ n \end{pmatrix}$$

Let the general vector $\mathbf{r}$ be $\begin{pmatrix} x \\ y \\ z \end{pmatrix}$, which gives

$$\begin{pmatrix} x \\ y \\ z \end{pmatrix} = \begin{pmatrix} x_1 \\ y_1 \\ z_1 \end{pmatrix} + t \begin{pmatrix} l \\ m \\ n \end{pmatrix}$$

Using the $\mathbf{i}$, $\mathbf{j}$, $\mathbf{k}$ components, we have

$$x = x_1 + tl$$
$$y = y_1 + tm$$
$$z = z_1 + tn$$

Finding t from each of these equations, we get

$$t = \frac{x - x_1}{l} = \frac{y - y_1}{m} = \frac{z - z_1}{n}$$

Hence, the three-dimensional cartesian equation of a straight line which passes through the point (x_1, y_1, z_1) in the direction $\begin{pmatrix} l \\ m \\ n \end{pmatrix}$ is

$$\frac{x - x_1}{l} = \frac{y - y_1}{m} = \frac{z - z_1}{n}$$

Example 2 Find the cartesian equation of the line PQ, where P is (2, 1, 7) and Q is (3, 8, 4).

SOLUTION

Let $\mathbf{p}$ and $\mathbf{q}$ be the position vectors of P and Q respectively. Then the direction of the line PQ is

$$\overrightarrow{PQ} = \mathbf{q} - \mathbf{p} = \begin{pmatrix} 3 \\ 8 \\ 4 \end{pmatrix} - \begin{pmatrix} 2 \\ 1 \\ 7 \end{pmatrix} = \begin{pmatrix} 1 \\ 7 \\ -3 \end{pmatrix}$$

Hence, given that the line passes through P(2, 1, 7), its **vector** equation is

$$\mathbf{r} = \begin{pmatrix} 2 \\ 1 \\ 7 \end{pmatrix} + t \begin{pmatrix} 1 \\ 7 \\ -3 \end{pmatrix}$$

Therefore, its cartesian equation is

$$\frac{x-2}{1} = \frac{y-1}{7} = \frac{z-7}{-3}$$

Note In Example 2, we could have used Q as the point on the line, in which case we would have obtained

$$\mathbf{r} = \begin{pmatrix} 3 \\ 8 \\ 4 \end{pmatrix} + t \begin{pmatrix} 1 \\ 7 \\ -3 \end{pmatrix}$$

leading to

$$\frac{x-3}{1} = \frac{y-8}{7} = \frac{z-4}{-3}$$

Example 3 For the line through $(4, 7, -1)$ in the direction $2\mathbf{i} - 3\mathbf{j} - 5\mathbf{k}$, find

a) its vector equation

b) its cartesian equation.

SOLUTION

a) The vector equation is

$$\mathbf{r} = \begin{pmatrix} 4 \\ 7 \\ -1 \end{pmatrix} + t \begin{pmatrix} 2 \\ -3 \\ -5 \end{pmatrix}$$

b) The cartesian equation is

$$\frac{x-4}{2} = \frac{y-7}{-3} = \frac{z+1}{-5}$$

which could be written as

$$\frac{x-4}{2} = \frac{7-y}{3} = \frac{-(1+z)}{5}$$

Example 4 Find the vector equation of the line

$$\frac{x-3}{4} = \frac{1-y}{2} = \frac{2z+7}{5}$$

SOLUTION

We **always start** by rearranging the cartesian equation in the form

$$\frac{x-x_1}{l} = \frac{y-y_1}{m} = \frac{z-z_1}{n}$$

which in this case gives

$$\frac{x-3}{4} = \frac{y-1}{-2} = \frac{z+\frac{7}{2}}{\frac{5}{2}}$$

Therefore, the vector equation of the line is

$$\mathbf{r} = \begin{pmatrix} 3 \\ 1 \\ -\frac{7}{2} \end{pmatrix} + t \begin{pmatrix} 4 \\ -2 \\ \frac{5}{2} \end{pmatrix}$$

Note

- The direction of a line is normally expressed in terms of integers. Hence, the vector equation in Example 4 would be given as

$$\mathbf{r} = \begin{pmatrix} 3 \\ 1 \\ -\frac{7}{2} \end{pmatrix} + s \begin{pmatrix} 8 \\ -4 \\ 5 \end{pmatrix}$$

where $s = \dfrac{t}{2}$ is also a parameter.

- It is neater to use a point on the line with integer coordinates, whereby this equation could be given as

$$\mathbf{r} = \begin{pmatrix} 7 \\ -1 \\ -1 \end{pmatrix} + \lambda \begin{pmatrix} 8 \\ -4 \\ 5 \end{pmatrix}$$

However, this further manipulation is not required in A-level examinations.

Example 5 Find the angle between the two lines

$$\frac{x-3}{4} = \frac{y-5}{2} = \frac{z-8}{-1} \quad \text{and} \quad \mathbf{r} = \begin{pmatrix} 3 \\ -1 \\ -2 \end{pmatrix} + t \begin{pmatrix} -7 \\ 4 \\ 3 \end{pmatrix}$$

SOLUTION

The required angle is between the **directions** of the two lines, which are

$$\begin{pmatrix} 4 \\ 2 \\ -1 \end{pmatrix} \quad \text{and} \quad \begin{pmatrix} -7 \\ 4 \\ 3 \end{pmatrix}$$

Using the scalar product in the form $\cos\theta = \dfrac{\mathbf{a} \cdot \mathbf{b}}{|\mathbf{a}|\,|\mathbf{b}|}$, where θ is the required angle, we have

$$\cos\theta = \frac{\begin{pmatrix} 4 \\ 2 \\ -1 \end{pmatrix} \cdot \begin{pmatrix} -7 \\ 4 \\ 3 \end{pmatrix}}{\sqrt{4^2 + 2^2 + (-1)^2} \times \sqrt{(-7)^2 + 4^2 + 3^2}}$$

$$= \frac{-28 + 8 - 3}{\sqrt{21} \times \sqrt{74}} = -\frac{23}{\sqrt{1554}}$$

The minus sign indicates that the angle between the two directions is obtuse. However, the angle between two lines would normally be taken to be acute. Therefore, the angle between the two lines is

$$\cos^{-1}\left(\frac{23}{\sqrt{1554}}\right) = 54.3° \quad \text{or} \quad 0.95 \text{ radians}$$

Note The scalar product of two vectors **a** and **b** is defined as

$$\mathbf{a} \cdot \mathbf{b} = |\mathbf{a}|\,|\mathbf{b}|\cos\theta$$

where θ is the angle between **a** and **b**. (See *Introducing Pure Mathematics*, pages 502–4, where examples are given of its application.)

Resolved part of a vector

We may consider the vector **a** to be composed of two parts: one in the direction of a vector **b**, and the other perpendicular to the direction of vector **b**.

In the diagram on the right, we have

$$\overrightarrow{OA} = \overrightarrow{OT} + \overrightarrow{TA}$$

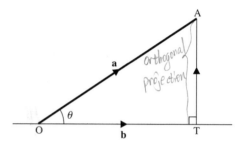

The magnitude of $\overrightarrow{OT}$ is the **resolved part** of vector **a** in the direction of vector **b**. That is,

$$OT = a\cos\theta$$

Using the scalar product $\mathbf{a} \cdot \mathbf{b} = ab\cos\theta$, we have

$$\textit{magnitude}\left(\frac{\mathbf{a} \cdot \mathbf{b}}{b}\right) = a\cos\theta = OT$$

Therefore, the resolved part of vector **a** in the direction of vector **b** is $\dfrac{\mathbf{a} \cdot \mathbf{b}}{b}$.

Note The resolved part is a **scalar**. The vector $\overrightarrow{OT}$ is

$$\frac{\mathbf{a} \cdot \mathbf{b}}{b}\frac{\mathbf{b}}{b} = \frac{(\mathbf{a} \cdot \mathbf{b})\mathbf{b}}{b^2}$$

Example 6 Find the resolved part of the vector of $2\mathbf{i} - 3\mathbf{j} + 6\mathbf{k}$ in the direction of $3\mathbf{i} + \mathbf{j} - 7\mathbf{k}$.

SOLUTION

The resolved part is

$$\frac{\begin{pmatrix} 2 \\ -3 \\ 6 \end{pmatrix} \cdot \begin{pmatrix} 3 \\ 1 \\ -7 \end{pmatrix}}{\sqrt{3^2 + 1^2 + (-7)^2}} = \frac{6 - 3 - 42}{\sqrt{9 + 1 + 49}} = -\frac{39}{\sqrt{59}}$$

Direction ratios

When one vector is a scalar multiple of another vector, the two vectors are parallel. For example, vector $\mathbf{a} = \begin{pmatrix} 3 \\ 4 \\ -5 \end{pmatrix}$ is parallel to vector $\mathbf{b} = \begin{pmatrix} -15 \\ -20 \\ 25 \end{pmatrix}$ since $\mathbf{b} = -5\mathbf{a}$.

The direction of a vector is specified by the ratios of the components in the **i**, **j** and **k** directions. These are called the **direction ratios** of the vector, and are normally expressed as integers.

For example, the direction ratios of the vector $28\mathbf{i} - 21\mathbf{j} - 14\mathbf{k}$ are $4 : -3 : -2$. Usually, these would be changed to $-4 : 3 : 2$.

Note Two lines which do not intersect and are not parallel are said to be **skew**.

Direction cosines

The angle which vector **a** makes with the **i**-axis is given by $\cos^{-1}\left(\dfrac{a_1}{a}\right)$, where a is the magnitude of vector **a**, and a_1 is the component of **a** in the **i**-direction. If θ_x is the angle which vector **a** makes with the **i**-axis, we have

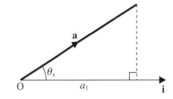

$$\cos \theta_x = \frac{a_1}{a}$$

Likewise for θ_y and θ_z, we have

$$\cos \theta_y = \frac{a_2}{a} \quad \text{and} \quad \cos \theta_z = \frac{a_3}{a}$$

These three values, $\dfrac{a_1}{a}, \dfrac{a_2}{a}$ and $\dfrac{a_3}{a}$, are known as the **direction cosines** of vector **a**. They represent another way of specifying the vector's direction.

> **Example 7** Find the direction cosines of the vector $3\mathbf{i} - 4\mathbf{j} + 5\mathbf{k}$, and find the angle which the vector makes with the z-axis.
>
> **SOLUTION**
>
> The direction cosines are given by $\dfrac{a_1}{a}, \dfrac{a_2}{a}$ and $\dfrac{a_3}{a}$, where $a_1 = 3$, $a_2 = -4$, $a_3 = 5$ and a represents the magnitude of the vector.
>
> The magnitude of $\begin{pmatrix} 3 \\ -4 \\ 5 \end{pmatrix}$ is
>
> $$\sqrt{3^2 + (-4)^2 + 5^2} = \sqrt{50} = 5\sqrt{2}$$
>
> Hence, the direction cosines are respectively
>
> $$\frac{3}{5\sqrt{2}} \qquad -\frac{4}{5\sqrt{2}} \qquad \frac{5}{5\sqrt{2}}$$
>
> If θ is the angle which the vector makes with the z-axis, we have
>
> $$\cos \theta = \frac{5}{5\sqrt{2}} = \frac{1}{\sqrt{2}}$$
>
> Therefore, the angle which the vector makes with the z-axis is $\dfrac{\pi}{4}$.

Exercise 6A

1 Find the vector equation of the line

a) through A(2, −7, 5) in the direction $3\mathbf{i} + 4\mathbf{j} - 7\mathbf{k}$
b) through B(4, 8, −6) in the direction $-2\mathbf{i} + 3\mathbf{j} + 6\mathbf{k}$
c) through P(7, 4, −1) in the direction $2\mathbf{i} - \mathbf{j} - 3\mathbf{k}$
d) through Q(−8, 1, −3) in the direction $\mathbf{i} + 3\mathbf{j} - 7\mathbf{k}$

2 Find the vector equation of the line through each pair of points.

a) A(4, 8, −2) and B(1, −3, 4)   b) C(−1, 8, 3) and D(2, −3, 9)
c) P(1, 7, −2) and B(−3, 4, 8)   d) R(3, −5, −9) and S(−2, −3, 7)

3 Find the cartesian equation of each line in Question **1.**

4 Find the vector equation of each of these lines.

a) $\dfrac{x-3}{4} = \dfrac{y+2}{3} = \dfrac{z-4}{-5}$ b) $\dfrac{x+2}{5} = \dfrac{y-1}{-7} = \dfrac{z+3}{-2}$

c) $\dfrac{x+5}{1} = \dfrac{2-y}{3} = \dfrac{z+4}{2}$ d) $\dfrac{2x-3}{4} = \dfrac{y-5}{3} = \dfrac{2-z}{1}$

e) $\dfrac{3x-5}{6} = \dfrac{y+2}{4} = \dfrac{2-z}{3}$

5 Find the acute angles between the lines with equations

a) $\mathbf{r} = 3\mathbf{i} + 4\mathbf{j} - 7\mathbf{k} + t(2\mathbf{i} - \mathbf{j} + 3\mathbf{k})$ and $\mathbf{r} = -2\mathbf{i} + 7\mathbf{j} + 2\mathbf{k} + t(3\mathbf{i} + 5\mathbf{j} - 3\mathbf{k})$

b) $\mathbf{r} = \begin{pmatrix} 1 \\ 2 \\ 4 \end{pmatrix} + t\begin{pmatrix} 7 \\ -2 \\ 3 \end{pmatrix}$ and $\mathbf{r} = \begin{pmatrix} -2 \\ 3 \\ 11 \end{pmatrix} + s\begin{pmatrix} 0 \\ 2 \\ 1 \end{pmatrix}$

6 Find the equation of the line AB where:

a) A is (2, 1, 4) and B is (4, 7, 5).
b) A is (−1, −4, 3) and B is (2, 8, 4).
c) A is (4, 1, −5) and B is (3, 2, −6).

7 Find the resolved part of $3\mathbf{i} - \mathbf{j} + 2\mathbf{k}$ in the direction $5\mathbf{i} + 3\mathbf{j} + 4\mathbf{k}$.

8 Find the resolved part of $4\mathbf{i} + 5\mathbf{j} - 2\mathbf{k}$ in the direction $\mathbf{j} - 7\mathbf{k}$.

9 Give for each vector **i)** its direction ratios and **ii)** its direction cosines.

a) $6\mathbf{i} + 12\mathbf{j} - 12\mathbf{k}$ b) $3\mathbf{i} - 4\mathbf{j} - 5\mathbf{k}$
c) $12\mathbf{i} + 8\mathbf{j} - 20\mathbf{k}$ d) $9\mathbf{i} - 18\mathbf{j} - 27\mathbf{k}$

10 Referred to a fixed origin O, the points P, Q and R have position vectors $(2\mathbf{i} + \mathbf{j} + \mathbf{k})$, $(5\mathbf{j} + 3\mathbf{k})$ and $(5\mathbf{i} - 4\mathbf{j} + 2\mathbf{k})$ respectively.

a) Find in the form $\mathbf{r} = \mathbf{a} + t\mathbf{b}$, an equation of the line PQ.
b) Show that the point S with position vector $(4\mathbf{i} - 3\mathbf{j} - \mathbf{k})$ lies on PQ.
c) Show that the lines PQ and RS are perpendicular.
d) Find the size of ∠PQR, giving your answer to 0.1°.   (EDEXCEL)

11 The lines l_1, l_2 and l_3 are given by

$$l_1: \quad \mathbf{r} = 10\mathbf{i} + \mathbf{j} + 9\mathbf{k} + \mu(3\mathbf{i} + \mathbf{j} + 4\mathbf{k})$$

$$l_2: \quad x = \frac{y+9}{2} = \frac{z-13}{-3}$$

$$l_3: \quad \mathbf{r} = -3\mathbf{i} - 5\mathbf{j} - 4\mathbf{k} + \lambda(4\mathbf{i} + 3\mathbf{j} + \mathbf{k})$$

where μ and λ are parameters.

a) Show that the point A(4, −1, 1) lies on both l_1 and l_2.
b) Rewrite the equation for l_2 in the form $\mathbf{r} = \mathbf{a} + v\mathbf{b}$, where v is a parameter.
c) Show that l_2 and l_3 intersect and find the coordinates of B, the point of intersection.

The lines l_1 and l_3 intersect at the point C(1, −2, −3).

d) Show that AC = BC.
e) Find the size of angle ACB, giving your answer in degrees to the nearest degree.
f) Write down the coordinates of the point D on AB such that CD is perpendicular to AB.

(EDEXCEL)

12 With respect to a fixed origin O, the lines l_1 and l_2 are given by the equations

$$l_1: \quad \mathbf{r} = (2\mathbf{i} + 3\mathbf{j} - 2\mathbf{k}) + \lambda(-2\mathbf{i} + 4\mathbf{j} + \mathbf{k})$$

$$l_2: \quad \mathbf{r} = (-6\mathbf{i} - 3\mathbf{j} + \mathbf{k}) + \mu(5\mathbf{i} + \mathbf{j} - 2\mathbf{k})$$

where λ and μ are scalar parameters.

a) Show that l_1 and l_2 meet and find the position vector of their point of intersection.
b) Find, to the nearest $0.1°$, the acute angle between l_1 and l_2. (EDEXCEL)

13 The line l passes through the points with position vectors $\mathbf{i} + 2\mathbf{j} + 3\mathbf{k}$ and $\mathbf{i} + 6\mathbf{j}$ relative to an origin O.

a) Find an equation for l in vector form.

The line m has equation $\mathbf{r} = 3\mathbf{i} + 6\mathbf{j} + \mathbf{k} + \lambda(\mathbf{i} - 2\mathbf{j} + 2\mathbf{k})$.

b) Find the acute angle between l and m, giving your answer to the nearest degree.

(EDEXCEL)

14 Two lines have vector equations

$$\mathbf{r} = (3\mathbf{i} + 2\mathbf{j} + 7\mathbf{k}) + \lambda(4\mathbf{i} - \mathbf{j} + 5\mathbf{k})$$

and

$$\mathbf{r} = (2\mathbf{i} + 6\mathbf{j} - 13\mathbf{k}) + \mu(-3\mathbf{i} + \mathbf{j} + a\mathbf{k})$$

where λ and μ are scalar parameters and a is a constant.

Given that these two lines intersect, find the position vector of the point of intersection and the value of a.   (AEB 98)

15 With respect to an origin O, the position vectors of the points L and M are $\mathbf{i} - \mathbf{j} + 3\mathbf{k}$ and $2\mathbf{i} - 4\mathbf{j} + 2\mathbf{k}$ respectively.

a) Write down the vector $\overrightarrow{LM}$.
b) Show that $|\overrightarrow{OL}| = |\overrightarrow{LM}|$.
c) Find $\angle$ OLM, giving your answer to the nearest tenth of a degree.   (EDEXCEL)

16 Two lines have equations

$$\mathbf{r} = (\mathbf{i} + 5\mathbf{j} + 2\mathbf{k}) + s(\mathbf{i} - 2\mathbf{j} + 3\mathbf{k}) \quad \text{and} \quad \mathbf{r} = (-\mathbf{i} - \mathbf{j} + 10\mathbf{k}) + t(3\mathbf{i} + 4\mathbf{j} - 5\mathbf{k})$$

i) Show that the lines meet, and find the point of intersection.
ii) Calculate the acute angle between the lines.   (OCR)

17 a) Find the angle between the vectors $2\mathbf{i} + 3\mathbf{j} + 6\mathbf{k}$ and $3\mathbf{i} + 4\mathbf{j} + 12\mathbf{k}$, giving your answer in radians.
b) The vectors $\mathbf{a}$ and $\mathbf{b}$ are non-zero.
  i) Given that $\mathbf{a} + \mathbf{b}$ is perpendicular to $\mathbf{a} - \mathbf{b}$, prove that $|\mathbf{a}| = |\mathbf{b}|$.
  ii) Given instead that $|\mathbf{a} + \mathbf{b}| = |\mathbf{a} - \mathbf{b}|$, prove that $\mathbf{a}$ and $\mathbf{b}$ are perpendicular.   (OCR)

18 The two lines

$$\frac{x + 11}{4} = \frac{y + 2}{1} = \frac{z + 6}{-2} \quad \text{and} \quad \frac{x - 6}{5} = \frac{y - 5}{4} = \frac{z + 20}{-8}$$

intersect. Find the coordinates of the point of intersection.   (OCR)

19 The points A and B have position vectors $7\mathbf{i} - 8\mathbf{j} + 7\mathbf{k}$ and $4\mathbf{i} + 7\mathbf{j} + 4\mathbf{k}$ respectively, and O is the origin.

i) Find, in vector form, an equation for the line passing through A and B.
ii) Find the position vector of the point P on the line AB such that OP is perpendicular to AB.
iii) Show that the line $\mathbf{r} = (8\mathbf{i} - 5\mathbf{j} + 2\mathbf{k}) + \lambda(\mathbf{i} - 10\mathbf{j} + 4\mathbf{k})$ does not intersect the line AB.

(OCR)

Vector product

The product of two vectors can be formed in two distinct ways. One of these, the **scalar product**, we have already met in *Introducing Pure Mathematics* (page 502), and in the present book on pages 97 and 98. The other is called the **vector product** (or sometimes the **cross product**).

The **vector product** of two vectors $\mathbf{a}$ and $\mathbf{b}$ is denoted by $\mathbf{a} \times \mathbf{b}$, and is defined as

$$\mathbf{a} \times \mathbf{b} = |\mathbf{a}|\,|\mathbf{b}|\sin\theta\,\hat{\mathbf{n}}$$

where θ is the angle measured in the anticlockwise sense between $\mathbf{a}$ and $\mathbf{b}$, and $\hat{\mathbf{n}}$ is a unit vector, such that $\mathbf{a}$, $\mathbf{b}$ and $\hat{\mathbf{n}}$ (in that order) form a right-handed set (see the diagram below).

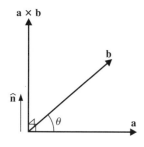
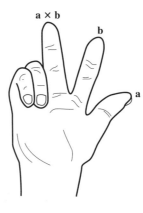

Some important properties of the vector product

The vector product is not commutative

Since $\mathbf{a} \times \mathbf{b} = ab \sin \theta \, \hat{\mathbf{n}}$, it follows that

$$\mathbf{b} \times \mathbf{a} = ab \sin(-\theta) \, \hat{\mathbf{n}} = -ab \sin \theta \, \hat{\mathbf{n}}$$

Therefore, we have

$$\mathbf{a} \times \mathbf{b} = -\mathbf{b} \times \mathbf{a}$$

which is known as the **anticommutative rule**.

The vector product of parallel vectors is zero

The angle, θ, between two parallel vectors, $\mathbf{a}$ and $\mathbf{b}$, is $0°$ or $180°$. Therefore, $\sin \theta = 0$, which gives

$$\mathbf{a} \times \mathbf{b} = \mathbf{0}$$

0 is called the **zero vector**. It is usually represented by an ordinary zero, 0, as below.

Likewise, $\mathbf{a} \times \mathbf{a} = 0$, since the angle between $\mathbf{a}$ and $\mathbf{a}$ is zero. Hence, we have the following important result:

$$\mathbf{i} \times \mathbf{i} = \mathbf{j} \times \mathbf{j} = \mathbf{k} \times \mathbf{k} = 0$$

Remember The scalar product $\mathbf{a} \cdot \mathbf{a} = a^2$.

The vector product of perpendicular vectors

Considering the unit vectors $\mathbf{i}$ and $\mathbf{j}$, we have

$$\mathbf{i} \times \mathbf{j} = 1 \times 1 \sin 90° \, \hat{\mathbf{n}} = \hat{\mathbf{n}}$$

Therefore, $\mathbf{i}, \mathbf{j}, \hat{\mathbf{n}}$ form a right-handed set.

But, by definition, $\mathbf{i}, \mathbf{j}, \mathbf{k}$ form a right-handed set. Therefore, $\hat{\mathbf{n}} = \mathbf{k}$. Hence, we have

$$\mathbf{i} \times \mathbf{j} = \mathbf{k} \qquad \mathbf{j} \times \mathbf{i} = -\mathbf{k}$$

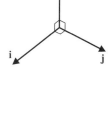

Similarly, we have

$$\mathbf{j} \times \mathbf{k} = \mathbf{i} \qquad \mathbf{k} \times \mathbf{j} = -\mathbf{i}$$
$$\mathbf{k} \times \mathbf{i} = \mathbf{j} \qquad \mathbf{i} \times \mathbf{k} = -\mathbf{j}$$

We notice (see diagram on the right) that these vector products are **positive** when the alphabetical order in which $\mathbf{i}, \mathbf{j}$ and $\mathbf{k}$ are taken is **clockwise**, but **negative** when this order is **anticlockwise**.

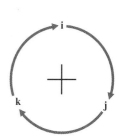

Remember For perpendicular vectors $\mathbf{a}$ and $\mathbf{b}$, the scalar product $\mathbf{a} \cdot \mathbf{b} = 0$.

The vector product in component form

Expressing $\mathbf{a}$ and $\mathbf{b}$ in their component form, we have $\mathbf{a} = a_1\mathbf{i} + a_2\mathbf{j} + a_3\mathbf{k}$ and $\mathbf{b} = b_1\mathbf{i} + b_2\mathbf{j} + b_3\mathbf{k}$. Therefore,

$$\mathbf{a} \times \mathbf{b} = (a_1\mathbf{i} + a_2\mathbf{j} + a_3\mathbf{k}) \times (b_1\mathbf{i} + b_2\mathbf{j} + b_3\mathbf{k})$$
$$= a_1\mathbf{i} \times b_1\mathbf{i} + a_2\mathbf{j} \times b_1\mathbf{i} + a_3\mathbf{k} \times b_1\mathbf{i} + a_1\mathbf{i} \times b_2\mathbf{j} + a_2\mathbf{j} \times b_2\mathbf{j} + a_3\mathbf{k} \times b_2\mathbf{j} +$$
$$+ a_1\mathbf{i} \times b_3\mathbf{k} + a_2\mathbf{j} \times b_3\mathbf{k} + a_3\mathbf{k} \times b_3\mathbf{k}$$

$$\Rightarrow \quad \mathbf{a} \times \mathbf{b} = a_1b_2\mathbf{k} - a_2b_1\mathbf{k} + a_3b_1\mathbf{j} - a_1b_3\mathbf{j} + a_2b_3\mathbf{i} - a_3b_2\mathbf{i}$$
$$= (a_2b_3 - a_3b_2)\mathbf{i} - (a_1b_3 - b_1a_3)\mathbf{j} + (a_1b_2 - a_2b_1)\mathbf{k} \quad [1]$$

From the definition of a 3×3 determinant on page 80, we obtain

$$\begin{vmatrix} \mathbf{i} & \mathbf{j} & \mathbf{k} \\ a_1 & a_2 & a_3 \\ b_1 & b_2 & b_3 \end{vmatrix} = \begin{vmatrix} a_2 & a_3 \\ b_2 & b_3 \end{vmatrix}\mathbf{i} - \begin{vmatrix} a_1 & a_3 \\ b_1 & b_3 \end{vmatrix}\mathbf{j} + \begin{vmatrix} a_1 & a_2 \\ b_1 & b_2 \end{vmatrix}\mathbf{k}$$
$$= (a_2b_3 - a_3b_2)\mathbf{i} - (a_1b_3 - b_1a_3)\mathbf{j} + (a_1b_2 - a_2b_1)\mathbf{k} \quad [2]$$

We note that the RHS of [1] and [2] are identical. Therefore, we have

$$\mathbf{a} \times \mathbf{b} = \begin{vmatrix} \mathbf{i} & \mathbf{j} & \mathbf{k} \\ a_1 & a_2 & a_3 \\ b_1 & b_2 & b_3 \end{vmatrix}$$

Example 8 Evaluate $(2\mathbf{i} + 3\mathbf{j} - \mathbf{k}) \times (7\mathbf{i} + 4\mathbf{j} + 2\mathbf{k})$.

SOLUTION

We use the result

$$\mathbf{a} \times \mathbf{b} = \begin{vmatrix} \mathbf{i} & \mathbf{j} & \mathbf{k} \\ a_1 & a_2 & a_3 \\ b_1 & b_2 & b_3 \end{vmatrix}$$

which gives

$$\begin{pmatrix} 2 \\ 3 \\ -1 \end{pmatrix} \times \begin{pmatrix} 7 \\ 4 \\ 2 \end{pmatrix} = \begin{vmatrix} \mathbf{i} & \mathbf{j} & \mathbf{k} \\ 2 & 3 & -1 \\ 7 & 4 & 2 \end{vmatrix}$$
$$= \mathbf{i}\begin{vmatrix} 3 & -1 \\ 4 & 2 \end{vmatrix} - \mathbf{j}\begin{vmatrix} 2 & -1 \\ 7 & 2 \end{vmatrix} + \mathbf{k}\begin{vmatrix} 2 & 3 \\ 7 & 4 \end{vmatrix}$$

Therefore, we have

$$\begin{pmatrix} 2 \\ 3 \\ -1 \end{pmatrix} \times \begin{pmatrix} 7 \\ 4 \\ 2 \end{pmatrix} = 10\mathbf{i} - 11\mathbf{j} - 13\mathbf{k}$$

Example 9 Evaluate $|\overrightarrow{AB} \times \overrightarrow{CD}|$, where A is $(6, -3, 0)$, B is $(3, -7, 1)$, C is $(3, 7, -1)$ and D is $(4, 5, -3)$. Hence find the shortest distance between AB and CD.

SOLUTION

First, we find $\overrightarrow{AB}$ and $\overrightarrow{CD}$:

$$\overrightarrow{AB} = \mathbf{b} - \mathbf{a} = \begin{pmatrix} 3 \\ -7 \\ 1 \end{pmatrix} - \begin{pmatrix} 6 \\ -3 \\ 0 \end{pmatrix} = \begin{pmatrix} -3 \\ -4 \\ 1 \end{pmatrix}$$

$$\overrightarrow{CD} = \mathbf{d} - \mathbf{c} = \begin{pmatrix} 4 \\ 5 \\ -3 \end{pmatrix} - \begin{pmatrix} 3 \\ 7 \\ -1 \end{pmatrix} = \begin{pmatrix} 1 \\ -2 \\ -2 \end{pmatrix}$$

Then, we find their vector product:

$$\overrightarrow{AB} \times \overrightarrow{CD} = \begin{pmatrix} -3 \\ -4 \\ 1 \end{pmatrix} \times \begin{pmatrix} 1 \\ -2 \\ -2 \end{pmatrix} = \begin{vmatrix} \mathbf{i} & \mathbf{j} & \mathbf{k} \\ -3 & -4 & 1 \\ 1 & -2 & -2 \end{vmatrix}$$

which gives

$$\overrightarrow{AB} \times \overrightarrow{CD} = 10\mathbf{i} - 5\mathbf{j} + 10\mathbf{k}$$

Therefore, we have

$$|\overrightarrow{AB} \times \overrightarrow{CD}| = \sqrt{10^2 + (-5)^2 + 10^2} = \sqrt{15}$$

The line which is the shortest distance between AB and CD is perpendicular to both AB and CD. So, if P and Q are general points on AB and CD respectively, and PQ is perpendicular to both AB and CD, we have

$$\overrightarrow{PQ} = k(\overrightarrow{AB} \times \overrightarrow{CD})$$

which gives

$$\begin{pmatrix} 6 \\ -3 \\ 0 \end{pmatrix} + t \begin{pmatrix} -3 \\ -4 \\ 1 \end{pmatrix} - \left[\begin{pmatrix} 4 \\ 5 \\ -3 \end{pmatrix} + s \begin{pmatrix} 1 \\ -2 \\ -2 \end{pmatrix} \right] = k \begin{pmatrix} 10 \\ -5 \\ 10 \end{pmatrix}$$

Hence, we have

$$2 - 3t - s = 10k$$
$$-8 - 4t + 2s = 5k$$
$$3 + t + 2s = 10k$$

Solving these simultaneous equations, we obtain $k = 0.4$, $s = 1$ and $t = -1$. Therefore, we have

$$\text{Shortest distance between AB and CD} = 0.4(\overrightarrow{AB} \times \overrightarrow{CD})$$
$$= 0.4 \times 15 = 6$$

Area of a triangle

Consider the triangle ABC whose sides are $\mathbf{a}$, $\mathbf{b}$ and $\mathbf{c}$, as shown in the diagram. From the definition of the vector product, we have

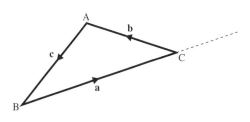

$$|\mathbf{a} \times \mathbf{b}| = |ab \sin \theta \, \hat{\mathbf{n}}|$$

where θ is the angle between $\mathbf{a}$ and $\mathbf{b}$.

However, the angle between $\mathbf{a}$ and $\mathbf{b}$ is $180° - C$, and $\sin(180° - C) = \sin C$. Therefore, we obtain

$$|\mathbf{a} \times \mathbf{b}| = |ab \sin(180° - C)\, \hat{\mathbf{n}}| = |ab \sin C \, \hat{\mathbf{n}}|$$

Since $\hat{\mathbf{n}}$ is a unit vector, $|\mathbf{a} \times \mathbf{b}| = ab \sin C$. Hence, we have

$$\text{Area of triangle ABC} = \tfrac{1}{2} ab \sin C = \tfrac{1}{2} |\mathbf{a} \times \mathbf{b}|$$

Similarly, we can show that the area of triangle ABC is given by

$$\tfrac{1}{2} bc \sin A = \tfrac{1}{2} |\mathbf{b} \times \mathbf{c}| \quad \text{and} \quad \tfrac{1}{2} ac \sin B = \tfrac{1}{2} |\mathbf{a} \times \mathbf{c}|$$

Generally, we have

$$\text{Area of a triangle} = \tfrac{1}{2}|\mathbf{a} \times \mathbf{b}| \quad \text{or} \quad \tfrac{1}{2}|\mathbf{b} \times \mathbf{c}| \quad \text{or} \quad \tfrac{1}{2}|\mathbf{a} \times \mathbf{c}|$$

Example 10 Find the area of triangle PQR where P is (4, 2, 5), Q is (3, −1, 6) and R is (1, 4, 2).

SOLUTION

First, we find any two sides (see diagram):

$$\overrightarrow{PR} = \mathbf{r} - \mathbf{p} = \begin{pmatrix} 1 \\ 4 \\ 2 \end{pmatrix} - \begin{pmatrix} 4 \\ 2 \\ 5 \end{pmatrix} = \begin{pmatrix} -3 \\ 2 \\ -3 \end{pmatrix}$$

$$\overrightarrow{PQ} = \mathbf{q} - \mathbf{p} = \begin{pmatrix} 3 \\ -1 \\ 6 \end{pmatrix} - \begin{pmatrix} 4 \\ 2 \\ 5 \end{pmatrix} = \begin{pmatrix} -1 \\ -3 \\ 1 \end{pmatrix}$$

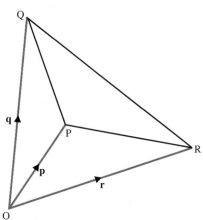

Then, we find their vector product:

$$\overrightarrow{PR} \times \overrightarrow{PQ} = \begin{vmatrix} \mathbf{i} & \mathbf{j} & \mathbf{k} \\ -3 & 2 & -3 \\ -1 & -3 & 1 \end{vmatrix}$$

which gives

$$\overrightarrow{PR} \times \overrightarrow{PQ} = -7\mathbf{i} + 6\mathbf{j} + 11\mathbf{k}$$
$$\Rightarrow \quad |\overrightarrow{PR} \times \overrightarrow{PQ}| = \sqrt{49 + 36 + 121} = \sqrt{206}$$

Therefore, we have

$$\text{Area of triangle PQR} = \tfrac{1}{2}\sqrt{206}$$

Equation of a plane

Equation in the form $\mathbf{r} = \mathbf{a} + t\mathbf{b} + s\mathbf{c}$

The position vector of **any** point on a plane can be expressed in terms of:

- **a**, the position vector of a point on the plane, and
- **b** and **c**, which are two **non-parallel** vectors **in** the plane.

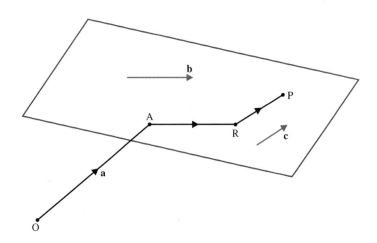

From the diagram, we see that the position vector of a point P on the plane is given by

$$\overrightarrow{OP} = \overrightarrow{OA} + \overrightarrow{AR} + \overrightarrow{RP}$$

where $\overrightarrow{AR}$ is parallel to vector **b**, and $\overrightarrow{RP}$ is parallel to vector **c**.

Hence, $\overrightarrow{AR} = t\mathbf{b}$ and $\overrightarrow{RP} = s\mathbf{c}$, for some parameters t and s.

The vector equation of a plane through point A is, therefore,

$$\mathbf{r} = \mathbf{a} + t\mathbf{b} + s\mathbf{c}$$

where **b** and **c** are non-parallel vectors in the plane, and t and s are parameters.

Equation in the form r . n = d

Given **n** is a vector perpendicular to the plane, we have

$$\mathbf{r} \cdot \mathbf{n} = (\mathbf{a} + t\mathbf{b} + s\mathbf{c}) \cdot \mathbf{n}$$
$$= \mathbf{a} \cdot \mathbf{n} + t\mathbf{b} \cdot \mathbf{n} + s\mathbf{c} \cdot \mathbf{n}$$

Since **b** and **c** are perpendicular to **n**, $\mathbf{b} \cdot \mathbf{n} = \mathbf{c} \cdot \mathbf{n} = 0$. Hence, we have

$$\mathbf{r} \cdot \mathbf{n} = \mathbf{a} \cdot \mathbf{n}$$

Therefore the vector equation of the plane is

$$\mathbf{r} \cdot \mathbf{n} = d$$

where d is a constant which determines the position of the plane.

Note

- If **n** is a unit vector, then d is the perpendicular distance of the plane from the origin.

- When d has the **same sign** for two planes, these planes are on the **same side** of the origin.

- When d has **opposite signs** for two planes, these planes are on **opposite sides** of the origin.

Cartesian form

In a similar way to finding the cartesian equation of a line, we take $\mathbf{r} \cdot \mathbf{n} = d$ and replace **r** by $x\mathbf{i} + y\mathbf{j} + z\mathbf{k}$, which gives the equation of a plane as

$$\begin{pmatrix} x \\ y \\ z \end{pmatrix} \cdot \mathbf{n} = d$$

Let $\mathbf{n} = a\mathbf{i} + b\mathbf{j} + c\mathbf{k}$, then the cartesian equation becomes

$$\begin{pmatrix} x \\ y \\ z \end{pmatrix} \cdot \begin{pmatrix} a \\ b \\ c \end{pmatrix} = d \quad \text{or} \quad ax + by + cz = d$$

$$(1, 2, 3)$$

direction vector $\begin{pmatrix} a \\ b \\ c \end{pmatrix}$

$$\frac{x-1}{a} = \frac{y-2}{b} = \frac{z-3}{c}$$

Example 11 Find the equation of the plane through (3, 2, 7) which is perpendicular to the vector $\begin{pmatrix} 1 \\ -5 \\ 8 \end{pmatrix}$, giving its equation **a)** in vector form, and **b)** in cartesian form.

[handwritten: $3-10+56=d$, $d=49$]
[handwritten: $x-5y+8z=d$]

SOLUTION

a) Using $\mathbf{r \cdot n = a \cdot n}$, we have

$$\mathbf{r} \cdot \begin{pmatrix} 1 \\ -5 \\ 8 \end{pmatrix} = \begin{pmatrix} 3 \\ 2 \\ 7 \end{pmatrix} \cdot \begin{pmatrix} 1 \\ -5 \\ 8 \end{pmatrix}$$

[handwritten: $x-5y+8z=49$]

Hence, the equation of the plane is $\mathbf{r} \cdot \begin{pmatrix} 1 \\ -5 \\ 8 \end{pmatrix} = 49$.

b) Replacing $\mathbf{r}$ by $x\mathbf{i} + y\mathbf{j} + z\mathbf{k}$, we get

$$\begin{pmatrix} x \\ y \\ z \end{pmatrix} \cdot \begin{pmatrix} 1 \\ -5 \\ 8 \end{pmatrix} = 49$$

Therefore, the cartesian equation is $x - 5y + 8z = 49$.

Note A plane is identified by

- a vector perpendicular to the plane, and
- a point on the plane.

Example 12 Find the unit vector perpendicular to the plane $2x + 3y - 7z = 11$.

SOLUTION

The vector $\begin{pmatrix} a \\ b \\ c \end{pmatrix}$ is perpendicular to the plane $ax + by + cz = d$.

Therefore, the vector perpendicular to the given plane is $\begin{pmatrix} 2 \\ 3 \\ -7 \end{pmatrix}$.

The magnitude of this vector is $\sqrt{2^2 + 3^2 + (-7)^2} = \sqrt{62}$.

Now, the unit vector perpendicular to the given plane must be of magnitude 1.

Therefore, the unit vector perpendicular to the given plane is $\begin{pmatrix} \dfrac{2}{\sqrt{62}} \\ \dfrac{3}{\sqrt{62}} \\ -\dfrac{7}{\sqrt{62}} \end{pmatrix}$.

[handwritten: unit vector in the direction of $\vec{a}$]
[handwritten: $\hat{a} = \dfrac{\vec{a}}{|\vec{a}|}$]
[handwritten: $\begin{pmatrix} x \\ y \\ z \end{pmatrix} \cdot \begin{pmatrix} \frac{2}{\sqrt{62}} \\ \frac{3}{\sqrt{62}} \\ -\frac{7}{\sqrt{62}} \end{pmatrix} = \dfrac{11}{\sqrt{62}}$]

Example 13 Find the equation of a plane through A(1, 4, 6), B(2, 7, 5) and C(−3, 8, 7).

SOLUTION

First, we find two vectors in the plane ABC: for example,

$$\overrightarrow{AB} = \mathbf{b} - \mathbf{a} = \begin{pmatrix} 2 \\ 7 \\ 5 \end{pmatrix} - \begin{pmatrix} 1 \\ 4 \\ 6 \end{pmatrix} = \begin{pmatrix} 1 \\ 3 \\ -1 \end{pmatrix}$$

$$\overrightarrow{AC} = \mathbf{c} - \mathbf{a} = \begin{pmatrix} -3 \\ 8 \\ 7 \end{pmatrix} - \begin{pmatrix} 1 \\ 4 \\ 6 \end{pmatrix} = \begin{pmatrix} -4 \\ 4 \\ 1 \end{pmatrix}$$

Result 1 To find the equation of the plane in the form $\mathbf{r} = \mathbf{a} + t\mathbf{b} + s\mathbf{c}$, we need to identify **one** point on the plane.

If we choose A(1, 4, 6), the equation of the plane ABC is

$$\mathbf{r} = \begin{pmatrix} 1 \\ 4 \\ 6 \end{pmatrix} + t \begin{pmatrix} 1 \\ 3 \\ -1 \end{pmatrix} + s \begin{pmatrix} -4 \\ 4 \\ 1 \end{pmatrix}$$

Note

- Instead of choosing A, we could have chosen B(2, 7, 5) or C(−3, 8, 7).

- Instead of $\begin{pmatrix} 1 \\ 3 \\ -1 \end{pmatrix}$, we could have used $\begin{pmatrix} -1 \\ -3 \\ 1 \end{pmatrix}$.

- Instead of $\begin{pmatrix} -4 \\ 4 \\ 1 \end{pmatrix}$, we could have used $\begin{pmatrix} 4 \\ -4 \\ -1 \end{pmatrix}$.

Result 2 To find the equation of the plane in the form $\mathbf{r} \cdot \mathbf{n} = d$, we need to find a vector perpendicular to the plane ABC.

A vector perpendicular to the plane ABC is given by $\overrightarrow{AB} \times \overrightarrow{AC}$, or any similar vector product of two vectors **in** the plane. (This follows from the definition of the vector product, page 102.)

Therefore, we have

$$\overrightarrow{AB} \times \overrightarrow{AC} = \begin{pmatrix} 1 \\ 3 \\ -1 \end{pmatrix} \times \begin{pmatrix} -4 \\ 4 \\ 1 \end{pmatrix} = \begin{vmatrix} \mathbf{i} & \mathbf{j} & \mathbf{k} \\ 1 & 3 & -1 \\ -4 & 4 & 1 \end{vmatrix}$$

which gives the vector perpendicular to the plane as $7\mathbf{i} + 3\mathbf{j} + 16\mathbf{k}$.

Hence, the vector equation of the plane ABC is

$$\mathbf{r} \cdot \begin{pmatrix} 7 \\ 3 \\ 16 \end{pmatrix} = \begin{pmatrix} 2 \\ 7 \\ 5 \end{pmatrix} \cdot \begin{pmatrix} 7 \\ 3 \\ 16 \end{pmatrix} = 14 + 21 + 80$$

$$\Rightarrow \quad \mathbf{r} \cdot \begin{pmatrix} 7 \\ 3 \\ 16 \end{pmatrix} = 115$$

Therefore, the cartesian equation is $7x + 3y + 16z = 115$.

Example 14 Find the angle between the planes $3x + 4y + 5z = 7$ and $x + 2y - 2z = 11$.

SOLUTION

The angle between the planes is the angle between the vectors perpendicular to the planes. That is, the angle between the two vectors

$$\begin{pmatrix} 3 \\ 4 \\ 5 \end{pmatrix} \quad \text{and} \quad \begin{pmatrix} 1 \\ 2 \\ -2 \end{pmatrix}$$

Using $\cos \theta = \dfrac{\mathbf{a} \cdot \mathbf{b}}{ab}$, where θ is the required angle, we have

$$\cos \theta = \frac{3 + 8 - 10}{5\sqrt{2} \times 3}$$

$$\Rightarrow \quad \theta = \cos^{-1}\left(\frac{1}{15\sqrt{2}}\right) = 87.3° \quad \text{(correct to 1 dp)}$$

Example 15 Find where the line from A(2, 7, 4), perpendicular to the plane Π, $3x - 5y + 2z + 2 = 0$, meets Π.

SOLUTION

Let T be the point where the line from A(2, 7, 4), perpendicular to the plane Π, $3x - 5y + 2z + 2 = 0$, meets Π.

The equation of AT is

$$\mathbf{r} = \begin{pmatrix} 2 \\ 7 \\ 4 \end{pmatrix} + t \begin{pmatrix} 3 \\ -5 \\ 2 \end{pmatrix}$$

Hence, T is the point where $\mathbf{r} = \begin{pmatrix} 2 \\ 7 \\ 4 \end{pmatrix} + t \begin{pmatrix} 3 \\ -5 \\ 2 \end{pmatrix}$ meets Π.

So, putting $x = (2 + 3t)$, $y = (7 - 5t)$ and $z = (4 + 2t)$ into the equation of the plane Π, we have

$$3(2 + 3t) - 5(7 - 5t) + 2(4 + 2t) + 2 = 0$$

$$\Rightarrow \quad 38t = 19 \quad \Rightarrow \quad t = \tfrac{1}{2}$$

Therefore, the point T is $(3\tfrac{1}{2}, 4\tfrac{1}{2}, 5)$

Example 16 Find the angle between the plane $3x + 4y - 5z = 6$ and the line

$$\mathbf{r} = \begin{pmatrix} 2 \\ 4 \\ 8 \end{pmatrix} + t \begin{pmatrix} 1 \\ 5 \\ -3 \end{pmatrix}$$

SOLUTION

The required angle is $90° - \theta$, where θ is the angle between the line and the vector perpendicular to the plane. That is,

Required angle =

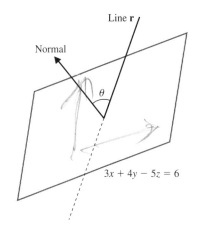

Line **r**

Normal

$3x + 4y - 5z = 6$

$$90° - \text{Angle between} \begin{pmatrix} 3 \\ 4 \\ -5 \end{pmatrix} \text{ and } \begin{pmatrix} 1 \\ 5 \\ -3 \end{pmatrix}$$

Using $\cos \theta = \dfrac{\mathbf{a} \cdot \mathbf{b}}{ab}$, we have

$$\cos \theta = \frac{3 + 20 + 15}{5\sqrt{2} \times \sqrt{35}}$$

which gives

$$\text{Required angle} = 90° - \cos^{-1} \left(\frac{38}{\sqrt{35} \times 5\sqrt{2}} \right)$$

$$= \sin^{-1} \left(\frac{38}{5\sqrt{70}} \right) = 65.3° \quad \text{(correct to 1 dp)}$$

Example 17 Find the equation of the plane containing the two lines

$$\mathbf{r} = \begin{pmatrix} 3 \\ 1 \\ 2 \end{pmatrix} + t \begin{pmatrix} -1 \\ 3 \\ -4 \end{pmatrix} \quad \text{and} \quad \mathbf{r} = \begin{pmatrix} -2 \\ -3 \\ 7 \end{pmatrix} + s \begin{pmatrix} 2 \\ -1 \\ 5 \end{pmatrix}$$

SOLUTION

The two vectors in the plane are the directions of the two lines, which are

$$\begin{pmatrix} -1 \\ 3 \\ -4 \end{pmatrix} \quad \text{and} \quad \begin{pmatrix} 2 \\ -1 \\ 5 \end{pmatrix}$$

Therefore, the vector perpendicular to this plane is

$$\begin{pmatrix} -1 \\ 3 \\ -4 \end{pmatrix} \times \begin{pmatrix} 2 \\ -1 \\ 5 \end{pmatrix} = \begin{vmatrix} \mathbf{i} & \mathbf{j} & \mathbf{k} \\ -1 & 3 & -4 \\ 2 & -1 & 5 \end{vmatrix} = 11\mathbf{i} - 3\mathbf{j} - 5\mathbf{k}$$

Hence, the equation of the plane is $11x - 3y - 5z = d$.

From the first line, we know that the point $(3, 1, 2)$ is in the plane. So, we have $d = 11 \times 3 - 3 \times 1 - 5 \times 2 = 20$.

Therefore, the equation of the plane containing the two lines is $11x - 3y - 5z = 20$.

Example 18 Find the equation of the common line (line of intersection) of the two planes

$$\Pi_1, \ 3x - y - 5z = 7 \quad \text{and} \quad \Pi_2, \ 2x + 3y - 4z = -2$$

SOLUTION

The vectors $\begin{pmatrix} 3 \\ -1 \\ -5 \end{pmatrix}$ and $\begin{pmatrix} 2 \\ 3 \\ -4 \end{pmatrix}$ are perpendicular to Π_1 and Π_2 respectively.

Therefore, $\begin{pmatrix} 3 \\ -1 \\ -5 \end{pmatrix} \times \begin{pmatrix} 2 \\ 3 \\ -4 \end{pmatrix}$ is perpendicular to both of these

perpendiculars, and hence is in the direction of the common line.

Therefore, the direction of the common line is

$$\begin{vmatrix} \mathbf{i} & \mathbf{j} & \mathbf{k} \\ 3 & -1 & -5 \\ 2 & 3 & -4 \end{vmatrix} = 19\mathbf{i} + 2\mathbf{j} + 11\mathbf{k}$$

To obtain the equation of the common line, we need to find a point on it.

Solving Π_1 and Π_2 will give only two equations to solve for three unknowns. So, we let $x = 0$ and solve the equations for the remaining two unknowns. However, if letting $x = 0$ causes problems because of the particular equations given, we may let either $y = 0$ or $z = 0$.

$$\Pi_1 \text{ is } 3x - y - 5z = 7 \qquad \text{When } x = 0, \ \Pi_1 \text{ gives } \quad -y - 5z = 7$$
$$\Pi_2 \text{ is } 2x + 3y - 4z = -2 \qquad \text{When } x = 0, \ \Pi_2 \text{ gives } \quad 3y - 4z = -2$$

Solving these simultaneous equations, we find $z = -1$, $y = -2$.

Therefore, the point $(0, -2, -1)$ lies on the common line, giving its equation as

$$\mathbf{r} = \begin{pmatrix} 0 \\ -2 \\ -1 \end{pmatrix} + t \begin{pmatrix} 19 \\ 2 \\ 11 \end{pmatrix}$$

Distance of a plane from the origin

Consider the equation of a plane in the form $\mathbf{r} \cdot \mathbf{n} = d$.

If $\mathbf{n}$ is a unit vector (usually denoted by $\hat{\mathbf{n}}$), then d is the perpendicular distance of the plane from the origin.

Example 19 Find the distance to the plane $3x + 4y - 5z = 21$ from the origin.

SOLUTION

The equation of the plane is

$$\mathbf{r} \cdot \begin{pmatrix} 3 \\ 4 \\ -5 \end{pmatrix} = 21$$

Changing this to the form $\mathbf{r} \cdot \hat{\mathbf{n}} = d$, where $\hat{\mathbf{n}}$ is a unit vector, we get

$$\mathbf{r} \cdot \begin{pmatrix} \dfrac{3}{5\sqrt{2}} \\[2mm] \dfrac{4}{5\sqrt{2}} \\[2mm] -\dfrac{5}{5\sqrt{2}} \end{pmatrix} = \dfrac{21}{5\sqrt{2}}$$

Therefore, the distance from the origin is $\dfrac{21}{5\sqrt{2}}$.

Distance of a plane from a point

Example 20 Find the distance from the point $(3, -2, 6)$ to the plane $3x + 4y - 5z = 21$.

SOLUTION

Method 1 First, we find the equation of the plane parallel to $3x + 4y - 5z = 21$ which passes through the point $(3, -2, 6)$. Then we find the distance of each plane from the origin. The difference between these distances is equal to the distance of the plane $3x + 4y - 5z = 21$ from $(3, -2, 6)$.

The equation of the plane parallel to $3x + 4y - 5z = 21$ through $(3, -2, 6)$ is

$$3x + 4y - 5z = (3 \times 3) - (2 \times 4) + (6 \times -5)$$

$$\Rightarrow \quad 3x + 4y - 5z = -29$$

Changing this to the form $\mathbf{r} \cdot \hat{\mathbf{n}} = d$, we get

$$\mathbf{r} \cdot \begin{pmatrix} \dfrac{3}{5\sqrt{2}} \\[2mm] \dfrac{4}{5\sqrt{2}} \\[2mm] -\dfrac{5}{5\sqrt{2}} \end{pmatrix} = -\dfrac{29}{5\sqrt{2}}$$

Therefore, the distance from the point $(3, -2, 6)$ to the plane is

$$\dfrac{29}{5\sqrt{2}} + \dfrac{21}{5\sqrt{2}} = \dfrac{50}{5\sqrt{2}} = 5\sqrt{2}$$

Method 2 Using the form $\mathbf{r} = \mathbf{a} + t\mathbf{b}$, the equation of the line perpendicular to $3x + 4y - 5z = 21$ through $(3, -2, 6)$ is

$$\mathbf{r} = \begin{pmatrix} 3 \\ -2 \\ 6 \end{pmatrix} + t \begin{pmatrix} 3 \\ 4 \\ -5 \end{pmatrix}$$

This line meets the plane $3x + 4y - 5z = 21$ when

$$3(3 + 3t) + 4(-2 + 4t) - 5(6 - 5t) = 21$$

$$\Rightarrow \quad 50t = 50 \quad \Rightarrow \quad t = 1$$

CHAPTER 6 VECTOR GEOMETRY

Using $\mathbf{r} = \mathbf{a} + t\mathbf{b}$ again, we see that the line meets the plane at (6, 2, 1).

The distance between the two points (3, −2, 6) and (6, 2, 1) is

$$\sqrt{3^2 + 4^2 + 5^2} = 5\sqrt{2}$$

Therefore, the distance from the point (3, −2, 6) to the plane is $5\sqrt{2}$.

Exercise 6B

1 Find $\mathbf{a} \times \mathbf{b}$ when

a) $\mathbf{a} = \begin{pmatrix} 1 \\ -4 \\ 3 \end{pmatrix}$ $\mathbf{b} = \begin{pmatrix} 2 \\ 3 \\ -1 \end{pmatrix}$
b) $\mathbf{a} = \begin{pmatrix} -3 \\ 4 \\ 5 \end{pmatrix}$ $\mathbf{b} = \begin{pmatrix} 2 \\ -3 \\ 4 \end{pmatrix}$

c) $\mathbf{a} = \begin{pmatrix} 4 \\ -4 \\ 2 \end{pmatrix}$ $\mathbf{b} = \begin{pmatrix} 1 \\ -5 \\ -3 \end{pmatrix}$
d) $\mathbf{a} = \begin{pmatrix} 1 \\ 4 \\ 6 \end{pmatrix}$ $\mathbf{b} = \begin{pmatrix} 3 \\ 2 \\ -5 \end{pmatrix}$

2 Find, in vector form, the equation of the plane through

a) A(4, 1, −5), B(2, −1, −6), C(−2, 3, 2)
b) P(2, 5, 3), Q(4, 1, −2), R(4, 3, 5)
c) D(4, 1, −3), E(2, 3, 2), F(−1, -3, 1)

3 Find, in cartesian form, the equation of the plane

a) $\mathbf{r} \cdot \begin{pmatrix} 3 \\ 1 \\ 7 \end{pmatrix} = 4$
b) $\mathbf{r} \cdot \begin{pmatrix} 2 \\ 4 \\ 3 \end{pmatrix} = 8$
c) $\mathbf{r} \cdot \begin{pmatrix} -1 \\ 5 \\ 3 \end{pmatrix} + 7 = 0$

4 Find the angle between each pair of planes.

a) $3x - y - 4z = 7$, $2x + 3y - z = 11$
b) $5x - 3y + z = 10$, $2x - y - z = 8$
c) $7x + 4y - 2z = 5$, $6x + 7y + z = 4$
d) $x - 2y - 9z = 1$, $x + 3y + 2z = 0$

5 Find the angle between the line

$$\mathbf{r} = \begin{pmatrix} 1 \\ 4 \\ 5 \end{pmatrix} + t \begin{pmatrix} 2 \\ -3 \\ 4 \end{pmatrix}$$

and the plane $2x + 4y - z = 7$.

6 Find the angle between the line

$$\mathbf{r} = \begin{pmatrix} 2 \\ -3 \\ 1 \end{pmatrix} + t \begin{pmatrix} 4 \\ 2 \\ -5 \end{pmatrix}$$

and the plane $3x - y + 2z = 11$.

7 Write the equation of the plane $3x + 4y - 5z = 20$ in the form $\mathbf{r} \cdot \hat{\mathbf{n}} = d$, where $\hat{\mathbf{n}}$ is a unit vector. Hence write down the distance from the plane to the origin.

8 The points A, B and C have position vectors $\mathbf{a} = \mathbf{i} + \mathbf{j} + 2\mathbf{k}$, $\mathbf{b} = 3\mathbf{i} + 2\mathbf{j} + 4\mathbf{k}$ and $\mathbf{c} = -\mathbf{i} + 4\mathbf{j} - 4\mathbf{k}$ respectively.

  a) Write down the vectors $\mathbf{b} - \mathbf{a}$ and $\mathbf{c} - \mathbf{a}$, and hence determine
   i) $(\mathbf{b} - \mathbf{a}) . (\mathbf{c} - \mathbf{a})$
   ii) $(\mathbf{b} - \mathbf{a}) \times (\mathbf{c} - \mathbf{a})$.

  b) Using the results from part **a**, or otherwise, find
   i) the cosine of the acute angle between the line AB and the line AC, giving your answer in an exact form
   ii) the area of triangle ABC, giving your answer in an exact surd form
   iii) a vector equation of the plane through A, B and C, giving your answer in the form $\mathbf{r} . \mathbf{n} = d$.   (AEB 98)

9 The plane Π_1 has vector equation

$$\mathbf{r} = (5\mathbf{i} + \mathbf{j}) + u(-4\mathbf{i} + \mathbf{j} + 3\mathbf{k}) + v(\mathbf{j} + 2\mathbf{k})$$

where u and v are parameters.

  a) Find a vector $\mathbf{n}_1$ normal to Π_1.

  The plane Π_2 has equation $3x + y - z = 3$.

  b) Write down a vector $\mathbf{n}_2$ normal to Π_2.
  c) Show that $4\mathbf{i} + 13\mathbf{j} + 25\mathbf{k}$ is normal to both $\mathbf{n}_1$ and $\mathbf{n}_2$.

  Given that the point (1, 1, 1) lies on both Π_1 and Π_2,

  d) write down an equation of the line of intersection of Π_1 and Π_2 in the form $\mathbf{r} = \mathbf{a} + t\mathbf{b}$, where t is a parameter.   (EDEXCEL)

10 The points A(24, 6, 0), B(30, 12, 12) and C(18, 6, 36) are referred to cartesian axes, origin O.

  a) Find a vector equation for the line passing through the points A and B.

  The point P lies on the line passing through A and B.

  b) Show that $\overrightarrow{CP}$ can be expressed as

$$(6 + t)\mathbf{i} + t\mathbf{j} + (2t - 36)\mathbf{k}$$

  where t is a parameter.

  c) Given that $\overrightarrow{CP}$ is perpendicular to $\overrightarrow{AB}$, find the coordinates of P.
  d) Hence, or otherwise, find the area of the triangle ABC, giving your answer to three significant figures.   (EDEXCEL)

11 The plane Π passes through the points A(−2, 3, 5), B(1, −3, 1) and C(4, −6, −7).
  a) Find $\overrightarrow{AC} \times \overrightarrow{BC}$.
  b) Hence, or otherwise, find the equation of the plane Π in the form $\mathbf{r} . \mathbf{n} = p$.

  The perpendicular from the point (25, 5, 7) to Π meets the plane at the point F.

  c) Find the coordinates of F.   (EDEXCEL)

12 The plane p has equation

$$\mathbf{r} = \mathbf{i} - \mathbf{j} + s(\mathbf{i} + \mathbf{k}) + t(\mathbf{j} - \mathbf{k})$$

and the line l has equation

$$\mathbf{r} = (\mathbf{i} - 2\mathbf{j} + \mathbf{k}) + \lambda(2\mathbf{i} - \mathbf{j})$$

i) Find a vector which is normal to p.

ii) Show that the acute angle between p and l is $\sin^{-1}(\frac{1}{5}\sqrt{15})$.   (OCR)

13 The planes P_1 and P_2 have equations

$$\mathbf{r}.(2\mathbf{i} - 3\mathbf{j} + \mathbf{k}) = 4 \quad \text{and} \quad \mathbf{r}.(\mathbf{i} + 2\mathbf{j} + 3\mathbf{k}) = 5$$

respectively. Find, in the form $\mathbf{r}.\mathbf{n} = d$, the equation of the plane which is perpendicular to both P_1 and P_2 and which passes through the point with position vector $3\mathbf{i} - \mathbf{j} + 2\mathbf{k}$.   (OCR)

14 The line l passes through the points with position vectors $\mathbf{i} - 8\mathbf{j} + 7\mathbf{k}$ and $7\mathbf{i} + 4\mathbf{j} + \mathbf{k}$. Find an equation of l in vector form.

The points A, B, C have position vectors $3\mathbf{i} + 5\mathbf{j} + 8\mathbf{k}$, $5\mathbf{i} + 6\mathbf{j} + 7\mathbf{k}$ and $4\mathbf{i} + 7\mathbf{j} + 5\mathbf{k}$ respectively.

i) Find the vector product $\overrightarrow{AB} \times \overrightarrow{AC}$. Hence or otherwise find the equation of the plane ABC.

ii) Show that the angle between l and the plane ABC is $24.5°$, correct to the nearest $0.1°$.

iii) Find the position vector of the point of intersection of l and the plane ABC.   (OCR)

15 The point A has position vector $2\mathbf{i} + 3\mathbf{j} + 5\mathbf{k}$ and the line l has equation

$$\mathbf{r} = -5\mathbf{i} + 6\mathbf{j} + 3\mathbf{k} + \lambda(2\mathbf{i} - 2\mathbf{j} - \mathbf{k})$$

i) Find the position vector of the point N on l such that AN is perpendicular to l.

ii) Show that the perpendicular distance from A to l is $\sqrt{26}$.

The points B and C have position vectors $-5\mathbf{i} + 6\mathbf{j} + 3\mathbf{k}$ and $6\mathbf{i} + 13\mathbf{j} - 7\mathbf{k}$ respectively, and the point D is the mid-point of BN.

iii) Show that the plane ANC is perpendicular to l.

iv) Find the acute angle between the planes ANC and ACD.   (OCR)

16 The line l has equation

$$\mathbf{r} = 5\mathbf{i} + 8\mathbf{j} + \mathbf{k} + t(\mathbf{i} + 8\mathbf{k})$$

and the plane P has equation

$$2x - 2y - z - 5 = 0$$

Find the position vector of the point at which l and P intersect.

Find also the acute angle between l and P, giving your answer to the nearest degree.   (OCR)

17 Consider the plane P_1, the line L_1 and the line L_2 given by the equations,

$$P_1: \quad x + 2y - z = 5$$

$$L_1: \quad \frac{x - 11}{-4} = \frac{y + 2}{2} = \frac{z + 8}{5}$$

$$L_2: \quad \frac{x - 1}{1} = \frac{y + 2}{-3}, \ z = 7$$

i) Show that L_1 and L_2 are coplanar.

ii) Find the equation of the plane, P_2, which contains L_1 and L_2.

iii) Find the equation of the line of intersection of the planes P_1 and P_2.   (NICCEA)

18 The points A, B and C have position vectors $(\mathbf{j} + 2\mathbf{k})$, $(2\mathbf{i} + 3\mathbf{j} + \mathbf{k})$ and $(\mathbf{i} + \mathbf{j} + 3\mathbf{k})$, respectively, relative to the origin O. The plane Π contains the points A, B and C.

a) Find a vector which is perpendicular to Π.
b) Find the area of $\triangle$ABC.
c) Find a vector equation of Π in the form $\mathbf{r} \cdot \mathbf{n} = p$.
d) Hence, or otherwise, obtain a cartesian equation of Π.
e) Find the distance of the origin O from Π.

The point D has position vector $(3\mathbf{i} + 4\mathbf{j} + \mathbf{k})$. The distance of D from Π is $\dfrac{1}{\sqrt{17}}$.

f) Using this distance, or otherwise, calculate the acute angle between the line AD and Π, giving your answer in degrees to one decimal place.   (EDEXCEL)

19 Given that

$$\mathbf{a} \times \mathbf{b} = \mathbf{i} \qquad \mathbf{b} \times \mathbf{c} = \mathbf{j} \qquad \mathbf{c} \times \mathbf{a} = \mathbf{k}$$

express

$$(\mathbf{a} + \mathbf{b}) \times (\mathbf{a} + 2\mathbf{b} + 3\mathbf{c})$$

in terms of $\mathbf{i}$, $\mathbf{j}$ and $\mathbf{k}$.   (NEAB)

20

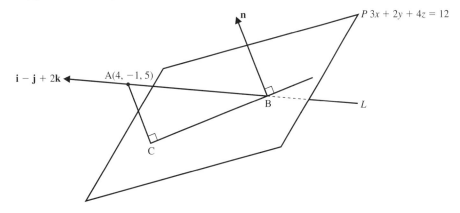

The figure above represents the line L and the plane P given by

$$L: \quad \mathbf{r} = 4\mathbf{i} - \mathbf{j} + 5\mathbf{k} + \lambda(\mathbf{i} - \mathbf{j} + 2\mathbf{k})$$
$$P: \quad 3x + 2y + 4z = 12$$

i) Find the coordinates of B, the point of intersection of the line L and the plane P.
ii) Write down a vector $\mathbf{n}$ which is perpendicular to the plane P.
iii) Calculate the vector $\mathbf{q}$ given by

$$\mathbf{q} = \mathbf{n} \times (\mathbf{i} - \mathbf{j} + 2\mathbf{k})$$

and mark it on a copy of the figure starting at B.

iv) Using $\mathbf{q}$, or otherwise, find the vector equation of the line BC which is the projection of the line L on the plane P.   (NICCEA)

21 Simplify

$$(\mathbf{a} + \mathbf{b}) \times (\mathbf{a} - \mathbf{b})$$

Given that $\mathbf{a}$ and $\mathbf{b}$ are non-zero vectors and that

$$(\mathbf{a} + \mathbf{b}) \times (\mathbf{a} - \mathbf{b}) = \mathbf{0}$$

write down the possible values of the angle between $\mathbf{a}$ and $\mathbf{b}$.   (NEAB)

22 The points A and B have position vectors $\mathbf{a} = \begin{pmatrix} -5 \\ 1 \\ 3 \end{pmatrix}$ and $\mathbf{b} = \begin{pmatrix} 2 \\ 4 \\ -1 \end{pmatrix}$ respectively, and the

plane Π has equation $\mathbf{r} \cdot \mathbf{n} = 1$, where $\mathbf{n} = \begin{pmatrix} 4 \\ -1 \\ 3 \end{pmatrix}$.

a) Show that B lies in Π and that A does not lie in Π.

b) Write down the vector $\overrightarrow{AB}$.

c) The angle between $\overrightarrow{AB}$ and $\mathbf{n}$ is θ. Find the value of θ, giving your answer correct to the nearest $0.1°$.

d) The point C lies in the plane Π and is such that $\overrightarrow{AC}$ is perpendicular to Π. Explain why $\overrightarrow{AC} = \lambda\mathbf{n}$ for some scalar parameter λ. By finding the value of λ, or otherwise, determine the position vector of C.

e) Find the shortest distance of the point A from Π.   (AEB 96)

23 The planes Π_1 and Π_2 have cartesian equations

$$x + 2y - z = 7 \quad \text{and} \quad 2x + y + z = -1$$

respectively.

a) Find the cartesian equations of the line of intersection of the planes Π_1 and Π_2 in the form

$$\frac{x - a}{l} = \frac{y - b}{m} = \frac{z - c}{n}$$

b) Find a cartesian equation of the plane Π_3 which contains the y-axis and which intersects Π_1 and Π_2 to form a prism.   (NEAB)

24 With respect to an origin O, the straight lines l_1 and l_2 have equations

$$l_1: \quad \mathbf{r} = p\mathbf{i} - 2\mathbf{j} + 2\mathbf{k} + \lambda(\mathbf{i} - \mathbf{k})$$
$$l_2: \quad \mathbf{r} = 3\mathbf{i} - \mathbf{j} + \mu(2\mathbf{i} + \mathbf{j} - 3\mathbf{k})$$

where λ and μ are scalar parameters and p is a scalar constant. The lines intersect at the point A.

a) Find the coordinates of A and show that $p = 2$.

The plane Π passes through A and is perpendicular to l_2.

b) Find a cartesian equation of Π.

c) Find the acute angle between the plane Π and the line l_1, giving your answer in degrees to one decimal place.   (EDEXCEL)

25 The point P has coordinates $(4, k, 5)$, where k is a constant. The line L has equation

$$\mathbf{r} = \begin{pmatrix} 1 \\ 0 \\ -4 \end{pmatrix} + t\begin{pmatrix} 1 \\ 2 \\ -2 \end{pmatrix}. \text{ The line } M \text{ has equation } \mathbf{r} = \begin{pmatrix} 4 \\ k \\ 5 \end{pmatrix} + t\begin{pmatrix} 7 \\ 3 \\ -4 \end{pmatrix}.$$

i) Show that the shortest distance from the point P to the line L is $\frac{1}{3}\sqrt{5(k^2 + 12k + 117)}$.

ii) Find (in terms of k) the shortest distance between lines L and M.

iii) Find the value of k for which the lines L and M intersect.

iv) When $k = 12$, show that the distances in parts **i** and **ii** are equal. In this case, find the equation of the line which is perpendicular to, and intersects, both L and M.   (MEI)

26 The planes Π_1 and Π_2 have equations

$$x + 2y - z = 3 \quad \text{and} \quad 3x + 4y - z = 1$$

respectively. Find

i) a vector which is parallel to both Π_1 and Π_2
ii) the equation of the plane which is perpendicular to both Π_1 and Π_2 and passes through the point $(3, -4, -5)$.   (NEAB)

27 The lines l_1 and l_2 have vector equations

$$\mathbf{r} = (2\lambda - 3)\mathbf{i} + \lambda\mathbf{j} + (1 - \lambda)\mathbf{k} \quad \text{and} \quad \mathbf{r} = (2 + 5\mu)\mathbf{i} + (1 + \mu)\mathbf{j} + (3 + 2\mu)\mathbf{k}$$

respectively, where λ and μ are scalar parameters.

a) Show that l_1 and l_2 intersect, stating the position vector of the point of intersection.
b) The vector $\mathbf{i} + a\mathbf{j} + b\mathbf{k}$ is perpendicular to both lines. Determine the value of the constants a and b.
c) Find a cartesian equation of the plane which contains l_1 and l_2.   (AEB 98)

28 The lines L_1 and L_2 have equations

$$\mathbf{r} = \begin{pmatrix} -3 \\ 0 \\ -15 \end{pmatrix} + s \begin{pmatrix} 0 \\ 2 \\ 1 \end{pmatrix}$$

and $\quad \mathbf{r} = \begin{pmatrix} 4 \\ -1 \\ 9 \end{pmatrix} + t \begin{pmatrix} 2 \\ 3 \\ 1 \end{pmatrix}$

respectively. Find direction ratios of a line which is perpendicular to both L_1 and L_2.

Verify that the plane Π_1, through the origin O, whose equation is

$$10x + y - 2z = 0$$

contains L_1. Find the equation of the plane Π_2 containing O and L_2. Show that the cartesian equations of the line L in which Π_1 and Π_2 intersect are

$$x = -\frac{y}{2} = \frac{z}{4}$$

Explain why L must be the common perpendicular of L_1 and L_2.   (NEAB)

29 A plane Π contains the points A$(1, -2, 1)$, B$(4, 0, 1)$ and C$(1, 0, 2)$.

a) i) Calculate the vector $\mathbf{n} = \overrightarrow{AB} \times \overrightarrow{AC}$.
ii) Explain why $\mathbf{n}$ is perpendicular to Π.
iii) Express the equation of Π in the form

$$\mathbf{r} \cdot \mathbf{n} = p$$

where p is a constant.
iv) The plane Π divides three-dimensional space into two regions. Show, with the aid of a diagram, that the region into which $\mathbf{n}$ is directed does not contain the origin.

b) A straight line L passes through the point D$(3, -1, 2)$ and has direction ratios $2 : 1 : 1$.
i) Write down a vector equation for L and verify that L passes through A.
ii) Show that the resolved part of the vector $\overrightarrow{DA}$ in the direction of $\mathbf{n}$ is -1.
iii) Write down two conclusions that can be drawn from this result about the position of D with respect to the plane Π.   (NEAB)

30 The planes Π_1 and Π_2 with equations

$$x + 2y + z + 2 = 0 \quad \text{and} \quad 2x + 3y + 2z - 1 = 0$$

respectively, meet in a line L. The point A has coordinates $(2, -2, 1)$.

a) i) Explain why the vector

$$\begin{pmatrix} 1 \\ 2 \\ 1 \end{pmatrix} \times \begin{pmatrix} 2 \\ 3 \\ 2 \end{pmatrix}$$

   is in the direction of L.

   ii) Hence find in the form $\mathbf{r} \cdot \mathbf{n} = a$ the equation of the plane which is perpendicular to L and contains A.

b) i) Explain why, for any constant λ, the plane Π_3 with equation

$$(x + 2y + z + 2) + \lambda(2x + 3y + 2z - 1) = 0$$

   contains L.

   ii) Hence, or otherwise, find the cartesian equation of the plane which contains L and the point A.   (NEAB)

31 The plane Π has equation $2x + y + 3z = 21$ and the origin is O. The line l passes through the point P(1, 2, 1) and is perpendicular to Π.

a) Find a vector equation of l.

The line l meets the plane Π at the point M.

b) Find the coordinates of M.

c) Find $\overrightarrow{OP} \times \overrightarrow{OM}$.

d) Hence, or otherwise, find the distance from P to the line OM, giving your answer in surd form.

The point Q is in the reflection of P in Π.

e) Find the coordinates of Q.   (EDEXCEL)

32 With respect to an origin O, the points A, B, C have position vectors $2\mathbf{i}$, $4\mathbf{j}$, $6\mathbf{k}$ respectively. The points P and Q are the mid-points of AB and BC respectively, and the point N has position vector $5\mathbf{i} + 6\mathbf{j} - 2\mathbf{k}$. The line l passes through P and N.

i) Find a vector equation of l and find the perpendicular distance from the point Q to l.

ii) Find a vector equation of the line of intersection of the planes ABC and OPQ, and find the acute angle between these two planes.

iii) Find the shortest distance between the lines OB and PQ.   (OCR)

33 The line l_1 passes through the point A, whose position vector is $\mathbf{i} - \mathbf{j} - 5\mathbf{k}$. and is parallel to the vector $\mathbf{i} - \mathbf{j} - 4\mathbf{k}$. The line l_2 passes through the point B, whose position vector is $2\mathbf{i} - 9\mathbf{j} - 14\mathbf{k}$, and is parallel to the vector $2\mathbf{i} + 5\mathbf{j} + 6\mathbf{k}$. The point P on l_1 and the point Q on l_2 are such that PQ is perpendicular to both l_1 and l_2.

i) Find the length of PQ.

ii) Find a vector perpendicular to the plane Π which contains PQ and l_2.

iii) Find the perpendicular distance from A to Π.   (OCR)

34 Let A, B, C, be the points $(2, 1, 0)$, $(3, 3, -1)$, $(5, 0, 2)$ respectively. Find $\overrightarrow{AB} \times \overrightarrow{AC}$.

Hence or otherwise obtain an equation for the plane containing A, B and C.   (SQA/CSYS)

35 The plane π has equation $\mathbf{r} \cdot (2\mathbf{i} - 3\mathbf{j} + 6\mathbf{k}) = 0$, and P and Q are the points with position vectors $7\mathbf{i} + 6\mathbf{j} + 5\mathbf{k}$ and $\mathbf{i} + 3\mathbf{j} - \mathbf{k}$ respectively. Find the position vector of the point in which the line passing through P and Q meets the plane π.

Find, in the form $ax + by + cz = d$, the equation of the plane which contains the line PQ and which is perpendicular to π. (OCR)

36 a) With the help of **Fig. 1** below and using, where appropriate, the notation in the figure, show that the volume of the tetrahedron OABC is

$$\frac{1}{6} |\mathbf{n} \cdot \mathbf{c}|$$

where $\mathbf{n} = \mathbf{a} \times \mathbf{b}$.

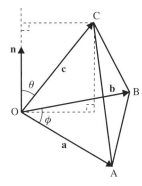
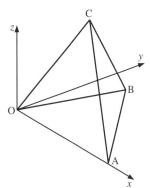

Fig. 1   Fig. 2

b) In the tetrahedron OABC, shown in **Fig. 2** above, the equation of the plane ABC is

$$12x + 4y + 5z = 48$$

i) Given that A is on the x-axis, find its coordinates.

The equation of the plane OBC is $-4x + 4y + z = 0$.

ii) Show that the equation of BC is

$$\mathbf{r} = \begin{pmatrix} 0 \\ -3 \\ 12 \end{pmatrix} + \lambda \begin{pmatrix} 1 \\ 2 \\ -4 \end{pmatrix}$$

iii) Given that B is in the xy-plane show that it is the point $(3, 3, 0)$.

The cartesian equation of AC is $\dfrac{x}{2} = \dfrac{2-y}{1} = \dfrac{8-z}{4}$.

iv) Find the coordinates of C.
v) Find the volume of this tetrahedron. (NICCEA)

Scalar triple product and its applications

The scalar triple product of $\mathbf{a}$, $\mathbf{b}$ and $\mathbf{c}$ is defined as $\mathbf{a} \cdot \mathbf{b} \times \mathbf{c}$.

Note We must calculate $\mathbf{a} \cdot \mathbf{b} \times \mathbf{c}$ as $\mathbf{a} \cdot (\mathbf{b} \times \mathbf{c})$. If we tried to calculate it as $(\mathbf{a} \cdot \mathbf{b}) \times \mathbf{c}$, we would have the vector product of a scalar and a vector, which, by definition, cannot exist.

Example 21 Calculate $\begin{pmatrix} 3 \\ 4 \\ 7 \end{pmatrix} \cdot \begin{pmatrix} 2 \\ 3 \\ -1 \end{pmatrix} \times \begin{pmatrix} 7 \\ 4 \\ 2 \end{pmatrix}$.

SOLUTION

We must calculate the **vector product first**:

$$\begin{pmatrix} 3 \\ 4 \\ 7 \end{pmatrix} \cdot \begin{pmatrix} 2 \\ 3 \\ -1 \end{pmatrix} \times \begin{pmatrix} 7 \\ 4 \\ 2 \end{pmatrix} = \begin{pmatrix} 3 \\ 4 \\ 7 \end{pmatrix} \cdot \left(\begin{pmatrix} 2 \\ 3 \\ -1 \end{pmatrix} \times \begin{pmatrix} 7 \\ 4 \\ 2 \end{pmatrix} \right)$$

$$= \begin{pmatrix} 3 \\ 4 \\ 7 \end{pmatrix} \cdot \begin{pmatrix} 10 \\ -11 \\ -13 \end{pmatrix}$$

Then we calculate the scalar product:

$$\begin{pmatrix} 3 \\ 4 \\ 7 \end{pmatrix} \cdot \begin{pmatrix} 10 \\ -11 \\ -13 \end{pmatrix} = 30 - 44 - 91 = -105$$

Therefore, we have

$$\begin{pmatrix} 3 \\ 4 \\ 7 \end{pmatrix} \cdot \begin{pmatrix} 2 \\ 3 \\ -1 \end{pmatrix} \times \begin{pmatrix} 7 \\ 4 \\ 2 \end{pmatrix} = -105$$

A quicker way to find $\mathbf{a} \cdot \mathbf{b} \times \mathbf{c}$ is as follows.

The vector product $\mathbf{b} \times \mathbf{c}$ is given by (see page 104)

$$\mathbf{b} \times \mathbf{c} = \begin{vmatrix} \mathbf{i} & \mathbf{j} & \mathbf{k} \\ b_1 & b_2 & b_3 \\ c_1 & c_2 & c_3 \end{vmatrix}$$

$$= \mathbf{i}(b_2 c_3 - b_3 c_2) - \mathbf{j}(b_1 c_3 - b_3 c_1) + \mathbf{k}(b_1 c_2 - b_2 c_1)$$

Therefore, the scalar triple product $\mathbf{a} \cdot \mathbf{b} \times \mathbf{c}$ is given by (see *Introducing Pure Mathematics*, page 503)

$$\mathbf{a} \cdot \mathbf{b} \times \mathbf{c} = a_1(b_2 c_3 - b_3 c_2) - a_2(b_1 c_3 - b_3 c_1) + a_3(b_1 c_2 - b_2 c_1)$$

That is,

$$\mathbf{a} \cdot \mathbf{b} \times \mathbf{c} = \begin{vmatrix} a_1 & a_2 & a_3 \\ b_1 & b_2 & b_3 \\ c_1 & c_2 & c_3 \end{vmatrix}$$

Applying this result to Example 21, we would have

$$\begin{pmatrix} 3 \\ 4 \\ 7 \end{pmatrix} \cdot \begin{pmatrix} 2 \\ 3 \\ -1 \end{pmatrix} \times \begin{pmatrix} 7 \\ 4 \\ 2 \end{pmatrix} = \begin{vmatrix} 3 & 4 & 7 \\ 2 & 3 & -1 \\ 7 & 4 & 2 \end{vmatrix}$$

$$= 3 \times 10 - 4 \times 11 + 7 \times -13$$

$$= -105$$

Coplanar vectors

We have

$$\mathbf{a}.\mathbf{b} \times \mathbf{c} = \mathbf{a}.(bc \sin \theta \, \hat{\mathbf{n}})$$
$$= abc \sin \theta \cos \phi$$

where θ is the angle between $\mathbf{b}$ and $\mathbf{c}$, and ϕ is the angle between $\mathbf{a}$ and $\hat{\mathbf{n}}$, which is perpendicular to the plane containing $\mathbf{b}$ and $\mathbf{c}$. Therefore, we get

$$\mathbf{a}.\mathbf{b} \times \mathbf{c} = abc \sin \theta \sin \psi$$

where $\psi = (90° - \phi)$ is the angle between $\mathbf{a}$ and the plane containing $\mathbf{b}$ and $\mathbf{c}$.

Hence, when $\mathbf{a}$, $\mathbf{b}$ and $\mathbf{c}$ are **coplanar** ($\mathbf{a}$, $\mathbf{b}$ and $\mathbf{c}$ lie in the same plane), we have

$$\mathbf{a}.\mathbf{b} \times \mathbf{c} = 0$$

Volume of a cuboid

Consider cuboid OBDCAQRS, which has adjacent edges $\overrightarrow{OA} = \mathbf{a}$, $\overrightarrow{OB} = \mathbf{b}$ and $\overrightarrow{OC} = \mathbf{c}$.

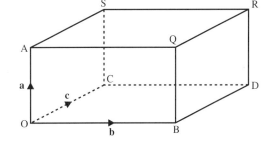

The volume of a cuboid is given by

Volume = Area of base × Perpendicular height

Therefore, the volume, V, of OBDCAQRS is

$$V = (b \times c) \times a = abc$$

Now, $\mathbf{b}$ and $\mathbf{c}$ are perpendicular to each other, and $\mathbf{a}$ is perpendicular to the plane containing them. Hence, we have

$$\mathbf{a}.\mathbf{b} \times \mathbf{c} = abc \sin 90° \sin 90° = abc$$

Therefore, the volume of a cuboid is given by

$$V = \mathbf{a}.\mathbf{b} \times \mathbf{c}$$

where the vectors $\mathbf{a}$, $\mathbf{b}$ and $\mathbf{c}$ represent three adjacent edges of the cuboid.

Note   Since the volume of any shape must be **positive**, we always use the **magnitude** of $\mathbf{a}.\mathbf{b} \times \mathbf{c}$ in volume calculations.

Volume of a parallelepiped

A parallelepiped is a polyhedron with six faces, each of which is a parallelogram.

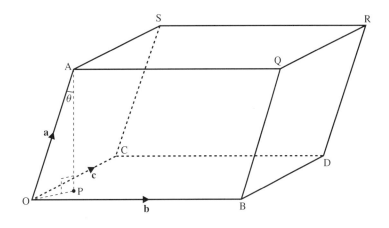

Consider the parallelepiped OBDCAQRS, which has adjacent edges $\overrightarrow{OA} = \mathbf{a}$, $\overrightarrow{OB} = \mathbf{b}$ and $\overrightarrow{OC} = \mathbf{c}$.

The volume of a parallelepiped is given by

$$\text{Volume} = \text{Area of base} \times \text{Perpendicular height}$$

Therefore, the volume, V, of OBDCAQRS is

$$V = |\mathbf{b} \times \mathbf{c}| \times \text{Perpendicular height}$$

Now, the perpendicular height, AP, is $|\mathbf{a}| \cos \theta$. Therefore, we have

$$V = |\mathbf{b} \times \mathbf{c}| \times |\mathbf{a}| \cos \theta$$
$$= |\mathbf{a}| \, |\mathbf{b} \times \mathbf{c}| \cos \theta$$

We note that this is identical to the scalar product $\mathbf{x} . \mathbf{y} = |\mathbf{x}| \, |\mathbf{y}| \cos \theta$, with $|\mathbf{x}| = |\mathbf{a}|$, $|\mathbf{y}| = |\mathbf{b} \times \mathbf{c}|$ and $\mathbf{b} \times \mathbf{c}$ having the same sense as PA. Therefore, the volume of a parallelepiped is given by

$$V = \mathbf{a} . \mathbf{b} \times \mathbf{c}$$

where the vectors $\mathbf{a}$, $\mathbf{b}$ and $\mathbf{c}$ represent three adjacent edges of the parallelepiped.

Example 22 Find the area of parallelogram ABCD, where A is $(3, 1, 7)$, B is $(2, 0, 4)$ and D is $(7, 2, -1)$.

SOLUTION

We have the adjacent sides

$$\overrightarrow{AB} = \mathbf{b} - \mathbf{a} = \begin{pmatrix} 2 \\ 0 \\ 4 \end{pmatrix} - \begin{pmatrix} 3 \\ 1 \\ 7 \end{pmatrix} = \begin{pmatrix} -1 \\ -1 \\ -3 \end{pmatrix}$$

$$\overrightarrow{AD} = \mathbf{d} - \mathbf{a} = \begin{pmatrix} 7 \\ 2 \\ -1 \end{pmatrix} - \begin{pmatrix} 3 \\ 1 \\ 7 \end{pmatrix} = \begin{pmatrix} 4 \\ 1 \\ -8 \end{pmatrix}$$

The area of parallelogram ABCD is $|\overrightarrow{AB} \times \overrightarrow{AD}|$, which gives

$$\text{Area} = \begin{vmatrix} \mathbf{i} & \mathbf{j} & \mathbf{k} \\ -1 & -1 & -3 \\ 4 & 1 & -8 \end{vmatrix}$$

$$= |11\mathbf{i} - 20\mathbf{j} + 3\mathbf{k}|$$

$$= \sqrt{11^2 + 20^2 + 3^2} = \sqrt{530}$$

Example 23 Find the volume of parallelepiped ABCDPQRS, where A is $(3, 1, 7)$, B is $(2, 0, 4)$, D is $(7, 2, -1)$ and P is $(8, 3, 11)$.

SOLUTION

The volume, V, of parallelepiped ABCDPQRS is given by

$$V = \overrightarrow{AP} . \overrightarrow{AB} \times \overrightarrow{AD}$$

We have

$$\overrightarrow{AP} = \mathbf{p} - \mathbf{a} = \begin{pmatrix} 8 \\ 3 \\ 11 \end{pmatrix} - \begin{pmatrix} 3 \\ 1 \\ 7 \end{pmatrix} = \begin{pmatrix} 5 \\ 2 \\ 4 \end{pmatrix}$$

Using $\overrightarrow{AB} \times \overrightarrow{AD} = \begin{pmatrix} 11 \\ -20 \\ 3 \end{pmatrix}$ from Example 22, we get

$$V = \overrightarrow{AP} \cdot \overrightarrow{AB} \times \overrightarrow{AD}$$

$$= \begin{pmatrix} 5 \\ 2 \\ 4 \end{pmatrix} \cdot \begin{pmatrix} 11 \\ -20 \\ 3 \end{pmatrix} = 55 - 40 + 12 = 27$$

Or we can use (see pages 122 and 124)

$$V = \begin{vmatrix} 5 & 2 & 4 \\ -1 & -1 & -3 \\ 4 & 1 & -8 \end{vmatrix} = 5(8+3) - 2(8+12) + 4(-1+4) = 27$$

Note $\overrightarrow{AB} \cdot \overrightarrow{AD} \times \overrightarrow{AP}$ could be used, since the order in which we select the three adjacent edges is not relevant, but each vector must be **away from, or towards, the same point** of the parallelepiped.

Volume of a tetrahedron

A tetrahedron is a polyhedron with four faces, each of which is a triangle. That is, it is a pyramid with a triangular base.

Consider the adjacent edges AD, AB and AC, represented by the vectors **a**, **b** and **c** respectively.

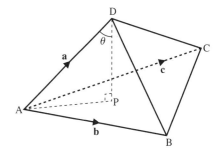

The volume of a tetrahedron is given by

$$\text{Volume} = \frac{1}{3} \times \text{Area of base} \times \text{Perpendicular height}$$

Therefore, the volume, V, of tetrahedron ABCD is

$$V = \frac{1}{3} \times \frac{1}{2} |\mathbf{b} \times \mathbf{c}| \times \text{Perpendicular height}$$

Now, the perpendicular height, DP, is $|\mathbf{a}|\cos\theta$. Therefore,

$$V = \frac{1}{6} |\mathbf{b} \times \mathbf{c}| \times |\mathbf{a}| \cos\theta = \frac{1}{6} |\mathbf{a}| \, |\mathbf{b} \times \mathbf{c}| \cos\theta$$

Because $\mathbf{b} \times \mathbf{c}$ has the same sense as PD, this gives

$$V = \frac{1}{6} \mathbf{a} \cdot \mathbf{b} \times \mathbf{c}$$

where **a**, **b** and **c** are the vectors representing three adjacent edges of the tetrahedron.

Therefore, the volume of a tetrahedron is one sixth of the volume of a parallelepiped.

Example 24 Find the volume of tetrahedron PQRS, where P is $(3, 4, 7)$, Q is $(-2, 1, 5)$, R is $(1, 3, -1)$ and S is $(-3, 6, 8)$.

SOLUTION

We have

$$\overrightarrow{PQ} = \mathbf{q} - \mathbf{p} = \begin{pmatrix} -5 \\ -3 \\ -2 \end{pmatrix} \quad \overrightarrow{PR} = \mathbf{r} - \mathbf{p} = \begin{pmatrix} -2 \\ -1 \\ -8 \end{pmatrix} \quad \overrightarrow{PS} = \mathbf{s} - \mathbf{p} = \begin{pmatrix} -6 \\ 2 \\ 1 \end{pmatrix}$$

Therefore, the volume, V, of tetrahedron PQRS is given by

$$V = \frac{1}{6} \times \overrightarrow{PQ} \cdot \overrightarrow{PR} \times \overrightarrow{PS}$$

$$= \frac{1}{6} \begin{vmatrix} -5 & -3 & -2 \\ -2 & -1 & -8 \\ -6 & 2 & 1 \end{vmatrix}$$

$$= \frac{1}{6}(-5 \times 15 + 3 \times -50 - 2 \times -10)$$

$$= -\frac{205}{6}$$

Therefore, the volume of tetrahedron PQRS is $\frac{205}{6}$.

Volume of a triangular prism

The volume of a triangular prism is given by

Volume = Area of base × Perpendicular height

By definition, the base is a triangle. So, we have

$$\text{Area of base} = \frac{1}{2}|\mathbf{b} \times \mathbf{c}|$$

Therefore, the volume, V, of the prism is

$$V = \frac{1}{2}|\mathbf{b} \times \mathbf{c}| \times \text{Perpendicular height}$$

$$= \frac{1}{2}|\mathbf{a}|\,|\mathbf{b} \times \mathbf{c}|$$

which gives

$$V = \frac{1}{2}\mathbf{a} \cdot \mathbf{b} \times \mathbf{c}$$

where $\mathbf{a}$ is the vector representing an edge of the prism, and $\mathbf{b}$ and $\mathbf{c}$ are the vectors representing two sides of its triangular base, adjacent to $\mathbf{a}$.

Volume of a pyramid

The volume of a pyramid is given by

$$V = \frac{1}{3} \times \text{Area of base} \times \text{Perpendicular height}$$

Taking the case of a rectangular (or parallelogram) base, we have

$$V = \frac{1}{3} |\mathbf{b} \times \mathbf{c}| \times \text{Perpendicular height}$$

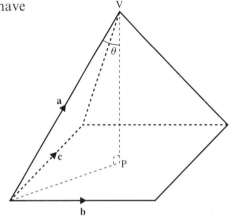

where **b** and **c** represent adjacent sides of the base, as shown in the diagram on the right.

From the diagram, we see that the perpendicular height is $|\mathbf{a}| \cos \theta$. Therefore, we have

$$V = \frac{1}{3} |\mathbf{a}| \, |\mathbf{b} \times \mathbf{c}| \cos \theta$$

Because $\mathbf{b} \times \mathbf{c}$ has the same sense as PV, this gives

$$V = \frac{1}{3} \mathbf{a} \cdot \mathbf{b} \times \mathbf{c}$$

Therefore, the volume of a pyramid with a rectangular (or parallelogram) base is one third of the volume of a parallelepiped.

Example 25 Find the volume of pyramid ABCDV, where ABCD is a parallelogram, and V is the vertex. A is $(2, 1, 5)$, B is $(3, 4, -2)$, D is $(5, 2, 3)$ and V is $(0, 6, 4)$.

SOLUTION

We have

$$\overrightarrow{AB} = \mathbf{b} - \mathbf{a} = \begin{pmatrix} 1 \\ 3 \\ -7 \end{pmatrix} \quad \overrightarrow{AD} = \mathbf{d} - \mathbf{a} = \begin{pmatrix} 3 \\ 1 \\ -2 \end{pmatrix} \quad \overrightarrow{AV} = \mathbf{v} - \mathbf{a} = \begin{pmatrix} -2 \\ 5 \\ -1 \end{pmatrix}$$

Therefore, the volume of pyramid ABCDV is

$$\frac{1}{3} \overrightarrow{AV} \cdot \overrightarrow{AB} \times \overrightarrow{AD} = \frac{1}{3} \begin{pmatrix} -2 \\ 5 \\ -1 \end{pmatrix} \cdot \left(\begin{pmatrix} 1 \\ 3 \\ -7 \end{pmatrix} \times \begin{pmatrix} 3 \\ 1 \\ -2 \end{pmatrix} \right)$$

$$= \frac{1}{3} \begin{vmatrix} -2 & 5 & -1 \\ 1 & 3 & -7 \\ 3 & 1 & -2 \end{vmatrix}$$

$$= \frac{1}{3} [-2(-6+7) - 5(-2+21) - 1(1-9)]$$

$$= \frac{1}{3} (-2 \times 1 + -5 \times 19 - 1 \times -8)$$

$$= \frac{1}{3} (-2 - 95 + 8)$$

Therefore, the volume of pyramid ABCDV is $\frac{89}{3}$.

Exercise 6C

1 Find the value of $\begin{pmatrix} 3 \\ -2 \\ 1 \end{pmatrix} \cdot \begin{pmatrix} 2 \\ 1 \\ 4 \end{pmatrix} \times \begin{pmatrix} 1 \\ 5 \\ -2 \end{pmatrix}$.

2 Find the value of $\begin{pmatrix} 2 \\ 4 \\ -5 \end{pmatrix} \cdot \begin{pmatrix} 3 \\ 8 \\ 2 \end{pmatrix} \times \begin{pmatrix} 2 \\ -3 \\ 6 \end{pmatrix}$.

3 Find the value of $\begin{pmatrix} -1 \\ 2 \\ 5 \end{pmatrix} \cdot \begin{pmatrix} 2 \\ 3 \\ 1 \end{pmatrix} \times \begin{pmatrix} 3 \\ 8 \\ 4 \end{pmatrix}$.

4 Find the volume of a parallelepiped ABCDEFGH,

$$\overrightarrow{AB} = \begin{pmatrix} 1 \\ 3 \\ 2 \end{pmatrix} \qquad \overrightarrow{AD} = \begin{pmatrix} -2 \\ 1 \\ -3 \end{pmatrix} \qquad \overrightarrow{AE} = \begin{pmatrix} 5 \\ 2 \\ 7 \end{pmatrix}$$

5 The figure on the right represents a cube with side of unit length.

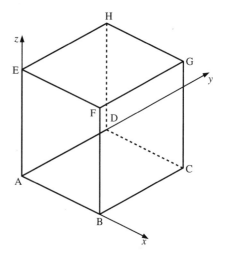

  i) Find $\overrightarrow{AB} \cdot \overrightarrow{AC}$.
  ii) Find a vector, using the letters in the diagram, which is equal to $\overrightarrow{EA} \times \overrightarrow{EH}$.
  iii) Find the value of λ in the following equation

$$\overrightarrow{EA} \times \overrightarrow{EC} = \lambda\overrightarrow{BD} \qquad \text{(NICCEA)}$$

6 The points A, B, C and D have coordinates $(3, 1, 2)$, $(5, 2, -1)$, $(6, 4, 5)$ and $(-7, 6, -3)$ respectively.

  a) Find $\overrightarrow{AC} \times \overrightarrow{AD}$.
  b) Find a vector equation of the line through A which is perpendicular to $\overrightarrow{AC}$ and $\overrightarrow{AD}$.
  c) Verify that B lies on this line.
  d) Find the volume of the tetrahedron ABCD.   (EDEXCEL)

7 The figure on the right shows a right prism with triangular ends ABC and DEF, and parallel edges AD, BE, CF. Given that A is $(2, 7, -1)$, B is $(5, 8, 2)$, C is $(6, 7, 4)$ and D is $(12, 1, -9)$,

  a) find $\overrightarrow{AB} \times \overrightarrow{AC}$
  b) find $\overrightarrow{AD} \cdot (\overrightarrow{AB} \times \overrightarrow{AC})$.
  c) Calculate the volume of the prism.   (EDEXCEL)

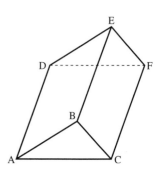

8 The points A, B and C have position vectors $\mathbf{a}\,\text{m}$, $\mathbf{b}\,\text{m}$ and $\mathbf{c}\,\text{m}$ respectively, relative to an origin O, where

$$\mathbf{a} = 3\mathbf{i} + 4\mathbf{j} + 5\mathbf{k} \qquad \mathbf{b} = 4\mathbf{i} + 6\mathbf{j} + 7\mathbf{k} \qquad \mathbf{c} = \mathbf{i} + 5\mathbf{j} + 3\mathbf{k}$$

a) Find $(\mathbf{b} - \mathbf{a}) \times (\mathbf{c} - \mathbf{a})$.

b) Hence, or otherwise, find the area of $\triangle$ABC and the volume of tetrahedron OABC.

c) Find an equation of the plane ABC in the form $\mathbf{r} \cdot \mathbf{n} = p$.

Given that the point D has position vector $(2\mathbf{i} + \mathbf{j} + 2\mathbf{k})\,\text{m}$,

d) find the coordinates of the point of intersection, E, of the OD with the plane ABC

e) find the acute angle between ED and the plane ABC.   (EDEXCEL)

7 Curve sketching and inequalities

And of the curveship lend a myth to God.
HART CRANE

Curve sketching

On page 306 of *Introducing Pure Mathematics*, there is a brief introduction to the use of asymptotes in curve sketching. We are now going to extend the procedure to more complex curves.

Remember An asymptote is a line which becomes a tangent to a curve as x or y tends to infinity.

We need to be able to find asymptotes if we want to sketch functions which are not trigonometric or polynomial.

Consider, for example, the curve

$$y = \frac{4x - 8}{x + 3}$$

As $y \to \pm\infty$, the denominator of this function must tend to zero. That is, as $x + 3 \to 0$, $x \to -3$. Hence, $x = -3$ is an asymptote.

To find the asymptote as $x \to \pm\infty$, we express the function as

$$y = \frac{4 - \dfrac{8}{x}}{1 + \dfrac{3}{x}}$$

As $x \to \pm\infty$, $\dfrac{3}{x} \to 0$, and $\dfrac{8}{x} \to 0$. Therefore, $y \to \dfrac{4}{1} = 4$. Hence, $y = 4$ is also an asymptote.

Notice that, as $x \to \pm\infty$, the largest terms in the numerator and the denominator are $4x$ and x respectively, and so $y \approx 4x \div x = 4$.

$x = -3$ is a **vertical asymptote**, as it is parallel to the y-axis, and $y = 4$ is a **horizontal asymptote**, as it is parallel to the x-axis.

To be able to sketch $y = \dfrac{4x - 8}{x + 3}$, we also need to find where it crosses the x- and y-axes:

$$\text{When } x = 0: \quad y = -\frac{8}{3}$$

$$\text{When } y = 0: \quad 4x - 8 = 0 \quad \Rightarrow \quad x = 2$$

To sketch the curve, we proceed as follows (see the diagram on the right):

- First, draw the asymptotes, using dashed lines.

- Next, mark the points where the curve crosses the axes.

- As the numerator **and** the denominator of the function each contain only a linear term in x, the curve cannot cross either asymptote.

- Considering the curve for $x > -3$, we see that it tends to $-\infty$ as x approaches -3 from values of x greater than -3. Hence, the curve tends to $+\infty$ as x approaches -3 from values of x less than -3.

- We can now complete the curve of $y = \dfrac{4x - 8}{x + 3}$.

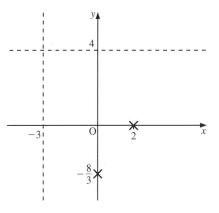

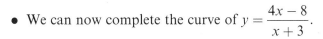

Example 1 Sketch $y = \dfrac{2x - 6}{x - 5}$.

SOLUTION

First, we find the asymptotes.

As $x \to \pm\infty$, $y \to 2$. That is, the horizontal asymptote is $y = 2$.

As $y \to \pm\infty$, $x - 5 \to 0$. Hence, $x = 5$ is the vertical asymptote.

Next, we find where the curve crosses the axes:

When $x = 0$: $\quad y = \dfrac{-6}{-5} = \dfrac{6}{5}$

When $y = 0$: $\quad 2x - 6 = 0 \quad \Rightarrow \quad x = 3$

We now complete the sketch of $y = \dfrac{2x - 6}{x - 5}$, which is shown below.

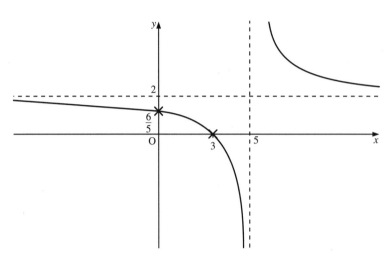

Curves with an oblique asymptote

For most curves, the value of y as $x \to \pm\infty$ will not be finite.

Consider, for example, the curve

$$y = x + \frac{1}{x}$$

As $x \to \pm\infty$, $\dfrac{1}{x} \to 0$ and thus $y \to x$. Therefore, $y = x$ is an asymptote to the curve.

This is called an **oblique asymptote** (sometimes an **inclined asymptote**), as $y = x$ is not parallel to either axis.

The other asymptote is $x = 0$ (the y-axis), as $\dfrac{1}{x} \to \infty$, when $x \to 0$.

When $x = 0$, y is not defined, thus the curve does not cross the y-axis. Thus, the y-axis is a vertical asymptote, as already shown.

We can now sketch the curve of $y = x + \dfrac{1}{x}$, as shown on the right.

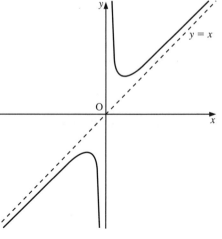

Example 2 Sketch $y = \dfrac{x^2 + 3x}{x + 1}$.

SOLUTION

Dividing $x^2 + 3x$ by $x + 1$, we obtain

$$y = x + 2 - \frac{2}{x + 1}$$

which gives the asymptotes as $x = -1$ and $y = x + 2$.

We now find where the curve crosses the axes:

$$\text{When } y = 0: \quad x^2 + 3x = 0 \quad \Rightarrow \quad x = 0 \quad \text{and} \quad -3$$
$$\text{When } x = 0: \quad y = 0$$

We now complete the sketch of $y = \dfrac{x^2 + 3x}{x + 1}$.

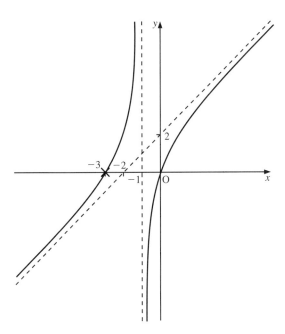

Sketching rational functions with a quadratic denominator

Curves with two vertical asymptotes

Take, for example, the curve $y = \dfrac{(x - 3)(2x - 5)}{(x + 1)(x + 2)}$.

When the denominator is a quadratic expression,

- there are **always two** vertical asymptotes, and
- the curve will normally cross the horizontal asymptote.

Hence, in addition to finding the asymptotes and the points where the curve crosses the axes, we need to establish where the curve crosses the horizontal asymptote.

Note The two vertical asymptotes could coincide, as in Example 4, on page 135.

Hence, there are four stages to sketching the given function.

1 To find the horizontal asymptote of $y = \dfrac{(x-3)(2x-5)}{(x+1)(x+2)}$, we express the function as

$$y = \frac{\left(1 - \dfrac{3}{x}\right)\left(2 - \dfrac{5}{x}\right)}{\left(1 + \dfrac{1}{x}\right)\left(1 + \dfrac{2}{x}\right)}$$

As $x \to \pm\infty$, $\dfrac{1}{x} \to 0$, and $y \to 2$. Therefore, the horizontal asymptote is $y = 2$.

2 To find the vertical asymptotes, we equate the denominator to zero, which gives

$$(x+1)(x+2) = 0$$

Hence, the vertical asymptotes are $x = -1$ and $x = -2$.

3 To find where the curve cuts the axes, we have

When $x = 0$: $\quad y = \dfrac{15}{2}$

When $y = 0$: $\quad x = 3 \quad$ and $\quad x = \dfrac{5}{2}$

4 To find where the curve crosses the horizontal asymptote, $y = 2$, we have

$$2 = \frac{(x-3)(2x-5)}{(x+1)(x+2)}$$

$$2(x^2 + 3x + 2) = 2x^2 - 11x + 15$$

$$\Rightarrow \quad x = \frac{11}{17}$$

To sketch the curve, we need to insert all four points, as well as the three asymptotes.

Note

- The curve can cross an axis or an asymptote **only** at the points found.

- If one branch of the curve goes to $+\infty$, the next branch must return from $-\infty$. The exception to this is when the two vertical asymptotes coincide as the result of a squared factor in the denominator. See Example 4 on page 135.

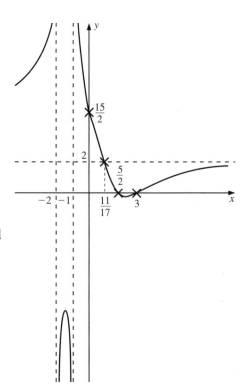

Example 3 Sketch $y = \dfrac{(x+1)(x-4)}{(x-2)(x-5)}$.

SOLUTION

The horizontal asymptote is $y = 1$.

The vertical asymptotes are $x = 2$ and $x = 5$.

The curve crosses the axes at $x = 0$, $y = -\frac{4}{10}$, and at $y = 0$, $x = -1, 4$.

The curve crosses the horizontal asymptote when $y = 1$, which gives

$$x^2 - 7x + 10 = x^2 - 3x - 4$$

$$\Rightarrow \quad x = \frac{7}{2}$$

We can now sketch the curve.

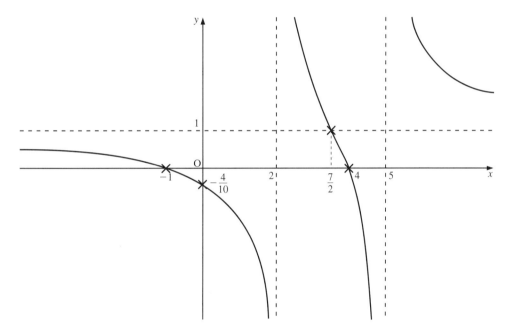

Example 4 Sketch the curve $y = \dfrac{(x-1)(3x+2)}{(x+1)^2}$.

SOLUTION

The horizontal asymptote is $y = 3$.

The vertical asymptotes are $x = -1$ (twice).

The curve crosses the axes at $x = 0$, $y = -2$, and at $y = 0$, $x = 1, -\frac{2}{3}$.

The curve crosses the horizontal asymptote when $y = 3$, which gives

$$3 = \frac{3x^2 - x - 2}{x^2 + 2x + 1}$$

$$3(x^2 + 2x + 1) = 3x^2 - x - 2$$

$$\Rightarrow \quad x = -\frac{5}{7}$$

Note Since $x = -1$ is a repeat asymptote, and the curve tends to $+\infty$ as x approaches the value of -1 from the right (that is, x tends to -1 from above), it also tends to $+\infty$ as x approaches the value of -1 from the left (that is, from below).

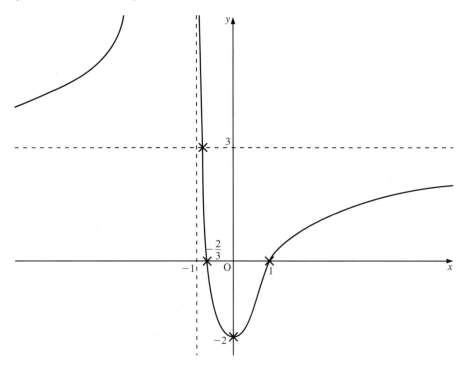

Curves without vertical asymptotes

Not all functions of the form $y = \dfrac{ax^2 + bx + c}{px^2 + qx + r}$ have vertical asymptotes. If the roots of $px^2 + qx + r = 0$ are not real, the curve will not have a vertical asymptote.

Example 5 Sketch the curve $y = \dfrac{2x^2 + 5x + 3}{4x^2 + 5x + 3}$, and find the range of possible values for y.

SOLUTION

To find the horizontal asymptote of $y = \dfrac{2x^2 + 5x + 3}{4x^2 + 5x + 3}$, we express the function as

$$y = \frac{2 + \dfrac{5}{x} + \dfrac{3}{x^2}}{4 + \dfrac{5}{x} + \dfrac{3}{x^2}}$$

As $x \to \infty$, $y \to \frac{1}{2}$. Therefore, $y = \frac{1}{2}$ is the horizontal asymptote.

For the vertical asymptotes, we have $4x^2 + 5x + 3 = 0$, which gives

$$x = \frac{-5 \pm \sqrt{-23}}{8}$$

These are not real. Therefore, the curve does not have a vertical asymptote.

To find where the curve cuts the axes, we have

When $y = 0$: $2x^2 + 5x + 3 = 0$

$(2x + 3)(x + 1) = 0$

$\Rightarrow \quad x = -1 \quad \text{and} \quad -\dfrac{3}{2}$

When $x = 0$: $y = 1$

The curve crosses the horizontal asymptote $y = \frac{1}{2}$ when

$$\frac{1}{2} = \frac{2x^2 + 5x + 3}{4x^2 + 5x + 3}$$

which gives

$$4x^2 + 5x + 3 = 4x^2 + 10x + 6$$

$$\Rightarrow \quad x = -\frac{3}{5}$$

We can now sketch the curve.

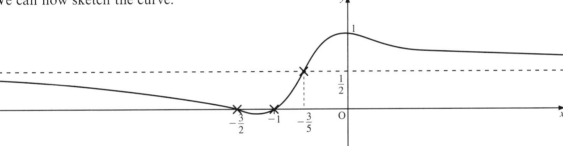

To find the range of values of y, we need to find the values for which

$y = \dfrac{2x^2 + 5x + 3}{4x^2 + 5x + 3}$ has real roots for x.

Cross-multiplying $y = \dfrac{2x^2 + 5x + 3}{4x^2 + 5x + 3}$, we obtain

$$4yx^2 + 5yx + 3y = 2x^2 + 5x + 3$$

$$\Rightarrow \quad (4y - 2)x^2 + (5y - 5)x + 3y - 3 = 0$$

From the quadratic formula, we know that $b^2 - 4ac \geqslant 0$ for the roots of x to be real. Therefore, we have

$$(5y - 5)^2 - 4(4y - 2)(3y - 3) \geqslant 0$$

$$\Rightarrow \quad 23y^2 - 22y - 1 \leqslant 0$$

$$\Rightarrow \quad (23y + 1)(y - 1) \leqslant 0$$

$$\Rightarrow \quad -\frac{1}{23} \leqslant y \leqslant 1$$

Therefore, the range of possible values of y is $-\frac{1}{23} \leqslant y \leqslant 1$.

Hence, the maximum value of y is 1, and the minimum value is $-\frac{1}{23}$.

Note   We could have used calculus to find these two stationary points.

Exercise 7A

Sketch the graph of each of the following functions.

1 $y = \dfrac{(x-3)(x-1)}{(x+2)(x-2)}$

2 $y = \dfrac{(2x-1)(x+4)}{(x-1)(x-2)}$

3 $y = \dfrac{(x+4)(x-5)}{(x-2)(x-3)}$

4 $y = \dfrac{(x+1)(2x+5)}{(x+2)(x-5)}$

5 $y = \dfrac{2x^2+3x-5}{x^2-x-2}$

6 $y = \dfrac{3x^2+4x+4}{x^2-2x-3}$

7 Find the range of values of

a) $y = \dfrac{4x^2-x-3}{2x^2-x-3}$

b) $y = \dfrac{x^2+x-1}{x^2+x-3}$

8 Find the equations of the three asymptotes of the curve

$$y = \frac{4x^2-5x+7}{21x^2-x-10} \qquad \text{(OCR)}$$

9 Find the equations of the asymptotes of the curve

$$y = \frac{x^2-x+1}{x+1} \qquad \text{(OCR)}$$

10 One of the two asymptotes of the curve

$$y = \frac{x^2+\lambda x+1}{x+2}$$

where λ is a constant, is $y = x+5$.

i) State the equation of the other asymptote.
ii) Find the value of λ. (OCR)

11 The curve C has equation

$$y = 10 + \frac{8}{x-2} - \frac{27}{x+2}$$

i) Write down the equations of the asymptotes of C.

ii) Find $\dfrac{d^2y}{dx^2}$.

iii) Show that C has one point of inflexion, and find the coordinates of this point. (OCR)

12 A curve has equation $y = \dfrac{x^2-5}{x^2+2x-11}$.

a) Determine the equations of the three asymptotes to the curve, giving each answer in an exact form.

b) Prove algebraically that there are no values of x for which $\frac{1}{2} < y < \frac{5}{6}$.

Hence, or otherwise, calculate the coordinates of the turning points on the curve. (AEB 98)

13 A curve has equation $y = \dfrac{x^2}{2x+1}$.

a) i) Write down the equation of the vertical asymptote to the curve, and determine the equation of the oblique asymptote.

ii) Use differentiation to determine the coordinates of the stationary points on the curve.

b) The region bounded by the curve, the x-axis between $x = 0$ and $x = 1$, and the line $x = 1$ is rotated through one revolution about the x-axis to form a solid with volume V.

Using the substitution $u = 2x + 1$, or otherwise, show that

$$V = \frac{\pi}{24}(4 - 3\ln 3) \qquad \text{(AEB 98)}$$

14 Let the function f be given by

$$f(x) = \frac{2x^3 - 7x^2 + 4x + 5}{(x-2)^2} \qquad x \neq 2$$

a) The graph of $y = f(x)$ crosses the y-axis at $(0, a)$. State the value of a.

b) For the graph of $f(x)$

i) write down the equation of the vertical asymptote,

ii) show algebraically that there is a non-vertical asymptote and state its equation.

c) Find the coordinates and nature of the stationary point of $f(x)$.

d) Show that $f(x) = 0$ has a root in the interval $-2 < x < 0$.

e) Sketch the graph of $y = f(x)$. (You must include on your sketch the results obtained in the first four parts of this question.)   (SQA/CSYS)

15 The curve C has equation

$$y = \frac{2x^2 + 6x + 1}{(x-1)(x+2)}$$

i) Express y in partial fractions.

ii) Deduce that

a) at every point of C, the gradient is negative

b) $y > 2$ for all $x > 1$.

iii) Write down the equations of the asymptotes of C.

iv) One of the asymptotes has a point in common with C. Determine the coordinates of this point.   (OCR)

16 A curve C is defined by the equations

$$x = \frac{1+t}{1-t} \qquad y = \frac{1+t^2}{1-t^2}$$

where t is a real parameter, $t \neq \pm 1$.

a) Find an expression for $\dfrac{dy}{dx}$ in terms of t, simplifying your answer as much as possible.

b) By eliminating t, prove that C has cartesian equation $y = \dfrac{x^2 + 1}{2x}$.

c) Write down the equations of the two asymptotes of C.

d) i) Prove algebraically that there are no values of x for which $-1 < y < 1$.

ii) Hence, or otherwise, determine the coordinates of the turning points of C.   (AEB 96)

Inequalities

In *Introducing Pure Mathematics* (pages 6 and 36), we found how to solve simple inequalities such as

$$4x + 7 > 3(x - 4) \quad \text{and} \quad x^2 - 7x + 10 \geqslant 0$$

We established that we can add and subtract as usual with an inequality symbol, as if it were an equals symbol. **But** to multiply or divide by a negative number, we must **change the sign of the inequality**. For example, we have

$$3 > 2 \quad \text{but} \quad -3 < -2$$
$$-2x > 4 \quad \Rightarrow \quad x < -2$$

Hence, an inequality such as $\dfrac{ax + b}{cx + d} > 2$ **cannot** be solved simply by multiplying both sides of the inequality by $cx + d$, since we do not know whether $cx + d$ is positive, giving

$$ax + b > 2(cx + d)$$

or negative, giving

$$ax + b < 2(cx + d)$$

To solve inequalities such as $\dfrac{ax + b}{cx + d} > k$, we can use either of the following two methods.

1 Multiply both sides of the inequality by $(cx + d)^2$, which we know must be positive.

2 Sketch $y = \dfrac{ax + b}{cx + d}$, solve $\dfrac{ax + b}{cx + d} = k$ and then, by comparing these two results, write down the solution to the inequality.

You should be able to use both methods, but the one which you prefer will probably depend on whichever is better, your algebraic skill or your graphical skill.

Example 6 Solve the inequality $\dfrac{5x - 9}{x + 3} > 2$.

SOLUTION

Method 1
Multiplying by $(x + 3)^2$, we obtain

$$\frac{5x - 9}{x + 3}(x + 3)^2 > 2(x + 3)^2$$

$$\Rightarrow \quad (5x - 9)(x + 3) > 2(x + 3)^2$$

$$\Rightarrow \quad (5x - 9)(x + 3) - 2(x + 3)^2 > 0$$

Noting that $(x + 3)$ is a factor, we factorise to obtain

$$(x + 3)[5x - 9 - 2(x + 3)] > 0$$
$$\Rightarrow \quad (x + 3)(3x - 15) > 0$$
$$\Rightarrow \quad (x + 3)(x - 5) > 0$$
$$\Rightarrow \quad x > 5 \quad \text{or} \quad x < -3$$

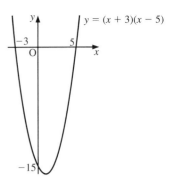

Method 2

Consider the curve $y = \dfrac{5x - 9}{x + 3}$.

The asymptotes are $x = -3$ and $y = 5$.

The curve cuts the axes at $\left(\dfrac{9}{5}, 0\right)$ and $(0, -3)$.

We can now sketch the curve of $y = \dfrac{5x - 9}{x + 3}$.

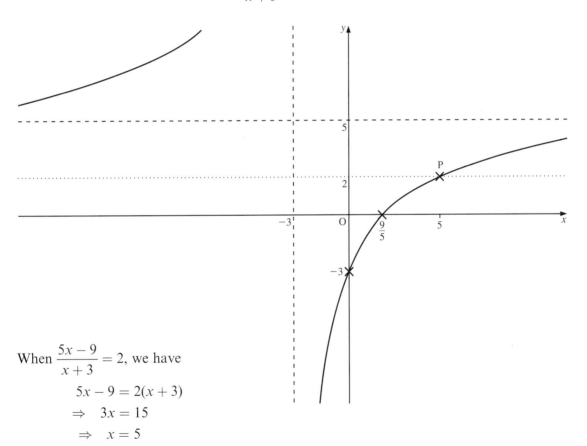

When $\dfrac{5x - 9}{x + 3} = 2$, we have

$$5x - 9 = 2(x + 3)$$
$$\Rightarrow \quad 3x = 15$$
$$\Rightarrow \quad x = 5$$

We insert the point P (5, 2) on the curve. Hence, we can see that

$$\frac{5x - 9}{x + 3} > 2$$

is satisfied by the part of the graph above the dotted line $y = 2$. That is,
where $x > 5$ or $x < -3$.

Example 7 Solve the inequality $\dfrac{(x+1)(x+4)}{(x-1)(x-2)} < 2$.

SOLUTION

Method 1

Multiplying by $(x-1)^2(x-2)^2$, we obtain

$$\frac{(x+1)(x+4)}{(x-1)(x-2)}(x-1)^2(x-2)^2 < 2(x-1)^2(x-2)^2$$

$$\Rightarrow \quad (x+1)(x+4)(x-1)(x-2) < 2(x-1)^2(x-2)^2$$

$$\Rightarrow \quad (x+1)(x+4)(x-1)(x-2) - 2(x-1)^2(x-2)^2 < 0$$

Noting that $(x-1)$ and $(x-2)$ are factors, we factorise to obtain

$$(x-1)(x-2)[(x+1)(x+4) - 2(x-1)(x-2)] < 0$$

$$\Rightarrow \quad (x-1)(x-2)[(x^2+5x+4-2x^2+6x-4] < 0$$

$$\Rightarrow \quad (x-1)(x-2)(-x^2+11x) < 0$$

$$\Rightarrow \quad (x-1)(x-2)(x^2-11x) > 0$$

$$\Rightarrow \quad (x-1)(x-2)x(x-11) > 0$$

Therefore, we have

$$\frac{(x+1)(x+4)}{(x-1)(x-2)} < 2$$

when $x > 11$, $1 < x < 2$, $x < 0$.

Method 2

Consider the curve of $y = \dfrac{(x+1)(x+4)}{(x-1)(x-2)}$.

The horizontal asymptote is $y = 1$.

The vertical asymptotes are $x = 1$ and $x = 2$.

The curve crosses the axes when $y = 0$, $x = -1$, -4, and when $x = 0$, $y = 2$.

When $y = 1$, we obtain

$$\frac{(x+1)(x+4)}{(x-1)(x-2)} = 1$$

$$\Rightarrow \quad x^2 + 5x + 4 = x^2 - 3x + 2$$

$$\Rightarrow \quad 8x = -2$$

$$\Rightarrow \quad x = -\frac{1}{4}$$

When $y = 2$, we obtain

$$\frac{(x+1)(x+4)}{(x-1)(x-2)} = 2$$

$$\Rightarrow \quad x^2 + 5x + 4 = 2(x^2 - 3x + 2)$$

$$\Rightarrow \quad 0 = x^2 - 11x$$

$$\Rightarrow \quad x = 0 \quad \text{and} \quad 11$$

Therefore, we have

$$\frac{(x+1)(x+4)}{(x-1)(x-2)} < 2$$

when $x > 11$, $1 < x < 2$, $x < 0$.

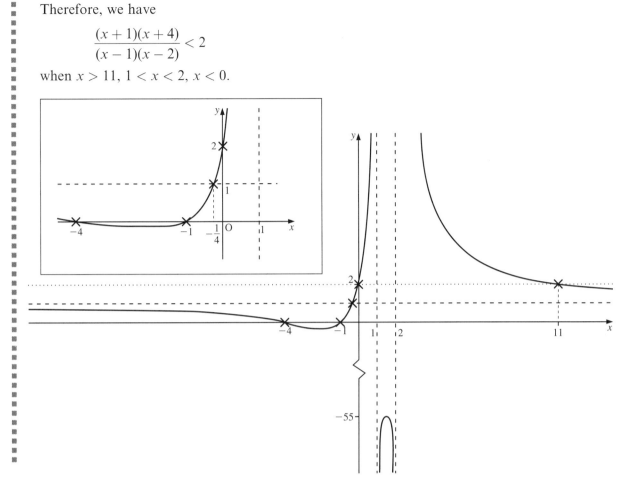

Inequalities involving modulus curves

In *Introducing Pure Mathematics* (page 95), we found how to solve simple modulus inequalities. Here, we consider modulus inequalities involving algebraic fractions.

Take, for example, the modulus inequality

$$\left|\frac{5x-9}{x+3}\right| > 2$$

We solve this by first solving

$$\frac{5x-9}{x+3} = +2 \quad \text{and} \quad \frac{5x-9}{x+3} = -2$$

and then deducing the required values of x from the sketch of the curve

$$y = \left|\frac{5x-9}{x+3}\right|$$

The sketch of $y = \left|\dfrac{5x-9}{x+3}\right|$ is obtained by sketching $y = \dfrac{5x-9}{x+3}$ and

reflecting in the x-axis that part of the curve below the x-axis.

Thus, to solve $\left| \dfrac{5x - 9}{x + 3} \right| > 2$, we proceed as follows.

First, we solve $\dfrac{5x - 9}{x + 3} = 2$, which gives $x = 5$ (as in Example 6, pages 140–1).

Next, we solve $\dfrac{5x - 9}{x + 3} = -2$, which gives

$$5x - 9 = -2(x + 3) \quad \Rightarrow \quad x = \tfrac{3}{7}$$

Then, we sketch $y = \dfrac{5x - 9}{x + 3}$.

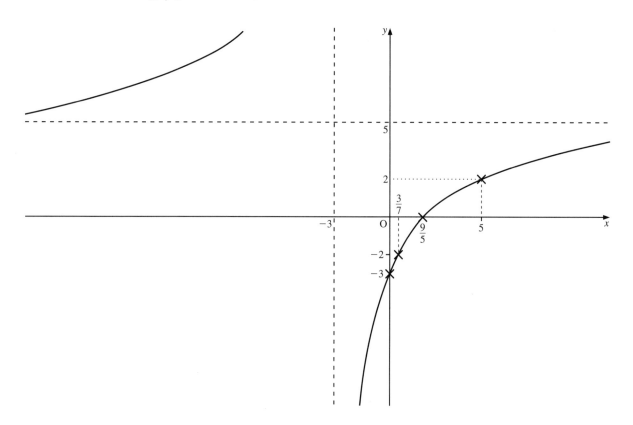

Finally, we sketch $y = \left| \dfrac{5x - 9}{x + 3} \right|$. (See top of page 145.)

Insert the point P where $\dfrac{5x - 9}{x + 3} = +2$, and the point Q where $\dfrac{5x - 9}{x + 3} = -2$.

Hence, we have

$$\left| \frac{5x - 9}{x + 3} \right| > 2$$

when $x > 5$ and $x < \tfrac{3}{7}$, **excluding** $x = -3$, where the curve is not defined.

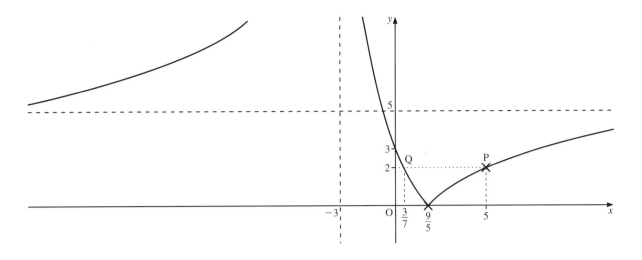

Exercise 7B

In Questions **1** to **4**, solve each of the inequalities for x.

1 a) $\dfrac{x+3}{x+2} < 2$

 b) $\dfrac{x+5}{x-3} > 1$

 c) $\dfrac{2x-1}{x+3} > 3$

 d) $\dfrac{3x+4}{x-5} > 2$

 e) $\dfrac{1-2x}{4x+2} > 2$

 f) $\dfrac{3+4x}{5x-1} > 3$

2 a) $\dfrac{(x-1)(x-2)}{(x+1)(x+2)} > 1$

 b) $\dfrac{(x+2)(x-5)}{(x-3)(x-2)} > 1$

 c) $\dfrac{(x-1)(x-4)}{(x+1)(x-5)} > 2$

 d) $\dfrac{(2x-1)(x-2)}{(x-3)(x+7)} > 2$

 e) $\dfrac{(x+1)(x+5)}{(x+2)(2x+3)} > 3$

3 a) $\left|\dfrac{x+3}{x+2}\right| > 1$

 b) $\left|\dfrac{x-1}{x+2}\right| > 2$

 c) $\left|\dfrac{x+3}{x-4}\right| > 2$

 d) $\left|\dfrac{2x-1}{x+5}\right| > 1$

 e) $\left|\dfrac{3x-1}{x+2}\right| > 2$

 f) $\left|\dfrac{x+2}{x+3}\right| \leqslant 1$

4 a) $\dfrac{x^2+x-3}{x^2+x-2} > 1$

 b) $\dfrac{2x^2+x-5}{2x^2+x-3} < 1$

 c) $\dfrac{x^2-x-2}{x^2+3x+2} > 1$

5 Find the complete set of values of x for which $\dfrac{x^2-12}{x} > 1$.   (EDEXCEL)

6 Given that $|x| \neq 1$, find the complete set of values of x for which $\dfrac{x}{x-1} > \dfrac{1}{x+1}$.   (EDEXCEL)

7 Find the set of values of x for which $x+2 > \dfrac{3}{2-x}$.   (EDEXCEL)

8 Find the set of values of x for which $\dfrac{x}{x+2} < \dfrac{2}{x-1}$.   (EDEXCEL)

9 Find the set of values of x for which $x < \dfrac{2x+5}{x-2}$.   (EDEXCEL)

10 For the curve with equation $y = \dfrac{2x^2 - x - 7}{x - 3}$, prove **algebraically** that there are no real values of x for which $3 < y < 19$.   (AEB 98)

11 $f(x) = \dfrac{3x - 6}{x(x + 6)}$ $x \in \mathbb{R},\ x \neq 0,\ x \neq -6$

a) Find the range of values of $f(x)$.

Hence, or otherwise, sketch the curve with equation $y = f(x)$. State the equations of any asymptotes and the coordinates of any turning points.

b) Use your graph to find the number of real roots of the equation
$$x^3 + 6x^2 - 3x + 6 = 0$$

c) On a separate diagram, sketch the curve with equation $y = |f(x)|$.   (EDEXCEL)

12 On the same diagram, sketch the graphs of
$$y = |x - 5| \quad \text{and} \quad y = |3x - 2|$$

distinguishing between them clearly.

Find the set of values of x for which $|x - 5| < |3x - 2|$.   (EDEXCEL)

13 Find the complete set of values of x for which $\dfrac{x - 2}{x + 1} < 3$.   (EDEXCEL)

14 Find the constants P, Q and R in the identity
$$\frac{x^2 + x + 2}{x - 1} \equiv Px + Q + \frac{R}{x - 1}$$

Hence write down the equation of the oblique asymptote of the curve C whose equation is
$$y = \frac{x^2 + x + 2}{x - 1}$$

Show that C does not intersect this asymptote.

The points $(-1, -1)$ and $(3, 7)$ are stationary points of C. Sketch C, indicating the asymptotes.   (NEAB)

15 a) Sketch the graph of $y = |2x + 3|$, giving the coordinates of the points where the graph meets the coordinate axes.

b) Hence, or otherwise, find the set of values of x for which $4x + 10 > |2x + 3|$.   (EDEXCEL)

8 Roots of polynomial equations

And the equation will come at last.

LOUIS MACNEICE

Roots of a quadratic equation

If α and β are the roots of a quadratic equation, $f(x) \equiv ax^2 + bx + c = 0$, then the equation must be of the form

$$f(x) = k(x - \alpha)(x - \beta) \quad \text{for some constant } k$$

Therefore, we have

$$k(x - \alpha)(x - \beta) \equiv ax^2 + bx + c$$
$$\Rightarrow \quad k(x^2 - [\alpha + \beta]x + \alpha\beta) \equiv ax^2 + bx + c$$

Equating the coefficients of x^2 gives: $\quad k = a$

Equating the coefficients of x gives: $\quad -k(\alpha + \beta) = b$

And equating the constants gives: $\quad k\alpha\beta = c$

Therefore, we obtain

$$\alpha + \beta = -\frac{b}{a} \quad \text{and} \quad \alpha\beta = \frac{c}{a}$$

Or

The **sum** of the roots is $-\dfrac{b}{a}$ and the **product** of the roots is $\dfrac{c}{a}$.

Example 1 In the equation $3x^2 - 7x + 11 = 0$, find

a) the sum of the roots
b) the product of the roots.

SOLUTION

a) Using $\alpha + \beta = -\dfrac{b}{a}$, we have

$$\text{Sum of the roots, } \alpha + \beta = -\frac{-7}{3} = +\frac{7}{3}$$

b) Using $\alpha\beta = \dfrac{c}{a}$, we have

$$\text{Product of the roots, } \alpha\beta = \frac{11}{3}$$

Conversely, we may write the quadratic equation as

$$x^2 - \text{(sum of roots)}x + \text{(product of roots)} = 0$$

Example 2 Find the equation whose roots have a sum of $\frac{1}{2}$ and a product of $-\frac{5}{2}$.

SOLUTION

Using $x^2 - \text{(sum of roots)}x + \text{(product of roots)} = 0$, we have

$$x^2 - \tfrac{1}{2}x - \tfrac{5}{2} = 0 \quad \text{or} \quad 2x^2 - x - 5 = 0$$

Example 3 The equation $3x^2 + 9x - 11 = 0$ has roots α and β. Find the equation whose roots are $\alpha + \beta$ and $\alpha\beta$.

SOLUTION

From $3x^2 + 9x - 11 = 0$, we have

$$\alpha + \beta = -3 \quad \text{and} \quad \alpha\beta = -\frac{11}{3}$$

The sum of the **new roots** is: $\quad \alpha + \beta + \alpha\beta = -3 - \dfrac{11}{3} = -\dfrac{20}{3}$

The product of the **new roots** is: $\quad (\alpha + \beta) \times \alpha\beta = -3 \times -\dfrac{11}{3} = 11$

Therefore, the new equation is

$$x^2 + \frac{20}{3}x + 11 = 0 \quad \text{or} \quad 3x^2 + 20x + 33 = 0$$

Example 4 The equation $4x^2 + 7x - 5 = 0$ has roots α and β. Find the equation whose roots are α^2 and β^2.

SOLUTION

From $4x^2 + 7x - 5 = 0$, we have

$$\alpha + \beta = -\frac{7}{4} \quad \text{and} \quad \alpha\beta = -\frac{5}{4}$$

The sum of the new roots is

$$\alpha^2 + \beta^2 = (\alpha + \beta)^2 - 2\alpha\beta$$

Substituting the above values in the RHS, we obtain

$$\alpha^2 + \beta^2 = \left(-\frac{7}{4}\right)^2 - 2 \times -\frac{5}{4} = \frac{89}{16}$$

The product of the new roots is $\alpha^2\beta^2 = (\alpha\beta)^2$. Substituting the value for $\alpha\beta$, we obtain

$$(\alpha\beta)^2 = \left(-\frac{5}{4}\right)^2 = \frac{25}{16}$$

Therefore, the new equation is

$$x^2 - \frac{89}{16}x + \frac{25}{16} = 0 \quad \text{or} \quad 16x^2 - 89x + 25 = 0$$

Roots of a cubic equation

In a similar manner, if α, β and γ are the roots of a cubic equation, $ax^3 + bx^2 + cx + d = 0$, then we have

$$ax^3 + bx^2 + cx + d \equiv k(x - \alpha)(x - \beta)(x - y)$$

$$\Rightarrow \quad ax^3 + bx^2 + cx + d \equiv k[x^3 - (\alpha + \beta + \gamma)x^2 + (\alpha\beta + \beta\gamma + \gamma\alpha)x - \alpha\beta\gamma]$$

Equating coefficients of x^2 gives: $\quad \alpha + \beta + \gamma = -\dfrac{b}{a}$

Equating coefficients of x gives: $\quad \alpha\beta + \beta\gamma + \gamma\alpha = \dfrac{c}{a}$

And equating the constants gives: $\quad \alpha\beta\gamma = -\dfrac{d}{a}$

Example 5 Find the cubic equation in x which has roots 4, 3 and -2.

SOLUTION

The sum of the roots is

$$\alpha + \beta + \gamma = 4 + 3 + (-2) = 5$$

The sum of the roots taken two at a time is

$$\alpha\beta + \beta\gamma + \gamma\alpha = 4 \times 3 + 3 \times -2 + (-2 \times 4) = -2$$

The product of the roots is

$$\alpha\beta\gamma = 4 \times 3 \times -2 = -24$$

Therefore, the equation is

$$x^3 - 5x^2 - 2x + 24 = 0$$

Example 6 The cubic equation $x^3 + 3x^2 - 7x + 2 = 0$ has roots α, β, γ. Find the value of $\alpha^2 + \beta^2 + \gamma^2$.

SOLUTION

From the cubic equation, we have

$$\alpha + \beta + \gamma = -3$$
$$\alpha\beta + \beta\gamma + \gamma\alpha = -7$$
$$\alpha\beta\gamma = -2$$

We now expand $(\alpha + \beta + \gamma)^2$ to obtain

$$\alpha^2 + \beta^2 + \gamma^2 = (\alpha + \beta + \gamma)^2 - 2(\alpha\beta + \beta\gamma + \gamma\alpha)$$

Substituting the values, we obtain

$$\alpha^2 + \beta^2 + \gamma^2 = (-3)^2 - 2 \times -7 = 23$$

Therefore, we have

$$\alpha^2 + \beta^2 + \gamma^2 = 23$$

Roots of a polynomial equation of degree n

From the properties of the roots of a quadratic equation and of a cubic equation, we see that in a polynomial equation of degree n, $ax^n + bx^{n-1} + cx^{n-2} + \ldots = 0$, the sum of the roots is $-\dfrac{b}{a}$ and the product of the roots is given by

$$(-1)^n \frac{\text{Last term}}{\text{First term}}$$

since the last term is the product of $-\alpha, -\beta, -\gamma, -\delta, \ldots$.

Example 7 The roots of $f(x) \equiv 4x^5 + 6x^4 - 3x^3 + 7x^2 - 11x - 3 = 0$ are $\alpha, \beta, \gamma, \delta$ and ε.

a) Find the product of the five roots.

b) i) Show that $x = 1$ is a root of the equation.

ii) Hence show that the sum of the roots other than 1 is $-\dfrac{5}{2}$.

SOLUTION

a) The sum of all five roots, $\alpha, \beta, \gamma, \delta$ and ε, is $-\dfrac{b}{a} = -\dfrac{6}{4} = -\dfrac{3}{2}$.

b) i) When $x = 1$, we have

$$f(1) = 4 + 6 - 3 + 7 - 11 - 3 = 0$$

Therefore, from the factor theorem, $x = 1$ is one root of the equation.

ii) The sum of all five roots is $-\dfrac{3}{2}$ (from part **a**). That is,

$$\alpha + \beta + \gamma + \delta + \varepsilon = -\frac{3}{2}$$

Putting $\varepsilon = 1$, we have

$$\alpha + \beta + \gamma + \delta + 1 = -\frac{3}{2} \quad \Rightarrow \quad \alpha + \beta + \gamma + \delta = -\frac{5}{2}$$

Therefore, the sum of the other four roots is $-\dfrac{5}{2}$.

Example 8 The equation $z^2 + (3 + i)z + p = 0$ has a root of $2 - i$. Find the value of p and the other root of the equation.

SOLUTION

Since $2 - i$ is a root, $z = 2 - i$ satisfies the equation. Therefore, we have

$$(2 - i)^2 + (3 + i)(2 - i) + p = 0$$
$$\Rightarrow \quad p = -10 + 5i$$

The sum of the roots, $\alpha + \beta = -\dfrac{b}{a}$, is $-(3 + i)$. Therefore, the other root is

$$-(3 + i) - (2 - i) = -5$$

Exercise 8A

1 Write down the sum and the product of the roots of each of the following equations.

a) $x^2 + 3x - 7 = 0$ **b)** $x^2 - 11x + 5 = 0$ **c)** $x^2 + 5x - 4 = 0$

d) $3x^2 + 11x + 2 = 0$ **e)** $x + 2 = \dfrac{5}{x}$ **f)** $2x^2 = 7 - 4x$

2 Write down the equation whose roots have the sum and the product given below.

a) Sum 7; product 15   **b)** Sum -3; product $+5$

c) Sum -2; product -4 **d)** Sum 5; product -11

3 If α, β, γ are the roots of the equation $x^3 - 5x + 3 = 0$, find the values of

a) $\alpha + \beta + \gamma$ **b)** $\alpha^2 + \beta^2 + \gamma^2$ **c)** $\alpha^3 + \beta^3 + \gamma^3 = (\alpha+\beta+\gamma)^3 - 3(\alpha+\beta+\gamma)(\alpha\beta+\beta\gamma+\alpha\gamma) + 3\alpha\beta\gamma$

4 The equation $2z^2 - (7 - 2i)z + q = 0$ has a root of $1 + i$. Find **i)** the value of q and **ii)** the other root of the equation.

5 The equation $3z^2 - (1 - i)z + t = 0$ has a root of $3 + 2i$. Find **i)** the value of t and **ii)** the other root of the equation.

6 Given that α, β, γ are the roots of the equation $x^3 + x^2 + 4x - 5 = 0$, find the cubic equation whose roots are $\beta\gamma$, $\gamma\alpha$ and $\alpha\beta$.   (WJEC)

7 Given the cubic equation $x^3 - 7x + q = 0$ has roots α, 2α and β, find the possible values of q.

                                                                 (WJEC)

8 The equation $3x^2 - 5x + 6 = 0$ has roots α and β. Without solving the given equation, find an equation with integer coefficients whose roots are $(\alpha + \beta)$ and $\alpha\beta$.   (EDEXCEL)

9 The roots of the equation $x^3 - 3x^2 - 3x - 7 = 0$ are α, β and γ.

a) Find the value of $\alpha^2 + \beta^2 + \gamma^2$.

b) Show that

$$\begin{vmatrix} 1 & \alpha & \beta \\ \alpha & 1 & \gamma \\ \beta & \gamma & 1 \end{vmatrix} = 0 \qquad \text{(NEAB)}$$

Equations with related roots

If α and β are the roots of $ax^2 + bx + c = 0$, then we can obtain the equation whose roots are 2α and 2β by making a substitution for x.

First, we express $ax^2 + bx + c = 0$ as

$$a(x - \alpha)(x - \beta) = 0$$

which gives

$$a(2x - 2\alpha)(2x - 2\beta) = 0$$

We obtain the required equation, whose roots are 2α and 2β, by putting $y = 2x$, which gives

$$a(y - 2\alpha)(y - 2\beta) = 0$$

Hence, replacing x by $\dfrac{y}{2}$ gives an equation whose roots are twice those of the original equation.

Example 9 Find the equation whose roots are 3α and 3β, where α and β are the roots of the equation $2x^2 - 5x + 3 = 0$.

SOLUTION

Replacing x by $\dfrac{y}{3}$ in $2x^2 - 5x + 3 = 0$, we obtain an equation in y whose roots for $\dfrac{y}{3}$ are the same as those for x: that is, α and β. Hence, the roots for y will be 3α and 3β.

Therefore, the required equation is

$$2\left(\frac{y}{3}\right)^2 - 5\left(\frac{y}{3}\right) + 3 = 0$$
$$\Rightarrow \quad 2y^2 - 15y + 27 = 0$$

If the equation is to be expressed in terms of x, it would be

$$2x^2 - 15x + 27 = 0$$

Example 10 Find the equation whose roots are α^2, β^2, γ^2, where α, β, γ are the roots of $3x^3 - 7x^2 + 11x - 5 = 0$.

SOLUTION

Replacing x by $\sqrt{y}$ in $3x^3 - 7x^2 + 11x - 5 = 0$, we obtain α, β, γ as the roots for $\sqrt{y}$. Hence, the roots for y are α^2, β^2, γ^2

Therefore, the equation in $\sqrt{y}$ is

$$3(\sqrt{y})^3 - 7(\sqrt{y})^2 + 11(\sqrt{y}) - 5 = 0$$
$$\Rightarrow \quad 3y\sqrt{y} + 11\sqrt{y} = 7y + 5$$

Squaring both sides, we have

$$9y^3 + 66y^2 + 121y = 49y^2 + 70y + 25$$

Therefore, the required equation is

$$9y^3 + 17y^2 + 51y - 25 = 0$$

Exercise 8B

1 The roots of the equation $x^2 + 7x + 11 = 0$ are α and β. Find the equation whose roots are 2α and 2β.

2 The roots of the equation $x^2 - 15x + 7 = 0$ are α and β. Find the equation whose roots are 3α and 3β.

3 The roots of the equation $3x^3 - 4x^2 + 8x - 7 = 0$ are α, β and γ. Find the equation whose roots are 2α, 2β and 2γ.

4 The roots of the equation $x^3 - 3x^2 - 11x + 5 = 0$ are α, β and γ. Find the equation whose roots are $\dfrac{\alpha}{2}, \dfrac{\beta}{2}$ and $\dfrac{\gamma}{2}$.

5 The roots of the equation $2x^2 + 3x + 17 = 0$ are α and β. Find the equation whose roots are α^2 and β^2.

6 The roots of the equation $3x^2 - 7x + 15 = 0$ are α and β. Find the equation whose roots are α^2 and β^2.

7 The equation $2x^2 + 7x + 3 = 0$ has roots α and β. Find the equation whose roots are

   a) $2\alpha, 2\beta$ **b)** $\dfrac{\alpha}{3}, \dfrac{\beta}{3}$ **c)** α^2, β^2 **d)** $\alpha + 2, \beta + 2$

8 The equation $3x^2 + 9x - 2 = 0$ has roots α and β. Find the equation whose roots are

   a) $4\alpha, 4\beta$ **b)** $\dfrac{\alpha}{2}, \dfrac{\beta}{2}$ **c)** α^2, β^2 **d)** $\alpha - 3, \beta - 3$

9 The roots of the equation $x^3 + 3x^2 + 5x + 7 = 0$ are α, β and γ. Find the equation whose roots are

   a) $3\alpha, 3\beta, 3\gamma$ **b)** $\alpha^2, \beta^2, \gamma^2$ **c)** $\alpha + 3, \beta + 3, \gamma + 3$

10 The roots of the equation $x^4 + 3x^3 + 7x^2 - 11x + 1 = 0$ are α, β, γ and δ. Find the equation whose roots are 3α, 3β, 3γ and 3δ.

11 The equation $x + 2 + \dfrac{3}{x} = 0$ has roots α and β. Find the equation whose roots are 5α and 5β.

12 The roots of the quadratic equation $x^2 - 3x + 4 = 0$ are α and β. Without solving the equation, find a quadratic equation, with integer coefficients, whose roots are $\dfrac{1}{\alpha}$ and $\dfrac{1}{\beta}$.   (EDEXCEL)

Complex roots of a polynomial equation

If $z \equiv x + iy$ is a root of a polynomial equation with **real coefficients**, then $\bar{z} \equiv x - iy$ is also a root of the polynomial equation, where $\bar{z}$ is the conjugate of z (see page 3).

Proof

Suppose z is a root of the polynomial

$$a_n z^n + a_{n-1} z^{n-1} + a_{n-2} z^{n-2} + \ldots + a_0 = 0$$

Then, taking the conjugate of both sides, we have

$$\overline{a_n z^n + a_{n-1} z^{n-1} + a_{n-2} z^{n-2} + \ldots + a_0} = 0$$

Using $\overline{z_1 + z_2} = \overline{z_1} + \overline{z_2}$, we obtain

$$\overline{a_n z^n} + \overline{a_{n-1} z^{n-1}} + \overline{a_{n-2} z^{n-2}} + \ldots + \overline{a_0} = 0$$

And using $\overline{z_1 z_2} = \overline{z_1}\,\overline{z_2}$, we obtain

$$\overline{a_n}\,\overline{z^n} + \overline{a_{n-1}}\,\overline{z^{n-1}} + \overline{a_{n-2}}\,\overline{z^{n-2}} + \ldots + \overline{a_0} = 0$$

which gives

$$\overline{a_n}\,(\bar{z})^n + \overline{a_{n-1}}\,(\bar{z})^{n-1} + \overline{a_{n-2}}\,(\bar{z})^{n-2} + \ldots + \overline{a_0} = 0$$

Since all the a_i are real, $\overline{a_i} = a_i$. Therefore, we have

$$a_n(\bar{z})^n + a_{n-1}(\bar{z})^{n-1} + a_{n-2}(\bar{z})^{n-2} + \ldots + a_0 = 0$$

Hence, $\bar{z}$ is also a root of the polynomial.

The complex roots of a polynomial with **real coefficients** always occur in **conjugate complex pairs**.

Note We found in Example 8 (page 150) that when a quadratic equation does **not have real coefficients**, the roots are **not conjugate complex pairs**. (In Example 8, they are $2 - i$ and -5.)

Example 11 Show that $4 - i$ is a root of the polynomial equation

$$f(z) \equiv z^3 - 6z^2 + z + 34 = 0$$

Hence find the other roots.

SOLUTION

To prove that $z = 4 - i$ is a root, we prove that $f(4 - i) = 0$. If $z = 4 - i$ is a root, then $z = 4 + i$ is also a root, since the roots occur as conjugate complex pairs.

Next, we find the quadratic with **real** coefficients which is a factor. We then divide $f(z)$ by this quadratic to find the other factor.

Substituting $z = 4 - i$ in $f(z) \equiv z^3 - 6z^2 + z + 34 = 0$, we have

$$f(4 - i) = (4 - i)^3 - 6(4 - i)^2 + (4 - i) + 34$$
$$= 52 - 47i - 90 + 48i + 4 - i + 34$$
$$= 0$$

Therefore, $4 - i$ is a root of $f(z) \equiv z^3 - 6z^2 + z + 34 = 0$. Hence, $4 + i$ is also a root.

If $z - (4 + i)$ and $z - (4 - i)$ are factors of the polynomial, so is

$$[z - (4 + i)][z - (4 - i)] = z^2 - 8z + 17$$

Dividing $z^3 - 6z^2 + z + 34 = 0$ by $z^2 - 8z + 17$, we obtain

$$f(z) = (z^2 - 8z + 17)(z + 2)$$

Therefore, the three roots of $f(z) \equiv z^3 - 6z^2 + z + 34 = 0$ are $4 + i$, $4 - i$ and -2.

Example 12 Show that $2 + i$ is a root of the polynomial equation

$$f(z) \equiv z^4 - 12z^3 + 62z^2 - 140z + 125 = 0$$

Hence find the other roots.

SOLUTION

As in Example 11, to prove that $z = 2 + i$ is a root, we prove that $f(2 + i) = 0$. If $z = 2 + i$ is a root, then $z = 2 - i$ is also a root.

Next, we find the quadratic with **real** coefficients which is a factor. We then divide $f(z)$ by this quadratic to find the other factors.

Substituting $z = 2 + i$ in $f(z) \equiv z^4 - 12z^3 + 62z^2 - 140z + 125 = 0$, we have

$$f(2 + i) = (2 + i)^4 - 12(2 + i)^3 + 62(2 + i)^2 - 140(2 + i) + 125$$
$$= -7 + 24i - 24 - 132i + 186 + 248i - 280 - 140i + 125$$
$$= 0$$

Therefore, $(2 + i)$ is a root of $f(z) \equiv z^4 - 12z^3 + 62z^2 - 140z + 125 = 0$. Hence, $(2 - i)$ is also a root.

If $z - (2 + i)$ and $z - (2 - i)$ are factors of the polynomial, so is

$$[z - (2 + i)][z - (2 - i)] = z^2 - 4z + 5$$

Dividing $z^4 - 12z^3 + 62z^2 - 140z + 125$ by $z^2 - 4z + 5$, we obtain

$$f(z) = (z^2 - 4z + 5)(z^2 - 8z + 25)$$

Using the quadratic formula, we find that the roots of $z^2 - 8z + 25 = 0$ are $4 \pm 3i$.

Therefore, the four roots of $f(z) \equiv z^4 - 12z^3 + 62z^2 - 140z + 125 = 0$ are $2 + i$, $2 - i$, $4 + 3i$ and $4 - 3i$.

Example 13 The roots of the equation $f(x) \equiv 2x^3 - 3x^2 + 7x - 19 = 0$ are α, β and γ. Show that

a) there is only one real root

b) the real root lies between $x = 2$ and $x = 3$

c) the real part of the two complex roots lies between $-\frac{1}{4}$ and $-\frac{3}{4}$.

SOLUTION

To show that a cubic equation has only one real root, we find the values of $f(x)$ at its turning points. Hence, we will be able to see which of the following curves is $f(x)$.

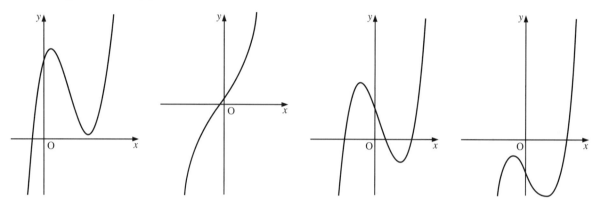

Note   When the values of $f(x)$ at its turning points are of opposite sign, $f(x) = 0$ has three real roots.

a) To find the values of $f(x)$ at its turning points, we differentiate $f(x)$:

$$f(x) \equiv 2x^3 - 3x^2 + 7x - 19$$
$$f'(x) = 6x^2 - 6x + 7$$

Hence, we have

$$6x^2 - 6x + 7 = 0$$
$$\Rightarrow \quad x = \frac{6 \pm \sqrt{36 - 168}}{12}$$

That is, $f'(x) = 0$ has no real roots. Hence, the cubic $f(x)$ has no turning points, which means that $f(x) = 0$ has only one real root.

b) We find that

$$f(2) = -1 \quad \text{and} \quad f(3) = +29$$

So, $f(x)$ has opposite signs at $x = 2$ and $x = 3$ and is continuous for $2 \leqslant x \leqslant 3$. Therefore, the real root of $f(x) = 0$ lies between $x = 2$ and $x = 3$.

c) Let the three roots of the equation be α, β, γ, where α is a real number between 2 and 3, and β and γ are complex numbers.

Since the roots of a polynomial with real coefficients occur in conjugate complex pairs, β and γ are conjugate complex numbers, which we will represent by $p + iq$ and $p - iq$.

Using $\alpha + \beta + \gamma = -\dfrac{b}{a}$, we find

$$\alpha + \beta + \gamma = \frac{3}{2}$$

which gives

$$\alpha + p + iq + p - iq = \frac{3}{2}$$

$$\Rightarrow \quad 2p = \frac{3}{2} - \alpha$$

Since $2 < \alpha < 3$, we therefore have

$$\frac{3}{2} - 3 < 2p < \frac{3}{2} - 2$$

$$\Rightarrow \quad -\frac{3}{2} < 2p < -\frac{1}{2}$$

$$\Rightarrow \quad -\frac{3}{4} < p < -\frac{1}{4}$$

Hence, the real part of each complex root lies between $-\frac{1}{4}$ and $-\frac{3}{4}$.

Exercise 8C

1 Solve the equation $x^4 - 5x^3 + 2x^2 - 5x + 1 = 0$, given that i is a root.

2 Solve the equation $3x^4 - x^3 + 2x^2 - 4x - 40 = 0$, given that 2i is a root.

3 Determine the number of real roots of the equation $2x^3 + x^2 = 3$.

4 Determine the number of real roots of the equation $2x^3 - 7x + 2 = 0$.

5 Determine the range of possible values of k if the equation $x^3 + 3x^2 = k$ has three real roots.

6 One root of the equation $z^4 - 5z^3 + 13z^2 - 16z + 10 = 0$ is $1 + i$. Find the other roots.

7 a) Show that one root of the equation $z^3 + 5z^2 - 56z + 110 = 0$ is $3 + i$.
  b) Find the other roots of the equation.

8 a) Show that one root of the equation $z^4 - 2z^3 + 6z^2 + 22z + 13 = 0$ is $2 - 3i$.
  b) i) Find the other roots of the equation.
    ii) Hence factorise $z^4 - 2z^3 + 6z^2 + 22z + 13$ into two quadratics, each of which has real coefficients.

9 The polynomial f(z) is defined by

$$f(z) \equiv z^4 - 2z^3 + 3z^2 - 2z + 2$$

  a) Verify that i is a root of the equation f(z) = 0.
  b) Find all the other roots of the equation f(z) = 0.   (EDEXCEL)

10 Given that $2 + i$ is a root of the equation $3x^3 - 14x^2 + 23x - 10 = 0$, find the other roots of the equation.   (WJEC)

11 One of the complex roots of $2z^4 - 13z^3 + 33z^2 - 80z - 50 = 0$ is $(1 - 3i)$, where $i^2 = -1$.

    i) State one other complex root.

    ii) Find the other two roots and plot all four on an Argand diagram.   (NICCEA)

12 Given that $3i$ is a root of the equation $3z^3 - 5z^2 + 27z - 45 = 0$, find the other two roots.   (OCR)

13 a) Verify that $z = 2$ is a solution of the equation $z^3 - 8z^2 + 22z - 20 = 0$.

    b) Express $z^3 - 8z^2 + 22z - 20$ as a product of a linear factor and a quadratic factor with real coefficients. Hence find **all** the solutions of $z^3 - 8z^2 + 22z - 20 = 0$.   (SQA/CSYS)

14 Two of the roots of a cubic equation, in which all the coefficients are real, are 2 and $1 + 3i$.

    i) State the third root.

    ii) Find the cubic equation, giving it in the form $z^3 + az^2 + bz + c = 0$.   (OCR)

15 Verify that $z = 1 + i$ is a solution of the equation $z^3 + 16z^2 - 34z + 36 = 0$.

    Write down a second solution of the equation.

    Hence find constants α and β such that

$$z^3 + 16z^2 - 34z + 36 = (z^2 - \alpha z + \alpha)(z + \beta)$$ (SQA/CSYS)

16 The roots of the equation $7x^3 - 8x^2 + 23x + 30 = 0$ are α, β, γ.

    a) Write down the value of $\alpha + \beta + \gamma$.

    b) Given that $1 + 2i$ is a root of the equation, find the other two roots.   (NEAB)

17 Derive expressions for the three cube roots of unity in the form $re^{i\theta}$. Represent the roots on an Argand diagram.

    Let ω denote one of the non-real roots. Show that the other non-real root is ω^2. Show also that

$$1 + \omega + \omega^2 = 0$$

    Given that

$$\alpha = p + q \qquad \beta = p + q\omega \qquad \gamma = p + q\omega^2$$

    where p and q are real,

    i) find, in terms of p, $\alpha\beta + \beta\gamma + \gamma\alpha$

    ii) show that $\alpha\beta\gamma = p^3 + q^3$

    iii) find a cubic equation, with coefficients in terms of p and q, whose roots are α, β, γ.

    (NEAB)

18 The polynomial $f(z)$ has real coefficients and one root of the equation $f(z) = 0$ is $5 + 4i$. Show that $z^2 - 10z + 41$ is a factor of $f(z)$.

    Given now that

$$f(z) = z^6 - 10z^5 + 41z^4 + 16z^2 - 160z + 656,$$

    solve the equation $f(z) = 0$, giving each root exactly in the form $a + ib$.   (OCR)

9 Proof, sequences and series

We must never assume that which is incapable of proof.
G. H. LEWES

We studied some aspects of proof in *Introducing Pure Mathematics* (pages 515–22). Here, we will examine **proof by induction**, including its application to divisibility, and will revisit **proof by contradiction**.

Proof by induction

Proof by induction is used when we are given a statement which applies to any natural number, n.

To prove a statement by induction, we proceed in two steps:

1 We assume that the statement is true for $n = k$, and then use this assumption to prove that it is true for $n = k + 1$.

2 We then prove the statement for $n = 1$.

Step 2 tells us that the statement is true for $n = 1$.

Step 1 then tells us that, when $k = 1$, the statement is true for $n = 2$.

Using step 1 again, when $k = 2$, the statement must be true for $n = 3$.

Using step 1 yet again, the statement is true for $n = 4$.

Similarly, step 1 can be repeated for $n = 5$, $n = 6$, and so on.

Therefore, the statement is true for all integer $n \, (\geqslant 1)$.

Example 1 Prove that $\sum_{r=1}^{n} r = \frac{1}{2} n(n + 1)$.

SOLUTION

We assume that the formula is true for $n = k$. Therefore, we have

$$\sum_{r=1}^{k} r = \frac{1}{2} k(k + 1)$$

We are trying to prove that $\sum_{r=1}^{n} r = \frac{1}{2} n(n + 1)$ is true for $n = k + 1$.

That is, we are trying to prove that $\sum_{r=1}^{k+1} r = \frac{1}{2}(k + 1)(k + 2)$.

159

We have

$$\sum_{r=1}^{k+1} r = \sum_{r=1}^{k} r + (k+1)\text{th term}$$

which gives

$$\sum_{r=1}^{k+1} r = \frac{1}{2}k(k+1) + k + 1$$

$$= \frac{1}{2}[k(k+1) + 2(k+1)]$$

$$= \frac{1}{2}(k+1)(k+2)$$

Therefore, $\sum_{r=1}^{n} r = \frac{1}{2}n(n+1)$ is true for $n = k+1$.

When $n = 1$:   LHS of the formula $= 1$

RHS of the formula $= \frac{1}{2} \times 1 \times 2 = 1$

Therefore, the formula is true for $n = 1$.

Therefore, $\sum_{r=1}^{n} r = \frac{1}{2}n(n+1)$ is true for all $n \geqslant 1$.

Note   In a mathematical proof by induction, it is vital that we **write these last four lines of the proof in full**.

Example 2 Prove that $\sum_{r=1}^{n} r.r! = (n+1)! - 1$.

SOLUTION

We assume that the formula is true for $n = k$, which gives

$$\sum_{r=1}^{k} r.r! = (k+1)! - 1$$

Therefore, we have

$$\sum_{r=1}^{k+1} r.r! = (k+1)! - 1 + (k+1)\text{th term}$$

$$= (k+1)! - 1 + (k+1)(k+1)!$$
$$= (k+1)!(1 + k + 1) - 1$$
$$= (k+2)(k+1)! - 1$$
$$= (k+2)! - 1$$

Therefore, the formula is true for $n = k+1$.

When $n = 1$:   LHS of $\displaystyle\sum_{r=1}^{n} r.r! = 1$

RHS of $\displaystyle\sum_{r=1}^{n} r.r! = (n+1)! - 1 = 2! - 1 = 1$

Therefore, the formula is true for $n = 1$.

Therefore, $\displaystyle\sum_{r=1}^{n} r.r! = (n+1)! - 1$ is true for all $n \geqslant 1$.

Example 3 Prove that $\dfrac{\mathrm{d}^n}{\mathrm{d}x^n}(e^x \sin x) = 2^{\frac{n}{2}} e^x \sin(x + \tfrac{1}{4} n\pi)$.

SOLUTION

We assume that the formula is true for $n = k$, which gives

$$\frac{\mathrm{d}^k}{\mathrm{d}x^k}(e^x \sin x) = 2^{\frac{k}{2}} e^x \sin(x + \tfrac{1}{4} k\pi)$$

Therefore, we have

$$\frac{\mathrm{d}^{k+1}}{\mathrm{d}x^{k+1}}(e^x \sin x) = \frac{\mathrm{d}}{\mathrm{d}x}\left(\frac{\mathrm{d}^k}{\mathrm{d}x^k}(e^x \sin x)\right) = \frac{\mathrm{d}}{\mathrm{d}x}[2^{\frac{k}{2}} e^x \sin(x + \tfrac{1}{4} k\pi)]$$

$$= 2^{\frac{k}{2}} e^x \sin(x + \tfrac{1}{4} k\pi) + 2^{\frac{k}{2}} e^x \cos(x + \tfrac{1}{4} k\pi)$$

$$= 2^{\frac{k}{2}} e^x [\sin(x + \tfrac{1}{4} k\pi) + \cos(x + \tfrac{1}{4} k\pi)]$$

Using $a \sin \theta + b \cos \theta = R \sin(\theta + \alpha)$, we obtain

$$\frac{\mathrm{d}^{k+1}}{\mathrm{d}x^{k+1}}(e^x \sin x) = 2^{\frac{k}{2}} e^x \sqrt{2} \sin[(x + \tfrac{1}{4} k\pi) + \tfrac{1}{4}\pi]$$

$$= 2^{\frac{1}{2}(k+1)} e^x \sin[x + \tfrac{1}{4}(k+1)\pi]$$

Therefore, $\dfrac{\mathrm{d}^n}{\mathrm{d}x^n}(e^x \sin x) = 2^{\frac{n}{2}} e^x \sin(x + \tfrac{1}{4} n\pi)$ is true for $n = k + 1$.

When $n = 1$: $\dfrac{\mathrm{d}}{\mathrm{d}x}(e^x \sin x) = e^x \sin x + e^x \cos x$

$$= \sqrt{2} e^x \sin(x + \tfrac{1}{4}\pi)$$

Therefore, the formula is true for $n = 1$.

Therefore, $\dfrac{\mathrm{d}^n}{\mathrm{d}x^n}(e^x \sin x) = 2^{\frac{n}{2}} e^x \sin(x + \tfrac{1}{4} n\pi)$ is true for all $n \geqslant 1$.

Example 4 If $A = \begin{pmatrix} 2 & 1 \\ 0 & 1 \end{pmatrix}$, prove that $A^n = \begin{pmatrix} 2^n & 2^n - 1 \\ 0 & 1 \end{pmatrix}$.

SOLUTION

We assume that the statement is true for $n = k$, which gives

$$A^k = \begin{pmatrix} 2^k & 2^k - 1 \\ 0 & 1 \end{pmatrix}$$

Therefore, we have

$$A^{k+1} = A^k \times A = \begin{pmatrix} 2^k & 2^k - 1 \\ 0 & 1 \end{pmatrix} \times \begin{pmatrix} 2 & 1 \\ 0 & 1 \end{pmatrix}$$

$$= \begin{pmatrix} 2^{k+1} & 2^k + 2^k - 1 \\ 0 & 1 \end{pmatrix}$$

$$\Rightarrow \quad A^{k+1} = \begin{pmatrix} 2^{k+1} & 2^{k+1} - 1 \\ 0 & 1 \end{pmatrix}$$

Therefore, the statement is true for $n = k + 1$.

When $n = 1$, the statement is true.

Therefore, if $A = \begin{pmatrix} 2 & 1 \\ 0 & 1 \end{pmatrix}$, $A^n = \begin{pmatrix} 2^n & 2^n - 1 \\ 0 & 1 \end{pmatrix}$ for all $n \geqslant 1$.

Divisibility

Proof by induction can also be used to prove that a term is divisible by a certain integer.

Example 5 Prove that $5^{2n} + 2^{2n-2}3^{n-1}$ is divisible by 13.

SOLUTION

Let $u_n = 5^{2n} + 2^{2n-2}3^{n-1}$. Therefore, we have

$$u_{n+1} = 5^{2(n+1)} + 2^{2(n+1)-2}3^{(n+1)-1}$$

Expressing u_{n+1} in the powers of u_n, we obtain

$$u_{n+1} = 5^2 5^{2n} + 2^2 2^{2n-2} 3^1 3^{n-1}$$
$$= 25 \times 5^{2n} + 12 \times 2^{2n-2}3^{n-1}$$

Adding u_n and u_{n+1}, we obtain

$$u_n + u_{n+1} = 26 \times 5^{2n} + 13 \times 2^{2n-2}3^{n-1}$$

Both 26 and 13 are divisible by 13. Therefore, since the sum of u_n and u_{n+1} is divisible by 13, either

> both u_n and u_{n+1} **are** divisible by 13, or
> both u_n and u_{n+1} **are not** divisible by 13.

When $n = 1$, $u_1 = 5^2 + 2^0 3^0 = 26$, which is divisible by 13.

Therefore, u_n is divisible by 13 for all integer $n \geqslant 1$.

It is not necessary to use simply $u_{n+1} + u_n$ as the term to be divisible by the required integer divisor. We can add, or subtract, any multiple of u_{n+1} and u_n, as long as that multiple is not the divisor, or a factor of the divisor.

In Example 5, we could have used $u_{n+1} - 12u_n = 13 \times 5^{2n}$. But obviously we could not use $13u_{n+1} - 13u_n$, which is divisible by 13, to prove anything about the divisibility of u_{n+1} or u_n.

Example 6 Prove that $3^{4n+2} + 5^{2n+1}$ is divisible by 14.

SOLUTION

Let $u_n = 3^{4n+2} + 5^{2n+1}$. Therefore, we have

$$u_{n+1} = 3^{4(n+1)+2} + 5^{2(n+1)+1}$$
$$= 3^4 3^{4n+2} + 5^2 5^{2n+1}$$
$$= 81 \times 3^{4n+2} + 25 \times 5^{2n+1}$$

Note   We are trying to prove divisibility by 14. But for the term in 5^{2n+1},

$$u_{n+1} + u_n \text{ gives } (25+1)5^{2n+1} \quad \text{and} \quad u_{n+1} - u_n \text{ gives } (25-1)5^{2n+1}$$

Neither $25 + 1 = 26$, nor $25 - 1 = 24$ are divisible by 14, and so are unhelpful. However, we can see that both $u_{n+1} + 3u_n$ **and** $u_{n+1} - 11u_n$ make the term in 5^{2n+1} divisible by 14, giving respectively $(25 + 3) = 28$ and $(25 - 11) = 14$.

We need to check that the term in 3^{4n+2} also satisfies this divisibility:

$$u_{n+1} - 11u_n = 81 \times 3^{4n+2} + 25 \times 5^{2n+1} - 11(3^{4n+2} + 5^{2n+1})$$
$$= 81 \times 3^{4n+2} + 25 \times 5^{2n+1} - 11 \times 3^{4n+2} - 11 \times 5^{2n+1}$$
$$= 70 \times 3^{4n+2} + 14 \times 5^{2n+1}$$

which is divisible by 14.

Therefore, either both u_{n+1} and u_n **are** divisible by 14, or both u_{n+1} and u_n **are not** divisible by 14.

When $n = 1$, $3^{4n+2} + 5^{2n+1} = 3^6 + 5^3 = 854$, which **is** divisible by 14.

Therefore, all u_n are divisible by 14.

Therefore, $3^{4n+2} + 5^{2n+1}$ is divisible by 14 for all $n \geqslant 1$.

Example 7 Prove that $7^n + 4^n + 1$ is divisible by 6.

SOLUTION

Let $u_n = 7^n + 4^n + 1$. Therefore, we have

$$u_{n+1} = 7^{n+1} + 4^{n+1} + 1$$
$$= 7 \times 7^n + 4 \times 4^n + 1$$

To eliminate the $+1$, we need to subtract u_n from u_{n+1}, giving

$$u_{n+1} - u_n = 6 \times 7^n + 3 \times 4^n$$

We cannot use $2u_{n+1} - 2u_n$, as this would involve multiplying by 2, which is a factor of 6, which we are trying to prove is a factor of the given expression. Hence, we need to show that 3×4^n, as well as 6×7^n, is divisible by 6:

$$u_{n+1} - u_n = 6 \times 7^n + 3 \times 4^n$$
$$= 6 \times 7^n + 3 \times 2^{2n}$$
$$= 6 \times 7^n + 6 \times 2^{2n-1}$$

which is divisible by 6.

▪
▪
▪
▪
▪
▪
▪
▪
▪
▪
▪
▪
▪

Therefore, either both u_{n+1} and u_n **are** divisible by 6, or both u_{n+1} and u_n **are not** divisible by 6.

When $n = 1$, $7^n + 4^n + 1 = 7 + 4 + 1 = 12$, which is divisible by 6.

Therefore, all u_n are divisible by 6.

Therefore, $7^n + 4^n + 1$ is divisible by 6 for all $n \geqslant 1$.

Proof by contradiction

Another way to prove that something is true is to assume that it is false, and then to arrive at a contradiction. (See also *Introducing Pure Mathematics*, pages 521–3.)

Suppose, for example, that we want to prove the statement

There is no biggest integer.

It seems obvious that there is no biggest whole number, but 'it seems obvious' is not a proper mathematical proof. One way to prove this statement is to assume that there **is** a biggest integer.

Call the biggest integer M. Then $M + 1$ must also be an integer. Now, $M + 1 > M$. But M was supposed to be the **biggest** integer. Therefore, we have a contradiction.

So our original assumption is false: there is no biggest integer.

One of the most beautiful proofs in all of mathematics concerns the statement

There are an infinite number of prime numbers

We suppose there are **not** an infinite number of prime numbers, and prove that this is nonsense.

Assume that there are a finite number of prime numbers. Then we can write them down as $\{p_1, p_2, \ldots, p_n\}$. The number $p_1 \times p_2 \times \ldots \times p_n + 1$ is not divisible by any of the prime numbers $\{p_1, p_2, \ldots, p_n\}$. This is nonsense because $\{p_1, p_2, \ldots, p_n\}$ was supposed to be a list of all the prime numbers. This contradiction tells us that our original assumption is wrong. Hence, there are infinitely many prime numbers.

Exercise 9A

1 Use proof by induction to prove that $\displaystyle\sum_{r=1}^{n} r^2 = \frac{1}{6} n(n + 1)(2n + 1)$.

2 Use proof by induction to prove that $\displaystyle\sum_{r=1}^{n} r^3 = \frac{1}{4} n^2 (n + 1)^2$.

3 Prove that $13^n - 6^{n-2}$ is divisible by 7.

4 Prove that $2^{6n} + 3^{2n-2}$ is divisible by 5.

5 It is given that $\phi(n) = 7^n(6n + 1) - 1$, for $n = 1, 2, 3, \ldots$

  i) Show that

$$\phi(n + 1) - \phi(n) = 7^n(36n + 48)$$

  ii) Hence prove by induction that $\phi(n)$ is divisible by 12 for every positive integer n.   (OCR)

6 Verify that $5^5 \equiv 1 \pmod{11}$. Hence find the remainder obtained on dividing 5^{1998} by 11.

                                                                      (OCR)

7 Use mathematical induction to prove that

$$\sum_{r=1}^{n} (r - 1)(3r - 2) = n^2(n - 1)$$

for all positive integers n.   (AEB 97)

8 $f(n) \equiv 24 \times 2^{4n} + 3^{4n}$, where n is a non-negative integer.

  a) Write down $f(n + 1) - f(n)$.
  b) Prove, by induction, that $f(n)$ is divisible by 5.   (EDEXCEL)

9 Prove by mathematical induction that $5^{2n} - 1$ is divisible by 24 for all positive integers n.

                                                              (WJEC)

10 Prove, by induction, that $\sum_{r=1}^{n} r(r + 3) = \dfrac{1}{3}n(n + 1)(n + 5)$, $n \in \mathbb{N}$.   (EDEXCEL)

11 Prove by induction that $\sum_{r=1}^{n} r^2 = \dfrac{1}{6}n(n + 1)(2n + 1)$.

  Find the sum of the squares of the first n positive odd integers.   (OCR)

12 Use induction to prove that

$$\sum_{r=1}^{n} r(r + 1) = \frac{1}{3}n(n + 1)(n + 2)$$

for all positive integers n.   (SQA/CSYS)

13 Prove by induction that

$$\sum_{r=1}^{n} r(r + 1)(r + 2) = \frac{1}{4}n(n + 1)(n + 2)(n + 3)$$ (NICCEA)

14 a) Write down an expression for the nth term of the series

$$\frac{1^2}{1 \times 3} + \frac{2^2}{3 \times 5} + \frac{3^2}{5 \times 7} + \frac{4^2}{7 \times 9} + \ldots$$

  b) Prove by induction, or otherwise, that the sum, S_n, of the first n terms of the above series is given by

$$S_n = \frac{n(n + 1)}{2(2n + 1)}$$ (NEAB)

15 Show by mathematical induction that

$$1 + 2.2 + 3.2^2 + \ldots + n2^{n-1} = (n - 1)2^n + 1$$

for all positive integer values of n.   (WJEC)

16 a) Use the results $\sum_{r=1}^{n} r = \frac{1}{2}n(n+1)$ and $\sum_{r=1}^{n} r^3 = \frac{1}{4}n^2(n+1)^2$ to find an expression, in terms of

n, for $\sum_{r=1}^{n} r(r-1)(r+1)$, factorising your answer as fully as possible.

b) Use mathematical induction to prove that

$$\sum_{r=2}^{n} \frac{1}{r(r-1)(r+1)} = \frac{1}{4} - \frac{1}{2n(n+1)}$$

for all positive integers $n \geq 2$.   (AEB 97)

17 Use mathematical induction to prove that

$$\sum_{r=1}^{n} r(r+1)(r+5) = \frac{1}{4}n(n+1)(n+2)(n+7)$$

for all positive integers n.   (AEB 98)

18 Show, by means of a counter-example, that the statement

$$\mathbf{a} \times \mathbf{b} = \mathbf{0} \text{ implies } \mathbf{a} = \mathbf{0} \text{ or } \mathbf{b} = \mathbf{0}$$

is false.

Find a unit vector $\mathbf{n}$ such that $\mathbf{n} \times \begin{pmatrix} 2 \\ -2 \\ 1 \end{pmatrix} = \mathbf{0}$.   (NEAB)

19 Prove by induction that

$$\sum_{r=1}^{n} \frac{2r+1}{r^2(r+1)^2} = 1 - \frac{1}{(n+1)^2}$$ (OCR)

20 Prove that there is no smallest positive rational number. [**Hint** Prove this by contradiction.]

21 Prove, by induction, that

$$\sum_{r=1}^{n} r^2(r-1) = \frac{1}{12}n(n-1)(n+1)(3n+2)$$ (EDEXCEL)

22 i) Show that, if $n = k+1$, then

$$\frac{(3n+2)(n-1)}{n(n+1)} = \frac{3k^3 + 5k^2}{k(k+1)(k+2)}$$

provided $k > 0$.

ii) Prove by induction

$$\sum_{r=2}^{n} \frac{4}{r^2-1} \equiv \frac{(3n+2)(n-1)}{n(n+1)}$$ (NICCEA)

23 Show that $\sum_{r=1}^{n} r(r+2) = \frac{n}{6}(n+1)(2n+7)$.

Using this result, or otherwise, find, in terms of n, the sum of the series

$$3\ln 2 + 4\ln 2^2 + 5\ln 2^3 + \ldots + (n+2)\ln 2^n$$

Express your answer in its simplest form.   (EDEXCEL)

24 Consider the sequence defined by the relationship $u_{n+1} = 5u_n + 2$ whose first term is $u_1 = 1$.

i) Show that the first four terms are 1, 7, 37, 187, ...

ii) Use the method of induction to prove that $u_n = \frac{1}{2}[3(5^{n-1}) - 1]$.   (NICCEA)

25 A sequence u_0, u_1, u_2, ... is defined by

$$u_0 = 2 \quad \text{and} \quad u_{n+1} = 1 - 2u_n \quad (n \geqslant 0)$$

a) Prove by induction that, for all $n \geqslant 0$,

$$u_n = \frac{1}{3}\{1 + 5(-2)^n\}$$

b) State, briefly giving a reason for your answer, whether the sequence is convergent.

(NEAB)

26 Prove by contradiction that if the sum of two numbers is greater than 50, then at least one of the original numbers must have been greater than 25.

27 Let

$$A = \begin{pmatrix} 1 & 0 \\ -1 & 2 \end{pmatrix}$$

Use induction to prove that, for all positive integers n,

$$A^n = \begin{pmatrix} 1 & 0 \\ 1 - 2^n & 2^n \end{pmatrix}$$

Determine whether or not this formula for A^n is also valid when $n = -1$.   (SQA/CSYS)

28 Prove by induction that, for every positive integer N,

$$\sum_{n=1}^{N} \frac{n4^n}{(n+4)!} = \frac{1}{6} - \frac{4^{N+1}}{(N+4)!}$$

Given that, for every positive integer N,

$$\frac{4^{N+1}}{(N+4)!} \leqslant \frac{1}{6}\left(\frac{4}{5}\right)^N$$

show that the infinite series

$$\frac{1 \times 4^1}{5!} + \frac{2 \times 4^2}{6!} + \frac{3 \times 4^3}{7!} + \ldots$$

is convergent, and give the sum to infinity.   (OCR)

29 Let u, v, w be positive integers. For each of the following, decide whether the statement is true or false. Where false, give a counter-example; where true, give a proof.

i) If u and v both divide w then $u + v$ divides w.
ii) If u divides both v and w then u divides $v + w$.
iii) If u divides v and v divides w then u divides $v + w$.

Write down the converse of statement **ii**, and determine whether or not this converse is true.

(SQA/CSYS)

Summation of series

As we have already seen on pages 159–61, proof by induction can be used to prove that a series has a known sum. Unfortunately, it is of no use when we do not know the sum in advance. Therefore, we will now introduce two other methods of summing a series: **applying standard formulae** and **differencing**.

Applying standard formulae

On pages 159 and 164, we found that

$$\sum_{r=1}^{n} r = \frac{1}{2}n(n+1)$$

$$\sum_{r=1}^{n} r^2 = \frac{1}{6}n(n+1)(2n+1)$$

$$\sum_{r=1}^{n} r^3 = \frac{1}{4}n^2(n+1)^2$$

We also have

$$\sum_{r=1}^{n} r^3 = \left(\sum_{r=1}^{n} r\right)^2$$

These four formulae can be used to find the sums of many series.

Note $\sum_{r=1}^{n} r$ is often expressed as $\sum_{1}^{n} r$.

Example 8 Find the sum of $\displaystyle\sum_{r=1}^{n} (4r^2 + 1)$.

SOLUTION

First, we split the given term into its parts, and then use the formulae above, as appropriate.

Note $\displaystyle\sum_{r=1}^{n} 1 = 1 + 1 + 1 + \ldots + 1 = n$ (total of n terms of 1)

Splitting the given term, we have

$$\sum_{r=1}^{n} (4r^2 + 1) = 4\sum_{r=1}^{n} r^2 + \sum_{r=1}^{n} 1$$

which gives

$$\sum_{r=1}^{n} (4r^2 + 1) = 4 \times \frac{1}{6}n(n+1)(2n+1) + n$$

$$= \frac{2}{3}n(n+1)(2n+1) + n$$

$$= \frac{1}{3}[2n(n+1)(2n+1) + 3n]$$

$$\Rightarrow \quad \sum_{r=1}^{n} (4r^2 + 1) = \frac{1}{3}n[2(n+1)(2n+1) + 3]$$

Therefore, we have

$$\sum_{r=1}^{n} (4r^2 + 1) = \frac{1}{3}n(4n^2 + 6n + 5)$$

Example 9 Find the sum of $\displaystyle\sum_{r=1}^{n} (2r^3 + 3r^2 + 1)$.

SOLUTION

Splitting the given term, we have

$$\sum_{r=1}^{n} (2r^3 + 3r^2 + 1) = \sum_{r=1}^{n} 2r^3 + \sum_{r=1}^{n} 3r^2 + \sum_{r=1}^{n} 1$$

$$= 2\sum_{r=1}^{n} r^3 + 3\sum_{r=1}^{n} r^2 + \sum_{r=1}^{n} 1$$

which gives

$$\sum_{r=1}^{n} (2r^3 + 3r^2 + 1) = 2 \times \frac{1}{4}n^2(n+1)^2 + 3 \times \frac{1}{6}n(n+1)(2n+1) + n$$

$$= \frac{n}{2}[n(n+1)^2 + (n+1)(2n+1) + 2]$$

Therefore, we have

$$\sum_{r=1}^{n} (2r^3 + 3r^2 + 1) = \frac{n}{2}(n^3 + 4n^2 + 4n + 3)$$

Example 10 Find the sum of $\displaystyle\sum_{r=n+1}^{2n} (4r^3 - 3)$.

SOLUTION

Splitting the given term, we have

$$\sum_{r=n+1}^{2n} (4r^3 - 3) = \sum_{r=1}^{2n} (4r^3 - 3) - \sum_{r=1}^{n} (4r^3 - 3)$$

which gives

$$\sum_{r=n+1}^{2n} (4r^3 - 3) = 4\sum_{1}^{2n} r^3 - 3\sum_{1}^{2n} 1 - \left(4\sum_{1}^{n} r^3 - 3\sum_{1}^{n} 1\right)$$

$$= 4 \times \frac{1}{4}(2n)^2(2n+1)^2 - 3 \times 2n - \left[4 \times \frac{1}{4}n^2(n+1)^2 - 3n\right]$$

$$= 4n^2(2n+1)^2 - 6n - n^2(n+1)^2 + 3n$$

$$= 4n^2(4n^2 + 4n + 1) - n^2(n^2 + 2n + 1) - 3n$$

$$= n^2(15n^2 + 14n + 3) - 3n$$

Therefore, we have

$$\sum_{r=n+1}^{2n} (4r^3 - 3) = 15n^4 + 14n^3 + 3n^2 - 3n$$

Example 11 Find $\displaystyle\sum_{r=1}^{8} (r^2 + 2)$.

SOLUTION

Splitting the given term, we have

$$\sum_{r=1}^{8} (r^2 + 2) = \sum_{1}^{8} r^2 + \sum_{1}^{8} 2$$

Using

$$\sum_{r=1}^{n} r^2 = \frac{1}{6}n(n + 1)(2n + 1)$$

and remembering that $\displaystyle\sum_{r=1}^{n} 1 = n$, therefore $\displaystyle\sum_{r=1}^{n} 2 = 2n$, we obtain

$$\sum_{r=1}^{n} (r^2 + 2) = \frac{1}{6}n(n + 1)(2n + 1) + 2n$$

Now, $n = 8$, therefore,

$$\sum_{r=1}^{8} (r^2 + 2) = \frac{1}{6} \times 8 \times 9 \times 17 + 16 = 220$$

Hence, we have

$$\sum_{r=1}^{8} (r^2 + 2) = 220$$

Exercise 9B

1 Find $\displaystyle\sum_{r=1}^{n} (2r^2 + 2r)$.

2 Find $\displaystyle\sum_{r=1}^{n} (2r^3 + r)$.

3 Find $\displaystyle\sum_{r=1}^{n} (r + 1)(r - 2)$.

4 Find $\displaystyle\sum_{r=1}^{n} (2r - 1)(r + 5)$.

5 a) Show that $\displaystyle\sum_{r=1}^{n} (2r - 1)(2r + 3) = \frac{1}{3}n(4n^2 + 12n - 1)$.

b) Hence find $\displaystyle\sum_{r=5}^{35} (2r - 1)(2r + 3)$.   (EDEXCEL)

6 Given that n is a positive integer, find $\displaystyle\sum_{r=1}^{n} (2r - 1)^3$, giving your answer in its simplest form.
     (EDEXCEL)

7 Show that $\displaystyle\sum_{r=1}^{n} r(2r + 1) = \frac{1}{6}n(n + 1)(4n + 5)$. Hence evaluate $\displaystyle\sum_{r=10}^{30} r(2r + 1)$.   (EDEXCEL)

8 Write down the sum

$$\sum_{n=1}^{2N} n^3$$

in terms of N, and hence find

$$1^3 - 2^3 + 3^3 - 4^3 + \ldots - (2N)^3$$

in terms of N, simplifying your answer.   (OCR)

Differencing

Some series can be summed using partial fractions (see *Introducing Pure Mathematics*, pages 280–89). The basis of this method is that most of the terms cancel out.

Example 12 Find $\displaystyle\sum_{r=1}^{n} \frac{1}{r(r+1)}$.

SOLUTION

First, we write $\dfrac{1}{r(r+1)}$ as the sum of partial fractions:

$$\frac{1}{r(r+1)} = \frac{1}{r} - \frac{1}{r+1}$$

Hence, we have

$$\sum_{r=1}^{n} \frac{1}{r(r+1)} = \sum_{r=1}^{n} \left(\frac{1}{r} - \frac{1}{r+1} \right)$$

$$= \left(\frac{1}{1} - \frac{1}{2} \right) + \left(\frac{1}{2} - \frac{1}{3} \right) + \left(\frac{1}{3} - \frac{1}{4} \right) + \ldots + \left(\frac{1}{n-1} - \frac{1}{n} \right) + \left(\frac{1}{n} - \frac{1}{n+1} \right)$$

We notice that all the terms except the first and the last cancel one another. Therefore, we have

$$\sum_{r=1}^{n} \frac{1}{r(r+1)} = \frac{1}{1} - \frac{1}{n+1} = \frac{n}{n+1}$$

Example 13   Find $\displaystyle\sum_{r=1}^{n} \frac{2}{r(r+1)(r+2)}$.

SOLUTION

First, we write $\dfrac{2}{r(r+1)(r+2)}$ as the sum of partial fractions:

$$\frac{2}{r(r+1)(r+2)} = \frac{1}{r} - \frac{2}{r+1} + \frac{1}{r+2}$$

Hence, we have

$$\sum_{r=1}^{n} \frac{2}{r(r+1)(r+2)} = \sum_{r=1}^{n} \left(\frac{1}{r} - \frac{2}{r+1} + \frac{1}{r+2} \right)$$

$$= \left(1 - \frac{2}{2} + \frac{1}{3} \right) + \left(\frac{1}{2} - \frac{2}{3} + \frac{1}{4} \right) + \left(\frac{1}{3} - \frac{2}{4} + \frac{1}{5} \right) +$$

$$+ \left(\frac{1}{4} - \frac{2}{5} + \frac{1}{6} \right) + \ldots + \left(\frac{1}{n-2} - \frac{2}{n-1} + \frac{1}{n} \right) +$$

$$+ \left(\frac{1}{n-1} - \frac{2}{n} + \frac{1}{n+1} \right) + \left(\frac{1}{n} - \frac{2}{n+1} + \frac{1}{n+2} \right)$$

Note   **Do not reduce** fractions to their lowest terms, since this obscures the cancellation which should occur.

We notice that almost all the terms cancel one another. We are left with

$$\sum_{r=1}^{n} \frac{2}{r(r+1)(r+2)} = 1 - \frac{2}{2} + \frac{1}{2} + \frac{1}{n+1} - \frac{2}{n+1} + \frac{1}{n+2}$$

$$= \frac{1}{2} - \frac{1}{n+1} + \frac{1}{n+2}$$

Example 14   Use the identity $r \equiv \frac{1}{2}[r(r+1) - (r-1)r]$ to find the

sum $\sum_{r=1}^{n} r$.

SOLUTION

Making the given substitution, we obtain

$$\sum_{r=1}^{n} r = \sum_{r=1}^{n} \frac{1}{2}[r(r+1) - (r-1)r]$$

$$= \frac{1}{2}(1 \times 2 - 0 \times 1) + \frac{1}{2}(2 \times 3 - 1 \times 2) + \frac{1}{2}(3 \times 4 - 2 \times 3) + \ldots$$

$$+ \frac{1}{2}[(n-1)n - (n-2)(n-1)] + \frac{1}{2}[n(n+1) - (n-1)n]$$

We notice that almost all the terms cancel one another. We are left with

$$\sum_{r=1}^{n} r = \frac{1}{2}[-0 \times 1 + n(n+1)]$$

$$= \frac{1}{2}n(n+1)$$

Note   This result was also found on pages 159–60, using a different method.

Exercise 9C

1 Verify the identity

$$\frac{2r-1}{r(r-1)} - \frac{2r+1}{r(r+1)} \equiv \frac{2}{(r-1)(r+1)}$$

Hence, using the method of differences, prove that

$$\sum_{r=2}^{n} \frac{2}{(r-1)(r+1)} = \frac{3}{2} - \frac{2n+1}{n(n+1)}$$

Deduce the sum of the infinite series

$$\frac{1}{1.3} + \frac{1}{2.4} + \frac{1}{3.5} + \cdots \frac{1}{(n-1)(n+1)} + \cdots \qquad \text{(AEB 98)}$$

2 Show that

$$\frac{1}{r(r+1)} - \frac{1}{(r+1)(r+2)} \equiv \frac{2}{r(r+1)(r+1)}$$

Hence, or otherwise, find a simplified expression for

$$\sum_{r=1}^{n} \frac{1}{r(r+1)(r+1)} \qquad \text{(WJEC)}$$

3 a) Express $\dfrac{1}{(2r-1)(2r+1)}$ in partial fractions.

b) Hence, or otherwise, show that

$$\sum_{r=n}^{r=2n} \frac{1}{(2r-1)(2r+1)} = \frac{an+b}{(2n-1)(4n+1)}$$

where a and b are integers to be found.

c) Determine the limit as $n \to \infty$ of $\displaystyle\sum_{r=n}^{r=2n} \frac{1}{(2r-1)(2r+1)}$. \qquad (NEAB)

4 Find the value of the constant A for which $(2r+1)^2 - (2r-1)^2 \equiv Ar$.

Use this result, and the method of differences, to prove that

$$\sum_{r=1}^{n} r = \frac{1}{2}n(n+1) \qquad \text{(AEB 96)}$$

5 Express $\dfrac{1}{(2r+1)(2r+3)}$ in partial fractions.

Hence find the sum of the series

$$\frac{1}{3 \times 5} + \frac{1}{5 \times 7} + \cdots + \frac{1}{(2n+1)(2n+3)}$$

Show that the series

$$\frac{1}{3 \times 5} + \frac{1}{5 \times 7} + \cdots + \frac{1}{(2n+1)(2n+3)} + \cdots$$

is convergent and state the sum to infinity. \qquad (OCR)

6 Verify that

$$\frac{1}{1+(n-1)x} - \frac{1}{1+nx} = \frac{x}{\{1+(n-1)x\}(1+nx)}$$

Hence show that, for $x \neq 0$,

$$\sum_{n=1}^{N} \frac{1}{\{1+(n-1)x\}(1+nx)} = \frac{N}{1+Nx}$$

Deduce that the infinite series

$$\frac{1}{1 \times \frac{3}{2}} + \frac{1}{\frac{3}{2} \times 2} + \frac{1}{2 \times \frac{5}{2}} + \dots$$

is convergent and find its sum to infinity.   (OCR)

7 Let $a_n = e^{-(n-1)x} - e^{-nx}$, where $x \neq 0$.

i) Find $\displaystyle\sum_{n=1}^{N} a_n$ in terms of N and x.

ii) Find the set of values of x for which the infinite series

$$a_1 + a_2 + a_3 + \dots$$

converges, and state the sum to infinity.   (OCR)

8 Given that

$$u_n = \frac{1}{\sqrt{(2n-1)}} - \frac{1}{\sqrt{(2n+1)}}$$

express $\displaystyle\sum_{n=25}^{N} u_n$ in terms of N.

Deduce the value of $\displaystyle\sum_{n=25}^{\infty} u_n$.   (OCR)

9 Show that

$$\frac{r}{(r+1)!} = \frac{1}{r!} - \frac{1}{(r+1)!} \qquad (r \in \mathbb{N})$$

Hence or otherwise, evaluate

i) $\displaystyle\sum_{r=1}^{n} \frac{r}{(r+1)!}$ **ii)** $\displaystyle\sum_{r=1}^{\infty} \frac{r+2}{(r+1)!}$

giving your answer to part **ii** in the terms of e.   (NEAB)

10 a) Show that

$$\frac{r+1}{r+2} - \frac{r}{r+1} \equiv \frac{1}{(r+1)(r+2)} \qquad r \in \mathbb{Z}^+$$

b) Hence, or otherwise, find

$$\sum_{r=1}^{n} \frac{1}{(r+1)(r+2)}$$

giving your answer as a single fraction in terms of n.   (EDEXCEL)

Convergence

As we found in geometric progressions, an infinite series is the sum of an infinite sequence of numbers (see *Introducing Pure Mathematics*, pages 248–50). For example, we have the infinite geometric progression

$$\frac{1}{2} + \frac{1}{2^2} + \ldots + \frac{1}{2^k} + \ldots$$

When we state that an infinite series $\sum\limits_{k=0}^{\infty} a_k$ **converges**, we mean that the sums $S_n = \sum\limits_{k=0}^{n} a_k$ have a **limit** as $n \to \infty$.

We say that an infinite series **diverges** if it does not converge.

When a series diverges, it could behave in one of the following ways.

- Diverge to $+\infty$; for example: $1 + 2 + 4 + 8 + 16 + \ldots$
- Diverge to $-\infty$; for example: $-1 - 2 - 4 - 8 - 16 - \ldots$
- Oscillate finitely; for example: $1 - 1 + 1 - 1 + 1 - \ldots$
- Oscillate infinitely: for example: $1 - 2 + 4 - 8 + 16 - \ldots$

D'Alembert's ratio test

D'Alembert's ratio test states that a series of the form $\sum\limits_{n=0}^{\infty} a_n$ converges when

$$\lim_{n \to \infty} \left| \frac{a_{n+1}}{a_n} \right| < 1$$

The test also states when $\lim\limits_{n \to \infty} \left| \dfrac{a_{n+1}}{a_n} \right|$ is greater than 1, the series diverges.

It does **not** imply anything when $\lim\limits_{n \to \infty} \left| \dfrac{a_{n+1}}{a_n} \right| = 1$.

Example 15 Prove that the series $\sum\limits_{n=0}^{\infty} \dfrac{x^n}{n!}$ converges for all real values of x.

SOLUTION

First, we find the ratio $\left| \dfrac{a_{n+1}}{a_n} \right|$. Then we find its limit as $n \to \infty$.

Hence, we have

$$\left| \frac{a_{n+1}}{a_n} \right| = \left| \frac{\dfrac{x^{n+1}}{(n+1)!}}{\dfrac{x^n}{n!}} \right|$$

$$= \left| \frac{x}{n+1} \right|$$

As $n \to \infty$, this ratio has a limit of zero regardless of the (real) value of x.

Therefore, the ratio test implies that the series converges for all real values of x.

Note The series $\displaystyle\sum_{k=0}^{\infty} \frac{x^k}{k!}$ is Maclaurin's expansion for e^x (see page 178) and is therefore known as the **exponential series**. That is,

$$e^x \equiv \sum_{k=0}^{\infty} \frac{x^k}{k!}$$

Example 16 Prove that the series $\displaystyle\sum_{n=1}^{\infty} \frac{1}{n}$ does not converge.

SOLUTION

Applying d'Alembert's ratio test, we obtain

$$\left| \frac{a_{n+1}}{a_n} \right| = \left| \frac{\dfrac{1}{n+1}}{\dfrac{1}{n}} \right| = \frac{n}{n+1}$$

which gives

$$\lim_{n \to \infty} \left| \frac{a_{n+1}}{a_n} \right| = \lim_{n \to \infty} \frac{n}{n+1} = 1$$

Thus, in this case, d'Alembert's ratio test **fails**, because it does **not** establish whether the series converges or diverges.

To prove that the series does not converge, we write out its first few terms:

$$\sum_{n=1}^{\infty} \frac{1}{n} = 1 + \frac{1}{2} + \frac{1}{3} + \frac{1}{4} + \frac{1}{5} + \frac{1}{6} + \frac{1}{7} + \frac{1}{8} + \dots$$

Now, the first term is greater than $\frac{1}{2}$.

The second term is $\frac{1}{2}$.

The sum of the next two terms is greater than $\frac{1}{4} + \frac{1}{4} = \frac{1}{2}$.

The sum of the next four terms is greater than $\frac{1}{8} + \frac{1}{8} + \frac{1}{8} + \frac{1}{8} = \frac{1}{2}$.

Similarly, the sum of the next eight terms is greater than eight times $\frac{1}{16}$, which is $\frac{1}{2}$.

This pattern keeps repeating. We can always increase the sum by more than $\frac{1}{2}$ by adding the next 2^k terms. Therefore, $\displaystyle\sum_{n=1}^{\infty} \frac{1}{n}$ exceeds any pre-assigned real number L. Hence, it cannot converge to L, and so it diverges.

Even though each term is less than the preceding term, and the terms tend to zero, the sum is not finite.

Maclaurin's series

Assuming that $f(x)$ can be expanded as a series in ascending positive integral powers of x, we can deduce the terms of the series, as shown below for $\sin x$, $\cos x$, e^x and $\ln(1 + x)$. These four expansions are needed frequently and therefore should be known.

Power series for $\sin x$

Let $\sin x = a_0 + a_1 x + a_2 x^2 + a_3 x^3 + \ldots$, where the a's are constants.

When $x = 0$, $\sin 0 = a_0$. But $\sin 0 = 0$, therefore $a_0 = 0$.

Differentiating $\sin x = a_1 x + a_2 x^2 + a_3 x^3 + \ldots$, we obtain

$$\cos x = a_1 + 2a_2 x + 3a_3 x^2 + 4a_4 x^3 + \ldots$$

When $x = 0$, $\cos 0 = a_1$. But $\cos 0 = 1$, therefore $a_1 = 1$.

Differentiating again, we obtain

$$-\sin x = 2a_2 + 3 \times 2a_3 x + 4 \times 3a_4 x^2 + 5 \times 4a_5 x^3 + \ldots$$

When $x = 0$, $\sin 0 = 2a_2 \Rightarrow a_2 = 0$.

Differentiating yet again, we obtain

$$-\cos x = 3 \times 2a_3 + 4 \times 3 \times 2a_4 x + 5 \times 4 \times 3a_5 x^2 + \ldots$$

When $x = 0$, $-\cos 0 = 3 \times 2a_3 \Rightarrow a_3 = -\dfrac{1}{3 \times 2 \times 1} = -\dfrac{1}{3!}$.

Repeating the differentiation, we obtain

$$a_4 = 0 \qquad a_5 = \frac{1}{5!} \qquad a_6 = 0 \qquad a_7 = -\frac{1}{7!}$$

Therefore, we have

$$\sin x = x - \frac{x^3}{3!} + \frac{x^5}{5!} - \frac{x^7}{7!} + \ldots + \frac{(-1)^n x^{2n+1}}{(2n+1)!} + \ldots$$

By d'Alembert's ratio test, this series converges for all real x.

Power series for $\cos x$

We can use the procedure for $\sin x$ to find the power series for $\cos x$. However, it is much easier to start from the expansion for $\sin x$. Hence, we have

$$\cos x = \frac{\mathrm{d}}{\mathrm{d}x} \sin x = \frac{\mathrm{d}}{\mathrm{d}x}\left(x - \frac{x^3}{3!} + \frac{x^5}{5!} - \frac{x^7}{7!} + \ldots \right)$$

which gives

$$\cos x = 1 - \frac{x^2}{2!} + \frac{x^4}{4!} - \frac{x^6}{6!} + \ldots + \frac{(-1)^n x^{2n}}{(2n)!} + \ldots$$

By d'Alembert's ratio test, this series is convergent for all real x.

Power series for e^x

Let $e^x = a_0 + a_1 x + a_2 x^2 + a_3 x^3 + \ldots$, where the a's are constants.

When $x = 0$, $e^0 = a_0$. But $e^0 = 1$, therefore $a_0 = 1$.

Differentiating $e^x = a_1 x + a_2 x^2 + a_3 x^3 + \ldots$, we obtain

$$e^x = a_1 + 2a_2 x + 3a_3 x^2 + 4a_4 x^3 + \ldots$$

When $x = 0$, $e^0 = a_1 \implies a_1 = 1$.

Differentiating again, we obtain

$$e^x = 2a_2 + 3 \times 2a_3 x + 4 \times 3a_4 x^2 + 5 \times 4a_5 x^3 + \ldots$$

When $x = 0$, $e^0 = 2a_2 \implies a_2 = \frac{1}{2}$.

Differentiating yet again, we obtain

$$e^x = 3 \times 2a_3 + 4 \times 3 \times 2a_4 x + 5 \times 4 \times 3a_5 x^2 + \ldots$$

When $x = 0$, $e^0 = 3 \times 2a_3 \implies a_3 = \dfrac{1}{3 \times 2 \times 1} = \dfrac{1}{3!}$.

Repeating the differentiation, we obtain

$$a_4 = \frac{1}{4!} \qquad a_5 = \frac{1}{5!} \qquad a_6 = \frac{1}{6!} \qquad a_7 = \frac{1}{7!}$$

Therefore, we have

$$e^x = 1 + x + \frac{x^2}{2!} + \frac{x^3}{3!} + \frac{x^4}{4!} + \frac{x^5}{5!} + \ldots$$

By d'Alembert's ratio test, this series converges for all real x.

Power series for $\ln(1 + x)$

Since $\ln 0$ is not finite, we cannot have a power series for $\ln x$. Instead, we use a power series for $\ln(1 + x)$.

Let $\ln(1 + x) = a_0 + a_1 x + a_2 x^2 + a_3 x^3 + \ldots$

When $x = 0$, $\ln 1 = a_0$. But $\log 1 = 0$, therefore $a_0 = 0$.

Differentiating $\ln(1 + x) = a_1 x + a_2 x^2 + a_3 x^3 + \ldots$, we obtain

$$\frac{1}{1 + x} = a_1 + 2a_2 x + 3a_3 x^2 + 4a_4 x^3 + \ldots$$

However, using the binomial theorem, we can expand $\dfrac{1}{1 + x}$ as $(1 + x)^{-1}$ to give $1 - x + x^2 - x^3 + x^4 - x^5 + \ldots$. Hence, we have

$$1 - x + x^2 - x^3 + x^4 - x^5 + \ldots \equiv a_1 + 2a_2 x + 3a_3 x^2 + 4a_4 x^3 + \ldots$$

Equating coefficients, we obtain $a_1 = 1$, $a_2 = -\frac{1}{2}$, $a_3 = \frac{1}{3}$, $a_4 = -\frac{1}{4}$, ...

Therefore, we have

$$\ln(1 + x) = x - \frac{x^2}{2} + \frac{x^3}{3} - \frac{x^4}{4} + \frac{x^5}{5} - \ldots$$

Using d'Alembert's ratio test, we obtain

$$\lim_{n \to \infty} \left| \frac{a_{n+1}}{a_n} \right| = \lim_{n \to \infty} \left| \frac{\frac{x^{n+1}}{n+1}}{\frac{x^n}{n}} \right| = \lim_{n \to \infty} \left| \frac{nx}{n+1} \right| = |x|$$

Thus, when $|x| < 1$, the series converges. By inspection, we notice that the expansion is valid when $x = 1$, but not when $x = -1$. Hence, we have

$$\ln(1 + x) = x - \frac{x^2}{2} + \frac{x^3}{3} - \frac{x^4}{4} + \frac{x^5}{5} - \ldots \quad \text{for } -1 < x \leqslant 1$$

Similarly, we have

$$\ln(1 - x) = -x - \frac{x^2}{2} - \frac{x^3}{3} - \frac{x^4}{4} - \frac{x^5}{5} - \ldots \quad \text{for } -1 \leqslant x < 1$$

Summary

The general result of this method for obtaining the power series of functions is known as **Maclaurin's series**, and is expressed as

$$f(x) = f(0) + xf'(0) + \frac{x^2}{2!} f''(0) + \frac{x^3}{3!} f'''(0) + \ldots$$

Exercise 9D

1 a) Show that the first two non-zero terms in the Maclaurin expansion of $\sin^{-1} x$ are given by

$$\sin^{-1} x = x + \frac{x^3}{6} + \ldots$$

b) By writing $x = \frac{1}{2}$, deduce an approximation to π as a rational fraction in its lowest terms.

c) The equation $\sin^{-1} x = 1.002x$ is satisfied by a small positive value of x. Find an approximation to this value, giving your answer correct to three decimal places. (WJEC)

2 i) Use Maclaurin's theorem to derive the series expansion for $\log_e(1 + x)$, where $-1 < x \leqslant 1$, giving the first three non-zero terms.

ii) If $\log_e(1 + x) \approx x(1 + ax)^b$ for small x, find the values of a and b so that the first three non-zero terms of the series expansions of the two sides agree. (NICCEA)

3 a) Find the first three derivatives of $(1 + x)^2 \cos x$.

b) Hence, or otherwise, find the expansion of $(1 + x)^2 \cos x$ in ascending powers of x up to and including the term in x^3. (EDEXCEL)

4 i) Use Maclaurin's theorem to derive the first three non-zero terms of the series expansion for $\sin x$.

ii) Show that, for sufficiently small x,

$$\sin x \approx x \left(1 - \frac{x^2}{15} \right)^{\frac{5}{2}}$$

iii) Show that when $x = \dfrac{\pi}{2}$ the error in using the approximation in part **ii** is about 0.2%.

(NICCEA)

5 Show that the first two non-zero terms of the Maclaurin series for $\ln(1 + x)$ are given by

$$\ln(1 + x) = x - \frac{x^2}{2} + \dots$$

a) Use the series to show that the equation $3 \ln(1 + x) = 100x^2$ has an approximate solution $x = 0.03$.

b) Taking $x = 0.03$ as a first approximation, obtain an improved value of the root by two applications of the Newton–Raphson method. Give your answer correct to six decimal places. (WJEC)

$$x_{n+1} = x_n - \frac{f(x_n)}{f'(x_n)}$$

6 Given that $y = (1 + \sin x)e^x$, find $\dfrac{dy}{dx}$ and show that $\dfrac{d^2y}{dx^2} = (1 + 2\cos x)e^x$.

Hence, or otherwise, prove that the Maclaurin series for y, in ascending powers of x, up to and including the term in x^2 is

$$1 + 2x + \frac{3}{2}x^2$$

The binomial expansion of $(1 + ax)^n$ also begins $1 + 2x + \dfrac{3}{2}x^2$. Find the value of the constants a and n. (AEB 97)

7 i) Use Maclaurin's theorem to find the values of A, B, C and D in the series expansion

$$\tan^{-1}x = A + Bx + Cx^2 + Dx^3 + \frac{x^5}{5} - \frac{x^7}{7} \dots$$

where $-1 < x < 1$.

ii) Find, using the binomial expansion, the first three non-zero terms of the series expansion, in ascending powers of u, for $\dfrac{1}{1 + u^2}$.

iii) Using the series in part **ii**, evaluate

$$\int_0^x \frac{1}{1 + u^2}\, du$$

as a series expansion in ascending powers of x.

iv) Explain briefly how the series expansion in part **i** can be derived from the result in part **iii**. (NICCEA)

8 Given that

$$y^2 = \sec x + \tan x \qquad -\frac{\pi}{2} < x < \frac{\pi}{2}, \quad y > 0$$

show that

a) $\dfrac{dy}{dx} = \dfrac{1}{2} y \sec x$

b) $\dfrac{d^2 y}{dx^2} = \dfrac{1}{4} y \sec x (\sec x + 2 \tan x)$

Given that x is small and that terms in x^3 and higher powers of x may be neglected, use Maclaurin's expansion to express y in the form $A + Bx + Cx^2$, stating the values of A, B and C.
(EDEXCEL)

9 Given that $f(x) = (1 + x) \ln (1 + x)$,

a) find the fifth derivative of $f(x)$

b) show that the first five non-zero terms in the Maclaurin expansion for $f(x)$ are

$$x + \frac{x^2}{2} - \frac{x^3}{6} + \frac{x^4}{12} - \frac{x^5}{20}$$

c) find, in terms of r, an expression for the rth term ($r \geqslant 2$) of the Maclaurin expansion for $f(x)$. (WJEC)

10 a) i) Given that $y = \ln(2 + x^2)$, find $\dfrac{dy}{dx}$ and show that

$$\frac{d^2 y}{dx^2} = \frac{4 - 2x^2}{(2 + x^2)^2}$$

ii) Deduce the Maclaurin series for $\ln(2 + x^2)$ in ascending powers of x, up to and including the term in x^2.

b) By writing $2 + x^2$ as $2(1 + \frac{1}{2} x^2)$ and using the series expansion

$$\ln(1 + t) = t - \frac{t^2}{2} + \frac{t^3}{3} - \dots$$

verify your result from part **a** and determine the next non-zero term in the series for $\ln(2 + x^2)$. (AEB 97)

11 i) Use Maclaurin's theorem to derive the first five terms of the series expansion for $(1 + x)^r$, where $-1 < x < 1$.

ii) Assuming that the series, obtained above, continues with the same pattern, sum the following infinite series

$$1 + \frac{1}{6} - \frac{1.2}{6.12} + \frac{1.2.5}{6.12.18} - \frac{1.2.5.8}{6.12.18.24} + \dots \qquad \text{(NICCEA)}$$

12 i) Use Maclaurin's theorem to derive the first five terms of the series expansion for e^x.

Consider the infinite series

$$\frac{1}{1!} + \frac{4}{2!} + \frac{7}{3!} + \frac{10}{4!} + \dots$$

ii) If the series continues with the same pattern, find an expression for the nth term.

iii) Sum the infinite series. (NICCEA)

Using power series

The series studied on pages 177–9 are used in a number of situations, including the two which are discussed below

Finding the limit of $\dfrac{f(x)}{g(x)}$ as $x \to 0$, when $f(0) = g(0) = 0$

If we simply insert $x = 0$, we obtain $\dfrac{f(0)}{g(0)} = \dfrac{0}{0}$, which means that we have proceeded incorrectly.

Example 17 Find the limit of $\dfrac{x - \sin x}{x^2(e^x - 1)}$ as $x \to 0$.

SOLUTION

To find such a limit, we expand the numerator and the denominator of the expression each as a power series in x and divide both by the lowest power of x present. Then we put $x = 0$.

Hence, we have

$$\frac{x - \sin x}{x^2(e^x - 1)} = \frac{x - \left(x - \dfrac{x^3}{3!} + \dfrac{x^5}{5!} - \dots\right)}{x^2\left(1 + x + \dfrac{x^2}{2!} + \dots - 1\right)} = \frac{\dfrac{x^3}{3!} - \dfrac{x^5}{5!} + \dots}{x^3 + \dfrac{x^4}{2!} + \dots}$$

Dividing the numerator and the denominator by x^3, we obtain

$$\frac{x - \sin x}{x^2(e^x - 1)} = \frac{\dfrac{1}{3!} - \dfrac{x^2}{5!} + \dots}{1 + \dfrac{x}{2!} + \dfrac{x^2}{3!} \dots}$$

Therefore, we have

$$\lim_{x \to 0} \frac{x - \sin x}{x^2(e^x - 1)} = \frac{\dfrac{1}{3!}}{1} = \frac{1}{6}$$

Example 18 Find the limit of $\dfrac{1 - \cos x}{\sin^2 x}$ as $x \to 0$.

SOLUTION

Expanding the numerator and the denominator each as a power series, we obtain

$$\frac{1 - \cos x}{\sin^2 x} = \frac{1 - \left(1 - \dfrac{x^2}{2!} + \dfrac{x^4}{4!} - \dots\right)}{\left(x - \dfrac{x^3}{3!} + \dfrac{x^5}{5!} - \dots\right)^2} = \frac{\dfrac{x^2}{2!} - \dfrac{x^4}{4!} + \dots}{x^2 - \dfrac{2x^4}{3!} + \dots}$$

Dividing the numerator and the denominator by x^2, we obtain

$$\frac{1 - \cos x}{\sin^2 x} = \frac{\dfrac{1}{2!} - \dfrac{x^2}{4!} + \dots}{1 - \dfrac{2x^2}{3!} + \dots}$$

Therefore, we have

$$\lim_{x \to 0} \frac{1 - \cos x}{\sin^2 x} = \frac{\dfrac{1}{2!}}{1} = \frac{1}{2}$$

L'Hôpital's rule

When evaluating the limits of some forms of $\dfrac{f(x)}{g(x)}$, the use of power series is not appropriate and so we apply l'Hôpital's rule, which states that if $f(a) = g(a) = 0$, and $g'(a) \neq 0$, then

$$\lim_{x \to a} \frac{f(x)}{g(x)} = \frac{f'(a)}{g'(a)}$$

If $g'(a) = 0$, we repeat the procedure until we find a derivative of $g(x)$ which is not zero when $x = a$.

Thus, if $f(a) = g(a) = 0$ and $g'(a) = 0$, but $g''(a) \neq 0$, then

$$\lim_{x \to a} \frac{f(x)}{g(x)} = \lim_{x \to a} \frac{f'(x)}{g'(x)} = \frac{f''(a)}{g''(a)}$$

Example 19 Find $\displaystyle\lim_{x \to 1} \frac{x^4 - 7x^3 + 8x^2 - 2}{x^3 + 5x - 6}$.

SOLUTION

We notice that both the numerator and the denominator are zero when $x = 1$. Hence, we have, after differentiating both the numerator and the denominator,

$$\lim_{x \to 1} \frac{x^4 - 7x^3 + 8x^2 - 2}{x^3 + 5x - 6} = \lim_{x \to 1} \frac{4x^3 - 21x^2 + 16x}{3x^2 + 5}$$

$$\Rightarrow \quad \lim_{x \to 1} \frac{x^4 - 7x^3 + 8x^2 - 2}{x^3 + 5x - 6} = -\frac{1}{8} = -0.125$$

Finding f(x) for small x

Example 20 Expand $\tan x$ as a power series in x as far as a term in x^5. Hence find the value of $\tan 0.001$ to 15 decimal places.

SOLUTION

We express $\tan x$ in terms of $\sin x$ and $\cos x$, and expand each as a power series. Hence, we have

$$\tan x = \frac{\sin x}{\cos x} = \frac{x - \dfrac{x^3}{3!} + \dfrac{x^5}{5!} - \ldots}{1 - \dfrac{x^2}{2!} + \dfrac{x^4}{4!} - \ldots}$$

$$= \frac{x - \dfrac{x^3}{6} + \dfrac{x^5}{120} - \ldots}{1 - \dfrac{x^2}{2} + \dfrac{x^4}{24} - \ldots}$$

We rearrange the above to give

$$\tan x = \left[x - \frac{x^3}{6} + \frac{x^5}{120} - \ldots\right]\left[1 - \left(\frac{x^2}{2} - \frac{x^4}{24} + \ldots\right)\right]^{-1}$$

and then we expand the second bracket, using the binomial theorem and ignoring terms in x^5 and higher, to obtain

$$\tan x = \left[x - \frac{x^3}{6} + \frac{x^5}{120} - \ldots\right]\left[1 + \frac{x^2}{2} - \frac{x^4}{24} + \ldots + \left(\frac{x^2}{2} - \frac{x^4}{24} + \ldots\right)^2 + \ldots\right]$$

$$= \left(x - \frac{x^3}{6} + \frac{x^5}{120} - \ldots\right)\left(1 + \frac{x^2}{2} - \frac{x^4}{24} + \frac{x^4}{4} - \ldots\right)$$

$$= \left(x - \frac{x^3}{6} + \frac{x^5}{120} - \ldots\right)\left(1 + \frac{x^2}{2} + \frac{5x^4}{24} + \ldots\right)$$

$$= x + \frac{x^3}{2} + \frac{5x^5}{24} - \frac{x^3}{6} - \frac{x^5}{12} + \frac{x^5}{120} + \ldots$$

Therefore, we have

$$\tan x = x + \frac{1}{3}x^3 + \frac{2}{15}x^5 + \ldots$$

Hence, $\tan 0.001$ is given by

$$\tan 0.001 = 0.001 + \frac{1}{3} \times 0.000\,000\,001 + \frac{2}{15} \times 0.000\,000\,000\,000\,001 + \ldots$$

That is,

$$\tan 0.001 = 0.001\,000\,000\,333\,333 \quad \text{to 15 dp}$$

Power series for more complicated functions

We can combine power series for simple functions to make power series for more complicated functions, as demonstrated in Examples 21 to 24.

Example 21 Find the power series for $\cos x^2$.

SOLUTION

The power series for $\cos x$ is

$$\cos x = 1 - \frac{x^2}{2!} + \frac{x^4}{4!} - \ldots + \frac{(-1)^n}{(2n)!} x^{2n} + \ldots$$

To obtain the power series for $\cos x^2$, we replace every x in the above series with x^2 to obtain

$$\cos x^2 = 1 - \frac{(x^2)^2}{2!} + \frac{(x^2)^4}{4!} - \ldots + \frac{(-1)^n}{(2n)!} (x^2)^{2n} + \ldots$$

$$= 1 - \frac{x^4}{2!} + \frac{x^8}{4!} - \ldots + \frac{(-1)^n}{(2n)!} x^{4n} + \ldots$$

Since the power series for $\cos x$ is valid for all real values of x, we know that the power series for $\cos x^2$ is valid for all values of x^2, i.e. for all real values of x.

Example 22 Find the power series for $\ln(1 + 3x)$, stating when the expansion is valid.

SOLUTION

In the expansion for $\ln(1 + x)$,

$$\ln(1 + x) = x - \frac{x^2}{2} + \frac{x^3}{3} - \ldots$$

we substitute $3x$ for x, which gives

$$\ln(1 + 3x) = (3x) - \frac{(3x)^2}{2} + \frac{(3x)^3}{3} -$$

$$= 3x - \frac{9}{2}x^2 + 9x^3 - \ldots$$

Since the expansion for $\ln(1 + x)$ is valid for $-1 < x \leqslant 1$, the expansion for $\ln(1 + 3x)$ is valid for $-1 < 3x \leqslant 1$, i.e. $-\frac{1}{3} < x \leqslant \frac{1}{3}$.

Therefore, we have

$$\ln(1 + 3x) = 3x - \frac{9}{2}x^2 + 9x^3 - \ldots \quad \text{for } -\frac{1}{3} < x \leqslant \frac{1}{3}$$

Example 23 Find the power series for $e^{4x}\sin 3x$, up to and including the term in x^4.

SOLUTION

Since we are asked for terms only up to x^4, we do not need to consider terms in higher powers of x.

The power series for e^x is

$$e^x = 1 + x + \frac{x^2}{2!} + \frac{x^3}{3!} + \frac{x^4}{4!} + \cdots$$

Therefore, the power series for e^{4x} is

$$e^{4x} = 1 + (4x) + \frac{(4x)^2}{2!} + \frac{(4x)^3}{3!} + \frac{(4x)^4}{4!} + \cdots$$

Similarly, using the power series for $\sin x$, and replacing x with $3x$, we obtain the power series expansion for $\sin 3x$:

$$\sin 3x = (3x) - \frac{(3x)^3}{3!} + \frac{(3x)^5}{5!} - \cdots$$

Therefore, the power series for $e^{4x}\sin 3x$ is

$$e^{4x}\sin 3x = \left[1 + (4x) + \frac{(4x)^2}{2!} + \frac{(4x)^3}{3!} + \frac{(4x)^4}{4!} + \cdots\right]\left[(3x) - \frac{(3x)^3}{3!} + \frac{(3x)^5}{5!} - \cdots\right]$$

$$= \left(1 + 4x + 8x^2 + \frac{32}{3}x^3 + \frac{32}{3}x^4 + \cdots\right)\left(3x - \frac{9}{2}x^3 + \cdots\right)$$

Ignoring terms in x^5 and higher powers, we obtain

$$e^{4x}\sin 3x = 3x + 12x^2 + 24x^3 - \frac{9}{2}x^3 + 32x^4 - 18x^4$$

Therefore, we have

$$e^{4x}\sin 3x = 3x + 12x^2 + \frac{39}{2}x^3 + 14x^4$$

Example 24 Find all the terms up to and including x^4 in the power series for $e^{\sin x}$.

SOLUTION

Using the power series for e^x, we obtain

$$e^{\sin x} = 1 + \frac{\sin x}{1!} + \frac{\sin^2 x}{2!} + \frac{\sin^3 x}{3!} + \cdots$$

We now apply the power series for $\sin x$. Since we are asked for terms only up to x^4, we can ignore terms in higher powers of x. Therefore, we have

$$e^{\sin x} = 1 + \frac{x - \frac{x^3}{3!} + \cdots}{1!} + \frac{\left(x - \frac{x^3}{3!} + \cdots\right)^2}{2!} + \frac{\left(x - \frac{x^3}{3!} + \cdots\right)^3}{3!} + \cdots$$

$$\Rightarrow \quad e^{\sin x} = 1 + x - \frac{x^3}{3!} + \frac{x^2 - \dfrac{2x^4}{3!}}{2!} + \frac{x^3}{3!} + \frac{x^4}{4!} + \cdots$$

which gives

$$e^{\sin x} = 1 + x + \frac{x^2}{2} - \frac{x^4}{8}$$

Exercise 9E

1 Find the power series of each of the following.

a) $\sin 2x$ **b)** $\cos 5x$ **c)** e^{8x}

d) $\ln(1 + x^2)$ **e)** $\ln(1 - 2x)$

2 Find the power series of each of the following, up to and including the term in x^4.

a) $\sin x^2$ **b)** $(1 + x)e^{3x}$ **c)** $(2 + x^2)\cos 3x$

d) $e^{\cos x}$ **e)** $\ln(1 + \cos x)$

3 Find out whether the following infinite series converge or diverge.

a) $\displaystyle\sum_{n=1}^{\infty} \frac{5^n}{n!}$ **b)** $\displaystyle\sum_{n=2}^{\infty} \frac{1}{2^n - 1}$ **c)** $\displaystyle\sum_{n=1}^{\infty} \frac{n^2}{2^n}$

4 Find the power series expansion of $\cos x^3$. Which values of x is this valid for?

5 Find the power series expansion of e^{2x^2}

6 You are told that $y = \displaystyle\sum_{n=0}^{\infty} \frac{nx^n}{3^n}$. When does this series converge?

7 Given that $|x| < 4$, find, in ascending powers of x up to and including the term in x^3, the series expansion of

a) $(4 - x)^{\frac{1}{2}}$ **b)** $(4 - x)^{\frac{1}{2}} \sin 3x$ (EDEXCEL)

8 a) Find the first four terms of the expansion, in ascending powers of x, of

$$(2 + 3x)^{-1} \qquad |x| < \tfrac{2}{3}$$

b) Hence, or otherwise, find the first four non-zero terms of the expansion, in ascending powers of x, of

$$\frac{\sin 2x}{2 + 3x} \qquad |x| < \tfrac{2}{3} \qquad \text{(EDEXCEL)}$$

9 $$\cos\left(2x + \frac{\pi}{3}\right) \equiv p\cos 2x + q\sin 2x$$

a) Find the exact values of the constants p and q.

Given that x is so small that terms in x^3 and higher powers of x are negligible,

b) show that $\cos\left(2x + \dfrac{\pi}{3}\right) = \dfrac{1}{2} - \sqrt{3}x - x^2$.   (EDEXCEL)

10 The function f is defined by

$$f(x) = e^{ax} - (1 + bx)^{\frac{1}{3}}$$

where a and b are positive constants and $|bx| < 1$.

a) Find, in terms of a and b, the coefficients of x, x^2 and x^3 in the expansion of $f(x)$ in ascending powers of x.

b) Given that the coefficient of x is zero and that the coefficient of x^2 is $\dfrac{3}{2}$,

   i) find the values of a and b

   ii) show that the coefficient of x^3 is $-\dfrac{3}{2}$.   (NEAB)

10 Hyperbolic functions

In the 1760s Johann Heinrich Lambert gave a very nice presentation in terms of the parametrization of the hyperbola, by analogy with such a treatment of the sine and cosine on the circle.
IVOR GRATTAN-GUINNESS

Definitions

The hyperbolic functions, of which there are six, are so named because they are related to the parametric equations for a hyperbola.

We begin with the two functions hyperbolic sine of x and hyperbolic cosine of x, which are written

$$\sinh x \quad \text{and} \quad \cosh x$$

They are defined by the relationships

$$\sinh x = \frac{1}{2}(e^x - e^{-x})$$

$$\cosh x = \frac{1}{2}(e^x + e^{-x})$$

In a similar manner to ordinary trigonometric functions, we have

$$\tanh x = \frac{\sinh x}{\cosh x} = \frac{e^x - e^{-x}}{e^x + e^{-x}}$$

$$\operatorname{cosech} x = \frac{1}{\sinh x}$$

$$\operatorname{sech} x = \frac{1}{\cosh x}$$

$$\coth x = \frac{1}{\tanh x}$$

By convention, we pronounce sinh as 'shine', tanh as 'than', (co)sech as '(co)sheck' and coth as 'coth'.

Example 1 Find **a)** $\sinh 2$ and **b)** $\operatorname{sech} 3$.

SOLUTION

a) Usually, you would use a calculator to find sinh values. Not all calculators operate in the same way, so you must first consult your calculator instructions to learn the **correct order** in which to press the hyperbolic (hyp) key, the sin key and, in this case, the 2 key. Your answer should be 3.6268...

Otherwise, you would have to evaluate sinh 2 using the relationship

$$\sinh 2 = \frac{1}{2}(e^2 - e^{-2})$$

and putting in the values of e^2 and e^{-2}, which you either obtain from tables or calculate from the exponential series.

b) Again, you would normally use a calculator with the relationship

$$\text{sech } 3 = \frac{1}{\cosh 3} = \frac{1}{10.0677} \quad \text{(to 4 dp)}$$

Therefore, sech 3 = 0.0993, to four decimal places.

Graphs of cosh x, sinh x and tanh x

$y = \cosh x$

We obtain the graph of $y = \cosh x$ (shown on the right) by finding the mean values of a few corresponding pairs of values of $y = e^x$ and $y = e^{-x}$, and then plotting these mean values.

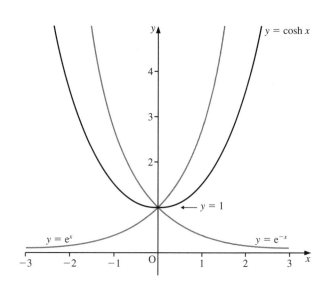

$y = \sinh x$

To produce the graph of $y = \sinh x$ (shown on the right), we find half the difference between a few corresponding pairs of values of $y = e^x$ and $y = e^{-x}$, and then plot these values.

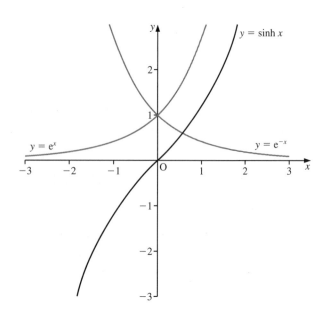

$y = \tanh x$

We have $\tanh x = \dfrac{\sinh x}{\cosh x}$, which gives

$$\tanh x = \frac{(e^x - e^{-x})}{(e^x + e^{-x})}$$

$$\Rightarrow \quad \tanh x = \frac{1 - e^{-2x}}{1 + e^{-2x}}$$

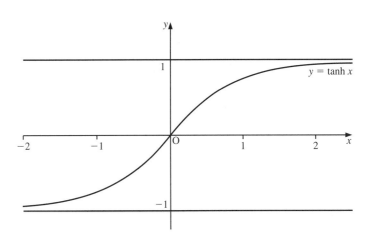

Therefore, $\tanh x < 1$ for all values of x, and as $x \to +\infty$, $\tanh x \to 1$.

Since $\tanh x = -\dfrac{1 - e^{2x}}{1 + e^{2x}}$, $\tanh x > -1$ for all values of x, and as $x \to -\infty$, $\tanh x \to -1$.

Hence, the graph of $y = \tanh x$ lies between the asymptotes $y = 1$ and $y = -1$.

Standard hyperbolic identities

From the exponential definitions for $\cosh x$ and $\sinh x$, we have

$$\cosh^2 x = \left[\frac{1}{2}(e^x + e^{-x})\right]^2$$

$$= \frac{1}{4}(e^{2x} + 2 + e^{-2x}) \qquad [1]$$

and

$$\sinh^2 x = \left[\frac{1}{2}(e^x - e^{-x})\right]^2$$

$$= \frac{1}{4}(e^{2x} - 2 + e^{-2x}) \qquad [2]$$

Hence, subtracting [2] from [1], we obtain

$$\cosh^2 x - \sinh^2 x = \frac{1}{4}(e^{2x} + 2 + e^{-2x}) - \frac{1}{4}(e^{2x} - 2 + e^{-2x}) = 1$$

Therefore, we have

$$\cosh^2 x - \sinh^2 x \equiv 1$$

Notice the similarity of this hyperbolic identity with the usual trigonometric identity $\cos^2 x + \sin^2 x \equiv 1$. See page 213 for Osborn's rule, which will help you to recall the standard hyperbolic identities.

Dividing $\cosh^2 x - \sinh^2 x \equiv 1$ by $\sinh^2 x$, we obtain

$$\frac{\cosh^2 x}{\sinh^2 x} - \frac{\sinh^2 x}{\sinh^2 x} \equiv \frac{1}{\sinh^2 x}$$

which gives

$$\coth^2 x - 1 \equiv \operatorname{cosech}^2 x$$

Similarly, dividing $\cosh^2 x - \sinh^2 x \equiv 1$ by $\cosh^2 x$, we obtain

$$\frac{\cosh^2 x}{\cosh^2 x} - \frac{\sinh^2 x}{\cosh^2 x} \equiv \frac{1}{\cosh^2 x}$$

which gives

$$1 - \tanh^2 x \equiv \operatorname{sech}^2 x$$

Differentiation of hyperbolic functions

To differentiate $\sinh x$ and $\cosh x$, we use their exponential definitions. Hence, for $\sinh x$, we have

$$\frac{\mathrm{d}}{\mathrm{d}x} \sinh x = \frac{\mathrm{d}}{\mathrm{d}x} \left[\frac{1}{2}(e^x - e^{-x}) \right] = \frac{1}{2}(e^x + e^{-x})$$

From the definitions, we know that

$$\frac{1}{2}(e^x + e^{-x}) = \cosh x$$

Therefore, we have

$$\frac{\mathrm{d}}{\mathrm{d}x} \sinh x = \cosh x$$

$$\frac{\mathrm{d}}{\mathrm{d}x} \cosh x = \frac{\mathrm{d}}{\mathrm{d}x} \left[\frac{1}{2}(e^x + e^{-x}) \right] = \frac{1}{2}(e^x - e^{-x})$$

From the definitions, we know that

$$\frac{1}{2}(e^x - e^{-x}) = \sinh x$$

Therefore, we have

$$\frac{\mathrm{d}}{\mathrm{d}x} \cosh x = \sinh x$$

To differentiate $\tanh x$, we use the identity

$$\tanh x \equiv \frac{\sinh x}{\cosh x}$$

which gives

$$\frac{\mathrm{d}}{\mathrm{d}x} \tanh x = \frac{\mathrm{d}}{\mathrm{d}x} \frac{\sinh x}{\cosh x}$$

$$= \frac{\cosh x \cosh x - \sinh x \sinh x}{\cosh^2 x} \quad \text{(using the quotient rule)}$$

$$= \frac{\cosh^2 x - \sinh^2 x}{\cosh^2 x}$$

$$= \frac{1}{\cosh^2 x} = \operatorname{sech}^2 x$$

Therefore, we have

$$\frac{\mathrm{d}}{\mathrm{d}x}\tanh x = \mathrm{sech}^2 x$$

To differentiate functions such as $\cosh ax$, again we use the exponential definitions. Hence, we have

$$\frac{\mathrm{d}}{\mathrm{d}x}\cosh ax = \frac{\mathrm{d}}{\mathrm{d}x}\left[\frac{1}{2}(\mathrm{e}^{ax} + \mathrm{e}^{-ax})\right]$$

$$= \frac{1}{2}(a\mathrm{e}^{ax} - a\mathrm{e}^{-ax})$$

From the exponential definitions, we note that

$$a\left[\frac{1}{2}(\mathrm{e}^{ax} - \mathrm{e}^{-ax})\right] = a\sinh ax$$

Therefore, we have

$$\frac{\mathrm{d}}{\mathrm{d}x}\cosh ax = a\sinh ax$$

Similarly, we have

$$\frac{\mathrm{d}}{\mathrm{d}x}\sinh ax = a\cosh ax$$

$$\frac{\mathrm{d}}{\mathrm{d}x}\tanh ax = a\,\mathrm{sech}^2 ax$$

Example 2 Find $\dfrac{\mathrm{d}y}{\mathrm{d}x}$ when $y = 3\cosh 3x + 5\sinh 4x + 2\cosh^4 7x$.

SOLUTION

To differentiate $\cosh^4 7x$, we express it as $(\cosh 7x)^4$ and apply the chain rule. Hence, we have

$$\frac{\mathrm{d}y}{\mathrm{d}x} = 9\sinh 3x + 20\cosh 4x + 2 \times 4 \times 7\sinh 7x\cosh^3 7x$$

$$= 9\sinh 3x + 20\cosh 4x + 56\sinh 7x\cosh^3 7x$$

Integration of hyperbolic functions

From the differentiation formulae given on pages 192–3, we deduce that

$$\int \cosh ax\,\mathrm{d}x = \frac{1}{a}\sinh ax + c$$

$$\int \sinh ax\,\mathrm{d}x = \frac{1}{a}\cosh ax + c$$

$$\int \mathrm{sech}^2 ax\,\mathrm{d}x = \frac{1}{a}\tanh ax + c$$

Example 3 Find $\displaystyle\int (2\sinh 4x + 9\,\mathrm{sech}^2 3x)\,\mathrm{d}x$.

SOLUTION

Splitting the given integral into two parts, we obtain

$$\int 2\sinh 4x\,\mathrm{d}x + \int 9\,\mathrm{sech}^2 3x\,\mathrm{d}x = \frac{2}{4}\cosh 4x + \frac{9}{3}\tanh 3x + c$$

$$= \tfrac{1}{2}\cosh 4x + 3\tanh 3x + c$$

Inverse hyperbolic functions

We define the inverses of the hyperbolic functions in a similar way to the inverses of the ordinary trigonometric functions. Hence, for example, if $y = \sinh^{-1}x$, then $\sinh y = x$. Likewise for $\cosh^{-1}x$, $\tanh^{-1}x$, $\mathrm{cosech}^{-1}x$, $\mathrm{sech}^{-1}x$ and $\coth^{-1}x$.

Sometimes, these functions are written as arsinh x, arcosh x etc.

Sketching inverse hyperbolic functions

The curve of $y = \sinh^{-1}x$ is obtained by reflecting the curve of $y = \sinh x$ in the line $y = x$.

To draw the curve with reasonable accuracy, we need to find the gradient of $y = \sinh x$ at the origin. Accordingly, we differentiate $y = \sinh x$, to obtain

$$\frac{\mathrm{d}y}{\mathrm{d}x} = \cosh x$$

Thus, at the origin, where $x = 0$, we have

$$\frac{\mathrm{d}y}{\mathrm{d}x} = \cosh 0 = 1$$

That is, the gradient of $y = \sinh x$ at the origin is 1.

We now proceed as follows:

- Draw the line $y = x$ as a dashed line.
- Sketch carefully the graph of $y = \sinh x$, remembering that $y = x$ is a tangent to $y = \sinh x$ at the origin.
- Reflect this sinh curve in the line $y = x$.

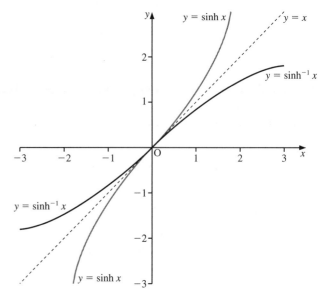

Similarly, we can sketch any other inverse hyperbolic function: that is, by reflecting the curve of the relevant hyperbolic function in the line $y = x$. In each case, we must find the gradient of the hyperbolic curve at the origin.

Take, for example, $y = \tanh x$, which gives

$$\frac{dy}{dx} = \text{sech}^2 x$$

At the origin, where $x = 0$, we have

$$\frac{dy}{dx} = \text{sech}^2 0 = \frac{1}{\cosh^2 0} = 1$$

That is, the gradient of $y = \tanh x$ at the origin is 1.

Also, we know that $y = \tanh x$ has asymptotes $y = 1$ and $y = -1$. Therefore, because $y = \tanh^{-1} x$ is the reflection of $y = \tanh x$ in $y = x$, $y = \tanh^{-1} x$ has asymptotes $x = 1$ and $x = -1$.

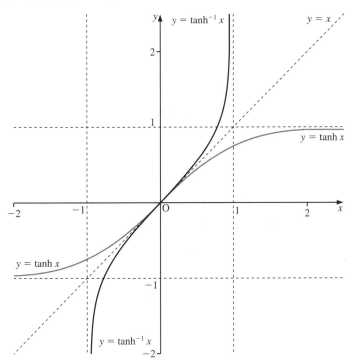

Example 4 Solve the equation $2 \cosh^2 x - \sinh x = 3$.

SOLUTION

Using the identity $\cosh^2 x - \sinh^2 x \equiv 1$, we obtain

$$2(1 + \sinh^2 x) - \sinh x - 3 = 0$$
$$\Rightarrow \quad 2 \sinh^2 x - \sinh x - 1 = 0$$

We now factorise this to obtain

$$(2 \sinh x + 1)(\sinh x - 1) = 0$$
$$\Rightarrow \quad \sinh x = 1 \quad \text{or} \quad -\tfrac{1}{2}$$
$$\Rightarrow \quad x = 0.8814 \quad \text{or} \quad -0.4812$$

Exercise 10A

1 Evaluate each of the following, giving your answer **i)** in terms of e and **ii)** correct to three significant figures.

   a) $\cosh 2$ **b)** $\sinh 3$ **c)** $\tanh 4$

2 Starting with the definitions of $\sinh x$ and $\cosh x$, prove each of the following identities.

   a) $\cosh (A + B) \equiv \cosh A \cosh B + \sinh A \sinh B$

   b) $\sinh (A - B) \equiv \sinh A \cosh B - \cosh A \sinh B$

   c) $\sinh A + \sinh B \equiv 2 \sinh \left(\dfrac{A + B}{2} \right) \cosh \left(\dfrac{A - B}{2} \right)$

3 Differentiate each of the following.

a) $\cosh 2x$	**b)** $\sinh 5x$	**c)** $\tanh 3x$
d) $2 \cosh 4x - 5 \sinh 3x$	**e)** $3 \cosh 2x + 6 \sinh 5x$	**f)** $\coth x$
g) $\operatorname{sech} x$	**h)** $3 \cosh^5 3x$	**i)** $2 \sinh^4 8x$
j) $\ln \cosh x$	**k)** $e^{\sinh 2x}$	**l)** $\ln \tanh 5x$

4 Integrate, with respect to x, each of the following.

   a) $\sinh 3x$ **b)** $\cosh 4x$ **c)** $\sinh \left(\dfrac{x}{3} \right)$

   d) $2 \cosh \left(\dfrac{x}{5} \right)$ **e)** $3 \cosh 5x - 2 \sinh \left(\dfrac{x}{2} \right)$ **f)** $\tanh 4x$

5 Solve each of these equations, giving your answer to three significant figures.

   a) $3 \sinh x + 2 \cosh x = 4$ **b)** $4 \cosh x - 8 \sinh x + 1 = 0$

   c) $\cosh x + 4 \sinh x = 3$ **d)** $3 \operatorname{sech} x - 2 = 5 \tanh x$

   e) $9 \cosh^2 x - 6 \sinh x = 17$ **f)** $3 \sinh^2 x + \cosh x - 2 = 0$

6 Find the values of x for which $8 \cosh x + 4 \sinh x = 7$, giving your answers as natural logarithms.   (EDEXCEL)

7 a) i) Write down an expression for $\tanh x$ in terms of e^x and e^{-x}.

     ii) Hence show that

$$1 - \tanh x = \frac{2e^{-2x}}{1 + e^{-2x}}$$

   b) Using the result in part **a ii**, evaluate

$$\int_0^\infty (1 - \tanh x) \, dx \quad \text{(NEAB)}$$

8 The curve C has equation $y = 5 \cosh x + 3 \sinh x$. Find the exact values of the coordinates of the turning point on C and determine its nature.   (EDEXCEL)

9 Show that, if x is real, $1 + \frac{1}{2}x^2 > x$.

Deduce that $\cosh x > x$.

The point P on the curve $y = \cosh x$ is such that its perpendicular distance from the line $y = x$ is a minimum. Show that the coordinates of P are $(\ln (1 + \sqrt{2}), \sqrt{2})$.   (NEAB)

10 Let $y = x \sinh x$.

i) Show that $\dfrac{d^2y}{dx^2} = x \sinh x + 2 \cosh x$, and find $\dfrac{d^4y}{dx^4}$.

ii) Write down a conjecture for $\dfrac{d^{2n}y}{dx^{2n}}$.

iii) Use induction to establish a formula for $\dfrac{d^{2n}y}{dx^{2n}}$.   (OCR)

11 Find the exact solution of the equation $2 \cosh x + \sinh x = 2$.   (OCR)

12 The curve C is defined parametrically by

$$x = t + \ln(\cosh t) \qquad y = \sinh t$$

i) Show that $\dfrac{dy}{dx} = e^{-t}\cosh^2 t$.

ii) Hence show that $\dfrac{d^2y}{dx^2} = e^{-2t}\cosh^2 t \,(2 \sinh t - \cosh t)$.

iii) Deduce that C has a point of inflexion where $t = \frac{1}{2} \ln 3$.   (OCR)

13 i) Show that

$$\frac{d}{dy}\left(\frac{1}{2} \sinh 4y + 4 \sinh 2y + 6y\right) = 16 \cosh^4 y$$

ii) Given that $x = 2 \sinh y$, show that

$$\sinh 2y = \frac{1}{2} x \sqrt{(x^2 + 4)}$$

and also that

$$\sinh 4y = \frac{1}{2} x(x^2 + 2)\sqrt{(x^2 + 4)}$$

iii) Use the results of parts **i** and **ii** to show that

$$\int (x^2 + 4)^{\frac{3}{2}} \, dx = \frac{1}{4} x(x^2 + 10)\sqrt{(x^2 + 4)} + 6 \sinh^{-1}\left(\frac{1}{2}x\right) + \text{constant} \qquad \text{(OCR)}$$

14 Consider the functions $y_1 = 7 + \sinh x$ and $y_2 = 5 \cosh x$ whose graphs are shown in the figure on the right.

i) Show, by solving the equation, that the solutions of $7 + \sinh x = 5 \cosh x$ are $-\log_e 2$ and $\log_e 3$.

ii) Show that the area bounded by the two graphs in the figure is $7 \log_e 6 - 10$.

   (NICCEA)

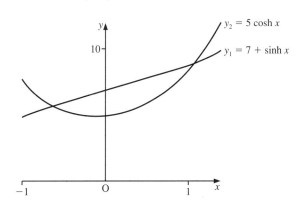

15 Let $I_n = \displaystyle\int \cosh^n x \, dx$. Show that

$$nI_n = \sinh x \cosh^{n-1} x + (n-1)I_{n-2}$$

Hence show that

$$\int_0^{\ln 2} \cosh^4 x \, dx = \frac{3}{8}\left(\frac{245}{128} + \ln 2\right) \qquad \text{(OCR)}$$

16 a) Show that $\dfrac{d}{dx}(\tanh x) = \operatorname{sech}^2 x$.

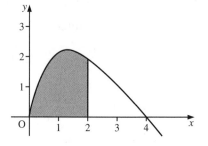

b) The diagram on the right shows a sketch of part of the curve whose equation is

$$y = 4\tanh x - x \qquad x \geqslant 0$$

  i) Find, correct to two decimal places, the coordinates of the stationary point on the curve.

  ii) Find, correct to four decimal places, the area of the shaded region bounded by the curve, the x-axis and the ordinate $x = 2$.

c) For large values of x, the curve is asymptotic to the line $y = mx + c$, where m and c are constants. State the values of m and c, and give a reason for your answer. (NEAB)

Logarithmic form

The inverse hyperbolic functions $\cosh^{-1} x$, $\sinh^{-1} x$ and $\tanh^{-1} x$ can all be expressed as logarithmic functions.

Expressing $\cosh^{-1} x$ as a logarithmic function

Let $\cosh^{-1} x = y$. We then have

$$x = \cosh y$$

$$\Rightarrow \quad x = \frac{1}{2}(e^y + e^{-y})$$

Multiplying throughout by $2e^y$, we obtain

$$2xe^y = e^{2y} + 1$$

$$\Rightarrow \quad e^{2y} - 2xe^y + 1 = 0$$

To solve this equation, we treat it as a quadratic in e^y, which gives

$$e^y = \frac{2x \pm \sqrt{4x^2 - 4}}{2}$$

$$\Rightarrow \quad e^y = x \pm \sqrt{x^2 - 1}$$

Taking the logarithms of both sides, we obtain

$$y = \ln(x \pm \sqrt{x^2 - 1})$$

That is, the **principal value** of $\cosh^{-1} x$ is $\ln(x + \sqrt{x^2 - 1})$.

Expressing the principal value in a different form, we obtain

$$\ln(x + \sqrt{x^2 - 1}) = \ln\left[\frac{(x + \sqrt{x^2 - 1})(x - \sqrt{x^2 - 1})}{x - \sqrt{x^2 - 1}}\right]$$

$$= \ln\left[\frac{x^2 - (x^2 - 1)}{x - \sqrt{x^2 - 1}}\right]$$

$$= \ln\left[\frac{1}{x - \sqrt{x^2 - 1}}\right]$$

$$= -\ln(x - \sqrt{x^2 - 1})$$

Hence, we have

$$\ln(x \pm \sqrt{x^2 - 1}) = \pm\ln(x + \sqrt{x^2 - 1})$$

which matches the symmetry of the graph of $\cosh x$.

Example 5 Find the value, in logarithmic form, of $\cosh^{-1}2$.

SOLUTION

Using $\cosh^{-1}x = \ln(x + \sqrt{x^2 - 1})$, we have

$$\cosh^{-1}2 = \ln(2 + \sqrt{3})$$

Example 6 Find the exact coordinates of the points where the line $y = 3$ cuts the graph of $y = \cosh x$.

SOLUTION

When $y = 3$, we have

$$x = \cosh^{-1}3$$
$$\Rightarrow \quad x = \ln(3 + \sqrt{8}) = \ln(3 + 2\sqrt{2})$$

By symmetry, the other value of x is $-\ln(3 + 2\sqrt{2})$.

Therefore, the two points are

$$(\ln(3 + 2\sqrt{2}), 3) \quad \text{and} \quad (-\ln(3 + 2\sqrt{2}), 3)$$

Expressing $\sinh^{-1}x$ as a logarithmic function

Let $y = \sinh^{-1}x$. We then have

$$x = \sinh y \quad \Rightarrow \quad x = \frac{1}{2}(e^y - e^{-y})$$

Multiplying throughout by $2e^y$, we obtain

$$2xe^y = e^{2y} - 1$$
$$\Rightarrow \quad e^{2y} - 2xe^y - 1 = 0$$

Treating this equation as a quadratic in e^y, we have

$$e^y = \frac{2x \pm \sqrt{4x^2 + 4}}{2} \quad \Rightarrow \quad e^y = x \pm \sqrt{x^2 + 1}$$

Taking the logarithms of both sides, we obtain

$$y = \ln(x \pm \sqrt{x^2 + 1})$$

The value of $\sinh^{-1} x$ can only be $\ln(x + \sqrt{x^2 + 1})$. We cannot have $\sinh^{-1} x = \ln(x - \sqrt{x^2 + 1})$, because $x < \sqrt{x^2 + 1}$, which would give the logarithm of a negative number, which is a complex number.

Hence, we have

$$\sinh^{-1} x = \ln(x + \sqrt{x^2 + 1})$$

Example 7 Find the value, in logarithmic form, of $\sinh^{-1} 3$.

SOLUTION

Using $\sinh^{-1} x = \ln(x + \sqrt{x^2 + 1})$, we have

$$\sinh^{-1} 3 = \ln(3 + \sqrt{10})$$

Expressing $\tanh^{-1} x$ as a logarithmic function

Let $y = \tanh^{-1} x$. We then have

$$x = \tanh y = \frac{\sinh y}{\cosh y}$$

$$\Rightarrow \quad x = \frac{\frac{1}{2}(e^y - e^{-y})}{\frac{1}{2}(e^y + e^{-y})}$$

Multiplying the numerator and the denominator by $2e^y$, we obtain

$$x = \frac{e^{2y} - 1}{e^{2y} + 1}$$

$$\Rightarrow \quad e^{2y} x + x = e^{2y} - 1$$

Therefore, we have

$$e^{2y} = \frac{1 + x}{1 - x} \quad \Rightarrow \quad y = \frac{1}{2} \ln\left(\frac{1 + x}{1 - x}\right)$$

Hence, the value of $\tanh^{-1} x$ is $\dfrac{1}{2} \ln\left(\dfrac{1 + x}{1 - x}\right)$, where $-1 < x < 1$.

Example 8 Find the value, in logarithmic form, of $\tanh^{-1} \frac{1}{2}$.

SOLUTION

Using $\tanh^{-1} x = \dfrac{1}{2} \ln\left(\dfrac{1 + x}{1 - x}\right)$, we have

$$\tanh^{-1} \frac{1}{2} = \frac{1}{2} \ln\left(\frac{\frac{3}{2}}{\frac{1}{2}}\right)$$

which gives $\tanh^{-1} \frac{1}{2} = \frac{1}{2} \ln 3$.

Example 9 Find the value, in logarithmic form, of $\operatorname{sech}^{-1}\frac{1}{2}$.

SOLUTION

Since $y = \operatorname{sech}^{-1}\frac{1}{2}$, we have

$$\operatorname{sech} y = \tfrac{1}{2}$$

$$\Rightarrow \quad \frac{1}{\cosh y} = \frac{1}{2}$$

$$\Rightarrow \quad \cosh y = 2$$

$$\Rightarrow \quad y = \cosh^{-1} 2$$

Using $\cosh^{-1} x = \ln(x + \sqrt{x^2 - 1})$, we have

$$\cosh^{-1} 2 = \ln(2 + \sqrt{3})$$

which gives $\operatorname{sech}^{-1}\frac{1}{2} = \ln(2 + \sqrt{3})$.

Summary

$$\cosh^{-1} x = \ln(x \pm \sqrt{x^2 - 1}) \quad x \geqslant 1 \quad \text{Plus sign gives the \textbf{principal value}}$$

$$\sinh^{-1} x = \ln(x + \sqrt{x^2 + 1})$$

$$\tanh^{-1} x = \frac{1}{2} \ln\left(\frac{1+x}{1-x}\right) \quad -1 < x < 1$$

Differentiation of inverse hyperbolic functions

$\sinh^{-1} x$

We have $y = \sinh^{-1} x$, therefore $\sinh y = x$.

Differentiating $\sinh y = x$, we obtain

$$\cosh y \, \frac{dy}{dx} = 1$$

$$\Rightarrow \quad \frac{dy}{dx} = \frac{1}{\cosh y} = \frac{1}{\sqrt{1 + \sinh^2 y}} = \frac{1}{\sqrt{1 + x^2}}$$

which gives

$$\frac{d}{dx} \sinh^{-1} x = \frac{1}{\sqrt{1 + x^2}}$$

Therefore, we have

$$\int \frac{dx}{\sqrt{1 + x^2}} = \sinh^{-1} x + c$$

We now take $y = \sinh^{-1}\left(\frac{x}{a}\right)$, giving $\sinh y = \frac{x}{a}$.

Differentiating $\sinh y = \dfrac{x}{a}$, we obtain

$$\cosh y \, \frac{dy}{dx} = \frac{1}{a}$$

$$\Rightarrow \quad \frac{dy}{dx} = \frac{1}{a \cosh y} = \frac{1}{a\sqrt{1 + \sinh^2 y}}$$

which gives

$$\frac{dy}{dx} = \frac{1}{a\sqrt{1 + \left(\dfrac{x}{a}\right)^2}} = \frac{1}{\sqrt{a^2 + x^2}}$$

That is, we have

$$\frac{d}{dx} \sinh^{-1}\left(\frac{x}{a}\right) = \frac{1}{\sqrt{a^2 + x^2}}$$

from which it follows that

$$\int \frac{dx}{\sqrt{a^2 + x^2}} = \sinh^{-1}\left(\frac{x}{a}\right) + c$$

$\cosh^{-1}x$

We have $y = \cosh^{-1}x$, therefore $\cosh y = x$.

Differentiating $\cosh y = x$, we obtain

$$\sinh y \, \frac{dy}{dx} = 1$$

$$\Rightarrow \quad \frac{dy}{dx} = \frac{1}{\sinh y} = \frac{1}{\sqrt{\cosh^2 y - 1}} = \frac{1}{\sqrt{x^2 - 1}}$$

which gives

$$\frac{d}{dx} \cosh^{-1}x = \frac{1}{\sqrt{x^2 - 1}}$$

Therefore, we have

$$\int \frac{dx}{\sqrt{x^2 - 1}} = \cosh^{-1}x + c$$

We now take $y = \cosh^{-1}\left(\dfrac{x}{a}\right)$, giving $\cosh y = \dfrac{x}{a}$.

Differentiating $\cosh y = \dfrac{x}{a}$, we obtain

$$\sinh y \, \frac{dy}{dx} = \frac{1}{a}$$

$$\Rightarrow \quad \frac{dy}{dx} = \frac{1}{a \sinh y} = \frac{1}{a\sqrt{\cosh^2 y - 1}}$$

which gives

$$\frac{dy}{dx} = \frac{1}{a\sqrt{\left(\frac{x}{a}\right)^2 - 1}} = \frac{1}{\sqrt{x^2 - a^2}}$$

That is, we have

$$\frac{d}{dx} \cosh^{-1}\left(\frac{x}{a}\right) = \frac{1}{\sqrt{x^2 - a^2}}$$

from which it follows that

$$\int \frac{dx}{\sqrt{x^2 - a^2}} = \cosh^{-1}\left(\frac{x}{a}\right) + c$$

$\tanh^{-1} x$

We have $y = \tanh^{-1}\left(\frac{x}{a}\right)$, therefore $\tanh y = \frac{x}{a}$.

Differentiating $\tanh y = \frac{x}{a}$, we obtain

$$\mathrm{sech}^2 y \frac{dy}{dx} = \frac{1}{a}$$

$$\Rightarrow \quad \frac{dy}{dx} = \frac{1}{a\,\mathrm{sech}^2 y} = \frac{1}{a(1 - \tanh^2 y)}$$

which gives

$$\frac{dy}{dx} = \frac{1}{a\left[1 - \left(\frac{x}{a}\right)^2\right]} = \frac{a}{a^2 - x^2}$$

That is, we have

$$\frac{d}{dx} \tanh^{-1}\left(\frac{x}{a}\right) = \frac{a}{a^2 - x^2}$$

from which it follows that

$$\int \frac{dx}{a^2 - x^2} = \frac{1}{a} \tanh^{-1}\left(\frac{x}{a}\right) + c$$

Note We can integrate $\dfrac{1}{a^2 - x^2}$ by partial fractions:

$$\int \frac{dx}{a^2 - x^2} = \frac{1}{2a} \int \left(\frac{1}{a + x} + \frac{1}{a - x}\right) dx = \frac{1}{2a} \ln\left(\frac{a + x}{a - x}\right) + c$$

This result is the logarithmic form of $\tanh^{-1}\left(\frac{x}{a}\right)$. Hence, it is unusual to use a function in $\tanh^{-1} x$ in differentiation or integration.

Example 10 Differentiate

a) i) $\sinh^{-1}\left(\dfrac{x}{3}\right)$ **ii)** $\sinh^{-1}4x$ **b)** $\cosh^{-1}\left(\dfrac{x}{5}\right)$

SOLUTION

a) Using $\dfrac{d}{dx}\sinh^{-1}\left(\dfrac{x}{a}\right) = \dfrac{1}{\sqrt{a^2 + x^2}}$, we have

i) $\qquad \dfrac{d}{dx}\sinh^{-1}\left(\dfrac{x}{3}\right) = \dfrac{1}{\sqrt{9 + x^2}}$

ii) $\qquad \dfrac{d}{dx}\sinh^{-1}4x = \dfrac{d}{dx}\sinh^{-1}\left(\dfrac{x}{\frac{1}{4}}\right) = \dfrac{1}{\sqrt{\frac{1}{16} + x^2}}$

which gives

$$\dfrac{d}{dx}\sinh^{-1}4x = \dfrac{4}{\sqrt{1 + 16x^2}}$$

b) Using $\dfrac{d}{dx}\cosh^{-1}\left(\dfrac{x}{a}\right) = \dfrac{1}{\sqrt{x^2 - a^2}}$, we have

$$\dfrac{d}{dx}\cosh^{-1}\left(\dfrac{x}{5}\right) = \dfrac{1}{\sqrt{x^2 - 25}}$$

Example 11 Find

a) $\displaystyle\int_0^2 \dfrac{1}{\sqrt{4 + x^2}}\,dx$ **b)** $\displaystyle\int_0^1 \dfrac{1}{\sqrt{4 + 3x^2}}\,dx$

SOLUTION

a) Using the first integral formula on page 202, we obtain

$$\int_0^2 \dfrac{1}{\sqrt{4 + x^2}}\,dx = \left[\sinh^{-1}\left(\dfrac{x}{2}\right)\right]_0^2$$

$$= \sinh^{-1}1 - \sinh^{-1}0 = \sinh^{-1}1 = \ln(1 + \sqrt{2})$$

Therefore, we have

$$\int_0^2 \dfrac{1}{\sqrt{4 + x^2}}\,dx = \ln(1 + \sqrt{2})$$

b) Before integrating, we must reduce the coefficient of x^2 to unity (as with inverse trigonometric functions), which gives

$$\int_0^1 \dfrac{1}{\sqrt{4 + 3x^2}}\,dx = \dfrac{1}{\sqrt{3}}\int_0^1 \dfrac{1}{\sqrt{\dfrac{4}{3} + x^2}}\,dx$$

$$= \dfrac{1}{\sqrt{3}}\int_0^1 \dfrac{1}{\sqrt{\left(\dfrac{2}{\sqrt{3}}\right)^2 + x^2}}$$

$$\Rightarrow \quad \int_0^1 \frac{1}{\sqrt{4+3x^2}} \, dx = \frac{1}{\sqrt{3}} \left[\sinh^{-1} \left(\frac{\sqrt{3}x}{2} \right) \right]_0^1$$

$$= \frac{1}{\sqrt{3}} \left[\sinh^{-1} \left(\frac{\sqrt{3}}{2} \right) - \sinh^{-1} 0 \right]$$

$$= \frac{1}{\sqrt{3}} \left[\ln \left(\frac{\sqrt{3}}{2} + \sqrt{\frac{3}{4}+1} \right) \right]$$

$$= \frac{1}{\sqrt{3}} \ln \left(\frac{\sqrt{3}+\sqrt{7}}{2} \right)$$

Therefore, we have

$$\int_0^1 \frac{1}{\sqrt{4+3x^2}} \, dx = \frac{1}{\sqrt{3}} \ln \left(\frac{\sqrt{3}+\sqrt{7}}{2} \right)$$

Example 12 Find $\int_3^6 \frac{1}{\sqrt{x^2-9}} \, dx$.

SOLUTION

Using the first integral formula on page 203, we obtain

$$\int_3^6 \frac{1}{\sqrt{x^2-9}} \, dx = \left[\cosh^{-1} \left(\frac{x}{3} \right) \right]_3^6$$

$$= \cosh^{-1} 2 - \cosh^{-1} 1 = \ln(2+\sqrt{3}) - 0$$

Therefore, we have

$$\int_3^6 \frac{1}{\sqrt{x^2-9}} \, dx = \ln(2+\sqrt{3})$$

Example 13 Find $\int \frac{1}{\sqrt{4x^2-8x-16}} \, dx$.

SOLUTION

Before integrating, we must

- Complete the square (as with inverse trigonometric functions).
- Reduce the coefficient of x^2 to unity.

Hence, we have

$$\sqrt{4x^2-8x-16} = \sqrt{4}\sqrt{x^2-2x-4}$$

$$= 2\sqrt{(x-1)^2-5}$$

which gives

$$\int \frac{1}{\sqrt{4x^2-8x-16}} \, dx = \frac{1}{2} \int \frac{dx}{\sqrt{(x-1)^2-5}}$$

$$= \frac{1}{2} \cosh^{-1} \left(\frac{x-1}{\sqrt{5}} \right) + c$$

We can express this result as

$$\frac{1}{2} \ln\left(\frac{x-1}{\sqrt{5}} + \sqrt{\frac{(x-1)^2}{5} - 1} \right) + c$$

which gives

$$\int \frac{1}{\sqrt{4x^2 - 8x - 16}} \, dx = \frac{1}{2} \ln\left(\sqrt{(x-1)^2 - 5} + x - 1 \right) - \frac{1}{2} \ln\sqrt{5} + c$$

$$= \frac{1}{2} \ln\left(\sqrt{x^2 - 2x - 4} + x - 1 \right) + c'$$

Exercise 10B

1 Differentiate each of the following with respect to x.

a) $\sinh^{-1} 5x$ **b)** $\cosh^{-1} 3x$ **c)** $\sinh^{-1}\sqrt{2}x$ **d)** $\cosh^{-1}\frac{3}{4}x$

e) $\sinh^{-1} x^2$ **f)** $\operatorname{sech}^{-1} x$ **g)** $\coth^{-1} x$

2 Find each of the following integrals.

a) $\displaystyle\int \frac{dx}{\sqrt{x^2 - 4}}$ **b)** $\displaystyle\int \frac{dx}{\sqrt{x^2 - 9}}$ **c)** $\displaystyle\int \frac{dx}{\sqrt{4x^2 - 25}}$

d) $\displaystyle\int \frac{dx}{\sqrt{9x^2 - 16}}$ **e)** $\displaystyle\int \frac{dx}{\sqrt{9 + x^2}}$ **f)** $\displaystyle\int \frac{dx}{\sqrt{16 + x^2}}$

g) $\displaystyle\int \frac{dx}{\sqrt{25 + 16x^2}}$ **h)** $\displaystyle\int \frac{dx}{\sqrt{9 + 25x^2}}$

3 Evaluate each of the following definite integrals, giving the exact value of your answer.

a) $\displaystyle\int_0^1 \frac{dx}{\sqrt{1 + x^2}}$ **b)** $\displaystyle\int_0^2 \frac{dx}{\sqrt{4 + x^2}}$ **c)** $\displaystyle\int_4^8 \frac{dx}{\sqrt{x^2 - 16}}$

d) $\displaystyle\int_0^2 \frac{dx}{\sqrt{4 + 3x^2}}$ **e)** $\displaystyle\int_{\frac{1}{5}}^1 \frac{dx}{\sqrt{25x^2 - 1}}$

4 Evaluate each of the following integrals, giving your answer in terms of logarithms.

a) $\displaystyle\int_1^2 \frac{dx}{\sqrt{25x^2 - 4}}$ **b)** $\displaystyle\int_1^2 \frac{dx}{\sqrt{4 + 9x^2}}$ **c)** $\displaystyle\int_3^4 \frac{dx}{\sqrt{(x-1)^2 - 3}}$

d) $\displaystyle\int_0^1 \frac{dx}{\sqrt{4(x+1)^2 + 5}}$ **e)** $\displaystyle\int_0^2 \frac{dx}{\sqrt{4 + 8x + x^2}}$ **f)** $\displaystyle\int_0^1 \frac{dx}{\sqrt{16x^2 + 20x + 35}}$

5 Given that $f(x) \equiv \dfrac{1}{\sqrt{(x^2 + 4x - 12)}}$,

 a) find $\displaystyle\int f(x)\,dx$.

 b) Hence find the exact value of $\displaystyle\int_6^{10} f(x)\,dx$, giving your answer as a single logarithm.

 (EDEXCEL)

6 a) Show that $\sinh^{-1}x = \ln(x + \sqrt{x^2 + 1})$.

 b) Evaluate $\displaystyle\int_{-1}^{0} \dfrac{dx}{\sqrt{x^2 + 2x + 2}}$. (WJEC)

7 a) Find $\displaystyle\int x\,\mathrm{sech}^2 x\,dx$.

 b) Find the general solution of the differential equation

$$\cosh x \, \frac{dy}{dx} - y\sinh x = x$$

 giving your answer in the form $y = f(x)$. (EDEXCEL)

8 $\qquad 4x^2 + 4x + 5 \equiv (px + q)^2 + r$

 a) Find the values of the constants p, q and r.

 b) Hence, or otherwise, find $\displaystyle\int \dfrac{1}{4x^2 + 4x + 5}\,dx$.

 c) Show that

$$\int \frac{2}{\sqrt{(4x^2 + 4x + 5)}}\,dx = \ln[(2x + 1) + \sqrt{(4x^2 + 4x + 5)}] + k$$

 where k is an arbitrary constant. (EDEXCEL)

9 a) Show that $\sinh^{-1}x = \ln(x + \sqrt{1 + x^2})$.

 b) Evaluate $\displaystyle\int_0^1 \dfrac{dx}{\sqrt{x^2 + 6x + 10}}$, giving your answer correct to four decimal places. (WJEC)

10 a) Express $4x^2 + 4x + 26$ in the form $(px + q)^2 + r$, where p, q and r are constants.

 b) Hence determine $\displaystyle\int \dfrac{1}{\sqrt{(4x^2 + 4x + 26)}}\,dx$. (EDEXCEL)

11 i) Find A, B and C such that

$$3x^2 + 24x + 23 \equiv A(x + B)^2 + C$$

 ii) Show that

$$\int \frac{dx}{\sqrt{3x^2 + 24x + 23}} = \frac{1}{\sqrt{3}}\cosh^{-1}\left(\frac{\sqrt{3}(x + 4)}{5}\right) + c \qquad \text{(NICCEA)}$$

12 Express $x^2 - 6x + 8$ in the form $(x - p)^2 - q^2$, for positive integers p and q.

 Hence evaluate $\displaystyle\int_4^5 \dfrac{dx}{\sqrt{(x^2 - 6x + 8)}}$ giving your answer in terms of natural logarithms.

 (AEB 97)

13 a) Simplify $(e^x + e^{-x})^2 - (e^x - e^{-x})^2$ and hence deduce that $\cosh^2 x - \sinh^2 x = 1$.

b) Given that $y = \operatorname{arsinh} x$, show that $\dfrac{dy}{dx} = \dfrac{1}{\sqrt{(x^2 + 1)}}$.

c) Find $\displaystyle\int \operatorname{arsinh} x \, dx$.   (EDEXCEL)

14 A curve has equation $y = x \sinh^{-1} x$.

i) Show that

$$\frac{d^2 y}{dx^2} = \frac{2 + x^2}{(1 + x^2)^{\frac{3}{2}}}$$

ii) Deduce that the curve has no point of inflexion.   (OCR)

15 Starting from the definition of cosh in terms of exponentials, show that

$$\cosh^{-1} x = \ln[x + \sqrt{(x^2 - 1)}]$$

Show that

$$\int_1^2 \frac{1}{\sqrt{(4x^2 - 1)}} \, dx = \frac{1}{2} \ln\left(\frac{4 + \sqrt{15}}{2 + \sqrt{3}}\right)   \text{(OCR)}$$

16 Given that $y = \tanh^{-1} x$, derive the result $\dfrac{dy}{dx} = \dfrac{1}{1 - x^2}$.

[No credit will be given for merely quoting the result from the *List of Formulae*.]

Show that $\displaystyle\int_0^{\frac{1}{4}} \tanh^{-1} 2x \, dx = \frac{1}{8} \ln \frac{27}{16}$.   (OCR)

17 i) Let $x = \sinh u$. By first expressing x in terms of exponentials, show that

$$\sinh^{-1} x = \ln[x + \sqrt{(x^2 + 1)}]$$

ii) By using an appropriate substitution, show that

$$\int \frac{1}{\sqrt{(x^2 + a^2)}} \, dx = \sinh^{-1}\left(\frac{x}{a}\right) + c$$

where a and c are constants ($a > 0$).

iii) Evaluate

$$\int_0^4 \frac{1}{\sqrt{(9x^2 + 4)}} \, dx$$

giving your answer in terms of a natural logarithm.   (OCR)

18 a) State the values of x for which $\cosh^{-1} x$ is defined.

b) A curve C is defined for these values of x by the equation $y = x - \cosh^{-1} x$.

i) Show that C has just one stationary point.

ii) Evaluate y at the stationary point, giving your answer in the form $p - \ln q$, where p and q are numbers to be determined.   (NEAB)

19 a) Using the substitution $u = e^x$, find $\int \operatorname{sech} x \, dx$.

b) Sketch the curve with equation $y = \operatorname{sech} x$.

The finite region R is bounded by the curve with equation $y = \operatorname{sech} x$, the lines $x = 2$, $x = -2$ and the x-axis.

c) Using your result from part **a**, find the area of R, giving your answer to three decimal places.   (EDEXCEL)

20 The diagram shows the curve with equation $y = \dfrac{1}{(x^2 + 4)^{\frac{1}{4}}}$.

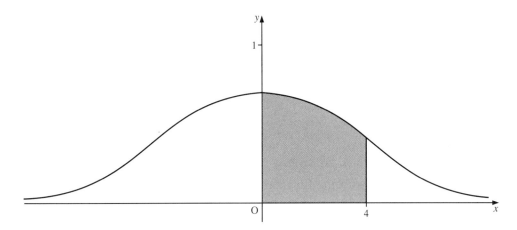

The finite region bounded by the curve, the x-axis, the y-axis and the line $x = 4$ is rotated through one full turn about the x-axis to form a solid of revolution.

Use integration to determine the volume of this solid, giving your answer in terms of π and a natural logarithm.   (AEB 98)

21 a) Use the definition of $\coth x$ in terms of exponential functions to prove that

$$\operatorname{arcoth} x = \frac{1}{2} \ln\left(\frac{x+1}{x-1}\right)$$

The function f is defined by $f(x) = \operatorname{arcoth}\left(\dfrac{x}{3}\right)$, $x^2 > 9$.

b) Show that f is odd.

c) Find $f'(x)$.

d) Expand $f(x)$ in a series of ascending powers of $\dfrac{1}{x}$ as far as the term in $\dfrac{1}{x^7}$ and state the coefficient of $\dfrac{1}{x^{2n+1}}$.

e) Hence, or otherwise, derive the expansion of $\dfrac{1}{9 - x^2}$ in a series of ascending powers of $\dfrac{1}{x}$ as far as the term in $\dfrac{1}{x^8}$ and state the coefficient of $\dfrac{1}{x^{2n}}$.   (EDEXCEL)

22 Starting from the definitions of $\sinh x$ and $\cosh x$ in terms of exponentials, show that for $|x| < 1$,

$$\operatorname{artanh} x = \frac{1}{2} \ln\left(\frac{1+x}{1-x}\right)$$

a) Expand $\operatorname{artanh} x$ as a series in ascending powers of x, as far as the term in x^5 and state the coefficient of x^{2n+1} in this expansion.

b) Solve the equation

$$3\operatorname{sech}^2 x + 4\tanh x + 1 = 0$$

giving any answers in terms of natural logarithms.

c) Sketch the graph of $y = \operatorname{artanh} x$ and evaluate the area of the finite region bounded by the curve with equation $y = \operatorname{artanh} x$ and the lines $x = \frac{1}{2}$ and $y = 0$.   (EDEXCEL)

23 a) Use integration by parts to show

$$\int x^2 \cosh x \, dx = x^2 \sinh x - 2x \cosh x + 2\sinh x + c$$

b) Consider the two curves whose equations are

$$y_1 = \sinh x \qquad y_2 = 2 - \cosh x$$

and which are shown in the figure on the right.

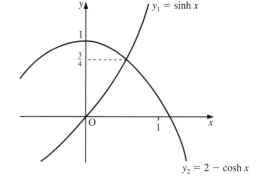

i) Show that they cross at the point $(\log_e 2, \frac{3}{4})$.

ii) Find the area bounded by the y-axis, the curve y_1 and the curve y_2.

iii) The area bounded by the y-axis, the line $y = \frac{3}{4}$ and the curve y_1 is rotated about the y-axis to form a solid of revolution. Show that its exact volume is

$$\frac{\pi}{4}[3(\log_e 2)^2 - 10\log_e 2 + 6]$$

$$\left[\text{The volume of revolution about the } y\text{-axis is given by } \pi \int x^2 \, dy. \right] \qquad \text{(NICCEA)}$$

'Double-angle' formulae

To integrate $\cosh^2 x$ and $\sinh^2 x$, we must express each in a form which contains $\cosh 2x$, in a similar manner to integrating $\cos^2 x$ and $\sin^2 x$ (see *Introducing Pure Mathematics*, pages 451–2).

To obtain the identity relating $\cosh 2x$ to $\cosh^2 x$, we have

$$\cosh 2x = \frac{1}{2}(e^{2x} + e^{-2x}) = \frac{1}{2}[(e^x + e^{-x})^2 - 2]$$

$$= \frac{1}{2}[4\cosh^2 x - 2]$$

which gives

$$\cosh 2x = 2\cosh^2 x - 1$$

To obtain the identity relating $\cosh 2x$ to $\sinh^2 x$, we take

$$\cosh 2x = 2\cosh^2 x - 1$$

and make the substitution $\cosh^2 x = 1 + \sinh^2 x$ to obtain

$$\cosh 2x = 2(1 + \sinh^2 x) - 1$$

which gives

$$\cosh 2x = 2\sinh^2 x + 1$$

Similarly, we have

$$\sinh 2x = \frac{1}{2}(e^{2x} - e^{-2x}) = \frac{1}{2}(e^x - e^{-x})(e^x + e^{-x})$$

which gives

$$\sinh 2x = 2\sinh x \cosh x$$

Hence, we see that $\displaystyle\int \cosh^2 ax \, dx$ is given by

$$\int \cosh^2 ax \, dx = \int \frac{1}{2}(\cosh 2ax + 1) \, dx$$

which gives

$$\int \cosh^2 ax \, dx = \frac{1}{4a}\sinh 2ax + \frac{x}{2} + c$$

Example 14 Using the substitution $x = 3\sinh u$, find the value of $\displaystyle\int \sqrt{9 + x^2} \, dx$.

SOLUTION

Differentiating the substitution $x = 3\sinh u$, we obtain

$$\frac{dx}{du} = 3\cosh u$$

$$\Rightarrow \quad dx = 3\cosh u \, du$$

Substituting for x and for dx in $\displaystyle\int \sqrt{9 + x^2} \, dx$, we have

$$\int \sqrt{9 + x^2} \, dx = \int \sqrt{9 + 9\sinh^2 u} \, (3\cosh u) \, du$$

$$= \int 9\cosh^2 u \, du$$

$$= \frac{9}{2}\int (\cosh 2u + 1) \, du$$

$$= \frac{9}{2}\left(\frac{1}{2}\sinh 2u + u\right) + c$$

Using $\sinh 2u = 2\sinh u \cosh u$, we obtain

$$\int \sqrt{9 + x^2}\, dx = \frac{9}{2}\sinh u \cosh u + \frac{9}{2}u + c$$

As the question involves an integral in terms of x, the answer must be given in terms of x.

Using $\cosh u = \sqrt{\sinh^2 u + 1}$ and $\sinh u = \frac{x}{3}$, we obtain

$$\int \sqrt{9 + x^2}\, dx = \frac{9}{2}\frac{x}{3}\sqrt{1 + \frac{x^2}{9}} + \frac{9}{2}\sinh^{-1}\left(\frac{x}{3}\right) + c$$

$$= \frac{1}{2}x\sqrt{9 + x^2} + \frac{9}{2}\ln\left(\frac{x}{3} + \sqrt{\frac{x^2}{9} + 1}\right) + c$$

Therefore, we have

$$\int \sqrt{9 + x^2}\, dx = \frac{x}{2}\sqrt{x^2 + 9} + \frac{9}{2}\ln\left(\sqrt{x^2 + 9} + x\right) + c'$$

Power series

On page 177, we used Maclaurin's series to find the power series for $\sin x$ and $\cos x$.

In a similar way, we can find the power series for $\sinh x$ and $\cosh x$.

Power series for $\sinh x$

Let $\sinh x = a_0 + a_1 x + a_2 x^2 + a_3 x^3 + \ldots$, where the a's are constants.

When $x = 0$, $\sinh 0 = a_0$. But $\sinh 0 = 0$, therefore $a_0 = 0$.

Differentiating $\sinh x = a_0 + a_1 x + a_2 x^2 + a_3 x^3 + \ldots$, we obtain

$$\cosh x = a_1 + 2a_2 x + 3a_3 x^2 + 4a_4 x^3 + \ldots$$

When $x = 0$, $\cosh 0 = a_1$. But $\cosh 0 = 1$, therefore $a_1 = 1$.

Differentiating again, we obtain

$$\sinh x = 2a_2 + 3 \times 2a_3 x + 4 \times 3a_4 x^2 + 5 \times 4a_5 x^3 + \ldots$$

When $x = 0$, $\sinh 0 = 2a_2 \quad \Rightarrow \quad a_2 = 0$.

Differentiating yet again, we obtain

$$\cosh x = 3 \times 2a_3 + 4 \times 3 \times 2a_4 x + 5 \times 4 \times 3a_5 x^2 + \ldots$$

When $x = 0$, $\cosh 0 = 3 \times 2a_3 \quad \Rightarrow \quad a_3 = \frac{1}{3!}$.

Repeating the differentiation, we obtain

$$a_4 = 0 \qquad a_5 = \frac{1}{5!} \qquad a_6 = 0 \qquad a_7 = \frac{1}{7!}$$

Hence, we have

$$\sinh x = x + \frac{1}{3!}x^3 + \frac{1}{5!}x^5 + \frac{1}{7!}x^7 + \dots$$

By d'Alembert's ratio test, this series converges for all real x.

Power series for cosh x

We can use the procedure for $\sinh x$ to find the power series for $\cosh x$. However, it is much easier to start from the expansion for $\sinh x$. Hence, we have

$$\cosh x = \frac{\mathrm{d}}{\mathrm{d}x} \sinh x = \frac{\mathrm{d}}{\mathrm{d}x}\left(x + \frac{1}{3!}x^3 + \frac{1}{5!}x^5 + \frac{1}{7!}x^7 + \dots\right)$$

which gives

$$\cosh x = 1 + \frac{1}{2!}x^2 + \frac{1}{4!}x^4 + \frac{1}{6!}x^6 + \dots$$

By d'Alembert's ratio rest, this series is convergent for all real x.

Osborn's rule

Taking the power series for $\cos \mathrm{i}x$, we have

$$\cos \mathrm{i}x = 1 - \frac{1}{2!}(\mathrm{i}x)^2 + \frac{1}{4!}(\mathrm{i}x)^4 - \dots$$

$$= 1 + \frac{1}{2!}x^2 + \frac{1}{4!}x^4 + \dots$$

which is the power series for $\cosh x$. Hence, we have

$$\cos \mathrm{i}x \equiv \cosh x$$

For $\sin \mathrm{i}x$, we have

$$\sin \mathrm{i}x = (\mathrm{i}x) - \frac{1}{3!}(\mathrm{i}x)^3 + \frac{1}{5!}(\mathrm{i}x)^5 - \dots$$

$$= \mathrm{i}\left(x + \frac{x^3}{3!} + \frac{x^5}{5!} + \dots\right)$$

which is the power series for $\mathrm{i}\sinh x$. Hence, we have

$$\sin \mathrm{i}x \equiv \mathrm{i}\sinh x$$

Since $\cos^2\theta + \sin^2\theta \equiv 1$ for any angle θ, we know that

$$\cos^2 \mathrm{i}x + \sin^2 \mathrm{i}x \equiv 1$$

which gives

$$\cosh^2 x + (\mathrm{i}\sinh x)^2 \equiv 1$$

Therefore, we have

$$\cosh^2 x - \sinh^2 x \equiv 1$$

The two identities

$$\cos^2 \theta + \sin^2 \theta \equiv 1 \quad \text{and} \quad \cosh^2 x - \sinh^2 x \equiv 1$$

are typical of the similarity between the standard ordinary trigonometric identities and the standard hyperbolic identities (see pages 191–2). Osborn's rule gives guidance on the similarity between such identities based on $\sin i x$ being equivalent to $i \sinh x$.

Osborn's rule states that to change a standard ordinary trigonometric identity into the equivalent standard hyperbolic identity, **change the sign** of the term which is the **product of two sines, and substitute the corresponding hyperbolic functions**.

Thus, for example,

$$\cos 2x \equiv 1 - 2\sin^2 x \quad \text{gives} \quad \cosh 2x \equiv 1 + 2\sinh^2 x$$

When applying the rule to $1 + \tan^2 x = \sec^2 x$, we treat $\tan^2 x$ as $\dfrac{\sin^2 x}{\cos^2 x}$. Hence, the equivalent hyperbolic identity is

$$1 - \tanh^2 x \equiv \text{sech}^2 x$$

Exercise 10C

1 Expand each of the following expressions up to and including the term in x^4.

    a) $\cosh 2x$ **b)** $\sinh 3x$

    c) $(1 + x)\cosh 5x$ **d)** $(1 + 2x)\sinh 6x$

2 By means of the substitution $x = 3\cosh\theta$, find $\displaystyle\int \sqrt{x^2 - 9}\,dx$.

3 Using the substitution $x = 4\sinh\theta$, find $\displaystyle\int \sqrt{16 + x^2}\,dx$.

4 Find $\displaystyle\int \sqrt{25 + x^2}\,dx$.   **5** Find $\displaystyle\int \sqrt{x^2 - 25}\,dx$.

6 Find $\displaystyle\int \frac{x^2}{\sqrt{x^2 - 4}}\,dx$.   **7** Find $\displaystyle\int \frac{x^2 + 2}{\sqrt{x^2 + 9}}\,dx$.

8 Use the substitution $x = 2\sinh u$ to find $\displaystyle\int \sqrt{(x^2 + 4)}\,dx$.   (OCR)

9 Use the definition $\cosh x = \frac{1}{2}(e^x + e^{-x})$ to prove that

$$\cosh A + \cosh B = 2\cosh \tfrac{1}{2}(A + B)\cosh \tfrac{1}{2}(A - B)$$

For $n \geqslant 0$, the function P_n is defined by

$$P_n(x) = 1 - (n + 1)\cosh nx + n\cosh(n + 1)x$$

    i) Evaluate $P_0(x)$.

ii) Show that

$$P_r(x) - P_{r-1}(x) = 2r \cosh rx (\cosh x - 1)$$

where $r \geqslant 1$.

Hence, or otherwise, find $\displaystyle\sum_{r=1}^{n} r \cosh rx$ for $x \neq 0$.   (NEAB)

10 i) Prove that

$$\sinh \{\log_e(x + \sqrt{x^2 + 1})\} \equiv x$$

ii) Show that

$$\int \frac{dx}{\sqrt{16x^2 + 9}} = \frac{1}{4} \log_e \{4x + \sqrt{16x^2 + 9}\} + c$$

iii) Show that

$$\frac{d}{dx}(x\sqrt{16x^2 + 9}) \equiv 2\sqrt{16x^2 + 9} - \frac{9}{\sqrt{16x^2 + 9}}$$

iv) Hence show that

$$\int_0^1 \sqrt{16x^2 + 9}\, dx = \frac{5}{2} + \frac{9}{8} \log_e 3$$ (NICCEA)

11 i) Show that

$$\int \frac{x^2}{\sqrt{16 - 9x^2}}\, dx = \frac{16}{27} \int \sin^2 \theta\, d\theta$$

ii) If $g(x)$ is a continuous function, show that

$$\int_0^{12} \frac{g(x)}{\sqrt{49 + 4x^2}}\, dx = \frac{1}{2} \int_0^{\log_e 7} g\left(\frac{7}{2} \sinh u\right) du$$ (NICCEA)

12 Find the general solution of the differential equation

$$(\sinh x)\frac{dy}{dx} + (2\cosh x)y = \sinh x$$ (NEAB)

13 a) The locus of a point (x, y) defined by the parametric equations

$$x = \cosh v$$
$$y = \sinh v$$

together with the point P at which $v = u$, where $u > 0$, is shown in the figure.

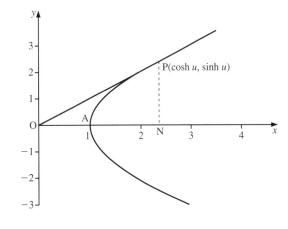

i) Show that the area bounded by the curve AP, the ordinate line PN and the x-axis is given by

$$\int_0^u \sinh^2 v\, dv$$

ii) Show that the area bounded by the curve AP, the straight line OP and the x-axis is $\frac{1}{2}u$.

b) Sketch the curve defined by $x = \cos\theta$, $y = \sin\theta$.

If the co-ordinates of P′ are $(\cos\phi, \sin\phi)$, shade a region whose area is $\frac{1}{2}\phi$. Comment on the similarities between the figure on page 215 and your sketch. (NICCEA)

14 Differentiate $\sqrt{(x^2 - 1)}$.

Show that

$$\int_1^{\frac{5}{4}} \cosh^{-1}x\,dx = a\ln 2 + b$$

where a and b are rational numbers to be determined. (NEAB)

15 The diagram on the right shows a region R in the x–y plane bounded by the curve $y = \sinh x$, the x-axis and the line AB which is perpendicular to the x-axis.

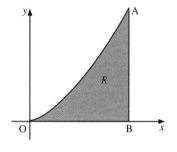

a) Given that $AB = \frac{4}{3}$, show that $OB = \ln 3$.

b) i) Show that

$$\cosh(\ln k) = \frac{k^2 + 1}{2k}$$

ii) Show that the area of the region R is $\frac{2}{3}$.

c) i) Show that

$$\int_0^{\ln 3} \sinh^2 x\,dx = \frac{1}{4}[\sinh(\ln 9) - \ln 9]$$

ii) Hence find, correct to three significant figures, the volume swept out when the region R is rotated through an angle of 2π radians about the x-axis. (NEAB)

16 a) Given that $u = \frac{1}{2}(e^y - e^{-y})$, prove that $y = \ln(u + \sqrt{(u^2 + 1)})$.

b) Using the substitution $x = \sinh\theta$, show that

$$\int \frac{x^2}{\sqrt{(1 + x^2)}}\,dx = \frac{1}{2}[x\sqrt{(1 + x^2)} - \ln(x + \sqrt{(1 + x^2)})] + k$$

where k is an arbitrary constant. (EDEXCEL)

17 The diagram shows a sketch of the curve defined by the parametric equations

$$x = \sinh t \quad y = \cosh t \quad t \geqslant 0$$

together with the tangent to the curve at the point P $(\sinh p, \cosh p)$. The curve meets the y-axis at the point Q and the tangent at P meets the y-axis at the point R.

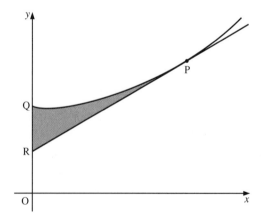

a) Show that the equation of the tangent to the curve at P is

$$y\cosh p - x\sinh p = 1$$

b) Given that R is the point $(0, \frac{1}{2})$, show that

$$p = \ln(2 + \sqrt{3})$$

c) Show that the area A (shown shaded in the diagram) bounded by QR, RP and the arc PQ is given by

$$A = \int_0^p \cosh^2 t \, dt - \frac{5\sqrt{3}}{4}$$

d) Hence find the value of A in the form

$$a \ln(2 + \sqrt{3}) + b\sqrt{3}$$

where a and b are rational numbers to be determined.   (NEAB)

18 a) Use the power series for $\sin x$ to show that, for small values of x,

$$\sin^3 x \approx x^3 \left(1 - \frac{x^2}{6} + \frac{x^4}{120} \right)^3$$

b) Hence, or otherwise, find the constants a, b, c in the approximation

$$\sin^3 x \approx ax^3 + bx^5 + cx^7$$

c) Find a similar approximation for $x^2 \sinh x$ for small values of x.

d) Show that

$$\lim_{x \to 0} \frac{x^2 \sinh x - \sin^3 x}{x^5} = \frac{2}{3} \qquad \text{(NEAB)}$$

11 Conics

That an extensive theory of the conics was obtained is eloquent testimony to the brilliance of Archimedes and Apollonius.
JEREMY J. GRAY

Generating conics

If we take a solid, right circular cone and, in any direction, cut a plane section through it, we obtain a curve which is a member of the class of curves known as **conics** or **conic sections**.

It follows that the shape of the curve so obtained is determined by the direction in which we make the cut: that is, on the inclination, θ, of the plane section to the axis, as the figure below shows.

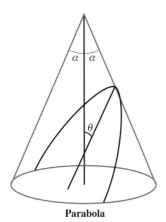

Parabola

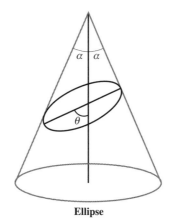

Ellipse

Hyperbola

Hence, with the cone standing on a horizontal plane, if we cut in a direction parallel to the slant height of the cone, whereby $\theta = \alpha$, we obtain a **parabola**.

If we cut in a direction for which $\alpha < \theta < \dfrac{\pi}{2}$, we obtain an **ellipse**.

If we cut in a direction, not through the vertex, for which $\theta < \alpha$, we obtain a **hyperbola**.

If we cut horizontally through the cone $\left(\text{that is, } \theta = \dfrac{\pi}{2}\right)$, we obtain a **circle**.

The study of the parabola, the ellipse and the hyperbola as sections of the same cone originated with the Greek geometer Apollonius, who flourished about 280 BC. They were not defined analytically as loci until the seventeenth century, largely due to the work of the renowned French mathematician René Descartes (1596–1650), and of the English mathematician John Wallis (1616–1703).

Conics as loci

Analytically we define a conic as the locus of a point which moves so that the ratio of its distance from a fixed point to its distance from a fixed line is constant.

The fixed point is called the **focus**, and the fixed line the **directrix**. The constant ratio is known as the **eccentricity** of the conic and is denoted by e.

Hence, in the figure on the right, where the point P is describing a conic, we have

$$PF = ePT$$

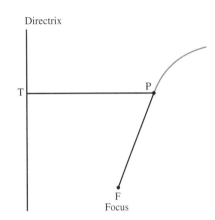

When $e = 1$, the conic is a **parabola**.
When $0 < e < 1$, the conic is an **ellipse**.
When $e > 1$, the conic is a **hyperbola**.

The circle (which we met in *Introducing Pure Mathematics*, pages 220–7) may be treated as the limiting case of an ellipse, in which $e = 0$ (see pages 222–6).

Parabola

Let the focus, F, be $(a, 0)$ and the directrix be $x = -a$. Then for the point $P(x, y)$, we have

$$PT = x + a \qquad PF = \sqrt{(x - a)^2 + y^2}$$

But $PT = PF$, since for a parabola $e = 1$. Hence, we obtain

$$(x - a)^2 + y^2 = (x + a)^2$$

which gives

$$y^2 = 4ax$$

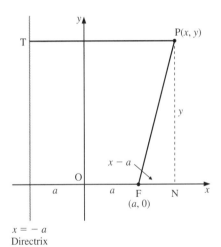

This is the **standard equation** for a parabola, an example of which is shown at bottom right.

Common parametric equations for the parabola $y^2 = 4ax$ are

$$x = at^2 \quad \text{and} \quad y = 2at$$

where t is the parameter.

The chord of a parabola through its focus, and perpendicular to its axis, is called the **latus rectum**. Thus, in the diagram on the right, CD is the latus rectum.

Half the length of this chord (FC or FD) is known as the **semi latus rectum**.

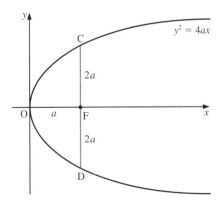

From the equation $y^2 = 4ax$, we see that the coordinates of C are $(a, 2a)$ and those of D are $(a, -2a)$. Hence, the length of the latus rectum is $4a$ and that of the semi latus rectum is $2a$. (See also pages 223 and 231.)

Note

- All quadratic curves are parabolas.
- All quadratic curves are similar.

Example 1 Find the focus and directrix of each of the parabolas

a) $y^2 = 32x$ **b)** $x^2 = 16(y + 1)$

SOLUTION

For a general parabola $y^2 = 4ax$, the focus is at $(a, 0)$ and the directrix is $x = -a$.

a) For the parabola $y^2 = 32x$, $a = 8$.

Therefore, the focus is $(8, 0)$, and the directrix is $x = -8$.

b) For the parabola $x^2 = 16(y + 1)$, the x- and y-axes have been interchanged and y has been translated to $y + 1$.

From the equation, $a = 4$. Thus the focus for the parabola $x^2 = 16y$ would be $(0, 4)$, which, after translation, becomes $(0, 3)$ for the parabola $x^2 = 16(y + 1)$.

Similarly, the directrix for $x^2 = 16y$, which would be $y = -4$, becomes $y = -5$ after translation.

Example 2 Find where the tangent to the parabola $y^2 = 8x$ at the point $(2t^2, 4t)$ meets the directrix.

SOLUTION

We have

$$x = 2t^2 \quad \Rightarrow \quad \frac{dx}{dt} = 4t$$

$$y = 4t \quad \Rightarrow \quad \frac{dy}{dt} = 4$$

which give

$$\frac{dy}{dx} = \frac{dy}{dt} \div \frac{dx}{dt} = \frac{1}{t}$$

Using $y - y_1 = m(x - x_1)$, we find the equation of the tangent at $(2t^2, 4t)$:

$$y - 4t = \frac{1}{t}(x - 2t^2)$$

$$\Rightarrow \quad yt = x + 2t^2 \qquad [1]$$

The directrix of the standard parabola $y^2 = 4ax$ is $x = -a$. Therefore, for the parabola $y^2 = 8x$, $a = 2$. Hence, the directrix is $x = -2$.

Substituting $x = -2$ in [1], we obtain

$$y = 2t - \frac{2}{t}$$

Therefore, the tangent meets the directrix at $\left(-2, 2t - \frac{2}{t}\right)$.

Example 3

a) Find the equation of the normal to the parabola $y^2 = 8x$ at the point $T(2t^2, 4t)$.

b) Find where the normals to the parabola at the points $P(2p^2, 4p)$ and $Q(2q^2, 4q)$ intersect.

Note When given the parametric equations for a parabola, it is much easier to stay with these equations. So, do **not** revert to the cartesian equation.

SOLUTION

a) We have

$$x = 2t^2 \quad \Rightarrow \quad \frac{dx}{dt} = 4t$$

$$y = 4t \quad \Rightarrow \quad \frac{dy}{dt} = 4$$

which give

$$\frac{dy}{dx} = \frac{dy}{dt} \div \frac{dx}{dt} = \frac{1}{t}$$

Therefore, the gradient of the normal is $-t$.

Using $y - y_1 = m(x - x_1)$, we find the equation of the normal at $(2t^2, 4t)$:

$$y - 4t = -t(x - 2t^2)$$
$$\Rightarrow \quad y + tx = 4t + 2t^3$$

b) To find the equation of the normal at the point P, we just substitute p for t. Therefore, the equation of the normal at the point P is

$$y + px = 4p + 2p^3 \qquad [1]$$

Similarly, the equation of the normal at Q is

$$y + qx = 4q + 2q^3 \qquad [2]$$

Subtracting [2] from [1], we find that these normals intersect when

$$px - qx = 4p + 2p^3 - 4q - 2q^3$$

Note In all such situations, where p and q are similarly considered, $(p - q)$ will be a factor. Therefore, we look for this factor, remove it and check that the answer is symmetrical in p and q.

Using $p^3 - q^3 = (p - q)(p^2 + pq + q^2)$, we obtain

$$(p - q)x = 4(p - q) + 2(p - q)(p^2 + pq + q^2)$$
$$\Rightarrow \quad x = 2(p^2 + pq + q^2 + 2) \qquad [3]$$

Substituting [3] into [1], we have

$$y = 4p + 2p^3 - 2p(p^2 + pq + q^2 + 2)$$
$$\Rightarrow \quad y = -2pq(p + q)$$

Therefore, the normals intersect at $(2(p^2 + pq + q^2 + 2), -2pq(p + q))$.

Exercise 11A

1 Find the focus and directrix of each of the following parabolas.

a) $y^2 = 16x$ b) $y^2 = 28x$ c) $x^2 = 8y$ d) $x^2 = -16y$

e) $y^2 + 12x = 0$ f) $(y + 1)^2 = 32x$ g) $(y - 2)^2 - 8(x - 3) = 0$

2 Find in cartesian form an equation of the parabola whose focus and directrix are respectively

a) $(3, 0), \quad x + 3 = 0$ b) $(4, 0), \quad x = -4$

c) $(0, 2), \quad y = -2$ d) $(0, -5), \quad y = 5$

3 Find the equation of the tangent to the parabola $y^2 = 20x$ at

a) the point $T(5t^2, 10t)$ b) the point $P(5p^2, 10p)$

c) the point $S(5, 10)$ d) the point $R(20, 20)$

4 a) Find the equation of the normal to the parabola $y^2 = 8x$ at the point $(2, 4)$.

    b) Find where this normal meets the parabola again.

Ellipse

Let the focus be $(ae, 0)$ and the directrix be $x = \dfrac{a}{e}$. Since, for an ellipse, e is less than 1, the directrix is further from the origin than the focus.

For the point $P(x, y)$, we have

$$PF = \sqrt{(x - ae)^2 + y^2} \qquad PT = \frac{a}{e} - x$$

Since the ratio of the distance of P from the focus to the distance of P from the directrix is e, we have

$$\frac{PF}{PT} = e \quad \Rightarrow \quad PF = ePT$$

which gives

$$\sqrt{(x - ae)^2 + y^2} = a - ex$$

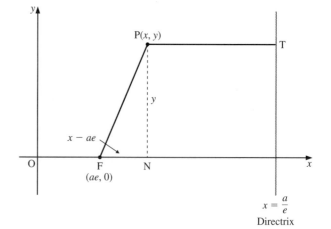

Squaring both sides, we obtain

$$(x - ae)^2 + y^2 = (a - ex)^2$$

$$\Rightarrow \quad x^2 - 2aex + a^2e^2 + y^2 = a^2 - 2aex + e^2x^2$$

$$\Rightarrow \quad x^2(1 - e^2) + y^2 = a^2(1 - e^2)$$

$$\Rightarrow \quad \frac{x^2}{a^2} + \frac{y^2}{a^2(1 - e^2)} = 1$$

We express this in the form

$$\frac{x^2}{a^2} + \frac{y^2}{b^2} = 1$$

where

$$b^2 = a^2(1 - e^2) \quad \Rightarrow \quad e^2 = 1 - \frac{b^2}{a^2}$$

Hence, the **standard equation** for an ellipse is

$$\frac{x^2}{a^2} + \frac{y^2}{b^2} = 1$$

where

$$e^2 = 1 - \frac{b^2}{a^2}.$$

Note   An ellipse is symmetrical with respect to its axes. Hence, it has two foci, one at $(ae, 0)$ and the other at $(-ae, 0)$, and two directrices, $x = \dfrac{a}{e}$ and $x = -\dfrac{a}{e}$.

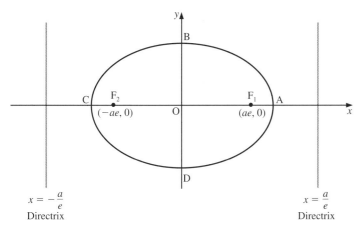

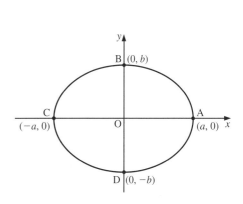

The longer axis, AC, is called the **major axis**, and the shorter axis, BD, is called the **minor axis**. We see that the length of the major axis is $2a$ and that of minor axis is $2b$.

A chord which passes through either focus, and which is perpendicular to the major axis, is called a **latus rectum**. Half the length of this chord is known as a **semi latus rectum**. (See also pages 219 and 231.)

The parametric equations for the ellipse $\dfrac{x^2}{a} + \dfrac{y^2}{b} = 1$ are

$$x = a\cos\theta \quad \text{and} \quad y = b\sin\theta$$

where θ is the **eccentric angle** of the ellipse. (Further discussion of this parameter and its use in projective geometry is beyond the scope of this book.)

Example 4

a) Find the eccentricity of the ellipse $\dfrac{x^2}{9} + \dfrac{y^2}{4} = 1$.

b) State the coordinates of its foci.

c) State the equations of its directrices.

SOLUTION

a) The general equation of an ellipse is $\dfrac{x^2}{a^2} + \dfrac{y^2}{b^2} = 1$. Hence, for the ellipse $\dfrac{x^2}{9} + \dfrac{y^2}{4} = 1$, we have $a = 3$, $b = 2$.

For an ellipse, $e^2 = 1 - \dfrac{b^2}{a^2}$, which in this case gives

$$e^2 = 1 - \frac{4}{9} = \frac{5}{9}$$

Therefore, the eccentricity is $\dfrac{\sqrt{5}}{3}$.

b) The foci are $(\pm ae, 0)$, which in this case gives $(\sqrt{5}, 0)$ and $(-\sqrt{5}, 0)$.

c) The directrices are $x = \pm\dfrac{a}{e}$, which in this case give

$$x = \pm\frac{3}{\frac{\sqrt{5}}{3}}$$

Therefore, its directrices are

$$x = \frac{9}{\sqrt{5}} \quad \text{and} \quad x = -\frac{9}{\sqrt{5}}$$

Example 5 Find the tangent and the normal to the ellipse $\dfrac{x^2}{a^2} + \dfrac{y^2}{b^2} = 1$ at the point $(a\cos\theta, b\sin\theta)$.

SOLUTION

We have

$$x = a\cos\theta \quad \Rightarrow \quad \frac{dx}{d\theta} = -a\sin\theta$$

$$y = b\sin\theta \quad \Rightarrow \quad \frac{dy}{d\theta} = b\cos\theta$$

which give

$$\frac{\mathrm{d}y}{\mathrm{d}x} = \frac{-b\cos\theta}{a\sin\theta}$$

Using $y - y_1 = m(x - x_1)$, we find the equation of the tangent:

$$y - b\sin\theta = \frac{-b\cos\theta}{a\sin\theta}(x - a\cos\theta)$$

$$\Rightarrow \quad a\sin\theta\, y + b\cos\theta\, x = ab(\cos^2\theta + \sin^2\theta)$$

$$\Rightarrow \quad a\sin\theta\, y + b\cos\theta\, x = ab$$

Note We can check this by taking $\dfrac{x^2}{a^2} + \dfrac{y^2}{b^2} = 1$, and replacing

x^2 by $(x \times \text{abscissa})$ and y^2 by $(y \times \text{ordinate})$

to obtain the equation of the tangent.

Hence, we have

$$\frac{xa\cos\theta}{a^2} + \frac{yb\sin\theta}{b^2} = 1$$

$$\Rightarrow \quad xb\cos\theta + ya\sin\theta = ab$$

as above.

To find the equation of the normal, we need its gradient, which is given by

$$\text{Gradient of normal} = -\frac{1}{\text{Gradient of tangent}}$$

$$= -\frac{1}{\dfrac{-b\cos\theta}{a\sin\theta}} = \frac{a\sin\theta}{b\cos\theta}$$

So, the equation of the normal is

$$y - b\sin\theta = \frac{a\sin\theta}{b\cos\theta}(x - a\cos\theta)$$

$$\Rightarrow \quad yb\cos\theta = xa\sin\theta - a^2\sin\theta\cos\theta + b^2\sin\theta\cos\theta$$

$$\Rightarrow \quad yb\cos\theta = xa\sin\theta + (b^2 - a^2)\sin\theta\cos\theta$$

Example 6 Find the area of the ellipse whose major axis is $2a$ and minor axis is $2b$.

SOLUTION

We have

$$\text{Area} = \int y\,\mathrm{d}x$$

To make the integration easier, we use the parametric equation for y, which gives

$$\int y\,\mathrm{d}x = \int b\sin\theta\,\mathrm{d}x$$

We cannot integrate a function in θ with respect to x. Therefore, we must convert the integration with respect to x to an integration with respect to θ. Hence, we have

$$\text{Area of ellipse} = \int_{-a}^{a} y\,dx = \int_{0}^{2\pi} b\sin\theta\,\frac{dx}{d\theta}\,d\theta$$

Using $x = a\cos\theta$, this gives

$$\text{Area} = \int_{0}^{2\pi} b\sin\theta(-a\sin\theta)\,d\theta$$

Because the ellipse is symmetrical about its axes, we can express the integral as

$$\text{Area} = 4\int_{0}^{\frac{\pi}{2}} ab\sin^2\theta\,d\theta$$

Using $\cos 2\theta = 1 - 2\sin^2\theta$, we obtain

$$\text{Area} = 4ab\int_{0}^{\frac{\pi}{2}} \frac{1}{2}(1 - \cos 2\theta)\,d\theta$$

$$= 4ab\left[\frac{1}{2}\theta - \frac{1}{4}\sin 2\theta\right]_{0}^{\frac{\pi}{2}}$$

Therefore, the area of an ellipse is πab.

Exercise 11B

1 Find the eccentricity, foci and directrices of each of the following ellipses.

a) $\dfrac{x^2}{16} + \dfrac{y^2}{9} = 1$

b) $\dfrac{x^2}{49} + \dfrac{y^2}{16} = 1$

c) $\dfrac{x^2}{25} + \dfrac{y^2}{16} = 1$

d) $\dfrac{x^2}{4} + \dfrac{y^2}{9} = 4$

e) $\dfrac{(x-1)^2}{25} + \dfrac{(y+2)^2}{9} = 1$

2 Find, in cartesian form, the equation of each ellipse with the focus and the directrix as given.

a) $(3,0),\quad x = 12$

b) $(2,0),\quad x = 18$

c) $(0,4),\quad y = 8$

d) $(0,3),\quad y = 15$

3 Find the equation of **a)** the tangent and **b)** the normal to the ellipse $\dfrac{x^2}{25} + \dfrac{y^2}{16} = 1$ at the point $(5\cos\theta, 4\sin\theta)$.

4 Sketch the curve given in polar coordinates by the equation

$$r = \frac{2a}{3 + 2\cos\theta}$$

Prove that this curve is an ellipse and identify its foci.

Hyperbola

Let the focus be $(ae, 0)$ and the directrix be $x = \dfrac{a}{e}$.

Since, for a hyperbola, e is greater than 1, the directrix is situated between the origin and the focus.

For the point $P(x, y)$, we have

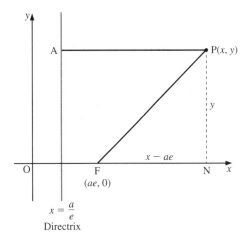

$$PF = \sqrt{(x - ae)^2 + y^2} \qquad PT = x - \frac{a}{e}$$

Since $\dfrac{PF}{PT} = e \quad \Rightarrow \quad PF = ePT$, we have

$$\sqrt{(x - ae)^2 + y^2} = e\left(x - \frac{a}{e}\right)$$

$$= ex - a$$

Squaring both sides, we obtain

$$(x - ae)^2 + y^2 = (ex - a)^2$$

$$\Rightarrow \quad x^2 - 2aex + a^2e^2 + y^2 = e^2x^2 - 2aex + a^2$$

$$\Rightarrow \quad x^2(1 - e^2) + y^2 = a^2(1 - e^2)$$

which gives

$$\frac{x^2}{a^2} + \frac{y^2}{a^2(1 - e^2)} = 1$$

But $e > 1$, therefore $a^2(1 - e^2)$ is negative.

Hence, the **standard equation** for a hyperbola is

$$\frac{x^2}{a^2} - \frac{y^2}{b^2} = 1$$

where

$$b^2 = a^2(e^2 - 1) \quad \text{or} \quad e^2 = 1 + \frac{b^2}{a^2}$$

As x and y become large, we have

$$\frac{x^2}{a^2} \to \frac{y^2}{b^2} \quad \Rightarrow \quad y \to \pm\frac{bx}{a}$$

Therefore, the asymptotes of a hyperbola are $y = \pm\dfrac{b}{a}x$

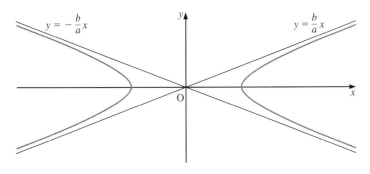

Common parametric equations for the hyperbola $\dfrac{x^2}{a^2} - \dfrac{y^2}{b^2} = 1$ are

$$x = a \sec \theta \quad \text{and} \quad y = b \tan \theta$$

where θ is the **eccentric angle** of the hyperbola. (Further discussion of this parameter is beyond the scope of this book.)

Alternatively, hyperbolic functions may be used (see pages 189–91). In this case, the parametric equations are

$$x = a \cosh \phi \quad \text{and} \quad y = b \sinh \phi$$

Like the ellipse, the hyperbola is symmetrical with respect to its axes. Hence, again there are two foci, one at $(ae, 0)$ and the other at $(-ae, 0)$, and two directrices, $x = \dfrac{a}{e}$ and $x = -\dfrac{a}{e}$.

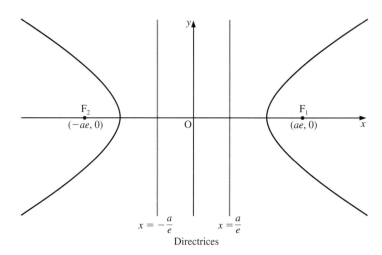

Rectangular hyperbola

When $a = b$, the asymptotes of the hyperbola are $y = x$ and $y = -x$, which are perpendicular to each other. Hence, such a hyperbola (shown on the right) is called a **rectangular hyperbola**.

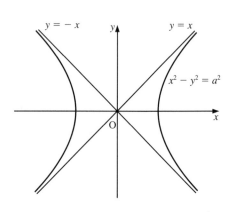

The general equation of a hyperbola, $\dfrac{x^2}{a^2} - \dfrac{y^2}{b^2} = 1$, becomes

$$x^2 - y^2 = a^2$$

for a rectangular hyperbola, as $a = b$. That is,

$$(x + y)(x - y) = a^2$$

Rotating the axes through $45°$ and designating the new axes (which are the asymptotes) X and Y, we transform this equation to

$$XY = \frac{a^2}{2}$$

where

$$X = \frac{1}{\sqrt{2}} (x - y) \quad \text{and} \quad Y = \frac{1}{\sqrt{2}} (x + y).$$

Hence, for a rectangular hyperbola, we have the equation

$$xy = c^2$$

where $c^2 = \dfrac{a^2}{2}$.

Common parametric equations for the rectangular hyperbola are

$$x = ct \quad \text{and} \quad y = \frac{c}{t}$$

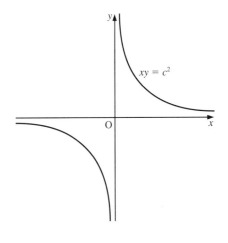

Example 7 Find the eccentricity of the hyperbola $\dfrac{x^2}{16} - \dfrac{y^2}{9} = 1$, and the coordinates of its foci.

SOLUTION

For a hyperbola $\dfrac{x^2}{a^2} - \dfrac{y^2}{b^2} = 1$, we have $e^2 = 1 + \dfrac{b^2}{a^2}$.

Therefore, for the hyperbola $\dfrac{x^2}{16} - \dfrac{y^2}{9} = 1$, we obtain

$$e^2 = 1 + \frac{9}{16} = \frac{25}{16}$$

Hence, the eccentricity is $\dfrac{5}{4}$.

When $\dfrac{x^2}{a^2} - \dfrac{y^2}{b^2} = 1$, the foci are $(\pm ae, 0)$. Therefore, for $\dfrac{x^2}{16} - \dfrac{y^2}{9} = 1$, the

foci are $\left(\pm \dfrac{5}{4} \times 4, 0 \right)$, giving $(\pm 5, 0)$.

Example 8 Find the equation of the tangent to $xy = c^2$ at the point

$\left(ct, \dfrac{c}{t} \right)$ Hence find the equation of the tangent to $xy = 16$ at the points

a) $(8, 2)$ and **b)** $\left(-12, -\dfrac{4}{3} \right)$.

SOLUTION

To find the equation of the tangent, we need its gradient. Hence, we have

$$x = ct \quad \Rightarrow \quad \frac{dx}{dt} = c$$

$$y = \frac{c}{t} \quad \Rightarrow \quad \frac{dy}{dt} = -\frac{c}{t^2}$$

which give

$$\frac{dy}{dx} = -\frac{1}{t^2}$$

Therefore, the equation of the tangent is

$$y - \frac{c}{t} = -\frac{1}{t^2}(x - ct)$$

$$\Rightarrow \quad t^2 y + x = 2ct$$

For the hyperbola $xy = 16$, $c = 4$. Therefore, the equation of the tangent at $\left(4t, \dfrac{4}{t} \right)$ is

$$t^2 y + x = 8t$$

a) At the point $(8, 2)$, we have $4t = 8$, which gives $t = 2$.

Therefore, the equation of the tangent at this point is

$$4y + x = 16$$

b) At the point $\left(-12, -\dfrac{4}{3} \right)$, we have $4t = -12$, which gives $t = -3$.

Therefore, the equation of the tangent at this point is

$$9y - x + 24 = 0$$

Polar equation of a conic

The polar equation (see pages 45–7) of a conic is formed by positioning the pole at the focus and keeping the directrix at $x = \dfrac{a}{e}$.

With reference to the diagram on the right, the locus of a general point, P, expressed in polar coordinates (r, θ) satisfies the condition (see page 219).

$$PF = e\,PT$$

where $PF = r$ and $PT = \dfrac{a}{e} - r\cos\theta$.

Hence, we have

$$r = e\left(\frac{a}{e} - r\cos\theta \right)$$

$$\Rightarrow \quad r(1 + e\cos\theta) = a$$

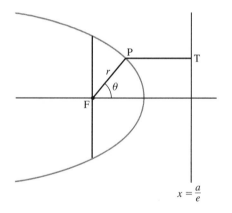

which gives the polar equation of a conic as

$$r = \frac{a}{1 + e\cos\theta}$$

This equation is valid for **all** conics whose pole and directrix are as specified above.

When $e = 0$, the conic is a circle.
When $0 < e < 1$, the conic is an ellipse.
When $e = 1$, the conic is a parabola.
When $e > 1$, the conic is a hyperbola.

We can express this equation in several different forms.

For example, when $x = d$ is used as the equation of the directrix, we have

$$r = \frac{ed}{1 + \cos\theta}$$

When the focus is at $(ae, 0)$ and the directrix is $x = \dfrac{a}{e}$, we have as the equation of the general conic

$$r = \frac{b^2}{a(1 + e\cos\theta)}$$

which gives

$$r = \frac{a(1 - e^2)}{1 + e\cos\theta} \quad \text{for an ellipse}$$

$$r = \frac{a(e^2 - 1)}{1 + e\cos\theta} \quad \text{for a hyperbola}$$

We can also derive a similar polar equation in terms of l, the length of the semi latus rectum (see also pages 219 and 223).

With reference to the diagram on the right, we have

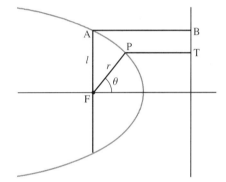

$$PT = ePT \quad \Rightarrow \quad r = e(AB - r\cos\theta)$$

The point A is on the conic, so we have

$$FA = e\,AB$$

FA is the semi latus rectum, so we obtain

$$l = e\,AB \quad \Rightarrow \quad AB = \frac{l}{e}$$

which gives

$$r = e\left(\frac{l}{e} - r\cos\theta\right)$$

$$\Rightarrow \quad r(1 + e\cos\theta) = l$$

That is, we have

$$r = \frac{l}{1 + e\cos\theta}$$

Note The distance between the directrix and the focus is $\dfrac{l}{e}$.

Exercise 11C

1 Find the eccentricity, foci and directrices of each of the following hyperbolas.

a) $\dfrac{x^2}{16} - \dfrac{y^2}{9} = 1$ **b)** $\dfrac{x^2}{49} - \dfrac{y^2}{16} = 1$ **c)** $\dfrac{x^2}{25} - \dfrac{y^2}{16} = 1$

d) $\dfrac{x^2}{4} - \dfrac{y^2}{9} = 4$ **e)** $\dfrac{(x-1)^2}{25} - \dfrac{(y+2)^2}{9} = 1$

2 Find, in cartesian form, the equation of each hyperbola with the focus and the directrix as given.

a) $(12, 0)$, $x = 3$ **b)** $(18, 0)$, $x = 2$ **c)** $(0, 8)$, $y = 4$ **d)** $(0, 15)$, $y = 3$

3 Find the equation of **a)** the tangent and **b)** the normal to the hyperbola $\dfrac{x^2}{25} - \dfrac{y^2}{16} = 1$ at the point $(5 \sec \theta, 4 \tan \theta)$.

Exercise 11D

1 Consider the parabola $y^2 = 4ax$.

  i) Show that the following parametric equations define a point on this parabola

$$x = at^2 \qquad y = 2at$$

  ii) Show that the tangent drawn to the parabola at the point $(at^2, 2at)$ has an equation given by

$$ty = x + at^2$$

Consider the points $P(ap^2, 2ap)$ and $Q(aq^2, 2aq)$, where $p \neq q$. Let M be the mid-point of PQ, and H be the intersection point of the tangents at P and Q.

  iii) Show that the line MH is parallel to the x-axis.   (NICCEA)

2 The equation of the curve C is $y^2 = 8x$. The point $P(2t^2, 4t)$ lies on C. The line through the point $(2, 0)$ perpendicular to the tangent to C at P intersects this tangent at the point Q.

  a) Find the coordinates of Q.

  b) Given that R is the mid-point of PQ, find the equation of the locus of R in cartesian form.   (WJEC)

3 The point P lies on the parabola with equation $y^2 = 4ax$, where a is a positive constant.

  a) Show that an equation of the tangent to the parabola at $P(ap^2, 2ap)$ is $py = x + ap^2$.

The tangents at the points $P(ap^2, 2ap)$ and $Q(aq^2, 2aq)$ $(p \neq q, p \neq 0, q \neq 0)$ meet at the point N.

  b) Find the coordinates of N.

Given further than N lies on the directrix of the parabola,

  c) write down a relationship between p and q.   (EDEXCEL)

4 The line with equation $y = mx + c$ is a tangent to the ellipse with equation $\dfrac{x^2}{a^2} + \dfrac{y^2}{b^2} = 1$.

  a) Show that $c^2 = a^2 m^2 + b^2$.

  b) Hence, or otherwise, find the equations of the tangents from the point $(3, 4)$ to the ellipse with equation $\dfrac{x^2}{16} + \dfrac{y^2}{25} = 1$.   (EDEXCEL)

5 An ellipse has equation $\dfrac{x^2}{a^2} + \dfrac{y^2}{b^2} = 1$, where a and b are positive constants and $a > b$.

  a) Find an equation of the tangent at the point $P(a \cos t, b \sin t)$.

  b) Find an equation of the normal at the point $P(a \cos t, b \sin t)$.

The normal at P meets the x-axis at the point Q. The tangent at P meets the y-axis at the point R.

c) Find, in terms of a, b and t, the coordinates of M, the mid-point of QR.

Given that $0 < t < \dfrac{\pi}{2}$,

d) show that, as t varies, the locus of M has equation $\left(\dfrac{2ax}{a^2 - b^2}\right)^2 + \left(\dfrac{b}{2y}\right)^2 = 1$.　　　(EDEXCEL)

6 The point $P(2\cos\theta, 3\sin\theta)$ lies on the ellipse $\dfrac{x^2}{4} + \dfrac{y^2}{9} = 1$.

a) Find the equation of the tangent to the ellipse at the point $P(2\cos\theta, 3\sin\theta)$, where $\theta \neq 0$.

b) Given that the tangent in part **a** passes through the point $(2, -6)$, show that

$$\cos\theta - 2\sin\theta = 1$$

c) Solve the equation in part **b** for $0° \leqslant \theta \leqslant 360°$ and deduce the coordinates of P.　　　(WJEC)

7 A curve C has equations

$$x = ct \quad y = \frac{c}{t} \quad t \neq 0$$

where c is a constant and t is a parameter.

a) Show that an equation of the normal to C at the point where $t = p$ is given by

$$py + cp^4 = p^3 x + c$$

b) Verify that this normal meets C again at the point at which $t = q$, where

$$qp^3 + 1 = 0 \quad \text{(EDEXCEL)}$$

8 The rectangular hyperbola C has equation $xy = c^2$, where c is a positive constant.

a) Show that the tangent to C at the point $P\left(cp, \dfrac{c}{p}\right)$ has equation

$$p^2 y = -x + 2cp$$

The point Q has coordinates $Q\left(cq, \dfrac{c}{q}\right)$, $q \neq p$. The tangents to C at P and Q meet at N. Given that $p + q \neq 0$,

b) show that the y-coordinate of N is $\dfrac{2c}{p + q}$.

The line joining N to the origin O is perpendicular to the chord PQ.

c) Find the numerical value of $p^2 q^2$.　　　(EDEXCEL)

9 The ellipse C has parametric equations

$$x = 2 + 3\cos\theta \quad y = 2\sin\theta$$

a) Obtain the cartesian equation of C and find the eccentricity of the ellipse.

b) Write down the coordinates of the foci.

c) Sketch C, stating the coordinates of its intersections with the axes.

The arc of the curve C between $\theta = 0$ and $\theta = \frac{1}{2}\pi$ is rotated through 2π about the x-axis.

d) Show that the area S of the resulting surface of revolution is given by

$$S = 4\pi \int_0^{\frac{\pi}{2}} \sin\theta(9 - 5\cos^2\theta)^{\frac{1}{2}}\,d\theta$$

Using the substitution $(\sqrt{5})\cos\theta = 3\sin u$, or otherwise, find the value of S, to two decimal places.   (EDEXCEL)

10 The curve C_1 is that arc of the hyperbola with equation

$$\frac{x^2}{9a^2} - \frac{y^2}{a^2} = 1 \quad a > 0$$

which contains the point $P(3a\cosh\theta, a\sinh\theta)$.

a) Show that the equation of the normal to C_1 at the point P can be written in the form

$$y\cosh\theta + 3x\sinh\theta = 10a\sinh\theta\cosh\theta$$

This normal meets the coordinate axes at A and B.

b) Show that, as θ varies, the locus C_2 of the mid-point of AB, is an arc of a hyperbola.

For each of the arcs C_1 and C_2

c) give the coordinates of any points of intersection with the coordinate axes and the equations of any asymptotes

d) find the eccentricity of the hyperbola and state the coordinates of the focus and the equation of the corresponding directrix.   (EDEXCEL)

11 The points $S\left(s, \dfrac{1}{s}\right)$ and $T\left(t, \dfrac{1}{t}\right)$ lie on the curve $xy = 1$ and the line ST passes through the point $(1, 2)$.

a) Show that $s + t = 1 + 2st$.

b) The tangents to the curve at S and T meet at the point P. Show that the locus of P is given by $y = 2 - 2x$.   (WJEC)

12 The figure on the right shows a parabola and a circle. The circle passes through the parabola's focus S, a point P on the parabola and the intersection point Q of the directrix and the tangent at P.

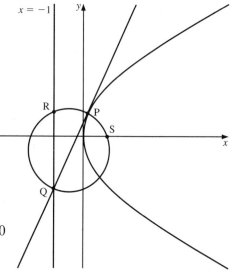

i) If the parabola has focus $S(1, 0)$ and directrix $x = -1$, show that its equation is $y^2 = 4x$.

Let the point P be given by $(t^2, 2t)$, where $t \neq 0$.

ii) Show that Q is the point $\left(-1, \dfrac{t^2 - 1}{t}\right)$.

iii) Verify that the focus S, the point P and the point Q lie on the circle with equation

$$tx^2 - t(t^2 - 1)x + ty^2 - (3t^2 - 1)y + t^3 - 2t = 0$$

iv) The circle intersects the directrix again at the point R. Find the coordinates of R.

v) Show that PR is parallel to the x-axis.   (NICCEA)

12 Further integration

Many a smale maketh a grate.
GEOFFREY CHAUCER

In *Introducing Pure Mathematics* (pages 433–8 and 445–7), we met integrals such as $\int x(x^2 + 1)^7 \, dx$, where we used the substitution $x^2 + 1 = u$, and $\int x e^{2x} \, dx$, where we integrated by parts.

We can extend these methods by using a greater variety of substitutions, including hyperbolic functions, to enable us to find integrals such as $\int \sqrt{9 + x^2} \, dx$.

To integrate more complicated expressions, we normally use the inverse function of a function rule given on page 294 of *Introducing Pure Mathematics*.

Inverse function of a function rule

$$\int f'(x)[f(x)]^n \, dx = \frac{1}{n+1} [f(x)]^{n+1} + c$$

It is usually quicker to differentiate by inspection the new expression than to obtain $f'(x)$ in the integrand, as shown in Examples 1 and 2.

Example 1 Find the constant k in

$$\int x(x^2 + 1)^7 \, dx = k(x^2 + 1)^8 + c$$

SOLUTION

Differentiating $(x^2 + 1)^8$, we obtain

$$\frac{d}{dx}(x^2 + 1)^8 = 8 \times 2x(x^2 + 1)^7 = 16x(x^2 + 1)^7$$

Therefore, we have

$$16 \int x(x^2 + 1)^7 \, dx = (x^2 + 1)^8 + c'$$

$$\Rightarrow \int x(x^2 + 1)^7 \, dx = \frac{1}{16}(x^2 + 1)^8 + c'$$

which gives $k = \frac{1}{16}$.

Example 2 Find the constant k in

$$\int \sin 2x \cos^7 2x \, dx = k \cos^8 2x + c$$

SOLUTION

Differentiating $\cos^8 2x$, we obtain

$$8 \times -2 \sin 2x \cos^7 2x = -16 \sin 2x \cos^7 2x$$

which gives $k = -\frac{1}{16}$.

Therefore, we have

$$\int \sin 2x \cos^7 2x \, dx = -\frac{1}{16} \cos^8 2x + c$$

Note In those cases where you experience difficulty in spotting the integral, use instead integration by substitution.

Integration by parts

Example 3 Evaluate $\int e^{3x} \cos 4x \, dx$.

SOLUTION

When faced with a product neither term of which will disappear after repeated differentiation, we usually use integration by parts until we obtain the integral with which we started as one of the terms on the right-hand side.

Hence, we have

$$\int e^{3x} \cos 4x \, dx = \frac{1}{3} e^{3x} \cos 4x - \int \frac{1}{3} e^{3x} \times -4 \sin 4x \, dx$$

$$= \frac{1}{3} e^{3x} \cos 4x + \frac{4}{3} \int e^{3x} \sin 4x \, dx$$

$$= \frac{1}{3} e^{3x} \cos 4x + \frac{4}{3} \left(\frac{1}{3} e^{3x} \sin 4x - \int \frac{1}{3} e^{3x} \times 4 \cos 4x \, dx \right)$$

$$= \frac{1}{3} e^{3x} \cos 4x + \frac{4}{9} e^{3x} \sin 4x - \frac{16}{9} \int e^{3x} \cos 4x \, dx$$

We now move the (original) integral on the RHS to the LHS:

$$\int e^{3x} \cos 4x \, dx + \frac{16}{9} \int e^{3x} \cos 4x \, dx = \frac{1}{3} e^{3x} \cos 4x + \frac{4}{9} e^{3x} \sin 4x + c$$

$$\Rightarrow \quad \frac{25}{9} \int e^{3x} \cos 4x \, dx = \frac{1}{3} e^{3x} \cos 4x + \frac{4}{9} e^{3x} \sin 4x + c$$

$$\Rightarrow \quad \int e^{3x} \cos 4x \, dx = \frac{9}{25} \left(\frac{1}{3} e^{3x} \cos 4x + \frac{4}{9} e^{3x} \sin 4x \right) + c'$$

Hence, we have

$$\int e^{3x} \cos 4x \, dx = \frac{3}{25} e^{3x} \cos 4x + \frac{4}{25} e^{3x} \sin 4x + c'$$

Integration of fractions

Example 4 Find $\int \dfrac{x^2 + 5}{x^2 + 9}\,dx$.

SOLUTION

When the power on the numerator is higher than, or equal to, the power on the denominator, we first divide the numerator by the denominator.

Hence, we have

$$\int \frac{x^2 + 5}{x^2 + 9}\,dx = \int \left(1 - \frac{4}{x^2 + 9}\right)dx$$

$$= x - \frac{4}{3}\tan^{-1}\left(\frac{x}{3}\right) + c$$

Example 5 Find $\int \dfrac{x^2 + 3x + 7}{x^2 + 2x + 4}\,dx$.

SOLUTION

Dividing, we obtain

$$\frac{x^2 + 3x + 7}{x^2 + 2x + 4} = 1 + \frac{x + 3}{x^2 + 2x + 4}$$

To integrate $\dfrac{x + 3}{x^2 + 2x + 4}$, we use

$$\int \frac{f'(x)}{f(x)}\,dx = \ln f(x) + c$$

(derived on page 422 of *Introducing Pure Mathematics*).

So, we need to obtain in the numerator a multiple of the differential of $x^2 + 2x + 4$. Hence, we convert

$$\frac{x + 3}{x^2 + 2x + 4} \quad \text{to} \quad \frac{\frac{1}{2}(2x + 2) + 2}{x^2 + 2x + 4}$$

which gives

$$\int \frac{x^2 + 3x + 7}{x^2 + 2x + 4}\,dx = \int \left(1 + \frac{x + 3}{x^2 + 2x + 4}\right)dx$$

$$= \int \left(1 + \frac{\frac{1}{2}(2x + 2)}{x^2 + 2x + 4} + \frac{2}{x^2 + 2x + 4}\right)dx$$

$$= \int 1\,dx + \int \frac{\frac{1}{2}(2x + 2)}{x^2 + 2x + 4}\,dx + 2\int \frac{dx}{(x + 1)^2 + 3}$$

Therefore, we have

$$\int \frac{x^2 + 3x + 7}{x^2 + 2x + 4}\,dx = x + \frac{1}{2}\ln(x^2 + 2x + 4) + \frac{2}{\sqrt{3}}\tan^{-1}\left(\frac{x + 1}{\sqrt{3}}\right) + c$$

Note When the denominator is the square root of a linear, or quadratic, function, a similar method is used.

Example 6 Find $\int \dfrac{x+3}{\sqrt{x^2+9}}\,\mathrm{d}x$.

SOLUTION

Ignoring the square-root sign, we note that the differential of x^2+9 is $2x$. Hence, we split the integral to obtain

$$\int \left(\frac{x}{\sqrt{x^2+9}} + \frac{3}{\sqrt{x^2+9}} \right) \mathrm{d}x$$

which gives

$$\sqrt{x^2+9} + 3\sinh^{-1}\left(\frac{x}{3}\right) + c$$

Therefore, we have

$$\int \frac{x+3}{\sqrt{x^2+9}}\,\mathrm{d}x = \sqrt{x^2+9} + 3\sinh^{-1}\left(\frac{x}{3}\right) + c$$

Example 7 Find $\int \dfrac{3x+8}{\sqrt{x^2+4x+9}}\,\mathrm{d}x$.

SOLUTION

Ignoring the square-root sign, we note that the differential of x^2+4x+9 is $2x+4$. Hence, we express the integral as

$$\int \frac{3x+8}{\sqrt{x^2+4x+9}}\,\mathrm{d}x = \frac{3}{2}\int \frac{2x+4}{\sqrt{x^2+4x+9}}\,\mathrm{d}x + \int \frac{2}{\sqrt{(x+2)^2+5}}\,\mathrm{d}x$$

$$= 3\sqrt{x^2+4x+9} + 2\sinh^{-1}\left(\frac{x+2}{\sqrt{5}}\right) + c$$

$$= 3\sqrt{x^2+4x+9} + 2\left[\ln(\sqrt{x^2+4x+9}+x+2) - \ln\sqrt{5}\right] + c$$

which gives

$$\int \frac{3x+8}{\sqrt{x^2+4x+9}} = 3\sqrt{x^2+4x+9} + 2\ln(\sqrt{x^2+4x+9}+x+2) + c'$$

Example 8 Find $\int \dfrac{x^2+3x+7}{\sqrt{x^2+2x+9}}\,\mathrm{d}x$.

SOLUTION

Proceeding as in Examples 4 and 6, we obtain

$$\int \frac{x^2+3x+7}{\sqrt{x^2+2x+9}}\,\mathrm{d}x = \int \left(\frac{x^2+2x+9}{\sqrt{x^2+2x+9}} + \frac{x-2}{\sqrt{x^2+2x+9}} \right) \mathrm{d}x$$

$$= \int \sqrt{x^2+2x+9}\,\mathrm{d}x + \int \frac{x-2}{\sqrt{x^2+2x+9}}\,\mathrm{d}x$$

Taking the first integral on the RHS, we have

$$\int \sqrt{x^2 + 2x + 9}\, dx = \int 1 \times \sqrt{x^2 + 2x + 9}\, dx$$

$$= x\sqrt{x^2 + 2x + 9} - \int x \times \frac{\frac{1}{2}(2x + 2)}{\sqrt{x^2 + 2x + 9}}\, dx$$

$$= x\sqrt{x^2 + 2x + 9} - \int \frac{x^2 + x}{\sqrt{x^2 + 2x + 9}}\, dx$$

$$= x\sqrt{x^2 + 2x + 9} - \int \frac{x^2 + 2x + 9 - x - 9}{\sqrt{x^2 + 2x + 9}}\, dx$$

$$= x\sqrt{x^2 + 2x + 9} - \int \sqrt{x^2 + 2x + 9}\, dx + \int \frac{x + 9}{\sqrt{x^2 + 2x + 9}}\, dx$$

which gives

$$2\int \sqrt{x^2 + 2x + 9}\, dx = x\sqrt{x^2 + 2x + 9} + \int \frac{x + 9}{\sqrt{x^2 + 2x + 9}}\, dx$$

$$\Rightarrow \quad \int \sqrt{x^2 + 2x + 9}\, dx = \frac{1}{2}x\sqrt{x^2 + 2x + 9} + \frac{1}{2}\int \frac{x + 9}{\sqrt{x^2 + 2x + 9}}\, dx$$

Hence, we obtain

$$\int \frac{x^2 + 3x + 7}{\sqrt{x^2 + 2x + 9}}\, dx = \frac{1}{2}x\sqrt{x^2 + 2x + 9} + \frac{1}{2}\int \left(\frac{x + 9}{\sqrt{x^2 + 2x + 9}} + \frac{2(x - 2)}{\sqrt{x^2 + 2x + 9}} \right) dx$$

$$= \frac{1}{2}x\sqrt{x^2 + 2x + 9} + \frac{1}{2}\int \left(\frac{\frac{3}{2}(2x + 2)}{\sqrt{x^2 + 2x + 9}} + \frac{2}{\sqrt{x^2 + 2x + 9}} \right) dx$$

$$= \frac{1}{2}x\sqrt{x^2 + 2x + 9} + \frac{3}{2}\sqrt{x^2 + 2x + 9} + \sinh^{-1}\left(\frac{x + 1}{2\sqrt{2}} \right) + c$$

$$= \frac{\sqrt{x^2 + 2x + 9}}{2}(3 + x) + \ln\left(\frac{\sqrt{x^2 + 2x + 9}}{2\sqrt{2}} + \frac{x + 1}{2\sqrt{2}} \right) + c$$

which gives

$$\int \frac{x^2 + 3x + 7}{\sqrt{x^2 + 2x + 9}}\, dx = \frac{1}{2}(3 + x)\sqrt{x^2 + 2x + 9} + \ln(\sqrt{x^2 + 2x + 9} + x + 1) + c'$$

Exercise 12A

1 Find each of these integrals.

a) $\displaystyle\int 2x(x^2 + 1)^5\, dx$

b) $\displaystyle\int x(x^2 - 1)^4\, dx$

c) $\displaystyle\int x^3(x^4 - 1)^7\, dx$

d) $\displaystyle\int x^2(1 - x^3)^4\, dx$

e) $\displaystyle\int \sin x \cos^5 x\, dx$

f) $\displaystyle\int \cosh x \sinh^4 x\, dx$

g) $\displaystyle\int \sinh 3x \cosh^4 3x\, dx$

h) $\displaystyle\int \sin^5 2x \cos 2x\, dx$

2 Find each of these integrals.

a) $\displaystyle\int e^x \cos x \, dx$

b) $\displaystyle\int e^x \sin 2x \, dx$

c) $\displaystyle\int e^{2x} \cos x \, dx$

d) $\displaystyle\int e^{3x} \cos 5x \, dx$

e) $\displaystyle\int e^{4x} \cosh 2x \, dx$

f) $\displaystyle\int e^{-7x} \sinh 3x \, dx$

3 Integrate each of the following with respect to x.

a) $\dfrac{x^2}{1+x^2}$

b) $\dfrac{x^2 - 4}{x^2 + 16}$

c) $\dfrac{2x - 5}{8x + 3}$

d) $\dfrac{3 + 7x}{5 - 4x}$

e) $\dfrac{2x - 1}{x^2 + 2x + 3}$

f) $\dfrac{x + 1}{x^2 + x + 1}$

g) $\dfrac{x - 1}{\sqrt{x^2 + x - 1}}$

h) $\dfrac{2x - 7}{\sqrt{2x^2 - 4x + 5}}$

i) $\dfrac{2x + 5}{\sqrt{1 - 4x - x^2}}$

j) $\dfrac{3x - 7}{\sqrt{2 - 5x - 3x^2}}$

4 a) Find $\displaystyle\int \dfrac{x + 1}{\sqrt{(1 - x^2)}} \, dx$.

b) Hence find the exact value of $\displaystyle\int_0^{\frac{1}{2}} \dfrac{x + 1}{\sqrt{(1 - x^2)}} \, dx$, giving your answer in the form $p + q\pi$, where $p, q \in \mathbb{R}$. (EDEXCEL)

5 If $x = 5\sin\theta - 3$, show that $16 - 6x - x^2 = 25\cos^2\theta$.

Hence, or otherwise, find

$$\int \dfrac{1}{\sqrt{(16 - 6x - x^2)}} \, dx \qquad \text{(OCR)}$$

6 i) Express

$$f(x) \equiv \dfrac{x^3 + 3x^2 + 8x + 26}{(x + 1)(x^2 + 9)}$$

in partial fractions of the form

$$a + \dfrac{b}{x + 1} + \dfrac{cx + d}{x^2 + 9}$$

ii) Hence show that

$$\int_0^3 f(x) \, dx = 3 + 4\ln 2 - \dfrac{\pi}{12} \qquad \text{(OCR)}$$

7 Express $y = \dfrac{7x^2 + 11x + 13}{(3x + 4)(x^2 + 9)}$ in partial fractions.

Hence show that

$$\int_0^3 y \, dx = \dfrac{1}{3}\ln 26 + \dfrac{\pi}{12} \qquad \text{(OCR)}$$

Reduction formulae

We need reduction formulae to facilitate the integration of functions whose integrals cannot otherwise be found directly.

An example of a reduction formula is

$$\int_0^{\frac{\pi}{2}} \sin^n x \, dx = \frac{n-1}{n} \int_0^{\frac{\pi}{2}} \sin^{n-2} x \, dx$$

which enables us to convert, for example, $\int_0^{\frac{\pi}{2}} \sin^6 x \, dx$ into $\int_0^{\frac{\pi}{2}} \sin^4 x \, dx$.

This we may further reduce to $\int_0^{\frac{\pi}{2}} \sin^2 x \, dx$, and hence to $\int_0^{\frac{\pi}{2}} \sin^0 x \, dx$, which is $\int_0^{\frac{\pi}{2}} 1 \, dx$, which we can integrate easily.

We usually obtain a reduction formula by changing the form of the integrand into a product which can be integrated by parts. But we must exercise discretion. For example, a possible product of $\int \sin^n x \, dx$ is $\int 1 \times \sin^n x \, dx$.

But this will not be helpful, as $\int 1 \, dx$ is x and $x \dfrac{d}{dx} \sin^n x$ is an awkward integrand. Thus, we must use

$$\int_0^{\frac{\pi}{2}} \sin^n x \, dx = \int_0^{\frac{\pi}{2}} \sin x \sin^{n-1} x \, dx$$

because we can integrate $\sin x$ easily.

Hence, we have

$$\int_0^{\frac{\pi}{2}} \sin x \sin^{n-1} x \, dx = \left[-\cos x \sin^{n-1} x \right]_0^{\frac{\pi}{2}} - \int_0^{\frac{\pi}{2}} -\cos x \times (n-1) \sin^{n-2} x \cos x \, dx$$

$$= 0 + (n-1) \int_0^{\frac{\pi}{2}} \sin^{n-2} x \cos^2 x \, dx$$

$$= (n-1) \int_0^{\frac{\pi}{2}} \sin^{n-2} x \, (1 - \sin^2 x) \, dx$$

$$= (n-1) \left(\int_0^{\frac{\pi}{2}} \sin^{n-2} x \, dx - \int_0^{\frac{\pi}{2}} \sin^n x \, dx \right)$$

We usually obtain the integral with which we started as one of the terms on the right-hand side. So, we take this integral to the left-hand side, which gives

$$n \int_0^{\frac{\pi}{2}} \sin^n x \, dx = (n-1) \int_0^{\frac{\pi}{2}} \sin^{n-2} x \, dx$$

$$\Rightarrow \int_0^{\frac{\pi}{2}} \sin^n x \, dx = \frac{n-1}{n} \int_0^{\frac{\pi}{2}} \sin^{n-2} x \, dx$$

Denoting $\int_0^{\frac{\pi}{2}} \sin^n x \, dx$ by I_n, we can express this reduction formula as

$$I_n = \left(\frac{n-1}{n} \right) I_{n-2}$$

Example 9 Use the reduction formula for $\int_0^{\frac{\pi}{2}} \sin^n x \, dx$ to evaluate $\int_0^{\frac{\pi}{2}} \sin^7 x \, dx$.

SOLUTION

In the reduction formula $I_n = \left(\dfrac{n-1}{n}\right) I_{n-2}$, we put $n = 7$, which gives

$$I_7 = \int_0^{\frac{\pi}{2}} \sin^7 x \, dx = \frac{6}{7} \int_0^{\frac{\pi}{2}} \sin^5 x \, dx$$

Using the formula again with $n = 5$, we obtain

$$I_5 = \frac{4}{5} \int_0^{\frac{\pi}{2}} \sin^3 x \, dx$$

which gives

$$I_7 = \frac{6}{7} \times \frac{4}{5} \int_0^{\frac{\pi}{2}} \sin^3 x \, dx$$

Repeating the procedure with $n = 3$, we have

$$I_7 = \frac{6}{7} \times \frac{4}{5} \times \frac{2}{3} \int_0^{\frac{\pi}{2}} \sin x \, dx$$

$$= \frac{6}{7} \times \frac{4}{5} \times \frac{2}{3} \left[-\cos x\right]_0^{\frac{\pi}{2}}$$

$$= \frac{6}{7} \times \frac{4}{5} \times \frac{2}{3} = \frac{16}{35}$$

Hence, we find $\int_0^{\frac{\pi}{2}} \sin^7 x \, dx = \dfrac{16}{35}$.

Similarly, we can find the reduction formula for $\int_0^{\frac{\pi}{2}} e^{ax} \cos^n x \, dx$, when a is not equal to 0. In this case, the integrand is already a product, and e^{ax} is a term which can be readily integrated. Therefore, we differentiate the term $\cos^n x$, which gives

$$\int_0^{\frac{\pi}{2}} e^{ax} \cos^n x \, dx = \left[\frac{1}{a} e^{ax} \cos^n x\right]_0^{\frac{\pi}{2}} - \int_0^{\frac{\pi}{2}} -\frac{1}{a} e^{ax} n \cos^{n-1} x \sin x \, dx$$

The new integrand is not in the form of $\int_0^{\frac{\pi}{2}} e^{ax} \cos^n x$, and therefore we must repeat the integration by parts, which gives

$$\int_0^{\frac{\pi}{2}} e^{ax} \cos^n x \, dx = \left[\frac{1}{a} e^{ax} \cos^n x\right]_0^{\frac{\pi}{2}} - \frac{n}{a} \int_0^{\frac{\pi}{2}} -e^{ax} \cos^{n-1} x \sin x \, dx$$

$$= -\frac{1}{a} + \frac{n}{a} \left\{ 0 - \frac{1}{a} \int_0^{\frac{\pi}{2}} \left[-(n-1) e^{ax} \cos^{n-2} x (1 - \cos^2 x) + e^{ax} \cos^n x \right] dx \right\}$$

$$= -\frac{1}{a} - \frac{n}{a^2} \int_0^{\frac{\pi}{2}} \left[-(n-1) e^{ax} (\cos^{n-2} x - \cos^n x) + e^{ax} \cos^n x \right] dx$$

$$= -\frac{1}{a} - \frac{n}{a^2} \int_0^{\frac{\pi}{2}} \left[n e^{ax} \cos^n x - (n-1) e^{ax} \cos^{n-2} x \right] dx$$

Hence, we have

$$(n^2 + a^2) \int_0^{\frac{\pi}{2}} e^{ax} \cos^n x \, dx = -a + n(n-1) \int_0^{\frac{\pi}{2}} e^{ax} \cos^{n-2} x \, dx$$

Example 10 If $I_n = \displaystyle\int_0^2 x^n e^{4x} \, dx$, show that

$$I_n = 2^{n-2} e^8 - \frac{n}{4} I_{n-1}$$

Hence find $\displaystyle\int_0^2 x^3 e^{4x} \, dx$.

SOLUTION

The reduction formula requires that the power of x is reduced. Therefore, we differentiate the term in x^n and integrate the term e^{4x}, obtaining

$$\int_0^2 x^n e^{4x} \, dx = \left[x^n \times \frac{e^{4x}}{4} \right]_0^2 - \int_0^2 n x^{n-1} \times \frac{e^{4x}}{4} \, dx$$

$$= \frac{2^n e^8}{4} - \frac{n}{4} \int_0^2 x^{n-1} e^{4x} \, dx$$

That is, we have

$$I_n = 2^{n-2} e^8 - \frac{n}{4} I_{n-1}$$

as required.

To find $\displaystyle\int_0^2 x^3 e^{4x} \, dx$, we first put $n = 3$, which gives

$$\int_0^2 x^3 e^{4x} \, dx = 2e^8 - \frac{3}{4} I_2$$

Then we put $n = 2$, which gives

$$\int_0^2 x^3 e^{4x} \, dx = 2e^8 - \frac{3}{4}\left(e^8 - \frac{2}{4} I_1 \right) = 2e^8 - \frac{3}{4} e^8 + \frac{3}{8} I_1$$

Finally, we put $n = 1$, which gives

$$\int_0^2 x^3 e^{4x} \, dx = \frac{5}{4} e^8 + \frac{3}{8}\left(\frac{1}{2} e^8 - \frac{1}{4} I_0 \right)$$

$$= \frac{5}{4} e^8 + \frac{3}{16} e^8 - \frac{3}{32} \int_0^2 e^{4x} \, dx$$

$$= \frac{5}{4} e^8 + \frac{3}{16} e^8 - \frac{3}{128} \left[e^{4x} \right]_0^2$$

Hence, we find

$$\int_0^2 x^3 e^{4x} \, dx = \frac{181}{128} e^8 - \frac{3}{128}$$

Example 11 If $I_n = \displaystyle\int_0^1 \frac{x^n}{\sqrt{x^2 - 1}}\,dx$, show that $I_n = \left(\dfrac{n-1}{n}\right)I_{n-2}$.

SOLUTION

When we separate I_n into a part to be integrated and a part to be differentiated, we must take account of the following:

- $\displaystyle\int \frac{1}{\sqrt{x^2 - 1}}\,dx = \cosh^{-1}x$ This result is unlikely to be helpful.

- $\dfrac{d}{dx}\left(\dfrac{1}{\sqrt{x^2 - 1}}\right) = \dfrac{x}{(x^2 - 1)^{\frac{3}{2}}}$ This result increases the power of the denominator, and so it also is unlikely to be helpful.

Hence, we avoid having to integrate or differentiate $\dfrac{1}{\sqrt{x^2 - 1}}$ by itself. We therefore separate $\dfrac{x^n}{\sqrt{x^2 - 1}}$ into $\dfrac{x}{\sqrt{x^2 - 1}} \times x^{n-1}$ because we can integrate $\dfrac{x}{\sqrt{x^2 - 1}}$ to give $\sqrt{x^2 - 1}$.

So, we have

$$I_n = \int_0^1 \frac{x^n}{\sqrt{x^2 - 1}}\,dx = \int_0^1 \frac{x}{\sqrt{x^2 - 1}} \times x^{n-1}\,dx$$

$$= \left[\sqrt{x^2 - 1} \times x^{n-1}\right]_0^1 - \int_0^1 \sqrt{x^2 - 1}(n-1)x^{n-2}\,dx$$

$$= -(n-1)\int_0^1 \frac{(x^2 - 1) \times x^{n-2}}{\sqrt{x^2 - 1}}\,dx$$

$$= -(n-1)\int_0^1 \left(\frac{x^n}{\sqrt{x^2 - 1}} - \frac{x^{n-2}}{\sqrt{x^2 - 1}}\right)dx$$

$$= -(n-1)(I_n - I_{n-2})$$

which gives

$$nI_n = (n-1)I_{n-2} \quad\Rightarrow\quad I_n = \left(\frac{n-1}{n}\right)I_{n-2}$$

as required.

Exercise 12B

1 If $\displaystyle\int_0^{\frac{\pi}{2}} \cos^n x\,dx = I_n$, prove that $nI_n = (n-1)I_{n-2}$, $(n > 1)$.

Evaluate

a) $\displaystyle\int_0^{\frac{\pi}{2}} \cos^6 x\,dx$ 　　　　　　　 **b)** $\displaystyle\int_0^{\frac{\pi}{2}} \cos^7 x\,dx$

2 Find the reduction formula for $\int x^n e^x \, dx$.

3 If $I_n = \int_0^1 x^n e^{-x} \, dx$, prove that

$$I_n = nI_{n-1} - e^{-1} \quad (n \geqslant 1)$$

Hence evaluate $\int_0^1 x^5 e^{-x} \, dx$.

4 If $I_n = \int_0^{\frac{\pi}{4}} \tan^n \theta \, d\theta$, prove that

$$I_n = \frac{1}{n-1} - I_{n-2} \quad (n > 1)$$

Hence evaluate

a) $\int_0^{\frac{\pi}{4}} \tan^4 \theta \, d\theta$ 　　　　　　　　**b)** $\int_0^{\frac{\pi}{4}} \tan^7 \theta \, d\theta$

5 If $I_n = \int_0^1 (\ln x)^n \, dx$, **a)** prove that $I_n = -nI_{n-1}$, and **b)** deduce that $I_n = (-1)^n n!$

6 Prove that

$$n \int \cosh^n x \, dx = \cosh^{n-1} x \sinh x + (n-1) \int \cosh^{n-2} x \, dx$$

Hence find $\int_0^1 \cosh^5 x \, dx$.

7 If $I_n = \int_0^1 x^n e^{x^2} \, dx$, prove that $I_n = \frac{1}{2} e - \frac{1}{2}(n-1)I_{n-2}$.

8 If $I_{m,n} = \int \frac{x^m}{(\ln x)^n} \, dx$, prove that

$$(n-1)I_{m,n} = -\frac{x^{m+1}}{(\ln x)^{n-1}} + (m+1)I_{m,n-1}$$

9 If $I_{m,n} = \int x^m (\ln x)^n \, dx$, prove that

$$(m+1)I_{m,n} = x^{(m+1)}(\ln x)^n - nI_{m,n-1}$$

10 Given that $I_n = \int_0^{\frac{\pi}{4}} \tan^n x \, dx$, show that

$$I_n = \frac{1}{n-1} - I_{n-2} \quad (n \geqslant 2)$$

Hence show that $I_4 = \frac{3\pi - 8}{12}$. 　　　　(WJEC)

11 a) Given that $I_n = \int \cosh^n x \, dx \; (n \geqslant 0)$, show that

$$nI_n = \sinh x \cosh^{n-1} x + (n-1)I_{n-2} \quad (n \geqslant 2)$$

b) Prove the identity

$$\cosh x - \operatorname{sech} x \equiv \sinh x \tanh x$$

Hence evaluate

$$\int_0^{\ln 3} (\operatorname{sech} x + \sinh x \tanh x)^3 \, dx \qquad \text{(AEB 98)}$$

12 Given that $I_n = \int_0^{\frac{\pi}{2}} x^n \cos x \, dx$, show that, for $n \geqslant 2$,

$$I_n = \left(\frac{\pi}{2}\right)^n - n(n-1)I_{n-2}$$

Hence find the area of the region enclosed by the curve $y = x^4 \cos x$, the x-axis and the lines $x = 0$ and $x = \frac{\pi}{2}$. (WJEC)

13 Given that $I_n = \int_0^{\frac{\pi}{2}} \sin^n\theta \, d\theta$, show that, for $n \geqslant 2$,

$$I_n = \left(\frac{n-1}{n}\right)I_{n-2}$$

Hence evaluate $\int_0^{\frac{\pi}{2}} \sin^5\theta \, d\theta$. (WJEC)

14 $$I_n = \int_0^1 x^{\frac{1}{2}n} e^{-\frac{1}{2}x} \, dx \quad n \geqslant 0$$

a) Show that $I_n = nI_{n-2} - e^{-\frac{1}{2}}$, $n \geqslant 2$.
b) Evaluate I_0 in terms of e.
c) Find, using the results of parts **a** and **b**, the value of I_4 in terms of e.
d) Show that the approximate value for I using Simpson's rule with three equally spaced ordinates is

$$\frac{1}{6}(2\sqrt{2}e^{-\frac{1}{4}} + e^{-\frac{1}{2}}) \qquad \text{(EDEXCEL)}$$

15 Consider $I_n = \int_0^{\frac{\pi}{2}} \frac{\sin 2n\theta}{\sin \theta} \, d\theta$, where n is a non-negative integer.

i) Using $\sin A - \sin B \equiv 2\cos\dfrac{A+B}{2}\sin\dfrac{A-B}{2}$, derive the reduction formula

$$I_n - I_{n-1} = \frac{2(-1)^{n-1}}{2n-1}$$

ii) Find $\int_0^{\frac{\pi}{2}} \frac{\sin 6\theta}{\sin \theta} \, d\theta$. (NICCEA)

16 Assuming the reduction formula

$$\int \tan^n x \, dx = \frac{1}{n-1} \tan^{n-1}x - \int \tan^{n-2}x \, dx$$

where $n \geqslant 2$, find the exact value of $\int_0^{\frac{\pi}{4}} \tan^5 x \, dx$. (NICCEA)

17 Given that $I_n = \int \sec^n x \, dx$,

a) show that

$$(n-1)I_n = \tan x \sec^{n-2} x + (n-2)I_{n-2} \qquad n \geqslant 2$$

b) Hence find the exact value of $\int_0^{\frac{\pi}{3}} \sec^3 x \, dx$, giving your answer in terms of natural logarithms and surds.   (EDEXCEL)

18 Find the value of each of the constants A, B and C for which

$$\frac{1}{1+x^3} \equiv \frac{A}{(1+x)} + \frac{Bx+C}{(1-x+x^2)}$$

Hence evaluate $\int_0^1 \frac{1}{(1+x^3)} \, dx$.

Given that $I_n = \int_0^1 (1+x^3)^n \, dx$, where n is an integer, show that

$$(3n+1)I_n = 2^n + 3nI_{n-1}$$

Hence evaluate $\int_0^1 (1+x^3)^{-2} \, dx$.   (EDEXCEL)

19 i) If $I_n = \int_0^1 x^n(1-x)^{\frac{1}{2}} \, dx$, prove that $(2n+3)I_n = 2nI_{n-1}$, where n is a positive integer.

ii) Show that $\int_0^1 x^3(1-x)^{\frac{1}{2}} \, dx = \frac{32}{315}$.   (NICCEA)

20 Given that $I_n = \int_0^1 x^n \cos \pi x \, dx$ for $n \geqslant 0$, show that

$$\pi^2 I_n + n(n-1)I_{n-2} + n = 0 \qquad \text{for } n \geqslant 2$$

Hence show that

$$\int_0^1 x^4 \cos \pi x \, dx = \frac{4(6-\pi^2)}{\pi^4} \qquad \text{(OCR)}$$

21 Show that

$$\frac{d}{dx}[x^{n-1}\sqrt{(16-x^2)}] = \frac{16(n-1)x^{n-2}}{\sqrt{(16-x^2)}} - \frac{nx^n}{\sqrt{(16-x^2)}}$$

Deduce, or prove otherwise, that if

$$I_n = \int_0^2 \frac{x^n}{\sqrt{(16-x^2)}} \, dx$$

then, for $n \geqslant 2$,

$$nI_n = 16(n-1)I_{n-2} - 2^n\sqrt{3}$$

Hence find the exact value of I_2.   (OCR)

22 Show that

$$\frac{d}{dt}[t(1 + t^4)^n] = (4n + 1)(1 + t^4)^n - 4n(1 + t^4)^{n-1}$$

The integral I_n is defined by $I_n = \int_0^1 (1 + t^4)^n \, dt$.

Show that $(4n + 1)I_n = 4nI_{n-1} + 2^n$.   (OCR)

23 Let $I_n = \int_0^1 \cosh^n x \, dx$.

i) By considering

$$\frac{d}{dx}(\sinh x \cosh^{n-1} x)$$

or otherwise, show that

$$nI_n = ab^{n-1} + (n - 1)I_{n-2}$$

where $a = \sinh(1)$ and $b = \cosh(1)$.

ii) Show that $I_4 = \frac{1}{8}(2ab^3 + 3ab + 3)$.   (OCR)

24 It is given that

$$I_n = \int_1^e x(\ln x)^n \, dx \qquad (n \geqslant 0)$$

By considering $\frac{d}{dx}[x^2(\ln x)^n]$, or otherwise, show that, for $n \geqslant 1$,

$$I_n = \frac{1}{2}e^2 - \frac{1}{2}nI_{n-1}$$

Hence find I_3, leaving your answer in terms of e.   (OCR)

25 For each non-negative integer n, let $I_n = \int \cos^n \theta \, d\theta$.

i) Show that if $n \geqslant 2$, then

$$nI_n = \sin \theta \cos^{n-1} \theta + (n - 1)I_{n-2}$$

ii) Show that $\int_0^{\frac{\pi}{3}} \cos^3 \theta \, d\theta = \frac{3\sqrt{3}}{8}$.   (NICCEA)

26 $$I_n = \int \frac{\sin nx}{\sin x} \, dx \qquad n > 0, \ n \in \mathbb{Z}$$

a) By considering $I_{n+2} - I_n$ or otherwise, show that

$$I_{n+2} = \frac{2\sin(n + 1)x}{n + 1} + I_n$$

b) Hence evaluate $\int_{\frac{\pi}{4}}^{\frac{\pi}{3}} \frac{\sin 6x}{\sin x} \, dx$, giving your answer in the form $p\sqrt{2} + q\sqrt{3}$, where p and q are rational numbers to be found.   (EDEXCEL)

27 a) Write down the values of $\cosh(\ln 2)$ and $\sinh(\ln 2)$.

b) For $n \geqslant 0$, the integral I_n is given by $I_n = \displaystyle\int_0^{\ln 2} \cosh^n x\, dx$.

i) By writing $\cosh^n x$ as $\cosh^{n-1} x \cosh x$, prove that, for $n \geqslant 2$,

$$nI_n = \frac{3 \times 5^{n-1}}{4^n} + (n-1)I_{n-2}$$

ii) Evaluate I_3.　　(AEB 96)

28　　　　$$I_n = \int_0^\pi \sin^{2n} x\, dx \qquad n \in \mathbb{N}$$

a) Calculate I_0 in terms of π.

b) Show that $I_n = \dfrac{(2n-1)}{2n} I_{n-1}, \; n \geqslant 1$.

c) Find I_3 in terms of π.

The figure on the right shows the curve with polar equation $r = a\sin^3\theta, \; 0 \leqslant \theta \leqslant \pi$, where a is a positive constant.

d) Using your answer to part **c**, or otherwise, calculate exactly the area bounded by this curve.　　(EDEXCEL)

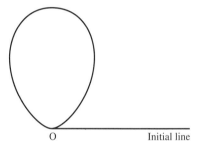

O 　　　　Initial line

29 a) Assuming the derivatives of $\sinh\theta$ and $\cosh\theta$, prove that

$$\frac{d}{d\theta}(\tanh\theta) = \mathrm{sech}^2\theta$$

b) Let I_r denote the integral $\displaystyle\int_0^{\ln 2} \tanh^{2r}\theta\, d\theta$ for integers $r \geqslant 0$.

i) Evaluate I_0.

ii) Show that $I_{r-1} - I_r = \dfrac{1}{2r-1}\left(\dfrac{3}{5}\right)^{2r-1}$.

iii) Hence prove that

$$\int_0^{\ln 2} \tanh^{2n}\theta\, d\theta = \ln 2 - \frac{5}{3}\sum_{r=1}^{n} \frac{1}{(2r-1)}\left(\frac{9}{25}\right)^r$$

iv) Deduce the sum of the infinite series

$$\sum_{r=1}^{\infty} \frac{1}{(2r-1)}\left(\frac{9}{25}\right)^r \qquad \text{(AEB 98)}$$

30 Let m and n be non-negative integers.

i) Determine $\displaystyle\int \sin\theta \cos^n\theta\, d\theta$.

ii) Show that

$$\int \sin^m\theta \cos^n\theta\, d\theta = -\frac{\sin^{m-1}\theta \cos^{n+1}\theta}{n+1} + \frac{m-1}{n+1}\int \sin^{m-2}\theta \cos^{n+2}\theta\, d\theta$$

where $m \geqslant 2$.

iii) If $I_{m,n} = \displaystyle\int_0^{\frac{\pi}{2}} \sin^m \theta \cos^n \theta \, d\theta$, show that

$$I_{m,n} = \frac{m-1}{m+n} I_{m-2,n}$$

where $m \geqslant 2$.

iv) Using the result in part **iii** and the similar result,

$$I_{m,n} = \frac{n-1}{m+n} I_{m,n-2}$$

where $n \geqslant 2$, show that

$$\int_0^{\frac{\pi}{2}} \sin^6 \theta \cos^4 \theta \, d\theta = \frac{3\pi}{512} \qquad \text{(NICCEA)}$$

31 $$I_n = \int \frac{x^n}{\sqrt{(1+x^2)}} \, dx$$

a) Show that $nI_n = x^{n-1}\sqrt{(1+x^2)} - (n-1)I_{n-2}, \; n \geqslant 2$.

The curve C has equation

$$y^2 = \frac{x^2}{\sqrt{(1+x^2)}} \qquad y \geqslant 0$$

The finite region R is bounded by C, the x-axis and the lines with equations $x = 0$ and $x = 2$. The region R is rotated through 2π radians about the x-axis.

b) Find the volume of the solid so formed, giving your answer in terms of π, surds and natural logarithms.

An estimate for the volume obtained in part **b** is found using Simpson's rule with three ordinates.

c) Find the percentage error resulting from using this estimate, giving your answer to three decimal places. (EDEXCEL)

Arc length

Cartesian form

Consider two points, P and Q, on a curve. P is the point (x, y) and Q is the point $(x + \delta x, y + \delta y)$.

Let s be the length of the arc from a point T, and δs the length of the arc PQ.

Since δs is very small, we can approximate the arc PQ to a straight line. Hence, using Pythagoras's theorem, we have

$$(\delta x)^2 + (\delta y)^2 = (\delta s)^2$$

Dividing by $(\delta x)^2$, we obtain

$$1 + \left(\frac{\delta y}{\delta x}\right)^2 = \left(\frac{\delta s}{\delta x}\right)^2$$

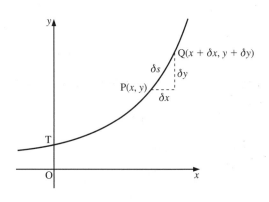

As $\delta x \to 0$, this gives

$$1 + \left(\frac{\mathrm{d}y}{\mathrm{d}x}\right)^2 = \left(\frac{\mathrm{d}s}{\mathrm{d}x}\right)^2$$

$$\Rightarrow \quad \frac{\mathrm{d}s}{\mathrm{d}x} = \sqrt{1 + \left(\frac{\mathrm{d}y}{\mathrm{d}x}\right)^2}$$

Therefore, we have

$$s = \int \sqrt{1 + \left(\frac{\mathrm{d}y}{\mathrm{d}x}\right)^2}\,\mathrm{d}x$$

Example 12 Find the length of the arc of the curve $x^3 = 3y^2$ from $x = 1$ to $x = 4$.

SOLUTION

Differentiating with respect to x, we have

$$3x^2 = 6y\frac{\mathrm{d}y}{\mathrm{d}x}$$

$$\Rightarrow \quad \frac{\mathrm{d}y}{\mathrm{d}x} = \frac{x^2}{2y}$$

which gives

$$\text{Arc length} = \int_1^4 \sqrt{1 + \left(\frac{\mathrm{d}y}{\mathrm{d}x}\right)^2}\,\mathrm{d}x$$

$$= \int_1^4 \sqrt{1 + \left(\frac{x^2}{2y}\right)^2}\,\mathrm{d}x$$

$$= \int_1^4 \sqrt{1 + \frac{x^4}{4y^2}}\,\mathrm{d}x$$

Substituting $y^2 = \frac{x^3}{3}$, we obtain

$$\text{Arc length} = \int_1^4 \sqrt{1 + \frac{3x^4}{4x^3}}\,\mathrm{d}x = \int_1^4 \sqrt{1 + \frac{3x}{4}}\,\mathrm{d}x$$

Putting $1 + \frac{3x}{4} = u^2$ and differentiating, we have

$$2u\frac{\mathrm{d}u}{\mathrm{d}x} = \frac{3}{4} \quad \Rightarrow \quad \frac{8}{3}u\,\mathrm{d}u = \mathrm{d}x$$

Substituting these in the original integral and changing the limits to $u = 2$ (from $x = 4$) and $u = \sqrt{\frac{7}{4}}$ (from $x = 1$), we obtain

$$\text{Arc length} = \int_{\sqrt{\frac{7}{4}}}^2 \frac{8}{3}u^2\,\mathrm{d}u = \left[\frac{8}{9}u^3\right]_{\sqrt{\frac{7}{4}}}^2$$

which gives

$$\text{Arc length} = \frac{64}{9} - \frac{8}{9}\left(\frac{7}{4}\right)^{\frac{3}{2}} = \frac{1}{9}(64 - 7\sqrt{7})$$

Parametric form

To obtain the parametric form, we divide $(\delta x)^2 + (\delta y)^2 = (\delta s)^2$ by $(\delta t)^2$, where t is the parameter, which gives

$$\left(\frac{\delta x}{\delta t}\right)^2 + \left(\frac{\delta y}{\delta t}\right)^2 = \left(\frac{\delta s}{\delta t}\right)^2$$

As $\delta x \to 0$, and thus $\delta t \to 0$, we have

$$\left(\frac{dx}{dt}\right)^2 + \left(\frac{dy}{dt}\right)^2 = \left(\frac{ds}{dt}\right)^2$$

$$\Rightarrow \quad \frac{ds}{dt} = \sqrt{\left(\frac{dx}{dt}\right)^2 + \left(\frac{dy}{dt}\right)^2}$$

which gives

$$s = \int \sqrt{\left(\frac{dx}{dt}\right)^2 + \left(\frac{dy}{dt}\right)^2}\, dt$$

We sometimes express this as

$$s = \int \sqrt{\dot{x}^2 + \dot{y}^2}\, dt$$

where $\dot{x} = \dfrac{dx}{dt}$ and $\dot{y} = \dfrac{dy}{dt}$

Note We use the dot notation only when the independent variable is t, and mostly when t represents time. Thus, $\dot{x}$ usually expresses speed and $\ddot{x}$ acceleration.

Example 13 Find the circumference of the circle $x^2 + y^2 = r^2$.

SOLUTION

The parametric equations for a circle are $x = r \cos \theta$, $y = r \sin \theta$. Therefore, we have

$$s = \int_0^{2\pi} \sqrt{\left(\frac{dx}{d\theta}\right)^2 + \left(\frac{dy}{d\theta}\right)^2}\, d\theta$$

Using just that part of the circle in the first quadrant and then multiplying by 4, we obtain

$$s = 4 \int_0^{\frac{\pi}{2}} \sqrt{\left(\frac{dx}{d\theta}\right)^2 + \left(\frac{dy}{d\theta}\right)^2}\, d\theta$$

which gives

$$s = 4 \int_0^{\frac{\pi}{2}} \sqrt{r^2 \sin^2\theta + r^2 \cos^2\theta}\, d\theta$$

$$= 4 \int_0^{\frac{\pi}{2}} r\, d\theta = 4 \Big[r\theta\Big]_0^{\frac{\pi}{2}} = 2\pi r$$

Polar form

To obtain the polar form, we consider two points, P and Q, on a curve which is expressed in its polar equation. P is the point (r, θ) and Q is the point $(r + \delta r, \theta + \delta\theta)$.

As $\delta\theta \to 0$, we can approximate TP to an arc of a circle of radius r, and hence of length $r\delta\theta$. Also, we can approximate TPQ to a right-angled triangle, for which, by Pythagoras's theorem,

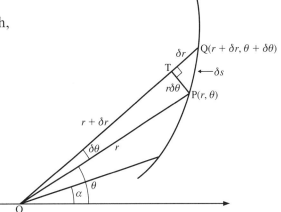

$$PQ^2 = TP^2 + TQ^2$$

Thus, we have

$$(r\delta\theta)^2 + (\delta r)^2 = (\delta s)^2$$

Dividing through by $(\delta\theta)^2$, we obtain

$$\left(\frac{\delta s}{\delta\theta}\right)^2 = \left(\frac{\delta r}{\delta\theta}\right)^2 + r^2$$

Therefore, as $\delta\theta \to 0$, we have

$$\frac{\mathrm{d}s}{\mathrm{d}\theta} = \sqrt{\left(\frac{\mathrm{d}r}{\mathrm{d}\theta}\right)^2 + r^2}$$

Hence, the length of the arc of a curve between the half-lines $\theta = \alpha$ and $\theta = \beta$ is given by

$$s = \int_\alpha^\beta \sqrt{r^2 + \left(\frac{\mathrm{d}r}{\mathrm{d}\theta}\right)^2}\, \mathrm{d}\theta$$

Example 14 Find the length of the arc of the curve $r = ae^{2\theta}$ between $\theta = 0$ and $\theta = \dfrac{\pi}{2}$.

SOLUTION

Differentiating $r = ae^{2\theta}$ with respect to θ, we have

$$\frac{\mathrm{d}r}{\mathrm{d}\theta} = 2ae^{2\theta}$$

Hence, the required arc length is given by

$$s = \int_0^{\frac{\pi}{2}} \sqrt{(ae^{2\theta})^2 + (2ae^{2\theta})^2}\, \mathrm{d}\theta$$

$$= \sqrt{5}a \int_0^{\frac{\pi}{2}} e^{2\theta}\, \mathrm{d}\theta$$

$$= \sqrt{5}a \left[\frac{1}{2}e^{2\theta}\right]_0^{\frac{\pi}{2}}$$

$$\Rightarrow \quad s = \frac{\sqrt{5}a}{2}(e^\pi - 1)$$

Area of a surface of revolution

Let A be the area of the surface formed by rotating the curve $y = f(x)$, between the lines $x = a$ and $x = b$, about the x-axis.

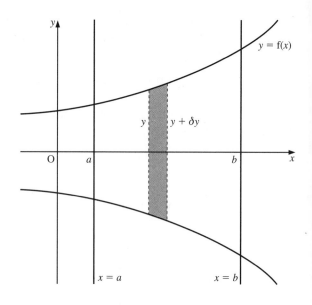

Let the curved surface area of the strip shown shaded be δA.

Treating the strip as being bounded by two cylinders, we have

$$2\pi y \delta s \leqslant \delta A \leqslant 2\pi(y + \delta y)\delta s$$

As $\delta x \to 0$, $\delta s \to 0$, so we have

$$\frac{\mathrm{d}A}{\mathrm{d}s} = 2\pi y$$

$$\Rightarrow \quad A = \int 2\pi y \, \mathrm{d}s$$

$$\Rightarrow \quad A = \int 2\pi y \frac{\mathrm{d}s}{\mathrm{d}x} \, \mathrm{d}x$$

which gives

$$A = \int 2\pi y \sqrt{1 + \left(\frac{\mathrm{d}y}{\mathrm{d}x}\right)^2} \, \mathrm{d}x$$

or, in parametric form,

$$A = \int 2\pi y \sqrt{\left(\frac{\mathrm{d}x}{\mathrm{d}t}\right)^2 + \left(\frac{\mathrm{d}y}{\mathrm{d}t}\right)^2} \, \mathrm{d}t \quad \text{or} \quad A = \int 2\pi y \sqrt{\dot{x}^2 + \dot{y}^2} \, \mathrm{d}t$$

Example 15 Find the surface area, A, of the sphere $x^2 + y^2 + z^2 = r^2$.

SOLUTION

The sphere $x^2 + y^2 + z^2 = r^2$ is obtained by rotating the circle $x^2 + y^2 = r^2$ about the x-axis. Hence, we have

$$A = \int 2\pi y \sqrt{1 + \left(\frac{\mathrm{d}y}{\mathrm{d}x}\right)^2} \, \mathrm{d}x$$

Differentiating $x^2 + y^2 = r^2$, we obtain

$$2x + 2y \frac{\mathrm{d}y}{\mathrm{d}x} = 0$$

which gives

$$A = \int_{-r}^{r} 2\pi\sqrt{r^2 - x^2} \times \sqrt{1 + \frac{x^2}{y^2}} \, \mathrm{d}x$$

$$= \int_{-r}^{r} 2\pi\sqrt{r^2 - x^2} \sqrt{\frac{y^2 + x^2}{y^2}} \, \mathrm{d}x = \int_{-r}^{r} 2\pi r \, \mathrm{d}x$$

By symmetry, the integral from $x = -r$ to $x = r$ is twice the integral from $x = 0$ to $x = r$. Therefore, we have

$$A = 2 \int_0^r 2\pi r \, dx = \left[4\pi r x \right]_0^r = 4\pi r^2$$

Hence, the surface area of a sphere is $4\pi r^2$.

Using the parametric form, $x = r\cos\theta$, $y = r\sin\theta$, for the rotated circle, we have

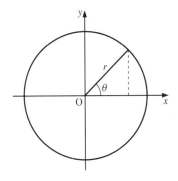

$$A = \int_{-\frac{\pi}{2}}^{\frac{\pi}{2}} 2\pi r \sin\theta \sqrt{r^2 \sin^2\theta + r^2 \cos^2\theta} \, d\theta$$

$$= 2 \int_0^{\frac{\pi}{2}} 2\pi r^2 \sin\theta \, d\theta = -4\pi r^2 \left[\cos\theta \right]_0^{\frac{\pi}{2}} = 4\pi r^2$$

Hence, the surface area of the sphere is $4\pi r^2$.

Exercise 12C

1 Find the length of the arc of $x^3 = y^2$ from $x = 0$ to $x = 3$.

2 Find the length of the arc of $x^3 = 6y^2$ from $x = 1$ to $x = 2$.

3 Find the length of the arc of the parabola $x = at^2$, $y = 2at$, between the points $(0, 0)$ and $(ap^2, 2ap)$.

4 Find the length of the arc of the cycloid $x = a(t + \sin t)$, $y = a(1 - \cos t)$, between the points $t = 0$ and $t = \pi$.

5 Find the length of the arc of the catenary $y = c \cosh\left(\dfrac{x}{c}\right)$, between the points where $x = 0$ and $x = c$.

6 Find the area of the surface generated by rotating about the x-axis each of the following.

a) Arc of the curve $x = 2t^3$, $y = 3t^2$, between the points where $t = 0$ and $t = 4$.
b) Arc of the curve $x = t^2$, $y = 2t$, between the points where $t = 0$ and $t = 2$.
c) Part of the asteroid $x = a\cos^3 t$, $y = a\sin^3 t$, which is above the x-axis.
d) Curve $y = 5x^{\frac{1}{2}}$, from $x = 4$ to $x = 9$.
e) Curve $y = \cosh x$, between $x = 0$ and $x = 1$.
f) Curve $y = e^{3x}$, from $x = 1$ to $x = 4$.

7 The diagram shows a wheel of radius a which rolls along the line Ox. The centre of the wheel is C and P is a point fixed on the rim of the wheel. Initially P is at O. When CP has rotated through an angle θ, show that the coordinates of P are

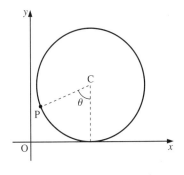

$$x = a(\theta - \sin\theta) \qquad y = a(1 - \cos\theta)$$

Hence find the length of the path of P when the wheel rolls through one complete revolution. (NEAB)

8 a) Find the values of the constants A and B for which

$$\frac{1+x^2}{1-x^2} \equiv A + \frac{B}{1-x^2}$$

b) The function f is defined for $-1 < x < 1$ by $f(x) = -\ln(1-x^2)$. The graph of the curve with equation $y = f(x)$ is shown.

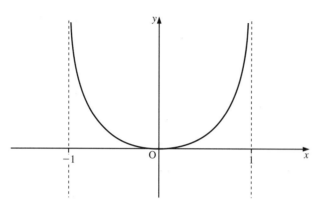

i) Find $\dfrac{dy}{dx}$ in terms of x. Hence show that

$$1 + \left(\frac{dy}{dx}\right)^2 = \left(\frac{1+x^2}{1-x^2}\right)^2$$

ii) Prove that the length of the arc of the curve from $x = 0$ to $x = \frac{3}{4}$ is $\ln 7 - \frac{3}{4}$. (AEB 96)

9 a) Using the definitions of the hyperbolic functions $\cosh x$ and $\sinh x$ given in the information booklet, show that

i) $\cosh^2 x - \sinh^2 x = 1$ **ii)** $2\cosh^2 x - 1 = \cosh 2x$ **iii)** $2\sinh x \cosh x = \sinh 2x$

b) Show that the length of the arc of the curve $y = x^2$ between the origin and the point $(1, 1)$ is

$$\frac{1}{4}(2\sqrt{5} + \sinh^{-1} 2).$$ (WJEC)

10 A curve C is given parametrically by

$$x = e^t \cos t \quad y = e^t \sin t \quad (0 \leqslant t \leqslant \pi)$$

Show that the length of C is $\sqrt{2}(e^\pi - 1)$.

Show also that in polar coordinates the equation of C is $r = e^\theta$ $(0 \leqslant \theta \leqslant \pi)$, and hence sketch C.

Find the area of the region bounded by C and the x-axis. (NEAB)

11 A curve is defined parametrically by

$$x = \frac{8}{3}t^{\frac{3}{2}} \quad y = t^2 - 2t + 4$$

The points A and B on the curve are defined by $t = 0$ and $t = 1$ respectively.

i) Find the length of the arc AB.

ii) Show that the area of the surface generated by one complete revolution of the arc AB about the y-axis is $\dfrac{256}{35}\pi$. (OCR)

12 The parametric equations of a curve are

$$x = a(t - \sin t) \qquad y = a(1 - \cos t)$$

where a is a positive constant. Show that

$$\left(\frac{dx}{dt}\right)^2 + \left(\frac{dy}{dt}\right)^2 = 4a^2 \sin^2(\tfrac{1}{2}t)$$

The arc of this curve between $t = 0$ and $t = 2\pi$ is rotated completely about the x-axis. Show that the area of the surface of revolution formed is

$$8\pi a^2 \int_0^{2\pi} [1 - \cos^2(\tfrac{1}{2}t)] \sin(\tfrac{1}{2}t) \, dt$$

and hence find this area.   (OCR)

13 The curve C is defined parametrically by

$$x = 3 + e^{-t}(\cos t + \sin t) \qquad y = 4 + e^{-t}(\cos t - \sin t)$$

Find the exact value, in terms of π and e, of the length of the arc of C from the point where $t = 0$ to the point where $t = \tfrac{1}{2}\pi$.

This arc is rotated about the x-axis through one revolution. Express the area of the surface generated as a definite integral. (You are not required to evaluate this integral.)   (OCR)

14 The parametric equations of a curve are

$$x = 3\cos\theta - \cos 3\theta \qquad y = 3\sin\theta - \sin 3\theta$$

Show that

$$\left(\frac{dx}{d\theta}\right)^2 + \left(\frac{dy}{d\theta}\right)^2 = 36\sin^2\theta$$

Hence find the length of the arc of the curve between the points given by $\theta = 0$ and $\theta = \tfrac{1}{2}\pi$.
     (OCR)

15 The arc of the curve $y = e^x$ from the point where $y = \tfrac{3}{4}$ to the point where $y = \tfrac{4}{3}$ is rotated through one revolution about the x-axis. Show that the area, S, of the surface generated is given by

$$S = 2\pi \int_{\frac{3}{4}}^{\frac{4}{3}} \sqrt{(1 + y^2)} \, dy$$

By using the substitution $y = \sinh u$, show that

$$S = \pi \left[\frac{185}{144} + \ln\left(\frac{3}{2}\right)\right] \qquad \text{(OCR)}$$

16 A curve C is defined parametrically by

$$x = 2(1 + t)^{\frac{3}{2}} \qquad y = 2(1 - t)^{\frac{3}{2}}$$

where $0 \leqslant t \leqslant 1$. Find

i) the length of C
ii) the area of the surface generated when C is rotated through one revolution about the x-axis.
     (OCR)

17 The curve C is defined parametrically by the equations $x = \frac{1}{3}t^3 - t$, $y = t^2$, where t is a parameter.

a) Show that $\left(\dfrac{dx}{dt}\right)^2 + \left(\dfrac{dy}{dt}\right)^2 = (t^2 + 1)^2$.

b) The arc of C between the points where $t = 0$ and $t = 3$ is denoted by L. Determine

   i) the length of L

   ii) the area of the surface generated when L is rotated through 2π radians about the x-axis.

   (AEB 98)

18

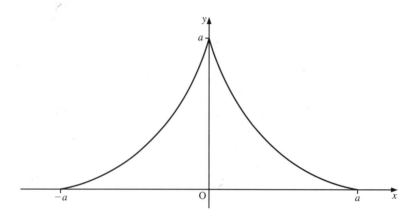

The figure above shows the curve C with parametric equations

$$x = a\cos^3 t \qquad y = a\sin^3 t \qquad 0 \leqslant t \leqslant \pi$$

where a is a positive constant.

The curve C is rotated through 2π radians about the x-axis. Show that the area of the surface

of revolution formed is $\dfrac{12\pi a^2}{5}$.   (EDEXCEL)

19 The arc of the curve $y = x^3$, between $x = 0$ and $x = 1$, is rotated through 2π radians about the x-axis. Determine the exact value of the surface area generated.   (AEB 98)

20 The curve C has the parametric equations

$$x = e^\theta \sin\theta \qquad y = e^\theta \cos\theta \qquad \text{for } 0 \leqslant \theta \leqslant \frac{\pi}{2}$$

a) Show that the area S of the surface generated when C is rotated through four right angles about the x-axis is given by

$$S = 2\sqrt{2}\pi \int_0^{\frac{\pi}{2}} e^{2\theta}\cos\theta \, d\theta$$

b) Find the value of S.   (WJEC)

21 a) i) Using only the definitions $\cosh\theta = \frac{1}{2}(e^\theta + e^{-\theta})$ and $\sinh\theta = \frac{1}{2}(e^\theta - e^{-\theta})$, prove the identity

$$\cosh^2\theta - \sinh^2\theta = 1$$

ii) Deduce a relationship between $\operatorname{sech}\theta$ and $\tanh\theta$.

b) A curve C has parametric representation $x = \operatorname{sech}\theta$, $y = \tanh\theta$.

i) Show that $\left(\dfrac{dx}{d\theta}\right)^2 + \left(\dfrac{dy}{d\theta}\right)^2 = \operatorname{sech}^2\theta$.

ii) The arc of the curve between the points where $\theta = 0$ and $\theta = \ln 7$ is rotated through one full turn about the x-axis. Show that the area of the surface generated is $\dfrac{36}{25}\pi$ square units.   (AEB 97)

22 a) Find $\displaystyle\int \cosh^2 t \sinh t\, dt$.

The curve C has parametric equations
$$x = \cosh^2 t \quad y = 2\sinh t \quad 0 \leqslant t \leqslant 2$$
The curve C is rotated through 2π radians about the x-axis.

b) Show that the area S of the curved surface generated is given by
$$S = 8\pi \int_0^2 \cosh^2 t \sinh t\, dt$$

c) Evaluate S to three significant figures.   (EDEXCEL)

Improper integrals

An **improper integral** is one which has either

- a limit of integration of $\pm\infty$, or
- an integrand which is infinite at one or other of its limits of integration, or between these limits.

In the first case, we replace $\pm\infty$ with n, say, and then find the limit of the integral as $n \to \pm\infty$. When this limit is **finite**, the integral **can be found** (see Example 16). When this limit is **not finite**, the integral **cannot be found** (see Example 17).

In the second case, we replace one or other of the limits of integration with p, say, and then find the limit of the integral as p tends to the value of the limit it has replaced (see Examples 18 and 19).

Example 16 Determine $\displaystyle\int_1^\infty \frac{1}{x^2}\, dx$.

SOLUTION

The upper limit is ∞, so we replace it with n, which gives

$$\int_1^\infty \frac{1}{x^2}\, dx = \lim_{n \to \infty} \int_1^n \frac{1}{x^2}\, dx$$

$$= \lim_{n \to \infty}\left[-\frac{1}{x}\right]_1^n = \lim_{n \to \infty}\left(-\frac{1}{n} + 1\right)$$

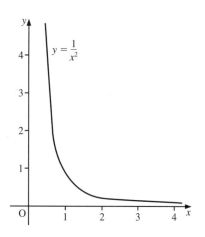

As $n \to \infty$, $\dfrac{1}{n} \to 0$, which gives

$$\lim_{n \to \infty} \left(-\frac{1}{n} + 1 \right) = 1$$

That is, we have

$$\int_1^\infty \frac{1}{x^2} \, dx = 1$$

This shows that the area under the curve $y = \dfrac{1}{x^2}$ is finite even though the boundary is of infinite length.

Example 17 Determine $\displaystyle\int_1^\infty \frac{1}{x} \, dx$.

SOLUTION

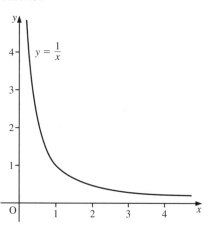

We have

$$\int_1^\infty \frac{1}{x} \, dx = \lim_{n \to \infty} \int_1^n \frac{1}{x} \, dx$$

$$= \lim_{n \to \infty} \Big[\ln x \Big]_1^n$$

$$= \lim_{n \to \infty} (\ln n - \ln 1)$$

which is not finite since $\lim\limits_{n \to \infty} \ln n$ is ∞.

This shows that the area under $y = \dfrac{1}{x}$ is not finite although the curve looks very similar to $y = \dfrac{1}{x^2}$, which has a finite area.

Example 18 Determine $\displaystyle\int_0^1 \frac{1}{\sqrt{x}} \, dx$.

SOLUTION

This is an improper integral since the integrand, $\dfrac{1}{\sqrt{x}}$, is infinite when $x = 0$. So, we replace the lower limit with p, which gives

$$\int_0^1 \frac{1}{\sqrt{x}} \, dx = \lim_{p \to 0} \int_p^1 \frac{1}{\sqrt{x}} \, dx$$

$$= \lim_{p \to 0} \left[2x^{\frac{1}{2}} \right]_p^1 = \lim_{p \to 0} (2 - 2\sqrt{p})$$

Since $\lim\limits_{p \to 0} 2\sqrt{p} = 0$, we have

$$\lim_{p \to 0} (2 - 2\sqrt{p}) = 2$$

That is, we have

$$\int_0^1 \frac{1}{\sqrt{x}} \, dx = 2$$

Example 19 Determine $\displaystyle\int_b^a \frac{dx}{\sqrt{(a-x)(x-b)}}$.

SOLUTION

At both limits the integrand is infinite, so we replace the upper limit with p and the lower limit with q and find the limit of the integral as $p \to a$ and $q \to b$. Hence, we have

$$\int_b^a \frac{dx}{\sqrt{(a-x)(x-b)}} = \lim_{p \to a}\lim_{q \to b}\int_q^p \frac{dx}{\sqrt{-ab+(a+b)x-x^2}}$$

$$= \lim_{p \to a}\lim_{q \to b}\int_q^p \frac{dx}{\sqrt{-[ab-(a+b)x+x^2]}}$$

$$= \lim_{p \to a}\lim_{q \to b}\int_q^p \frac{dx}{\sqrt{-\left[\left(x-\dfrac{a+b}{2}\right)^2-\left(\dfrac{a-b}{2}\right)^2\right]}}$$

$$= \lim_{p \to a}\lim_{q \to b}\int_q^p \frac{dx}{\sqrt{\left(\dfrac{a-b}{2}\right)^2-\left(x-\dfrac{a+b}{2}\right)^2}}$$

which gives

$$\int_a^b \frac{dx}{\sqrt{(a-x)(x-b)}} = \lim_{p \to a}\lim_{q \to b}\left[\sin^{-1}\frac{x-\dfrac{a+b}{2}}{\dfrac{a-b}{2}}\right]_q^p$$

$$= \lim_{p \to a}\lim_{q \to b}\left[\sin^{-1}\frac{2x-(a+b)}{a-b}\right]_q^p$$

$$= \lim_{p \to a}\lim_{q \to b}\left\{\sin^{-1}\left[\frac{2p-(a+b)}{a-b}\right]-\sin^{-1}\left[\frac{2q-(a+b)}{a-b}\right]\right\}$$

$$= \sin^{-1}1 - \sin^{-1}(-1) = \frac{\pi}{2}+\frac{\pi}{2}$$

That is, we have

$$\int_a^b \frac{dx}{\sqrt{(a-x)(x-b)}} = \pi$$

Summation of series

On pages 196–7 of *Introducing Pure Mathematics*, we noted that the definite integral $\displaystyle\int_a^b f(x)\,dx$ is an area bounded by the curve $y = f(x)$, the x-axis, and the lines $x = a$ and $x = b$.

We arrived at this result by dividing the given area into a series of infinitesimally narrow 'rectangular' strips of equal width, δx, and summing their areas, $y\delta x$: that is, $f(x)\delta x$.

If we divide the interval $a \leqslant x \leqslant b$ into n equal strips, the x-coordinates of the strips are

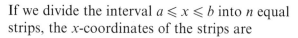

$$a, \quad a + \frac{b-a}{n}, \quad a + 2\left(\frac{b-a}{n}\right), \quad a + 3\left(\frac{b-a}{n}\right), \quad \ldots, \quad b$$

Hence, the sum of the areas of all n rectangles 'inside' the integral, shown on Figure A, is

$$f(a)\frac{b-a}{n} + f\left(a + \frac{b-a}{n}\right)\frac{b-a}{n} + f\left(a + 2\frac{b-a}{n}\right)\frac{b-a}{n} + \ldots$$

$$+ f\left(a + [n-1]\frac{b-a}{n}\right)\frac{b-a}{n}$$

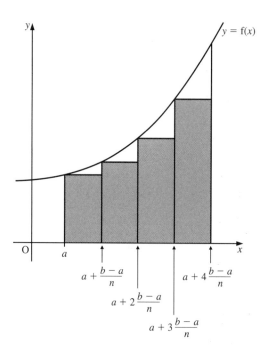

Figure A

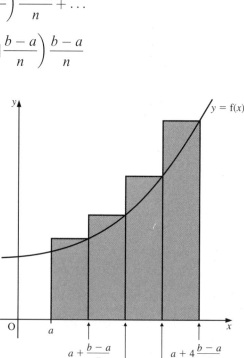

Figure B

And the sum of the areas of all n rectangles 'outside' the integral, shown in Figure B, is

$$\mathrm{f}\left(a+\frac{b-a}{n}\right)\frac{b-a}{n}+\mathrm{f}\left(a+2\frac{b-a}{n}\right)\frac{b-a}{n}+\mathrm{f}\left(a+3\frac{b-a}{n}\right)\frac{b-a}{n}+\ldots$$

$$+\,\mathrm{f}\left(a+[n-1]\frac{b-a}{n}\right)\frac{b-a}{n}+\mathrm{f}(b)\frac{b-a}{n}$$

The actual value of the definite integral is between these two values. As n tends to infinity, these two sums tend to the same value.

Hence, we have

$$\int_a^b \mathrm{f}(x)\,\mathrm{d}x=\lim_{n\to\infty}\left[\mathrm{f}(a)\frac{b-a}{n}+\mathrm{f}\left(a+\frac{b-a}{n}\right)\frac{b-a}{n}+f\left(a+2\frac{b-a}{n}\right)\frac{b-a}{n}+\ldots\right]$$

which gives

$$\int_a^b \mathrm{f}(x)\,\mathrm{d}x=\lim_{n\to\infty}\sum_{r=0}^{n-1}\mathrm{f}\left(a+r\frac{b-a}{n}\right)\frac{b-a}{n}$$

That is, the integral is the limit as n tends to infinity of the sum of the series.

This method may be used to find **upper** and **lower bounds** of integrals and series.

Consider, for example, the curve $y=\dfrac{1}{x^2}$.

The areas of the rectangles 'inside' $\displaystyle\int_1^\infty \frac{1}{x^2}\,\mathrm{d}x$ are

$$\frac{1}{2^2},\quad \frac{1}{3^2},\quad \frac{1}{4^2},\quad \frac{1}{5^2},\quad \cdots$$

The areas of the rectangles 'outside' the integral are

$$\frac{1}{1^2},\quad \frac{1}{2^2},\quad \frac{1}{3^2},\quad \frac{1}{4^2},\quad \cdots$$

Therefore, we have

$$\frac{1}{1^2}+\frac{1}{2^2}+\frac{1}{3^2}+\frac{1}{4^2}+\ldots \;>\; \int_1^\infty \frac{1}{x^2}\,\mathrm{d}x$$

$$>\;\frac{1}{2^2}+\frac{1}{3^2}+\frac{1}{4^2}+\frac{1}{5^2}+\ldots$$

Since $\displaystyle\int_1^\infty \frac{1}{x^2}\,\mathrm{d}x=1$ (see page 260), this gives

$$\sum_{r=1}^\infty \frac{1}{r^2} \;>\; 1 \;>\; \sum_{r=2}^\infty \frac{1}{r^2}$$

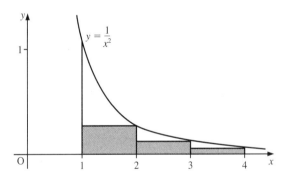

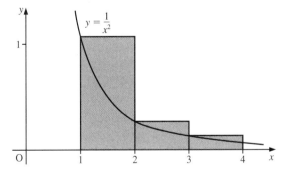

Example 20 Prove that

$$\frac{4}{441} < \frac{1}{7^3} + \frac{1}{8^3} + \dots + \frac{1}{20^3} < \frac{91}{7200}$$

SOLUTION

Since we require terms $\frac{1}{r^3}$, we take the curve $y = \frac{1}{x^3}$.

The first two rectangles 'outside' the curve, areas $\frac{1}{7^3}$ and $\frac{1}{8^3}$, are shown in Figure C.

We could continue in a similar manner until we obtain

$$\frac{1}{7^3} + \frac{1}{8^3} + \dots + \frac{1}{20^3} > \int_7^{21} \frac{1}{x^3}\, dx$$

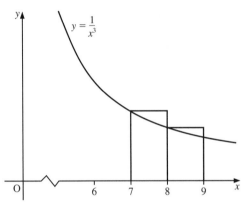

Figure C

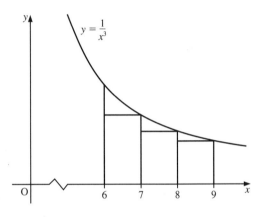

Figure D

Using rectangles 'inside' the curve, as shown in Figure D, we obtain

$$\frac{1}{7^3} + \dots + \frac{1}{20^3} < \int_6^{20} \frac{1}{x^3}\, dx$$

Evaluating both integrals, we obtain

$$\int_7^{21} \frac{1}{x^3}\, dx = \left[-\frac{1}{2x^2} \right]_7^{21} = -\frac{1}{2 \times 21^2} + \frac{1}{98} = \frac{4}{441}$$

$$\int_6^{20} \frac{1}{x^3}\, dx = \left[-\frac{1}{2x^2} \right]_6^{20} = -\frac{1}{800} + \frac{1}{72} = \frac{91}{7200}$$

Therefore, we have

$$\frac{4}{441} < \frac{1}{7^3} + \frac{1}{8^3} + \dots + \frac{1}{20^3} < \frac{91}{7200}$$

as required.

Exercise 12D

Find the value, where it exists, of each of the following.

1 $\int_0^1 \dfrac{1}{x^{\frac{1}{3}}}\, dx$

2 $\int_0^1 \dfrac{1}{x^{\frac{3}{2}}}\, dx$

3 $\int_1^2 \dfrac{1}{(1-x)^2}\, dx$

4 $\int_0^\infty \dfrac{x}{1+x^2}\, dx$

5 $\int_0^\infty \dfrac{1}{x^{\frac{1}{2}}}\, dx$

6 $\int_0^\infty \dfrac{x}{x^{\frac{4}{3}}}\, dx$

7 $\int_{-a}^\infty \dfrac{1}{x^2 - a^2}\, dx$

8 $\int_0^\infty \dfrac{1}{x^2 + a^2}\, dx$

9 $\int_{-2}^2 \dfrac{1}{x+2}\, dx$

10 $\int_0^{\frac{\pi}{2}} \tan x\, dx$

11 a) Use integration by parts to find $\int x \ln x\, dx$.

  b) Explain why $\int_0^1 x \ln x\, dx$ exists, and obtain the value of this integral.   (NEAB)

12 a) Write down the value of $\displaystyle\lim_{x \to \infty} \dfrac{x}{2x+1}$.

  b) Evaluate

$$\int_1^\infty \left(\frac{1}{x} - \frac{2}{2x+1} \right) dx$$

  giving your answer in the form $\ln k$, where k is a constant to be determined.   (NEAB)

13 a) Find (in terms of the constant k) the limit of $\dfrac{\cos\left(\frac{1}{2}\pi x^k\right)}{\ln x}$ as $x \to 1$.

  b) i) Explain in detail how $\displaystyle\sum_{r=1}^n \dfrac{r}{n^2 + r^2}$ is related to the area under the curve $y = \dfrac{x}{1+x^2}$

  between $x = 0$ and $x = 1$.

  $\left(\text{You should include a diagram. You may assume that } \dfrac{x}{1+x^2} \text{ is an increasing function for } 0 \leqslant x \leqslant 1.\right)$

  ii) Evaluate the limit $L = \displaystyle\lim_{n \to \infty} \sum_{r=1}^n \dfrac{r}{n^2 + r^2}$.

  iii) Show that $L < \displaystyle\sum_{r=1}^n \dfrac{r}{n^2 + r^2} < L + \dfrac{1}{2n}$.   (MEI)

14 i) Find the exact value of $\displaystyle\int_0^1 \frac{1}{1+x}\,\mathrm{d}x$.

ii)

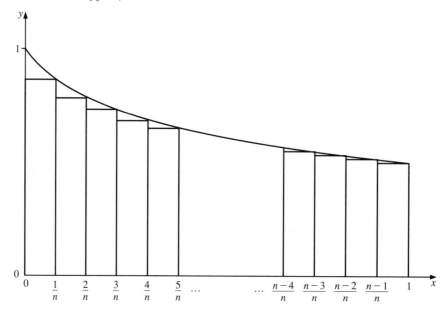

The graph of $y = \dfrac{1}{1+x}$, for $0 \leqslant x \leqslant 1$, is shown in the diagram together with n rectangles, each of width $\dfrac{1}{n}$. Show that the total area of all n rectangles is

$$\frac{1}{n}\left\{ \frac{1}{1+\frac{1}{n}} + \frac{1}{1+\frac{2}{n}} + \frac{1}{1+\frac{3}{n}} + \ldots + \frac{1}{2} \right\}$$

iii) State the limit, as $n \to \infty$, of the expression in part **ii**.

iv) By considering an appropriate graph, find the limit, as $n \to \infty$, of

$$\frac{1}{n}\left\{ \frac{1}{1+\left(\frac{1}{n}\right)^2} + \frac{1}{1+\left(\frac{2}{n}\right)^2} + \frac{1}{1+\left(\frac{3}{n}\right)^2} + \ldots + \frac{1}{2} \right\} \qquad \text{(OCR)}$$

15 Prove by induction, or otherwise, that

$$\sum_{r=1}^{n} r^3 = \frac{1}{4}n^2(n+1)^2$$

The diagram shows a sketch of the graph of

$$y = x^3 \quad (0 \leqslant x \leqslant 1)$$

The area of the region between the curve and the x-axis is divided up into n strips, each of width $\dfrac{1}{n}$, by lines drawn parallel to the y-axis. Show that the area A of the rth strip BCDE, shown in the diagram, satisfies the inequalities

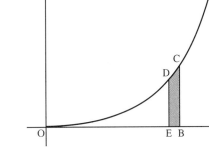

$$\frac{(r-1)^3}{n^4} < A_r < \frac{r^3}{n^4}$$

Hence show that the sum S of the areas of all n strips satisfies

$$\frac{1}{4}\left(\frac{n-1}{n}\right)^2 < S < \frac{1}{4}\left(\frac{n+1}{n}\right)^2$$

Deduce the value of the integral $\displaystyle\int_0^1 x^3 \, dx$.   (NEAB)

16 In this question, you may assume the following three results:

A) $\dfrac{\ln w}{w} \to 0$ as $w \to \infty$.

B) $y = \dfrac{\ln x}{x^{\frac{3}{2}}}$ is a decreasing function for $x > 2$.

C) $\displaystyle\int_a^b x^k \ln x \, dx = \frac{b^{k+1} \ln b - a^{k+1} \ln a}{k+1} - \frac{b^{k+1} - a^{k+1}}{(k+1)^2}$, where $0 < a < b$ and $k \neq -1$.

i) By substituting $w = \dfrac{1}{\sqrt{x}}$ into the result in A, show that $\sqrt{x}\ln x \to 0$ as $x \to 0$.

ii) Show that $\dfrac{\ln x}{\sqrt{x}} \to 0$ as $x \to \infty$.

iii) Explain why $\displaystyle\int_0^1 \frac{\ln x}{\sqrt{x}} \, dx$ is an improper integral, and evaluate this integral.

iv) Draw a diagram to show that $\displaystyle\sum_{r=3}^{n} \frac{\ln r}{r^{\frac{3}{2}}} < \int_2^n \frac{\ln x}{x^{\frac{3}{2}}} \, dx$, and write down a similar integral I for which $\displaystyle\sum_{r=3}^{n} \frac{\ln r}{r^{\frac{3}{2}}} > I$.

v) Deduce that the infinite series $\displaystyle\sum_{r=3}^{\infty} \frac{\ln r}{r^{\frac{3}{2}}}$ is convergent, and show that

$$3.57 < \sum_{r=3}^{\infty} \frac{\ln r}{r^{\frac{3}{2}}} < 3.81 \qquad \text{(MEI)}$$

13 Numerical methods

Which is so small that it scarcely admits of calculation.
DAVID HUME

Solution of polynomial equations

Most equations cannot be solved using algebraic procedures which give exact solutions, and so we have to turn to numerical methods to solve them.

While there are several, distinct numerical methods available to use, they all have one property in common: if we repeatedly apply any of the methods to a problem, we will normally be able to obtain the solution to any desired degree of accuracy.

Initially, we need to determine an interval in which the root lies. Hence, generally, to find $f(x) = 0$, we find $f(\alpha)$ and $f(\beta)$. If these are of opposite sign, and $f(x)$ is continuous between α and β, then $f(x) = 0$ has a root for some x satisfying $\alpha < x < \beta$.

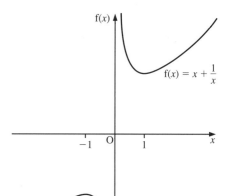

If $f(x)$ is not continuous, it may be as in the graph on the right, where $f(1)$ and $f(-1)$ are of opposite sign, and $f(x) \neq 0$ for any value between -1 and 1.

Example 1 Find an approximate value for the root of $f(x) \equiv x^3 + 5x - 9 = 0$.

SOLUTION

We have

$$f(1) = 1 + 5 - 9 = -3$$

$$f(2) = 8 + 10 - 9 = 9$$

We know that $f(x)$ is continuous for $1 < x < 2$. Hence, there is a root of $f(x) = 0$ for a value of x between 1 and 2.

To find the value of the root more accurately, we could repeat this method, finding $f(1.1)$, $f(1.2)$, $f(1.3)$, and so on, noting that the values of $f(x)$ change sign between 1.3 and 1.4, and then finding $f(1.31)$, $f(1.32)$, etc.

The method used in Example 1 is time-consuming, although with a sensible choice of values of x it can be reasonably effective in finding a solution without too many unnecessary calculations.

The procedures which are normally used to solve polynomial equations such as that in Example 1 are **interval bisection**, **linear interpolation**, the **Newton–Raphson method**, and **iteration**.

Interval bisection

As the name suggests, if we know that there is a root of $f(x) = 0$ between $x = \alpha$ and $x = \beta$, we try $x = \dfrac{(\alpha + \beta)}{2}$. The sign of $f(x)$ determines which side of $\dfrac{(\alpha + \beta)}{2}$ the root lies.

The method is repeated until we obtain the same answer to the degree of accuracy required.

Example 2 Find, by interval bisection, an approximate value for the root of $f(x) \equiv x^3 + 5x - 9 = 0$, correct to two significant figures.

SOLUTION

$$f(1) = 1 + 5 - 9 = -3$$

$$f(2) = 8 + 10 - 9 = 9$$

Therefore, the root lies between $x = 1$ and $x = 2$.

We now put $x = 1.5$, which gives

$$f(1.5) = 1.875$$

We note that $f(1.5)$ and $f(1)$ are of opposite sign. Therefore, the root lies between $x = 1$ and $x = 1.5$.

We continue to bisect the interval in which we know the root lies, until we obtain the required accuracy. Hence, we have the following results.

$f(1.25) = -0.796\,875$
$f(1.25)$ and $f(1.5)$ of opposite sign: root between $x = 1.25$ and 1.5

$f(1.375) = 0.474\,609\,375$
$f(1.25)$ and $f(1.375)$ of opposite sign: root between $x = 1.25$ and 1.375

$f(1.3125) = -0.176\,513\,67$
$f(1.3125)$ and $f(1.375)$ of opposite sign: root between $x = 1.3125$ and 1.375

$f(1.343\,75) = 0.145\,111$
$f(1.343\,75)$ and $f(1.3125)$ of opposite sign: root between $x = 1.3125$ and $1.343\,75$

Only now are we able to state that the solution of $f(x) = 0$ is 1.3 to two significant figures.

Interval bisection is a very long and generally slow method. Also, it fails if the graph of $f(x)$ is not continuous over the interval in question, as in the case of the graph of $f(x) = x + \dfrac{1}{x}$ on page 268.

(The actual value of the solution is $1.329\,744\,122$ to ten significant figures.)

Linear interpolation

A more efficient method of progressing from f(1) = −3 and f(2) = 9 is to deduce that the root of f(x) ≡ x^3 + 5x − 9 = 0 is likely to be much nearer to 1 than to 2, since |f(2)| > |f(1)|.

This intuitive approach is formalised in **linear interpolation**, where the two points (1, −3) and (2, 9) are joined by a straight line and the x-value of the point on this line is calculated.

Using similar triangles, with the root at $x = 1 + p_1$, we have

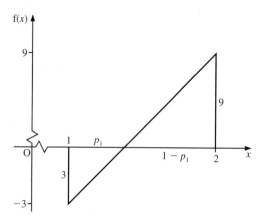

$$\frac{p_1}{3} = \frac{1 - p_1}{9} \quad \Rightarrow \quad p_1 = \frac{1}{4}$$

Therefore, a better approximation to the root of f(x) = 0 is 1.25, which gives

$$f(1.25) = -0.796\,875$$

Hence, the root is between 1.25 and 2.

Using similar triangles again, we have

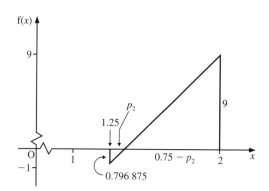

$$\frac{p_2}{0.796\,875} = \frac{0.75 - p_2}{9}$$

$$\Rightarrow \quad 9.796\,875p_2 = 0.75 \times 0.796\,875$$

$$\Rightarrow \quad p_2 = 0.061\,004\,784$$

Therefore, the second approximation to the root of f(x) = 0 is 1.311 004 784, which gives

$$f(1.311\,004\,784) = -0.191\,708\,181$$

Hence, the root is between 1.311 004 784 and 2.

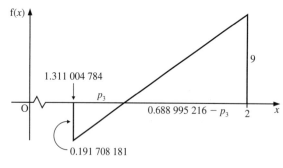

Repeating the procedure again (see figure above), we obtain

$$\frac{p_3}{0.191\,708\,181} = \frac{0.688\,995\,216 - p_3}{9}$$

$$\Rightarrow \quad 9.191\,708\,181p_3 = 0.688\,995\,216 \times 0.191\,708\,181$$

$$\Rightarrow \quad p_3 = 0.014\,370\,127$$

Therefore, the third approximation to the root is 1.325 374 912, which gives

$$f(1.325\,374\,912) = -0.044\,947\,145$$

Hence, the root is between 1.325 374 912 and 2.

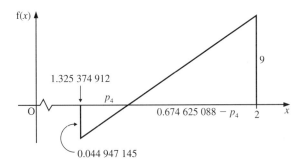

Repeating the procedure yet again (see figure above), we obtain

$$\frac{p_4}{0.044\,947\,145} = \frac{0.674\,625\,088 - p_4}{9}$$

$$\Rightarrow \quad 9.044\,947\,145 p_4 = 0.674\,625\,088 \times 0.044\,947\,145$$

$$\Rightarrow \quad p_4 = 0.003\,352\,421\,099$$

Therefore, the fourth approximation 1.328 727 333 is to the root.

Both the fourth and third approximations are 1.33 correct to two decimal places. To check that this is the correct answer to two decimal places, we find $f(1.335)$:

$$f(1.335) = 0.054\,27$$

which has the opposite sign to $f(1.326)$.

Hence, the root is 1.33 correct to two decimal places.

Although linear interpolation is much quicker than interval bisection, it still does not take into account the shape of the graph of $f(x)$ between the starting points.

The procedure which does is the Newton–Raphson method.

Newton–Raphson method

If α is an approximate value for the root of $f(x) = 0$, then $\alpha - \dfrac{f(\alpha)}{f'(\alpha)}$ is generally a better approximation.

Consider the graph of $y = f(x)$. Draw the tangent at P, where $x = \alpha$, and let the tangent meet the x-axis at T.

We see that the x-value at T is closer than α is to the x-value at N, where the graph cuts the axis.

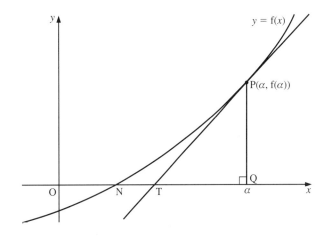

Using triangle PTQ, we have

$$\text{Gradient of tangent} = \frac{PQ}{QT}$$

$$\Rightarrow \quad f'(\alpha) = \frac{f(\alpha)}{QT}$$

$$\Rightarrow \quad QT = \frac{f(\alpha)}{f'(\alpha)}$$

The x-value of the point T is

$$\alpha - QT = \alpha - \frac{f(\alpha)}{f'(\alpha)}$$

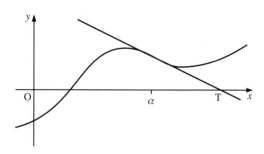

Figure A

which is a better approximation to the root of $f(x) = 0$. When the root of $f(x) = 0$ is not close to α, the method may fail. For example, in Figure A, the next x-value found is at T, which is further from the root than α is. And in Figure B, $f'(\alpha) = 0$, which is unhelpful.

In its iterative form, the Newton–Raphson method gives

$$\alpha_{n+1} = \alpha_n - \frac{f(\alpha_n)}{f'(\alpha_n)}$$

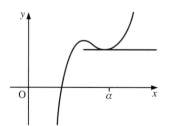

Figure B

Example 3 Use the Newton–Raphson method, with an initial value of $x = 1$, to find a root of $f(x) \equiv x^3 + 5x - 9$ to three significant figures.

SOLUTION

Let α be the required root.

Differentiating $f(x) \equiv x^3 + 5x - 9$, we have

$$f'(x) = 3x^2 + 5$$

Putting $\alpha_1 = 1$, we obtain

$$f(\alpha_1) = 1 + 5 - 9 = -3$$
$$f'(\alpha_1) = 3 + 5 = 8$$

Using Newton–Raphson, we have

$$\alpha_{n+1} = \alpha_n - \frac{f(\alpha_n)}{f'(\alpha_n)}$$

which gives

$$\alpha_2 = \alpha_1 - \frac{f(\alpha_1)}{f'(\alpha_1)} = 1 + \frac{3}{8} = 1.375$$

Hence, we have

$$\alpha_3 = 1.375 - \frac{f(1.375)}{f'(1.375)} = 1.375 - \frac{0.474\,609\,375}{10.671\,875} = 1.330\,5271$$

which then gives

$$\alpha_4 = 1.330\,5271 - \frac{f(1.330\,5271)}{f'(1.530\,5271)}$$

$$= 1.330\,5271 - \frac{0.008\,070\,770\,27}{10.310\,907} = 1.329\,744\,36$$

α_3 and α_4 are now so close together that we can say that the root is 1.33 to three significant figures. (The root is actually 1.329 744 to seven significant figures. Were we to repeat the procedure a few more times, we would find that the root is 1.329 744 122, correct to ten significant figures.)

So, the root of $f(x) \equiv x^3 + 5x - 9$ is 1.33, correct to three significant figures.

Iteration

An iterative process is one which is repeated several times, following exactly the same procedure each time.

It provides yet another way of obtaining the solution to an equation $f(x) = 0$, in which we rearrange the equation to create an **iterative formula** of the form

$$x_{n+1} = g(x_n)$$

where x_{n+1} is a closer approximation than x_n to the solution of $f(x) = 0$.

For example, we can rearrange $x^3 + 5x - 9 = 0$ as

$$x^3 = 9 - 5x \quad \Rightarrow \quad x = \sqrt[3]{9 - 5x}$$

from which we can obtain the iterative formula

$$x_{n+1} = (9 - 5x_n)^{\frac{1}{3}}$$

Alternatively, we can rearrange $x^3 + 5x - 9 = 0$ as

$$x^3 = 9 - 5x \quad \Rightarrow \quad x^2 = \frac{9}{x} - 5 \quad \Rightarrow \quad x = \sqrt{\frac{9}{x} - 5}$$

from which we can obtain the iterative formula

$$x_{n+1} = \sqrt{\frac{9}{x_n} - 5}$$

Naturally, some iterative procedures produce an accurate solution more quickly than others, and some iterative procedures fail quickly.

For example, using $x_{n+1} = \sqrt{\frac{9}{x_n} - 5}$, with $x_1 = 1$, we obtain

$$x_2 = \sqrt{\frac{9}{1} - 5} = 2$$

$$x_3 = \sqrt{\frac{9}{2} - 5} = \sqrt{-0.5} \quad \text{which does not exist.}$$

How to decide which iterative formula to use is beyond the scope of the A-level syllabuses, and therefore of this book. All the iterations you will meet at this level result in one of the patterns shown next.

Example 4 Use $x_{n+1} = \dfrac{1}{6}(x_n^3 + 1)$ to find a solution to

$f(x) = x^3 - 6x + 1 = 0$.

SOLUTION

As in the previous methods, we need to determine an interval in which the root lies.

Putting $x = 2$ and $x = 3$ in $f(x) = x^3 - 6x + 1$, we obtain

$$f(2) = -3 \quad \text{and} \quad f(3) = 10$$

Therefore, there is a root of $f(x) = 0$ for a value of x between 2 and 3.

(Also, since $f(0) = 1$ and $f(1) = -4$, there is another root of $f(x) = 0$ for a value of x between 0 and 1.)

Using $x_{n+1} = \dfrac{1}{6}(x_n{}^3 + 1)$, with $x_1 = 2$, we have

$$x_2 = \frac{1}{6}(2^3 + 1) = 1.5$$

which gives

$$x_3 = \frac{1}{6}(1.5^3 + 1) = 0.729\,1666$$

$$x_4 = 0.231\,28$$

We see that these values of x are not converging to the required root.

Alternatively, starting at $x_1 = 3$, we have

$$x_2 = \frac{1}{6}(3^3 + 1) = \frac{28}{6} = 4\tfrac{2}{3}$$

$$x_3 = \frac{1}{6}[(4\tfrac{2}{3})^3 + 1] = 17.1049$$

We see that these values of x are not converging to the root either.

In Example 4, we note that starting below the root sends the iteration to the smaller root, whereas starting above the root sends the iteration off to infinity. We can graphically represent these results by a **staircase diagram**, as shown on the right.

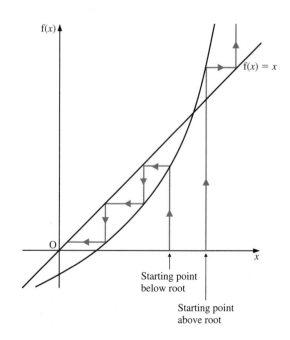

$f(x)$

$f(x) = x$

O

x

Starting point
below root

Starting point
above root

Example 5 Starting with $x = -1$, use the iteration $x_{n+1} = \dfrac{1}{6}x_n^2 - 2$ to find the root to three significant figures.

SOLUTION

Using $x_{n+1} = \dfrac{1}{6}x_n^2 - 2$ with $x_1 = -1$, we have

$$x_2 = \tfrac{1}{6} - 2 = -1\tfrac{5}{6} = -1.833\,33$$

$$x_3 = \tfrac{1}{6}(-1.833\,33)^2 - 2 = -1\tfrac{95}{216} = -1.4398$$

$$x_4 = -1.654\,488\,883 \qquad x_5 = -1.543\,777\,756$$

$$x_6 = -1.602\,791\,707 \qquad x_7 = -1.571\,843\,124$$

$$x_8 = -1.588\,218\,199 \qquad x_9 = -1.579\,593\,825$$

Therefore, the root is -1.58, correct to three significant figures.

Eventually, we would find that the root is $-1.582\,575\,695$.

The result in Example 5 is represented graphically by a pattern which spirals **into** the root, as shown on the right. Hence, it is called a **cobweb diagram**.

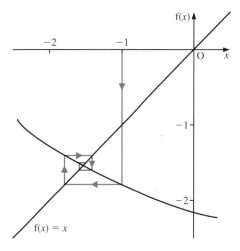

Two other patterns which you are likely to meet are shown below.

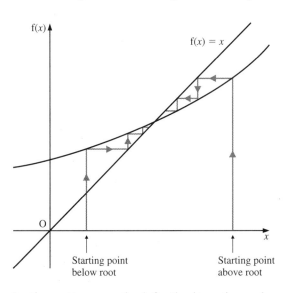

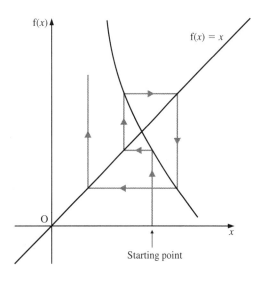

In the pattern on the left, the iterative values step directly **into** the root from above (or below). The pattern on the right spirals **out from** the root.

Exercise 13A

1 Show that a root of the equation $x^3 = 7 - 5x$ lies in the interval $1 < x < 2$. Use linear interpolation to find this root correct to two decimal places.

2 Show that a root of the equation $xe^{3x} = 12$ lies in the interval $0 < x < 1$. Use linear interpolation to find this root correct to two decimal places.

3 Show that a root of the equation $x^3 - 4x = 5$ lies in the interval $2 < x < 3$. Use interval bisection to find this root correct to two decimal places.

4 Show that a root of the equation $3x^3 - e^x = 0$ lies in the interval $0 < x < 1$. Use interval bisection to find this root correct to two decimal places.

5 Show that a root of the equation $2x^3 - e^{\frac{1}{2}x} = 0$ lies in the interval $0 < x < 1$. Use linear interpolation to find this root correct to two decimal places.

6 Show that a root of the equation $\sin \dfrac{\pi x}{2} = 3x - 1$ lies in the interval $0 < x < 1$. Use linear interpolation to find this root correct to two decimal places.

7 Using the Newton–Raphson method, find the real root of $x^3 + 3x - 7 = 0$ correct to two decimal places.

8 An iterative formula for solving a cubic is

$$x_{n+1} = \frac{3}{x_n^2} - 4$$

a) Take $x_1 = 4$ and calculate $x_2, x_3, x_4, x_5, x_6, x_7$ and x_8. For each iteration, write down the first five digits.
b) What is the solution, correct to three dp?
c) How many iterations are required to find this solution?
d) By replacing x_{n+1} and x_n with x, show that this value is a solution of the equation
$$x^3 + 4x^2 - 3 = 0$$

9 It is given that the equation $e^{0.5x} + x^2 - 3.5x = 0$ has **exactly** one root in the interval $[0, 1]$.

Apply the bisection method three times to obtain a more accurate determination of the interval containing the root.

Calculate the minimum number of **further** applications of the bisection method required to make the length of the interval less than 10^{-5}.   (SQA/CSYS)

10 $f(x) \equiv \tan x + 1 - 4x^2$ $-\dfrac{\pi}{2} < x < \dfrac{\pi}{2}$

a) Show that the equation $f(x) = 0$ has a root α in the interval $[1.42, 1.44]$
b) Use linear interpolation once on the interval $[1.42, 1.44]$ to find an estimate of α, giving your answer to three decimal places.
c) Show that the equation $f(x) = 0$ has another root β in the interval $[0.6, 0.7]$.

d) Use the iteration
$$x_{n+1} = \tfrac{1}{2}(1 + \tan x_n)^{\frac{1}{2}} \qquad x_0 = 0.65$$
to find β to three decimal places. (EDEXCEL)

11 $\qquad$ $f(x) \equiv 2^x - x^3$

a) Show that a root, α, of the equation $f(x) = 0$ lies in the interval $1.3 < \alpha < 1.4$.

b) Taking 1.37 as your starting value, apply the Newton–Raphson procedure once to $f(x)$ to obtain a second approximation to this root. Give your answer to three decimal places.
(EDEXCEL)

12 Show that the equation
$$e^x + x - 3 = 0$$
has a root between 0 and 1. Use the Newton–Raphson method to solve the equation, giving your answers correct to five decimal places. Record your values of $x_0, x_1, x_2, \ldots$ to as many decimal places as your calculator will allow. (WJEC)

13 Given that x is measured in radians and $f(x) \equiv \sin x - 0.4x$,

a) find the values of $f(2)$ and $f(2.5)$ and deduce that the equation $f(x) = 0$ has a root α in the interval $[2, 2.5]$

b) use linear interpolation once on the interval $[2, 2.5]$ to estimate a value for α, giving your answer to two decimal places

c) using 2.1 as a first approximation to α, use the Newton–Raphson process once to find a second approximation to α, giving your answer to two decimal places. (EDEXCEL)

14 The equation $x^3 + 3x^2 - 1 = 0$ has a root between 0 and 1. Use the Newton–Raphson method, with initial approximation 0.5, to find this root correct to two decimal places.

Give a clear reason why it would be impossible to use the Newton–Raphson method with initial approximation 0. (OCR)

15 Use the Newton–Raphson method to find, correct to three decimal places, the root of the equation $x^3 - 10x = 25$ which is close to 4. (OCR)

16 $\qquad$ $f(x) \equiv \cosh x - x^3$

a) Show that the equation $f(x) = 0$ has one root, α, between 1 and 2.

A second root, β, of the equation $f(x) = 0$ lies close to 6.14.

b) Apply the Newton–Raphson procedure once to $f(x)$ to obtain a second approximation to β, giving your answer to three decimal places. (EDEXCEL)

17 $\qquad$ $f(x) \equiv e^x - 2x^2$

a) Show that the equation $f(x) = 0$ has a root α in the interval $[-1, 0]$ and a root β in the interval $[1, 2]$.

b) Use linear interpolation once on the interval $[1, 2]$ to find an approximation to β, giving your answer to two decimal places.

c) Apply the Newton–Raphson process twice to $f(x)$, starting with -0.5, to find an approximation to α, giving your final answer as accurately as you think is appropriate.
(EDEXCEL)

18 a) Solve $x = 0.5 + \sin x$ by **each** of the following two methods.

  i) An iterative method, other than the Newton–Raphson method, starting with $x_1 = 1.5$. Give a solution which is correct to five significant figures.

  ii) The Newton–Raphson method, **applied once only**, starting from $x_1 = 1.5$.

  b) Calculate the gradient of $0.5 + \sin x$, where $x = 1.5$. Comment on its relevance to **one** of the methods used in part **a**.   (NEAB/SMP 16–19)

19 Given that $f(\theta) = \theta - \sqrt{(\sin \theta)}$, $0 < \theta < \frac{1}{2}\pi$, show that

  a) the equation $f(\theta) = 0$ has a root lying between $\frac{1}{4}\pi$ and $\frac{3}{10}\pi$

  b) $f'(\theta) = 1 - \dfrac{\cos \theta}{2\sqrt{(\sin \theta)}}$

  c) Taking $\frac{3}{10}\pi$ as a first approximation to this root of the equation $f(\theta) = 0$, use the Newton–Raphson procedure once to find a second approximation, giving your answer to two decimal places.

  d) Show that $f'(\theta) = 0$ when $\sin \theta = \sqrt{5} - 2$.   (EDEXCEL)

20 The figure shows the line with equation $y = 5x$ and the curve with equation $y = e^x$. They meet where $x = \alpha$ and $x = \beta$. Approximate values for α and β are 0.2 and 2.5 respectively.

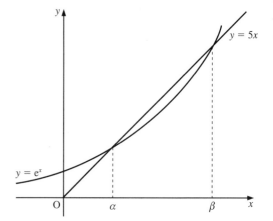

  a) The iterative formula $a_{n+1} = \frac{1}{5}e^{a_n}$ is used to find a more accurate approximation for α. Taking $a_1 = 0.2$ use the iterative formula to obtain a_2, a_3, a_4 and a_5, giving your answers to four decimal places.

The Newton–Raphson process is used to find a more accurate approximation for β.

  b) Taking $f(x) \equiv e^x - 5x$ and a first approximation to β of 2.5, apply the Newton–Raphson process once to obtain a second approximation, giving your answer to three decimal places.

  c) Explain, with the aid of a diagram, why the Newton–Raphson process fails if the first approximation used for β is $\ln 5$.   (EDEXCEL)

21 a) The cubic equation

$$x^3 - 9x + 3 = 0$$

has a root that lies between 0 and 1. Use the Newton–Raphson method with starting value $x_0 = 0.5$ to find this root, giving your answer correct to six decimal places.

  b) A rearrangement of the equation

$$x + 3 = 2\tan x$$

gives the iterative formula

$$x_{n+1} = \tan^{-1}\left(\frac{x_n + 3}{2}\right)$$

By considering the condition for convergence, show that this iterative formula can be used to find any root of the equation.   (WJEC)

22 The diagram below shows part of the graph of the function f, where

$$f(x) = \frac{6}{4x^3 - 12x^2 + 9x + 3}$$

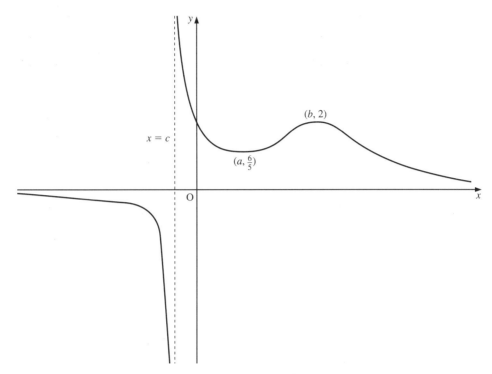

a) The graph of f has a minimum turning point at $(a, \frac{6}{5})$ and a maximum turning point at $(b, 2)$. Use calculus to obtain the values of a and b.

b) The line $x = c$ is a vertical asymptote to the graph of f.

  i) Write down an equation which c must satisfy.

  ii) Use Newton's method, with $x_0 = -0.2$, to find an approximation to the value of c correct to four decimal places.

$\left[\text{Newton's method uses the iteration } x_{n+1} = x_n - \dfrac{\text{p}(x_n)}{\text{p}'(x_n)} \text{ to produce successive approximations} \right.$

$\left. \text{to a solution of the equation } \text{p}(x) = 0. \right]$ (SQA/CSYS)

23 The equation $f(x) = 0$ has a root at $x = a$, which is known to be close to $x = x_0$. By drawing a suitable graph to illustrate this situation, derive the formula for the first iteration of the Newton–Raphson method of solution of $f(x) = 0$. Hence explain how the general formula is obtained.

It is known that the equation $f(x) = 0$, where

$$f(x) = 3x^5 - 8x^2 + 4$$

has three distinct real roots of which two are positive.

Use the Newton–Raphson method with starting value -1 to determine the negative root correct to three decimal places.

It is known that the other two roots lie in the narrow interval $[0.75, 1.25]$. Use a diagram to explain why the Newton–Raphson method may be difficult to use in the determination of these roots.

It is proposed to determine the root near $x = 0.75$ using simple iteration with the iterative scheme

$$x_{n+1} = \frac{3x_n^4}{8} + \frac{1}{2x_n}$$

Show that this **may** be suitable to obtain a solution in the neighbourhood of $x = 0.75$.

Using $x = 0.75$ as a starting value and recording successive iterates to three decimal places, use simple iteration to determine this root to two decimal places.

The third root is known to lie in the interval $[1.2, 1.25]$. Use three applications of the bisection method to determine a more accurate estimate of the interval in which this root lies.

(SQA/CSYS)

Evaluation of areas under curves

When we need to find the area under a curve but are unable to integrate the function, we have to use a numerical method. The two most common numerical techniques are the **trapezium rule** and **Simpson's rule**.

Trapezium rule

We can find the area under a curve by drawing equally spaced lines parallel to the y-axis. These will produce a number of trapezia of equal widths, as the figure shows.

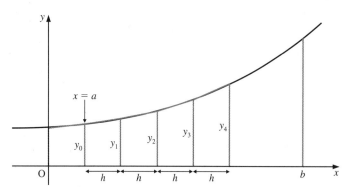

If we divide the x-axis from $x = a$ to $x = b$ into n equal intervals, then we will obtain n trapezia.

Let the y-values of the curve at these x-values be $y_0, y_1, \ldots, y_n$, as shown.

The area of the first trapezium is $\frac{1}{2}h(y_0 + y_1)$, where h is the width of each strip.

The area of the second trapezium is $\frac{1}{2}h(y_1 + y_2)$.

Hence, the total area of the trapezia is

$$\frac{1}{2}h(y_0 + y_1) + \frac{1}{2}h(y_1 + y_2) + \ldots + \frac{1}{2}h(y_{n-1} + y_n)$$

By collecting like terms, we obtain the **trapezium rule**, which is

$$\text{Area} \approx \frac{h}{2}[y_0 + y_n + 2(y_1 + y_2 + \ldots + y_{n-1})]$$

where h is the width of a strip and y_0 and y_n are the first and last ordinates.

Example 6 Find, by the trapezium rule, an approximate value for $\int_1^7 e^x \, dx$. Use six intervals.

SOLUTION

First, we divide the x-axis from $x = 1$ to $x = 7$ (the limits of the integral) into six strips (as requested).

Hence, the x-values of these points are $x = 1, 2, 3, 4, 5, 6, 7$, as the figure shows.

The corresponding y-values are $e^1, e^2, e^3, e^4, e^5, e^6, e^7$.

Therefore, using the trapezium rule, we have

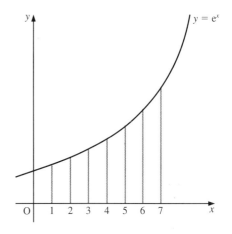

$$\text{Area} \approx \frac{h}{2}[y_0 + y_n + 2(y + y_2 + \ldots + y_{n-1})]$$

$$\approx \frac{1}{2}[e^1 + e^7 + 2(e^2 + e^3 + e^4 + e^5 + e^6)]$$

which gives

$$\text{Area} = 1183.590\,416 \quad \text{or} \quad 1183.6 \quad \text{to 1 dp}$$

Note

- The accurate answer to Example 6 is $e^7 - e^1$, which is $1093.914\,877$ or 1093.9 to one decimal place.
- The answer obtained by the trapezium rule can be made more accurate by using more strips of smaller width.

Simpson's rule

The trapezium rule is rarely very accurate because we usually use too small a number of trapezia to approximate the area to be found.

We obtain a better approximation by imposing a known, integrable quadratic curve which passes through points on the original curve.

Simpson's rule is based upon the use of a quadratic curve which passes through three consecutive points. Thus, Simpson's rule finds the approximate value for a **pair** of strips.

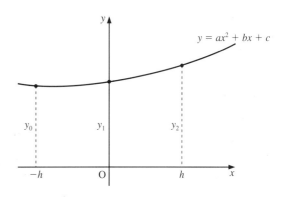

Consider the quadratic curve $y = ax^2 + bx + c$, passing through three consecutive points, (h, y_2), $(0, y_1)$ and $(-h, y_0)$, as shown on the right.

When $x = 0$, $y = y_1 \quad \Rightarrow \quad c = y_1$ [1]

When $x = h$, $y = y_2 \quad \Rightarrow \quad y_2 = ah^2 + bh + y_1$ [2]

When $x = -h$, $y = y_0 \quad \Rightarrow \quad y_0 = ah^2 - bh + y_1$ [3]

Adding [2] and [3], we obtain

$$y_0 + y_2 = 2ah^2 + 2y_1 \qquad\qquad [4]$$

Using integration to find the area under the quadratic curve, we have

$$\text{Area of pair of strips} = \int_{-h}^{h} (ax^2 + bx + c)\, dx$$

$$= \left[\frac{ax^3}{3} + \frac{bx^2}{2} + cx \right]_{-h}^{h}$$

$$= \frac{ah^3}{3} + \frac{bh^2}{2} + ch - \left(-\frac{ah^3}{3} + \frac{bh^2}{2} - ch \right)$$

$$= \frac{2ah^3}{3} + 2ch$$

Substituting from [1] and [4], we obtain

$$\text{Area of pair of strips} = \frac{h(y_0 + y_2 - 2y_1)}{3} + 2y_1 h$$

$$= \frac{h}{3}(y_0 + 4y_1 + y_2)$$

Using a number of such pairs of strips, we have

$$\text{Total area} \approx \frac{h}{3}(y_0 + 4y_1 + y_2) + \frac{h}{3}(y_2 + 4y_3 + y_4) + \frac{h}{3}(y_4 + 4y_5 + y_6) + \dots$$

$$\approx \frac{h}{3}(y_0 + 4y_1 + y_2 + y_2 + 4y_3 + y_4 + y_4 + 4y_5 + y_6 + \dots)$$

By factorising, we obtain Simpson's rule, which is

$$\text{Area} \approx \frac{h}{3}[y_0 + y_n + 4(y_1 + y_3 + y_5 + \dots) + 2(y_2 + y_4 + y_6 + \dots)]$$

Or

$$\text{Area} \approx \frac{1}{3} \times \text{Strip width (First + Last + 4 × Sum of odds + 2 × Sum of evens)}$$

Note There **must** always be an **even** number of strips. That is, *n* **must be even**.

Example 7 Find, by Simpson's rule, an approximate value for $\int_{1}^{7} e^x\, dx$. Use six intervals.

SOLUTION

First, we divide the *x*-axis from $x = 1$ to $x = 7$ (the limits of the integral) into six strips (as requested).

Note Since we are using Simpson's rule, we ensure that we use an **even** number of strips.

Hence, the *x*-values of these points are $x = 1, 2, 3, 4, 5, 6, 7$. (See top figure on page 281.)

The corresponding y-values are e^1, e^2, e^3, e^4, e^5, e^6, e^7.

Therefore, using Simpson's rule, we have

$$\text{Area} \approx \frac{1}{3}[e^1 + e^7 + 4(e^2 + e^4 + e^6) + 2(e^3 + e^5)]$$

which gives

$$\text{Area} = 1099.337\,61 \quad \text{or} \quad 1099.3 \quad \text{to 1 dp}$$

Note On page 281, we gave the accurate value of this area as 1093.9 (to 1 dp). Hence, the value obtained by Simpson's rule gives a better approximation than that obtained by the trapezium rule.

Exercise 13B

1 Using five equally spaced ordinates, estimate the value of each of the following to four decimal places by means of **i)** the trapezium rule, and **ii)** Simpson's rule.

a) $\displaystyle\int_1^5 x^2 \, dx$ **b)** $\displaystyle\int_2^6 x^3 \, dx$ **c)** $\displaystyle\int_0^6 \sin\frac{x}{4} \, dx$ **d)** $\displaystyle\int_1^4 e^{\sin x} \, dx$

2 Using six strips, find an estimate for $\displaystyle\int_2^5 x^x \, dx$ by means of **i)** Simpson's rule, and **ii)** the trapezium rule.

3 a) Show that the length of the arc, s, of the curve with equation $y = \cosh x$ between $x = 0$ and $x = 2$ is given by

$$s = \int_0^2 \cosh x \, dx$$

b) Obtain an estimate to this integral by using Simpson's rule with five equally spaced ordinates, giving your answer to four decimal places.
c) Find the exact value of s.
d) Determine the percentage error which results from using the estimate for s calculated in part **b** rather than the exact value obtained in part **c**, giving your answer to one significant figure. (EDEXCEL)

4 $$I_n = \int_0^1 x^{\frac{1}{2}n} e^{-\frac{1}{2}x} \, dx \qquad n \geqslant 0$$

a) Show that $I_n = nI_{n-2} - 2e^{-\frac{1}{2}}$, $n \geqslant 2$.
b) Evaluate I_0 in terms of e.
c) Find, using the results of parts **a** and **b**, the value of I_4 in terms of e.
d) Show that the approximate value for I_1 using Simpson's rule with three equally spaced ordinates is

$$\frac{1}{6}(2\sqrt{2}e^{-\frac{1}{4}} + e^{-\frac{1}{2}}) \qquad \text{(EDEXCEL)}$$

5
$$A = \int_2^4 \frac{1}{\sqrt{(4x^2 - 9)}} \, dx$$

a) Using five equally spaced ordinates, obtain estimates for A, to four decimal places, by means of
   i) the trapezium rule
   ii) Simpson's rule.

b) Find
$$\int \frac{1}{\sqrt{(4x^2 - 9)}} \, dx$$
and hence evaluate A, giving your answer to four decimal places.

c) Which of your estimates in part **a** is the more accurate? Give a reason for your answer.

(EDEXCEL)

6
$$I_n = \int \frac{x^n}{\sqrt{(1 + x^2)}} \, dx$$

a) Show that $nI_n = x^{n-1}\sqrt{(1 + x^2)} - (n - 1)I_{n-2}$, $n \geqslant 2$.

The curve C has equation
$$y^2 = \frac{x^2}{\sqrt{(1 + x^2)}} \qquad y \geqslant 0$$

The finite region R is bounded by C, the x-axis and the lines with equations $x = 0$ and $x = 2$. The region R is rotated through 2π radians about the x-axis.

b) Find the volume of the solid so formed, giving your answer in terms of π, surds and natural logarithms.

An estimate for the volume obtained in part **b** is found using Simpson's rule with three ordinates.

c) Find the percentage error resulting from using this estimate, giving your answer to three decimal places.   (EDEXCEL)

7 For $0 < x < \pi$, the curve C has equation $y = \ln(\sin x)$. The region of the plane bounded by C, the x-axis and the lines $x = \dfrac{\pi}{4}$ and $x = \dfrac{\pi}{2}$ is rotated through 2π radians about the x-axis.

Show that the surface area of the solid generated in this way is given by S, where
$$S = 2\pi \int_{\frac{\pi}{4}}^{\frac{\pi}{2}} \left| \frac{\ln(\sin x)}{\sin x} \right| \, dx$$

Use the trapezium rule with four ordinates (three strips) to find an approximate value for S, giving your answer to three decimal places.   (AEB 97)

8 Use the trapezium rule, with six intervals, to estimate the value of
$$\int_0^3 \ln(1 + x) \, dx$$
showing your working. Give your answer correct to three significant figures.

Hence write down an approximate value for
$$\int_0^3 \ln \sqrt{(1 + x)} \, dx \qquad \text{(OCR)}$$

9 Use the trapezium rule with five intervals to estimate the value of

$$\int_0^{0.5} \sqrt{(1+x^2)}\,dx$$

showing your working. Give your answer correct to two decimal places.

By expanding $(1+x^2)^{\frac{1}{2}}$ in powers of x as far as the term in x^4, and integrating term by term, obtain a second estimate for the value of

$$\int_0^{0.5} \sqrt{(1+x^2)}\,dx$$

giving this answer also correct to two decimal places.   (OCR)

10 Derive Simpson's rule with two strips for evaluating an approximation to $\int_{-h}^{h} f(x)\,dx$.

Use Simpson's composite rule with **four** strips to obtain an estimate of $\int_2^3 \cos(x-2)\ln x\,dx$.

(Use five decimal place arithmetic in your calculation.)   (SQA/CSYS)

11 Use the composite trapezium rule with **four** sub-intervals to obtain an approximation to the definite integral

$$\int_0^{\frac{1}{2}} x\sin(\pi x)\,dx$$

(Give your final answer to four decimal places.)   (SQA/CSYS)

12 Use the trapezium rule, with four intervals, to estimate the value of

$$\int_1^2 \sqrt{\left(x - \frac{1}{x}\right)}\,dx$$

showing your working and giving your answer correct to two decimal places.

The diagram shows part of the graph of $y = \sqrt{\left(x - \dfrac{1}{x}\right)}$.

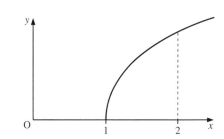

i) State, with a reason, whether this use of the trapezium rule gives an underestimate or an overestimate of the value of $\int_1^2 \sqrt{\left(x - \dfrac{1}{x}\right)}\,dx$.

ii) State, without further calculation, whether increasing the number of intervals in the trapezium rule from four to eight would lead to a larger or a smaller estimate for $\int_1^2 \sqrt{\left(x - \dfrac{1}{x}\right)}\,dx$. Give a reason for your answer.   (OCR)

Step-by-step solution of differential equations

First-order differential equations

Most differential equations cannot be solved exactly, but need a step-by-step approach.

These methods depend on drawing lines parallel to the y-axis, distance h apart.

h is called the **step length**.

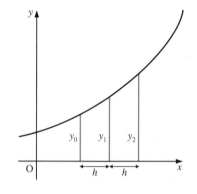

Single-step approximation

The linear approximation

$$\left(\frac{\mathrm{d}y}{\mathrm{d}x}\right)_0 \approx \frac{y_1 - y_0}{h}$$

is commonly used in the step-by-step solution of first-order differential equation. It is known as **Euler's method**, after Léonard Euler (1707–83), the prolific Swiss mathematician. We derive it as follows.

With reference to the figure on the right, $P(x_0, y_0)$ is a point on the curve $y = f(x)$ and $Q(x_1, y_1)$ is another point on the curve close to P, where $x_1 - x_0 = h$ and h is small.

We see that the gradient of the chord PQ is approximately the same as the gradient of the tangent at P. Hence, we have

$$\text{Gradient of PQ} = \frac{y_1 - y_0}{h}$$

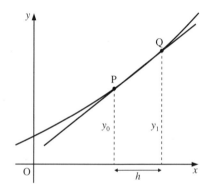

which gives

$$\text{Gradient of tangent at P} = \left(\frac{\mathrm{d}y}{\mathrm{d}x}\right)_0 \approx \frac{y_1 - y_0}{h}$$

Naturally, the accuracy of this method depends on the size of the step length, h.

Example 8 Use a step length of 0.1 to find $y(0.3)$ for $\dfrac{\mathrm{d}y}{\mathrm{d}x} = \ln(x + y)$, given that $y = 2$ when $x = 0$.

SOLUTION

Using $\left(\dfrac{\mathrm{d}y}{\mathrm{d}x}\right)_0 \approx \dfrac{y_1 - y_0}{h}$, we obtain

$$y_1 \approx y_0 + h\left(\frac{\mathrm{d}y}{\mathrm{d}x}\right)_0$$

which means that

$$y \text{ at new value of } x \text{ (i.e. when } x \text{ is } 0.1) =$$

$$= y \text{ at original value of } x + h \times \frac{\mathrm{d}y}{\mathrm{d}x} \text{ at original value of } x$$

Hence, we have

$$y(0.1) \approx 2 + 0.1\ln(0 + 2)$$

$$\Rightarrow \quad y(0.1) \approx 2.0693$$

We repeat this procedure with the values obtained for y and $\dfrac{dy}{dx}$ when $x = 0.1$ now being treated as the original values, and the new value for y being found for $x = 0.2$. Thus, we obtain

$$y(0.2) \approx y(0.1) + h\left(\frac{dy}{dx}\right)_{x=0.1}$$

$$\Rightarrow \quad y(0.2) \approx 2.0693 + 0.1\ln(0.1 + 2.0693)$$

$$\Rightarrow \quad y(0.2) \approx 2.1467$$

Repeating again, we have

$$y(0.3) \approx y(0.2) + h\left(\frac{dy}{dx}\right)_{x=0.2}$$

$$\Rightarrow \quad y(0.3) \approx 2.1467 + 0.1\ln(0.2 + 2.1467)$$

$$\Rightarrow \quad y(0.3) \approx 2.2320$$

Example 9 Use a step length of 0.2 to find $y(1.4)$ for $\dfrac{dy}{dx} = e^{\cos x}$ given that $y = 3$ when $x = 1$.

SOLUTION

Using $\left(\dfrac{dy}{dx}\right)_0 \approx \dfrac{y_1 - y_0}{h}$, we obtain

$$y_1 \approx y_0 + h\left(\frac{dy}{dx}\right)_0$$

which means that

$$y \text{ at new value of } x \text{ (i.e. when } x \text{ is 1.2)} =$$

$$= y \text{ at original value of } x + h \times \frac{dy}{dx} \text{ at original value of } x$$

Hence, we have

$$y(1.2) \approx 3 + 0.2\,e^{\cos 1}$$

$$\Rightarrow \quad y(1.2) \approx 3.3433$$

We repeat this procedure with the values obtained for y and $\dfrac{dy}{dx}$ when $x = 1.2$ now being treated as the original values, and the new value for y being found for $x = 1.4$. Thus, we obtain

$$y(1.4) \approx y(1.2) + h\left(\frac{dy}{dx}\right)_{x=1.2}$$

$$\Rightarrow \quad y(1.4) \approx 3.3433 + 0.2\,e^{\cos 1.2}$$

$$\Rightarrow \quad y(1.4) \approx 3.6307$$

Double-step approximation

A better approximation is given by

$$\left(\frac{dy}{dx}\right)_0 \approx \frac{y_1 - y_{-1}}{2h}$$

which uses a double step, as shown in the figure on the right.

We see that the gradient of the chord TQ is a better approximation to the gradient of the tangent at P than that obtained with the single step.

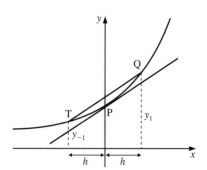

We have

$$\text{Gradient of chord TQ} = \frac{y_1 - y_{-1}}{2h}$$

which gives

$$\text{Gradient of tangent at P} = \left(\frac{dy}{dx}\right)_0 \approx \frac{y_1 - y_{-1}}{2h}$$

Example 10 Using $\left(\dfrac{dy}{dx}\right)_0 \approx \dfrac{y_1 - y_{-1}}{2h}$, and using a step length of 0.1, find y when $x = 1.2$ for

$$\frac{dy}{dx} = \frac{3x^2 - y^2}{2xy}$$

given that $y = 2$ at $x = 1$.

SOLUTION

Since we are required to use the double-step approximation, we need to know the values of y at two values of x.

To find the second value of y, we use the single-step method. As we are given the y-value when $x = 1$, the original value of x is 1, and the new value of x is 1.1. Hence, we have

$$\left(\frac{dy}{dx}\right)_0 \approx \frac{y_1 - y_0}{h} \quad \Rightarrow \quad y_1 \approx y_0 + h\left(\frac{dy}{dx}\right)_0$$

which gives

$$y(1.1) \approx y(1) + 0.1\left(\frac{dy}{dx}\right)_{x=1}$$

When $x = 1$ and $y = 2$, we have

$$\frac{dy}{dx} = \frac{3x^2 - y^2}{2xy} = \frac{3 - 4}{2 \times 1 \times 2} = -\frac{1}{4}$$

which gives

$$y(1.1) \approx 2 + 0.1 \times -\frac{1}{4}$$

$$\Rightarrow \quad y(1.1) \approx 1.975$$

Now we have two values for y, we can use the double-step approximation

$$\left(\frac{dy}{dx}\right)_0 \approx \frac{y_1 - y_{-1}}{2h}$$

$$\Rightarrow \quad y_1 \approx y_{-1} + 2h\left(\frac{dy}{dx}\right)_0$$

which gives

$$y(1.2) \approx y(1) + 2h\left(\frac{dy}{dx}\right)_{x=1.1}$$

$$\Rightarrow \quad y(1.2) \approx 2 + 2 \times 0.1 \times \frac{3 \times 1.1^2 - 1.975^2}{2 \times 1.1 \times 1.975}$$

$$\Rightarrow \quad y(1.2) \approx 2 + 2 \times 0.1 \times -0.062\,284$$

Therefore, when $x = 1.2$, $y = 1.9875$, correct to 4 dp.

Second-order differential equations of the form $\dfrac{d^2y}{dx^2} = f(x, y)$

With reference to the figure on the right, P is a point at $x = -\frac{1}{2}h$ on the curve of $\dfrac{dy}{dx}$ against x, and Q is a point at $x = \frac{1}{2}h$ on the same curve, where h is small.

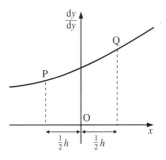

We see that the gradient of the chord PQ is approximately the same as the gradient of the tangent at $x = 0$. Hence, we have

$$\text{Gradient of PQ} = \frac{\left(\dfrac{dy}{dx}\right)_{\frac{1}{2}h} - \left(\dfrac{dy}{dx}\right)_{-\frac{1}{2}h}}{h}$$

That is, we have

$$\text{Gradient of tangent} = \left(\frac{d^2y}{dx^2}\right)_0 \approx \frac{\left(\dfrac{dy}{dx}\right)_{\frac{1}{2}h} - \left(\dfrac{dy}{dx}\right)_{-\frac{1}{2}h}}{h}$$

which gives

$$\left(\frac{d^2y}{dx^2}\right)_0 \approx \frac{\dfrac{y_1 - y_0}{h} - \dfrac{y_0 - y_{-1}}{h}}{h}$$

$$\Rightarrow \quad \left(\frac{d^2y}{dx^2}\right)_0 \approx \frac{y_1 - 2y_0 + y_{-1}}{h^2}$$

To solve numerically a second-order differential equation, we need either the values of y at two different values of x, or one value of y and one value of $\dfrac{dy}{dx}$.

Example 11

$$\frac{d^2y}{dx^2} = xe^{\cos y}$$

Using a step length of 0.1, find y when $x = 1.3$, given that $y = 1$ when $x = 1$ and $y = 1.2$ when $x = 1.1$.

SOLUTION

We use

$$\left(\frac{d^2y}{dx^2}\right)_0 \approx \frac{y_1 - 2y_0 + y_{-1}}{h^2}$$

with

y_0 as the value when $x = 1.1$

y_1 as the value when $x = 1.1 + h = 1.2$

y_{-1} as the value when $x = 1.1 - h = 1$

Hence, we have

$$\left(\frac{d^2y}{dx^2}\right)_{x=1.1} \approx \frac{y(1.2) - 2y(1.1) + y(1)}{0.1^2}$$

$$\Rightarrow \quad \left(\frac{d^2y}{dx^2}\right)_{x=1.1} \approx \frac{y(1.2) - 2.4 + 1}{0.01}$$

When $x = 1.1$ and $y = 1.2$, we have

$$\frac{d^2y}{dx^2} = 1.1\,e^{\cos 1.2} = 1.580\,38$$

which gives

$$\frac{y(1.2) - 1.4}{0.01} \approx 1.580\,38$$

$$\Rightarrow \quad y(1.2) \approx 2.4 - 1 + 0.015\,8038$$

$$\Rightarrow y(1.2) \approx 1.4158$$

We repeat this procedure, using

$$\left(\frac{d^2y}{dx^2}\right)_0 \approx \frac{y_1 - 2y_0 + y_{-1}}{h^2}$$

with

y_0 as the value when $x = 1.2$

y_1 as the value when $x = 1.2 + h = 1.3$

y_{-1} as the value when $x = 1.2 - h = 1.1$

Hence, we obtain

$$\left(\frac{d^2y}{dx^2}\right)_{x=1.2} \approx \frac{y(1.3) - 2y(1.2) + y(1.1)}{0.1^2}$$

For $x = 1.2$ and $y = 1.4158$, this gives

$$\left(\frac{d^2y}{dx^2}\right)_{x=1.2} = 1.4003 \approx \frac{y(1.3) - 2.8316 + 1.2}{0.01}$$

$$\Rightarrow \quad y(1.3) \approx 2.8316 - 1.2 + 0.014\,003$$

$$\Rightarrow \quad y(1.3) \approx 1.6456$$

Therefore, when $x = 1.3$, $y = 1.6456$, correct to 4 dp.

Example 12

$$\frac{d^2y}{dx^2} = 1 + x\cos y + \sin y \cos y$$

Using a step length of 0.05, find y when $x = 1.1$, given that $\dfrac{dy}{dx} = 1$ and $y = 0$ when $x = 1$.

SOLUTION

Because we are given y and $\dfrac{dy}{dx}$ at **only one** value of x, we need to use a first-order step-by-step approximation to find a second value for y.

We know the value of y when $x = 1$, so $x = 1$ becomes the original value for x. We require a step length of 0.05, hence we use

$$\left(\frac{d^2y}{dx^2}\right)_0 \approx \frac{y_1 - 2y_0 + y_{-1}}{h^2}$$

with

y_0 as the value when $x = 1$

y_1 as the value when $x = 1 + h = 1.05$

y_{-1} as the value when $x = 1 - h = 0.95$

The most accurate first-order step-by-step method is the double-step approximation

$$\left(\frac{dy}{dx}\right)_0 \approx \frac{y_1 - y_{-1}}{2h}$$

which gives

$$1 \approx \frac{y(1.05) - y(0.95)}{0.1}$$

$$\Rightarrow \quad 0.1 \approx y(1.05) - y(0.95) \qquad [1]$$

Using $\left(\dfrac{d^2y}{dx^2}\right)_0 \approx \dfrac{y_1 - 2y_0 + y_{-1}}{h^2}$, with $x = 1$ and $y = 0$, we obtain

$$\left(\frac{d^2y}{dx^2}\right)_{x=1} \approx \frac{y(1.05) - 2 \times 0 + y(0.95)}{h^2}$$

When $x = 1$ and $y = 0$, we have

$$\left(\frac{d^2y}{dx^2}\right)_{x=1} = 1 + 1\cos 0 + \sin 0\cos 0 = 2$$

which gives

$$2 \approx \frac{y(1.05) + y(0.95)}{0.0025}$$

$$\Rightarrow \quad 0.005 \approx y(1.05) + y(0.95) \qquad [2]$$

Hence, adding [1] and [2], we obtain $y(1.05) \approx 0.0525$.

We have now two values of y, namely $y(1)$ and $y(1.05)$, so we are able to use

$$\left(\frac{d^2y}{dx^2}\right)_0 \approx \frac{y_1 - 2y_0 + y_{-1}}{h^2}$$

to find y when $x = 1.1$

Thus, we have

$$\left(\frac{d^2y}{dx^2}\right)_{x=1.05} \approx \frac{y(1.1) - 2y(1.05) + y(1)}{0.05^2}$$

When $x = 1.05$ and $y = 0.0525$, we also have

$$\left(\frac{d^2y}{dx^2}\right)_{x=1.05} = 1 + 1.05\cos 0.0525 + \sin 0.0525 \cos 0.0525$$

$$= 2.100\,957$$

which gives

$$2.100\,957 \approx \frac{y(1.1) - 2 \times 0.0525 + 0}{0.05^2}$$

$$\Rightarrow \quad y(1.1) \approx 0.0025 \times 2.1009\,57 + 0.105$$

$$\Rightarrow \quad y(1.1) \approx 0.1103$$

Therefore, when $x = 1.1$, $y = 0.1103$, correct to 4 dp.

Taylor's series

The other main method for solving differential equations numerically is to use Taylor's series (the derivation of which is beyond the scope of this book):

$$f(x) = f(a) + (x - a)f'(a) + \frac{(x-a)^2}{2!}f''(a) + \frac{(x-a)^3}{3!}f'''(a) + \dots$$

We use this series to find values of $f(x)$, or y, near a given value of $f(x)$ (see Example 12). Its most common application is in the special case when $a = 0$, which gives

$$f(x) = f(0) + xf'(0) + \frac{x^2}{2!}f''(0) + \frac{x^3}{3!}f'''(0) + \dots$$

Notice that this is the same as Maclaurin's series, which we studied on pages 177–9. In the numerical solution of differential equations, when we refer to a series we always mean Taylor's series, though it is rarely seen in its full form.

Example 13 Expand f(x) up to terms in x^4, where

$$\frac{d^2 y}{dx^2} + x\frac{dy}{dx} + y = 0$$

given that $y = 1$ and $\frac{dy}{dx} = 0$ at $x = 0$. Hence find y when $x = 0.01$, giving your answer to 11 decimal places.

SOLUTION

Differentiating $\frac{d^2 y}{dx^2} + x\frac{dy}{dx} + y = 0$, we obtain

$$\frac{d^3 y}{dx^3} + \frac{dy}{dx} + x\frac{d^2 y}{dx^2} + \frac{dy}{dx} = 0$$

$$\Rightarrow \quad \frac{d^3 y}{dx^3} + x\frac{d^2 y}{dx^2} + 2\frac{dy}{dx} = 0$$

Differentiating again, we obtain

$$\frac{d^4 y}{dx^4} + \frac{d^2 y}{dx^2} + x\frac{d^3 y}{dx^3} + 2\frac{d^2 y}{dx^2} = 0$$

$$\Rightarrow \quad \frac{d^4 y}{dx^4} + x\frac{d^3 y}{dx^3} + 3\frac{d^2 y}{dx^2} = 0$$

But f(0) = 1 and f'(0) = 0 (given), so we have

From $\frac{d^2 y}{dx^2} + x\frac{dy}{dx} + y = 0$:   f''(0) = −1

From $\frac{d^3 y}{dx^3} + x\frac{d^2 y}{dx^2} + 2\frac{dy}{dx} = 0$:   f'''(0) = 0

From $\frac{d^4 y}{dx^4} + x\frac{d^3 y}{dx^3} + 3\frac{d^2 y}{dx^2} = 0$:   f''''(0) = 3

which give

$$f(x) = 1 - \frac{x^2}{2!} + \frac{3x^4}{4!}$$

Therefore, substituting $x = 0.01$, we obtain

$$f(0.01) = 1 - 0.000\,05 + \frac{1}{8} \times 0.000\,000\,01$$

That is, $y = 0.999\,950\,001\,25$, correct to 11 dp as the next term is 10^{-12}.

Example 14 Expand y up to terms in $(x - 1)^3$, where

$$\frac{d^2 y}{dx^2} + y\frac{dy}{dx} = x$$

given that $y = 0$ and $\frac{dy}{dx} = 1$ at $x = 1$. Hence find y when $x = 1.01$, giving your answer to six decimal places.

SOLUTION

As the values of y and $\dfrac{dy}{dx}$ are given when $x = 1$, we must use the full version of Taylor's series and obtain a solution for y in powers of $x - 1$.

Differentiating $\dfrac{d^2y}{dx^2} + y\dfrac{dy}{dx} = x$, we obtain

$$\frac{d^3y}{dx^3} + \frac{dy}{dx} \times \frac{dy}{dx} + y\frac{d^2y}{dx^2} - 1 = 0$$

$$\Rightarrow \quad \frac{d^3y}{dx^3} + \left(\frac{dy}{dx}\right)^2 + y\frac{d^2y}{dx^2} - 1 = 0$$

But $f(1) = 0$ and $f'(1) = 1$ (given), so we have

From $\dfrac{d^2y}{dx^2} + y\dfrac{dy}{dx} = x$: $\quad f''(1) = 1$

From $\dfrac{d^3y}{dx^3} + \left(\dfrac{dy}{dx}\right)^2 + y\dfrac{d^2y}{dx^2} - 1 = 0$: $\quad f'''(1) = 0$

which give

$$f(x) = (x - 1) + \frac{(x - 1)^2}{2!}$$

(Note that since $f'''(1) = 0$, there is no term in $(x - 1)^3$.)

When $x = 1.01$, we obtain

$$f(0.1) = 0.01 + 0.00005$$

Therefore, $y = 0.010050$, correct to 6 dp as the next term is 10^{-8}.

Exercise 13C

In Questions **1** to **4**, find the Taylor's series solution for y up to and including terms in x^4:

1 $\dfrac{dy}{dx} = y^3 + x^8$, for which $y = 1$, when $x = 0$.

2 $\dfrac{dy}{dx} = x^2y + xy^2$, for which $y = 2$, when $x = 0$.

3 $\dfrac{d^2y}{dx^2} + x\dfrac{dy}{dx} + 4y = 0$, for which $\dfrac{dy}{dx} = 1$ and $y = 0$ when $x = 0$.

4 $\dfrac{d^2y}{dx^2} + y\dfrac{dy}{dx} + 2x^2y = 0$, for which $\dfrac{dy}{dx} = 0$ and $y = 1$ when $x = 0$.

Hence find y correct to nine decimal places when $x = 0.01$.

5 Given that

$$\frac{d^2y}{dx^2} = x^3 + 2y^3$$

and that $y = 0$ and $\frac{dy}{dx} = 1$ when $x = 0$, expand y as a power series in $(x - 1)$. Hence find y, correct to four decimal places, when **a)** $x = 1.1$, and **b)** $x = 0.9$.

6 Given that y satisfies the differential equation

$$\frac{d^2y}{dx^2} - 4y\frac{dy}{dx} = 0$$

and that $y = 0$ at $x = 0$, and $\frac{dy}{dx} = 2$ at $x = 0$, use the Taylor series method to find a series for y in ascending powers of x up to, and including, the term in x^3.　　(EDEXCEL)

7 Obtain the Taylor polynomial of degree two for the function $\sin x$ near $x = \frac{\pi}{4}$. Estimate the value of $\sin 46°$ using the first-degree approximation.　　(SQA/CSYS)

8 Obtain the Taylor polynomial of degree two, in the form $f(0.5 + h) = c_0 + c_1 h + c_2 h^2$ for the function $f(x) = \frac{1}{7x - 4}$ near $x = 0.5$.

State, with a reason, whether $f(x)$ is sensitive to small changes in the value of x in the neighbourhood of $x = 0.5$.　　(SQA/CSYS)

9
$$\frac{d^2y}{dx^2} + x\frac{dy}{dx} + 3y = 0$$

where $y = 1$ at $x = 0$ and $\frac{dy}{dx} = 2$ at $x = 0$.

Find y as a series in ascending powers of x, up to and including the term in x^3.　　(EDEXCEL)

10 Given that y satisfies the differential equation $\frac{dy}{dx} = (x + y)^3$, and $y = 1$ at $x = 0$,

a) find expressions for $\frac{d^2y}{dx^2}$ and $\frac{d^3y}{dx^3}$.

b) Hence, or otherwise, find y as a series in ascending powers of x up to and including the term in x^3.

c) Use your series to estimate the value of y at $x = -0.1$, giving your answer to one decimal place.　　(EDEXCEL)

11 Obtain the series solution in ascending powers of x, up to and including the term in x^3, of the differential equation

$$\frac{d^2y}{dx^2} + y\frac{dy}{dx} - 4y = 0$$

given that $y = 3$ and $\frac{dy}{dx} = 2$ at $x = 0$.　　(EDEXCEL)

12 $$\frac{dy}{dx} = y(xy - 1), \quad y = 1 \text{ at } x = 0$$

a) Use the approximation of $\frac{y_1 - y_0}{h} \approx \left(\frac{dy}{dx}\right)_0$ to estimate the value of y at $x = 0.1$.

b) Using a step length of 0.1 with the approximation $\frac{y_2 - y_0}{2h} \approx \left(\frac{dy}{dx}\right)_1$ and your answer from part **a**, estimate the value of y at $x = 0.2$.

c) Using a step length of 0.1 again and by repeating the application of the approximation used in part **b**, estimate the value of y at $x = 0.3$. (EDEXCEL)

13 The function $y(x)$ satisfies the differential equation

$$\frac{dy}{dx} = f(x, y)$$

where $f(x, y) = (x^2 + y^2)^{\frac{1}{2}}$, and $y(0) = 1$.

a) Use the Euler formula

$$y_{r+1} = y_r + h f(x_r, y_r)$$

with $h = 0.1$ to obtain an approximation to $y(0.1)$.

b) Use the improved Euler formula

$$y_{r+1} = y_r + h f(x_r, y_r)$$

together with your answer to part **a** to obtain an approximation to $y(0.2)$, giving your answer correct to three decimal places. (NEAB)

14 The motion of one point of a turbine blade is given by

$$\frac{dx}{dt} = 4y + 3 \qquad \frac{dy}{dt} = 5 - 4x$$

Initially, $x = 2$, $y = 0$.

a) Use a step-by-step method with $dt = 0.05$ to estimate its position one tenth of a second later.

b) Find a second-order equation, in x and t only, which gives the displacement x at any time t.

c) Write down a first-order differential equation in x and y only. Solve this equation by an exact method, leaving your solution in the form $f(y) = g(x)$. (NEAB/SMP 16–19)

15 The function $y(x)$ satisfies the differential equation

$$\frac{dy}{dx} = f(x, y)$$

where $f(x, y) = 2 + \frac{y}{x}$ and $y(1) = 1$.

a) Use the Euler formula

$$y_{r+1} = y_r + h f(x_r, y_r)$$

with $h = 0.05$ to obtain an approximate value for $y(1.2)$, giving your answer correct to three decimal places.

b) i) Show that the integrating factor for the above differential equation is $\dfrac{1}{x}$.

ii) Solve the differential equation to find y in terms of x, and use it to show that $y(1.2) = 1.638$, correct to three decimal places.

c) Hence find, correct to one decimal place, the percentage error in using Euler's formula in the evaluation of $y(1.2)$.   (NEAB)

16 The variable y satisfies the differential equation $\dfrac{\mathrm{d}y}{\mathrm{d}x} = x^2 + y^2$, and $y = 0$ at $x = 0.5$.

Use the approximation $\left(\dfrac{\mathrm{d}y}{\mathrm{d}x}\right)_0 \approx \dfrac{y_1 - y_0}{h}$ with step length $h = 0.01$ to estimate the values of y at $x = 0.51$, $x = 0.52$ and $x = 0.53$, giving your answers to four decimal places.   (EDEXCEL)

17 a) The differential equation
$$\frac{\mathrm{d}^2 x}{\mathrm{d}t^2} - 4\frac{\mathrm{d}x}{\mathrm{d}t} + 3x = 0$$
can be written as two simultaneous first-order differential equations.

i) If one of these equations is $v = \dfrac{\mathrm{d}x}{\mathrm{d}t}$, write down the other equation.

ii) Use a step-by-step method with two steps of $\mathrm{d}t = 0.05$ to estimate the value of x at $t = 0.1$, given that at $t = 0$, $x = 0$ and $v = 2$.

b) i) Find the general solution of the differential equation
$$\frac{\mathrm{d}^2 x}{\mathrm{d}t^2} - 4\frac{\mathrm{d}x}{\mathrm{d}t} + 3x = 0$$

ii) Find the particular solution if $x = 0$ and $\dfrac{\mathrm{d}x}{\mathrm{d}t} = 2$ at $t = 0$. Hence calculate the value of x when $t = 0.1$, giving your answer to two decimal places.   (NEAB/SMP 16–19)

18 The equation $f(x) = 0$ has a root at $x = a$, which is known to be close to $x = x_0$. Use the Taylor series expansion of $f(x)$ about $x = x_0$ to derive the formula for the Newton–Raphson method of solution of $f(x) = 0$

It is known that the equation $f(x) = 0$, where
$$f(x) = x^4 - 6x^2 + 2x + 1$$
has four distinct roots of which two are positive.

Show that exactly one root of the equation lies in the interval $[2, 3]$.

Use the Newton–Raphson method to determine this root correct to two decimal places.

It is proposed to determine the other positive root using simple iteration. Show that the equation can be rearranged to give the iterative scheme
$$x_{n+1} = \frac{x_n^3}{6} + \frac{1}{3} + \frac{1}{6x_n}$$
and that this **may** be suitable to obtain a solution in the interval $[0.5, 1]$.

Using $x_0 = 0.5$ as a starting value, and recording the successive iterates to three decimal places, use simple iteration to determine this root to two decimal places.

State the order of convergence of the iterative scheme used and explain how the data from the iterative process can be seen to agree with this.   (SQA/CSYS)

19 Derive Euler's method for the approximate solution of the differential equation

$$\frac{dy}{dx} = f(x, y)$$

subject to the initial condition $y(x_0) = y_0$.

The differential equation $\dfrac{dy}{dx} = (x^2 + y)e^{-2x}$ with $y(1) = 2$ is to be solved.

Use Euler's method with step lengths of 0.1 and 0.05 to obtain two approximations to the solution of this equation at $x = 1.2$. Perform the calculations using four decimal place accuracy.

Assuming that the difference in the two estimates for $y(1.2)$ is due entirely to the truncation error, estimate the size of this error in the calculation with step size 0.05. Hence give a better estimate of $y(1.2)$ to an appropriate degree of accuracy.

The predictor–corrector method of solution where Euler's method is used as the predictor and the trapezium rule as the corrector (with **one** corrector application on each step) is to be used to approximate the solution of the above equation at $x = 1.2$. Use step length 0.1 and perform the calculation using four decimal place accuracy.   (SQA/CSYS)

20 The solution of the differential equation

$$x\frac{dy}{dx} = (y + 1)^2 - \cos x \qquad y(1) = 0$$

is required at $x = 1.15$. Obtain an approximation to this solution using Euler's method with step size 0.05. Perform the calculation using three decimal place accuracy.

If a step size 0.01 had been used in this calculation, by what factor would you expect the truncation error to be reduced?   (SQA/CSYS)

14 Matrices

Mathematics is not a book confined within a cover and bound between brazen clasps, whose contents it needs only patience to ransack.
JAMES JOSEPH SYLVESTER

A matrix stores mathematical information in a concise way. The information is written down in a rectangular array of rows and columns of terms, called **elements** or **entries**, each of which has its own precise position in the array.

$\begin{pmatrix} 4 \\ 8 \\ 7 \end{pmatrix}$ is a matrix, but its meaning depends on the context.

As in Chapter 6, it could represent a vector, meaning $4\mathbf{i} + 8\mathbf{j} + 7\mathbf{k}$. In football, it could represent the number of goals scored by three different clubs. In a shop, it could represent the number of packets of three different items bought.

Notation

We normally represent matrices by bold capital letters. For example,

$$\mathbf{M} = \begin{pmatrix} 4 & 11 & 5 \\ 1 & 4 & 2 \\ 1 & 2 & 1 \end{pmatrix}$$

Example 1, on page 300, illustrates an application of this notation.

The order of a matrix

The order of a matrix is its shape. For example, the matrix $\begin{pmatrix} 6 & -2 & 7 \\ 4 & 3 & -5 \end{pmatrix}$

has order 2×3, since its elements are arranged in two rows and three columns.

When stating the order of a matrix, we must **always give first the number of rows**, followed by the number of columns.

$\begin{pmatrix} 4 \\ 8 \\ 7 \end{pmatrix}$ is a **column matrix** and has order 3×1, since its elements are arranged

in three rows and only one column.

The matrix $(4 \quad 8 \quad 7)$ has order 1×3 and is a **row matrix**.

When the number of rows and the number of columns are equal, the matrix is called a **square matrix**.

Note $(4, 8, 7)$ with the numbers separated by commas is a point. $(4 \quad 8 \quad 7)$ with no commas is a matrix.

Addition and subtraction of matrices

Only when two matrices are of the **same order** can we add them or subtract them.

To add two matrices of the same order, we proceed as follows, element by element:

$$\begin{pmatrix} a & b & c \\ d & e & f \\ g & h & i \end{pmatrix} + \begin{pmatrix} p & q & r \\ s & t & u \\ v & w & x \end{pmatrix} = \begin{pmatrix} a+p & b+q & c+r \\ d+s & e+t & f+u \\ g+v & h+w & i+x \end{pmatrix}$$

We subtract two matrices of the same order in a similar way.

We **cannot** evaluate $\begin{pmatrix} a \\ b \end{pmatrix} + \begin{pmatrix} c & d \\ e & f \end{pmatrix}$ because the matrices are **not of the same order**.

Multiplication of matrices

Multiplying a matrix by a number

To multiply a matrix by, for example, k, we multiply **every** element of the matrix by k. Hence, we have

$$k \begin{pmatrix} a & b & c \\ d & e & f \\ g & h & i \end{pmatrix} = \begin{pmatrix} ka & kb & kc \\ kd & ke & kf \\ kg & kh & ki \end{pmatrix}$$

Example 1 Find $3\mathbf{A} + 2\mathbf{B}$ when $\mathbf{A} = \begin{pmatrix} 4 & 7 & -1 \\ 8 & 1 & 5 \end{pmatrix}$ and $\mathbf{B} = \begin{pmatrix} 3 & 2 & 4 \\ -1 & -3 & 2 \end{pmatrix}$.

SOLUTION

We have

$$3\mathbf{A} + 2\mathbf{B} = 3 \begin{pmatrix} 4 & 7 & -1 \\ 8 & 1 & 5 \end{pmatrix} + 2 \begin{pmatrix} 3 & 2 & 4 \\ -1 & -3 & 2 \end{pmatrix}$$

Multiplying out the RHS, we obtain

$$3\mathbf{A} + 2\mathbf{B} = \begin{pmatrix} 12 & 21 & -3 \\ 24 & 3 & 15 \end{pmatrix} + \begin{pmatrix} 6 & 4 & 8 \\ -2 & -6 & 4 \end{pmatrix}$$

which gives

$$3\mathbf{A} + 2\mathbf{B} = \begin{pmatrix} 18 & 25 & 5 \\ 22 & -3 & 19 \end{pmatrix}$$

Multiplying one matrix by another

We **cannot** multiply **any** matrix by **any other** matrix.

To allow multiplication, the orders of the two matrices concerned must **conform** to the following rule:

> The number of columns in the first matrix must be the same as the number of rows in the second matrix.

For example, if the first matrix has order 3×3, the second must have order $3 \times$ something, as in the case of **A** and **B** below, which we will multiply together:

$$\mathbf{A} = \begin{pmatrix} 2 & 3 & 1 \\ 0 & -2 & 3 \\ 0 & 2 & 3 \end{pmatrix} \qquad \mathbf{B} = \begin{pmatrix} 1 & 2 & 0 \\ 1 & -2 & 1 \\ 0 & 2 & 1 \end{pmatrix}$$

To multiply **A** by **B**, we start by taking the first row of matrix **A**, $(2 \quad 3 \quad 1)$,

and the first column of matrix **B**, $\begin{pmatrix} 1 \\ 1 \\ 0 \end{pmatrix}$.

We then multiply the first element of the row by the first element of the column, the second element of the row by the second element of the column, and the third element of the row by the last element of the column. We then add up these three products.

This gives the element in the top left-hand corner of the matrix **AB**, which is

$$2 \times 1 + 3 \times 1 + 1 \times 0 = 5$$

So, we have

$$\mathbf{AB} = \begin{pmatrix} 5 & ? & ? \\ ? & ? & ? \\ ? & ? & ? \end{pmatrix}$$

Next, we take the second row of matrix **A**, $(0 \quad -2 \quad 3)$, and the first column

of matrix **B**, $\begin{pmatrix} 1 \\ 1 \\ 0 \end{pmatrix}$.

Again, we multiply each element of the row by the corresponding element of the column and add up the products.

This gives the second element of the first column of matrix **AB**, which is

$$0 \times 1 - 2 \times 1 + 3 \times 0 = -2$$

So, now we have

$$\mathbf{AB} = \begin{pmatrix} 5 & ? & ? \\ -2 & ? & ? \\ ? & ? & ? \end{pmatrix}$$

We repeat the procedure on the second and third columns of matrix **B**, eventually obtaining

$$\mathbf{AB} = \begin{pmatrix} 5 & 0 & 4 \\ -2 & 10 & 1 \\ 2 & 2 & 5 \end{pmatrix}$$

(Notice that at each stage it looks as if we are finding a scalar dot product of two vectors.)

Generally, the product **PQ** produces a matrix which has the **same number of rows** as **P**, and the **same number of columns** as **Q**. Hence, if **P** has order $p \times t$ and **Q** has order $t \times q$, then **PQ** has order $p \times q$.

Multiplication is not commutative

It is important to note that the multiplication of two matrices is **not commutative**. That is,

$$\mathbf{AB} \neq \mathbf{BA}$$

Therefore, we must ensure that we write the matrices in the **correct sequence**. (See Exercise 14A, Question 1, page 306.)

Also, to avoid ambiguity when referring to the product of **A** and **B**, we must **specify their sequence**. For example, in the case of **AB**, we say either that **A premultiplies B** or that **B postmultiplies A**.

There are, however, three exceptions to the non-commutative law:

- Multiplication of a zero matrix by a non-zero matrix of the same order (see page 304).
- Multiplication of a square matrix by its inverse (see page 304).
- Multiplication of a square matrix by the identity matrix of the same order (see page 303).

We also note the following:

- If **AB** exists, **BA** does not necessarily exist.
- The matrix $\mathbf{A}^2$ is $\mathbf{A} \times \mathbf{A}$, which can only exist if **A** is a square matrix.

Multiplication is associative

We find that for **any** matrices **A**, **B** and **C**, which are conformable for multiplication,

$$\mathbf{A}(\mathbf{BC}) = (\mathbf{AB})\mathbf{C}$$

provided their **sequence is not changed**.

Known as the **associative law**, this allows us to decide whether we start the multiplication with the first pair of matrices or the second pair. Consequently, we can refer to the product **ABC** without ambiguity.

Determinant of a matrix

As stated on page 81, determinants always consist of a square array of elements. It follows, therefore, that **only a square matrix** can have a determinant.

From our definition of a determinant, we see that it is the scalar representation of its originating square matrix, and gives the value associated with that matrix.

If **A** is a square matrix, we can find the determinant of **A**, denoted by det **A** or |**A**|, by the method shown on pages 80–1.

Determinant of the product of two matrices

The determinant of the product **AB** is the same as the product of the determinant of **A** and that of **B**:

$$\det(\mathbf{AB}) = \det \mathbf{A} \times \det \mathbf{B}$$

Identity matrices and zero matrices

An **identity matrix** is any square matrix all of whose elements in the leading diagonal are 1, and all of whose other elements are zeros. It is denoted by **I**. Hence,

$$\mathbf{I} = \begin{pmatrix} 1 & 0 \\ 0 & 1 \end{pmatrix}$$

is known as the 2×2 identity matrix, and

$$\mathbf{I} = \begin{pmatrix} 1 & 0 & 0 \\ 0 & 1 & 0 \\ 0 & 0 & 1 \end{pmatrix}$$

is known as the 3×3 identity matrix.

When we multiply **I** by any square matrix **M** of the same order as **I**, **I** behaves as unity. That is,

$$\mathbf{IM} = \mathbf{MI} = \mathbf{M}$$

Zero matrices

When all the elements of a matrix are zeros, it is known as a **zero matrix**, and is denoted by **0**.

A zero matrix may have **any** order and therefore is not unique. For example,

$$\mathbf{0} = \begin{pmatrix} 0 \\ 0 \end{pmatrix} \qquad \mathbf{0} = \begin{pmatrix} 0 & 0 \\ 0 & 0 \end{pmatrix}$$

We can multiply any non-zero matrix by a zero matrix provided the zero matrix is **conformable for multiplication**. For example,

$$\begin{pmatrix} 5 & -2 \\ -4 & 3 \end{pmatrix} \begin{pmatrix} 0 \\ 0 \end{pmatrix} = \begin{pmatrix} 0 \\ 0 \end{pmatrix}$$

and $\begin{pmatrix} 0 & 0 \\ 0 & 0 \end{pmatrix} \begin{pmatrix} 3 & 2 \\ 4 & 5 \end{pmatrix} = \begin{pmatrix} 0 & 0 \\ 0 & 0 \end{pmatrix} = \begin{pmatrix} 3 & 2 \\ 4 & 5 \end{pmatrix} \begin{pmatrix} 0 & 0 \\ 0 & 0 \end{pmatrix}$

Generally, we have

$$\mathbf{0M} = \mathbf{0} \quad \text{and} \quad \mathbf{N0} = \mathbf{0}$$

Also, from the second example, we note that when $\mathbf{0}$ and $\mathbf{M}$ have the **same order**

$$\mathbf{0M} = \mathbf{0} = \mathbf{M0}$$

which is one of the three exceptions to the non-commutative laws discussed on page 302.

When we multiply together two **non-zero matrices**, we can get a **zero matrix** as the result. For example,

$$\begin{pmatrix} 5 & 2 \\ 10 & 4 \end{pmatrix} \begin{pmatrix} 2 & 4 \\ -5 & -10 \end{pmatrix} = \begin{pmatrix} 0 & 0 \\ 0 & 0 \end{pmatrix}$$

Inverse matrices

If $\mathbf{M}$ is a square matrix, its **inverse**, denoted by $\mathbf{M}^{-1}$, is defined by

$$\mathbf{MM}^{-1} = \mathbf{M}^{-1}\mathbf{M} = \mathbf{I}$$

Contrary to the non-commutative law discussed on page 302, we note that the order in which we multiply $\mathbf{M}$ and $\mathbf{M}^{-1}$ does not matter, which means that $\mathbf{M}^{-1}$, if it exists, is unique.

The inverse of a square matrix, $\mathbf{M}$, exists when $\det \mathbf{M} \neq 0$. That is, when $\mathbf{M}$ is said to be **non-singular**. When $\det \mathbf{M} = 0$, $\mathbf{M}$ is said to be **singular**.

The minor determinant

The **minor determinant** of an element of a matrix is the determinant of the matrix formed by deleting the row and column containing that element.

For example, the minor determinant of the middle element, 2, of the

matrix $\begin{pmatrix} 5 & 6 & 9 \\ 7 & 2 & 1 \\ 3 & 4 & 8 \end{pmatrix}$ is the determinant of the matrix $\begin{pmatrix} 5 & 9 \\ 3 & 8 \end{pmatrix}$, which is

$$\begin{vmatrix} 5 & 9 \\ 3 & 8 \end{vmatrix} = 13$$

Finding the inverse of a 3 × 3 matrix

We proceed in the following order:

1 Find the value of the determinant, Δ, of the matrix.

2 Find the value of the minor determinant of each of the elements.

3 Form a new matrix from the minor values, inserting them in the positions corresponding to the elements from which they were derived. Also insert a minus sign at each odd-numbered place, counting on from the top left entry of the matrix. These minor values with their associated signs (+ or −) are called the **cofactors** of the elements of the original matrix.

4 Find the transpose of the result.

Hence, we have

$$\begin{pmatrix} a & b & c \\ d & e & f \\ g & h & i \end{pmatrix}^{-1} = \frac{1}{\Delta} \begin{pmatrix} A & -B & C \\ -D & E & -F \\ G & -H & I \end{pmatrix}^{T}$$

where $A, B, C, \ldots$ are the minor determinants of the elements $a, b, c, \ldots$ respectively.

Example 2 Find the inverse of **M**, where $\mathbf{M} = \begin{pmatrix} 1 & 2 & 5 \\ 2 & 3 & 4 \\ 1 & 1 & 2 \end{pmatrix}$

SOLUTION

First, we calculate det **M**, which gives

$$\det \mathbf{M} = 1(6 - 4) - 2(4 - 4) + 5(2 - 3) = -3$$

Next, we calculate the minor determinants, obtaining

$$\begin{array}{ccc} 2 & 0 & -1 \\ -1 & -3 & -1 \\ -7 & -6 & -1 \end{array}$$

Then, we insert those minor values in their appropriate positions, together with their associated signs (+ or −), to form the matrix to be transposed.

(For example, the minor value of element 4 is $\begin{vmatrix} 1 & 2 \\ 1 & 1 \end{vmatrix} = -1$. This is

inserted three places from the top left corner of the matrix, for which the associated sign is minus, giving $-(-1) = +1$. Thus, $+1$ is the cofactor of element 4.)

Hence, we have

$$\mathbf{M}^{-1} = \frac{1}{-3} \begin{pmatrix} 2 & -0 & -1 \\ +1 & -3 & +1 \\ -7 & +6 & -1 \end{pmatrix}^{T}$$

We obtain the transpose by reflecting the matrix in its leading diagonal (see page 84), giving

$$\mathbf{M}^{-1} = -\frac{1}{3}\begin{pmatrix} 2 & 1 & -7 \\ 0 & -3 & 6 \\ -1 & 1 & -1 \end{pmatrix} = \begin{pmatrix} -\frac{2}{3} & -\frac{1}{3} & \frac{7}{3} \\ 0 & 1 & -2 \\ \frac{1}{3} & -\frac{1}{3} & \frac{1}{3} \end{pmatrix}$$

Exercise 14A

1 Evaluate **PQ** and **QP**, where

$$\mathbf{P} = \begin{pmatrix} 6 & 4 \\ 2 & 3 \end{pmatrix} \quad \text{and} \quad \mathbf{Q} = \begin{pmatrix} 1 & -2 \\ 2 & 3 \end{pmatrix}$$

What do you conclude from your results, and why has it happened?

2 Find the inverse of each of the following.

a) $\begin{pmatrix} 3 & 4 \\ 4 & 5 \end{pmatrix}$
    b) $\begin{pmatrix} 2 & 7 \\ 1 & 4 \end{pmatrix}$
    c) $\begin{pmatrix} 1 & -2 & 1 \\ 3 & -1 & 5 \\ -1 & 4 & 0 \end{pmatrix}$

d) $\begin{pmatrix} 4 & 11 & 5 \\ 1 & 4 & 2 \\ 1 & 2 & 1 \end{pmatrix}$
    e) $\begin{pmatrix} 3 & 4 & -2 \\ 2 & -1 & 5 \\ -3 & 4 & 1 \end{pmatrix}$

3 Find the inverse of the matrix $\begin{pmatrix} -1 & 0 & 1 \\ 2 & 0 & 1 \\ k & -1 & 0 \end{pmatrix}$ in terms of k.   (NICCEA)

4 Given the matrix $\mathbf{A} = \begin{pmatrix} \cos\theta & -\sin\theta \\ \sin\theta & \cos\theta \end{pmatrix}$, show by induction that

$$\mathbf{A}^n = \begin{pmatrix} \cos n\theta & -\sin n\theta \\ \sin n\theta & \cos n\theta \end{pmatrix}$$

for all positive integers n.   (WJEC)

5 a) Calculate the inverse of the matrix

$$\mathbf{A}(x) = \begin{pmatrix} 1 & x & -1 \\ 3 & 0 & 2 \\ 1 & 1 & 0 \end{pmatrix} \quad x \neq \frac{5}{2}$$

The image of the vector $\begin{pmatrix} a \\ b \\ c \end{pmatrix}$ when transformed by the matrix $\begin{pmatrix} 1 & 3 & -1 \\ 3 & 0 & 2 \\ 1 & 1 & 0 \end{pmatrix}$ is the vector $\begin{pmatrix} 4 \\ 3 \\ 5 \end{pmatrix}$.

b) Find the values of a, b and c.   (EDEXCEL)

6 Given that the matrix $\mathbf{A} = \begin{pmatrix} 5 & 2 & 3 \\ 3 & 2 & 1 \\ 2 & 5 & 2 \end{pmatrix}$ and that the determinant of $\mathbf{A} = 20$, find $\mathbf{A}^{-1}$.

(WJEC)

7 The matrices $\mathbf{A}$ and $\mathbf{C}$ are given by

$$\mathbf{A} = \begin{pmatrix} 1 & 1 & 1 \\ 1 & 2 & 2 \\ 2 & 1 & 3 \end{pmatrix} \qquad \mathbf{C} = \begin{pmatrix} 1 & 0 & 2 \\ 3 & 1 & 0 \\ 1 & 1 & 1 \end{pmatrix}$$

Find the matrix $\mathbf{B}$ satisfying $\mathbf{BA} = \mathbf{C}$. (WJEC)

8 Let matrix $\mathbf{A} = \begin{pmatrix} 0 & 1 \\ -2 & 3 \end{pmatrix}$ and $\mathbf{I}$ be the unit matrix $\begin{pmatrix} 1 & 0 \\ 0 & 1 \end{pmatrix}$.

i) Show that $\mathbf{A}^2 = 3\mathbf{A} - 2\mathbf{I}$.
ii) By writing $\mathbf{A}^3 = \mathbf{A} \times \mathbf{A}^2$ and using part **i**, show that $\mathbf{A}^3 = 7\mathbf{A} - 6\mathbf{I}$.
iii) For positive n, use the method of induction to prove that

$$\mathbf{A}^n = (2^n - 1)\mathbf{A} + (2 - 2^n)\mathbf{I} \qquad \text{(NICCEA)}$$

9 The matrix $\mathbf{A}$ is given by

$$\mathbf{A} = \begin{pmatrix} 1 & a & 0 \\ -1 & 1 & 0 \\ a & 5 & 1 \end{pmatrix}$$

where $a \neq -1$.

i) Find $\mathbf{A}^{-1}$.
ii) Given that $a = 2$, find the coordinates of the point which is mapped onto the point with coordinates $(1, 2, 3)$ by the transformation represented by $\mathbf{A}$. (OCR)

10 The matrix $\mathbf{A}$ is given by

$$\mathbf{A} = \begin{pmatrix} 2 & -1 & 1 \\ 0 & 3 & 1 \\ 1 & 1 & a \end{pmatrix}$$

where $a \neq 1$. Find the inverse of $\mathbf{A}$.

Hence, or otherwise, find the point of intersection of the three planes with equations

$$2x - y + z = 0$$
$$3y + z = 0$$
$$x + y + az = 3 \qquad \text{(OCR)}$$

11 Matrices $\mathbf{A}$ and $\mathbf{B}$ are given by

$$\mathbf{A} = \begin{pmatrix} 1 & 0 & 0 \\ 1 & -1 & 0 \\ 1 & 0 & a \end{pmatrix} \quad \text{and} \quad \mathbf{B} = \begin{pmatrix} 1 & 1 & 1 \\ 0 & 1 & -1 \\ 0 & 0 & 2 \end{pmatrix}$$

where $a \neq 0$.

i) Find the inverse of $\mathbf{A}$.

ii) Given that

$$\mathbf{B}^{-1} = \begin{pmatrix} 1 & -1 & -1 \\ 0 & 1 & \frac{1}{2} \\ 0 & 0 & \frac{1}{2} \end{pmatrix}$$

find the matrix $\mathbf{C}$ such that $\mathbf{ABC} = \mathbf{I}$, where $\mathbf{I}$ is the identity matrix.   (OCR)

12 It is given that

$$\mathbf{A} = \begin{pmatrix} 1 & -1 & 2 \\ 1 & 1 & 3 \\ a & 0 & 5 \end{pmatrix}$$

where $a \neq 2$.

i) Show that $\mathbf{A}$ has an inverse, and find it.

ii) It is given that

$$\mathbf{A}\begin{pmatrix} x_1 \\ x_2 \\ x_3 \end{pmatrix} = \mathbf{B}\begin{pmatrix} y_1 \\ y_2 \\ y_3 \end{pmatrix}$$

where

$$\mathbf{B} = \begin{pmatrix} 0 & 1 & 1 \\ 1 & 0 & 1 \\ 1 & 1 & 0 \end{pmatrix}$$

Find x_2 in terms of y_1, y_2, y_3 and a.   (OCR)

13 Let $\mathbf{A} = \begin{pmatrix} 1 & 1 & 1 \\ 1 & 1 & 1 \\ 1 & 1 & 1 \end{pmatrix}$ and $\mathbf{B} = \begin{pmatrix} 1 & -1 & -1 \\ -1 & 1 & -1 \\ -1 & -1 & 1 \end{pmatrix}$.

a) Determine whether or not $\mathbf{AB} = \mathbf{BA}$.

b) Verify that $\mathbf{A}^2 + 3\mathbf{B}^2 = 12\mathbf{I}$, where $\mathbf{I}$ is the 3×3 identity matrix.

c) Find $\mathbf{AB}$, $\mathbf{AB}^2$ and $\mathbf{AB}^3$ as multiples of $\mathbf{A}$, and make a conjecture about a general result for $\mathbf{AB}^n$. Use induction to prove your conjecture.

d) It is given that $\mathbf{B}$ is invertible, with inverse of the form

$$\mathbf{B}^{-1} = \begin{pmatrix} x & y & z \\ z & x & y \\ y & z & x \end{pmatrix}$$

Write down a system of linear equations which x, y and z must satisfy, and hence find the values of x, y and z.

e) Verify that $\mathbf{B}^2 - \mathbf{B}$ is a multiple of $\mathbf{I}$, and hence find $\mathbf{B}^{-1}$ in the form $r\mathbf{B} + s\mathbf{I}$ where r, s are real numbers. Hence check your answer to part **d**.   (SQA/CSYS)

14 a) Given that $\mathbf{A} = \begin{pmatrix} 1 & 1 & 2 \\ 0 & 2 & 1 \\ 1 & 0 & 2 \end{pmatrix}$, find $\mathbf{A}^2$.

b) Using $\mathbf{A}^3 = \begin{pmatrix} 10 & 9 & 23 \\ 5 & 9 & 14 \\ 9 & 5 & 19 \end{pmatrix}$, show that $\mathbf{A}^3 - 5\mathbf{A}^2 + 6\mathbf{A} - \mathbf{I} = 0$.

c) Deduce that $\mathbf{A}(\mathbf{A} - 2\mathbf{I})(\mathbf{A} - 3\mathbf{I}) = \mathbf{I}$.

d) Hence find $\mathbf{A}^{-1}$. (EDEXCEL)

15 Given that $\mathbf{A} = \begin{pmatrix} 1 & 0 & 0 \\ 0 & 2 & 1 \\ 0 & 0 & 1 \end{pmatrix}$, use matrix multiplication to find

a) $\mathbf{A}^2$ **b)** $\mathbf{A}^3$

c) Prove by induction that

$$\mathbf{A}^n = \begin{pmatrix} 1 & 0 & 0 \\ 0 & 2^n & 2^n - 1 \\ 0 & 0 & 1 \end{pmatrix} \quad n \geqslant 1$$

d) Find the inverse of $\mathbf{A}^n$. (EDEXCEL)

Transformations

A number of transformations of a two-dimensional plane onto a two-dimensional plane, $\mathbb{R}^2$, and of a three-dimensional space onto a three-dimensional space, $\mathbb{R}^3$, may be represented by a matrix $\mathbf{M}$, where

$$\mathbf{M} \begin{pmatrix} x \\ y \\ z \end{pmatrix} = \begin{pmatrix} x_1 \\ y_1 \\ z_1 \end{pmatrix}$$

means that the image of (x, y, z) under the transformation, T, is (x_1, y_1, z_1).

Linear transformations

T is described as a **linear transformation** of n-dimensional space (where $n = 2, 3, \ldots$) when it has the properties

$$T(\lambda \mathbf{x}) = \lambda T(\mathbf{x}) \quad \text{and} \quad T(\lambda \mathbf{x} + \mu \mathbf{y}) = \lambda T(\mathbf{x}) + \mu T(\mathbf{y})$$

where λ and μ are arbitrary constants.

We may represent a linear transformation by a matrix. For example, in three dimensions, we might represent T by the matrix

$$\mathbf{M} = \begin{pmatrix} a & b & c \\ d & e & f \\ g & h & i \end{pmatrix}$$

Hence, to find, under T, the image of the point with position vector **i**, we calculate

$$\begin{pmatrix} a & b & c \\ d & e & f \\ g & h & i \end{pmatrix} \begin{pmatrix} 1 \\ 0 \\ 0 \end{pmatrix} = \begin{pmatrix} a \\ d \\ g \end{pmatrix}$$

So, under T, the image of the point $(1, 0, 0)$ is (a, d, g), which we can see is the first column of **M**.

To find which type of transformation is represented by a matrix, we find the images of the vectors $(1, 0, 0)$ and $(0, 1, 0)$ and $(0, 0, 1)$. Common linear transformations are rotations about the origin, reflections in lines through the origin, stretches and shears.

We can represent a linear transformation of $\mathbb{R}^2$, which is the xy-plane, by a matrix. Such a matrix will have order 2×2.

For example, let T be an anticlockwise rotation of two-dimensional space.

The rotation, centred at the origin, is through angle θ.

Then the vector $\begin{pmatrix} 1 \\ 0 \end{pmatrix}$, which is **i**, transforms to the vector $\begin{pmatrix} \cos \theta \\ \sin \theta \end{pmatrix}$, and the vector $\begin{pmatrix} 0 \\ 1 \end{pmatrix}$,

which is **j**, transforms to the vector $\begin{pmatrix} -\sin \theta \\ \cos \theta \end{pmatrix}$.

The matrix for T is then given by

$$\mathbf{M} = \begin{pmatrix} \cos \theta & -\sin \theta \\ \sin \theta & \cos \theta \end{pmatrix}$$

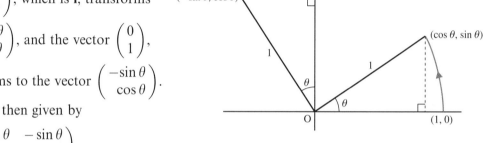

So, to find the matrix representing this transformation, we find the images of $(1, 0)$ and $(0, 1)$, which become the two columns of the matrix.

In three dimensions, we find the images of the points $(1, 0, 0)$, $(0, 1, 0)$ and $(0, 0, 1)$, which are the vertices of the unit cube. In vector form, these are the images of the vectors **i**, **j** and **k**. As seen below and on page 311, these become the columns of the matrix representing the transformation.

Example 3 Find the matrix **M** representing an enlargement, scale factor 2, with the origin as the centre of enlargement.

SOLUTION

The images of the vertices of the unit cube are

$$(1, 0, 0) \rightarrow (2, 0, 0)$$

$$(0, 1, 0) \rightarrow (0, 2, 0)$$

$$(0, 0, 1) \rightarrow (0, 0, 2)$$

Hence, we have

$$\mathbf{M} = \begin{pmatrix} 2 & 0 & 0 \\ 0 & 2 & 0 \\ 0 & 0 & 2 \end{pmatrix}$$

Example 4 Find the matrix **M** representing a reflection in the line $y = x$ in the xy-plane.

SOLUTION

The images of the vertices of the unit cube are

$$(1, 0, 0) \rightarrow (0, 1, 0)$$

$$(0, 1, 0) \rightarrow (1, 0, 0)$$

$$(0, 0, 1) \rightarrow (0, 0, 1)$$

Hence, we have

$$\mathbf{M} = \begin{pmatrix} 0 & 1 & 0 \\ 1 & 0 & 0 \\ 0 & 0 & 1 \end{pmatrix}$$

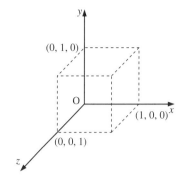

Example 5 Find the matrix **M** representing a shear in the yz-plane, in which $(0, 1, 0)$ is invariant and $(0, 0, 1)$ moves to $(0, 2, 1)$.

SOLUTION

The images of the vertices of the unit cube are

$$(1, 0, 0) \rightarrow (1, 0, 0)$$

$$(0, 1, 0) \rightarrow (0, 1, 0)$$

$$(0, 0, 1) \rightarrow (0, 2, 1)$$

Hence, we have

$$\mathbf{M} = \begin{pmatrix} 1 & 0 & 0 \\ 0 & 1 & 2 \\ 0 & 0 & 1 \end{pmatrix}$$

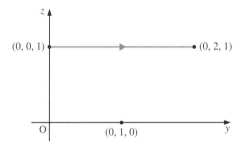

Example 6 Find the image of the line $y = 7x$ under the transformation whose matrix is $\begin{pmatrix} 4 & -1 \\ 2 & 5 \end{pmatrix}$.

SOLUTION

To find the image of a line (or a plane), we first obtain the general point on the line (or plane), and then obtain the image of this general point.

The general point on the line $y = 7x$ is $(t, 7t)$.

The image of this point is given by

$$\begin{pmatrix} 4 & -1 \\ 2 & 5 \end{pmatrix} \begin{pmatrix} t \\ 7t \end{pmatrix} = \begin{pmatrix} -3t \\ 37t \end{pmatrix}$$

Hence, we have $x = -3t$, $y = 37t$.

So, to find the desired line, we eliminate t, obtaining

$$37x + 3y = 0$$

Therefore, the image of the line $y = 7x$ is $37x + 3y = 0$.

Example 7 The transformation T is the composite transformation of

i) a one-way stretch in the x-direction, scale factor 3
ii) a one-way stretch in the y-direction, scale factor 9
iii) a one-way stretch in the z-direction, scale factor 3
iv) a reflection in the xy-plane.

Find the matrix $\mathbf{N}$ representing the composite transformation.

SOLUTION

Transformations **iii** and **iv** can be combined to give a one-way stretch in the z-direction of scale factor -3.

After all four transformations have taken place, the images of the vertices of the unit cube are

$$(1, 0, 0) \rightarrow (3, 0, 0)$$
$$(0, 1, 0) \rightarrow (0, 9, 0)$$
$$(0, 0, 1) \rightarrow (0, 0, -3)$$

Hence, we have

$$\mathbf{N} = \begin{pmatrix} 3 & 0 & 0 \\ 0 & 9 & 0 \\ 0 & 0 & -3 \end{pmatrix}$$

Invariant points and lines

An **invariant point** of the transformation T is a point which is unchanged by that transformation. That is, $T(\mathbf{x}) = \mathbf{x}$.

For example, the only points which are unchanged by reflection in the line $y = x$ are the points on the line $y = x$ itself. Therefore, the **only** invariant points in this transformation are on the line $y = x$.

Reflection in the line $y = x$ does not affect the line $y = x$. In addition, the line $y = -x$ maps onto itself. These are the **only** two lines which map onto themselves. Both lines pass through the origin.

We say that these are the **invariant lines** of the transformation which is a reflection in the line $y = x$. The **only** invariant lines are $y = x$ and $y = -x$.

We notice that some points which are not invariant points are on an invariant line. For example, the point $(1, -1)$, which is on the line $y = -x$, is reflected to the point $(-1, 1)$, which is still on the same invariant line $y = -x$.

All invariant lines of a transformation which can be represented by a matrix, other than those with an invariant plane, pass through the origin. If the transformation is represented by the identity matrix, all lines are invariant.

Example 8 Find **a)** the invariant points and **b)** the invariant lines of the transformation whose matrix is $\begin{pmatrix} 4 & -1 \\ 2 & 5 \end{pmatrix}$.

SOLUTION

a) The invariant points are the points (x, y) which satisfy

$$\begin{pmatrix} 4 & -1 \\ 2 & 5 \end{pmatrix} \begin{pmatrix} x \\ y \end{pmatrix} = \begin{pmatrix} x \\ y \end{pmatrix}$$

From this, we obtain the following simultaneous equations:

$$4x - y = x \quad \Rightarrow \quad 3x = y \qquad [1]$$
$$2x + 5y = y \quad \Rightarrow \quad 2x = -4y \qquad [2]$$

Substituting [1] into [2], we have

$$2x = -4(3x)$$
$$\Rightarrow \quad 2x = -12x$$
$$\Rightarrow \quad x = 0$$

The only solution to equations [1] and [2] is $x = y = 0$.

Therefore, the origin $(0, 0)$ is the only invariant point under this transformation.

b) The line $y = mx + c$ is invariant if points on it map onto points on the same line, but not necessarily onto the same points.

Thus, the general point, (t, mt), on the line $y = mx$ should map onto another point, (T, mT), on the line. So, we must solve the equation

$$\begin{pmatrix} 4 & -1 \\ 2 & 5 \end{pmatrix} \begin{pmatrix} t \\ mt \end{pmatrix} = \begin{pmatrix} T \\ mT \end{pmatrix}$$

Multiplying out the LHS, we obtain

$$\begin{pmatrix} (4 - m)t \\ (2 + 5m)t \end{pmatrix} = \begin{pmatrix} T \\ mT \end{pmatrix}$$

Therefore, we have the simultaneous equations

$$(4 - m)t = T$$
$$(2 + 5m)t = mT$$

which give

$$\frac{4 - m}{2 + 5m} = \frac{1}{m}$$

Cross-multiplying, we obtain

$$4m - m^2 = 2 + 5m$$
$$\Rightarrow \quad m^2 + m + 2 = 0$$

This equation has no real roots, and so the transformation has no invariant line.

Consider the anticlockwise rotation by $\dfrac{\pi}{2}$ about the origin in $\mathbb{R}^2$. Every line is rotated, and so there are no invariant lines. Also, there is only one invariant point, namely $(0, 0)$.

Any rotation (except by the angle $0°$ or $180°$) in two-dimensional space has no invariant lines. For example, we can see from the figure on the right that the image line can never lie along the object line, unless $\theta = 0°$ or $180°$.

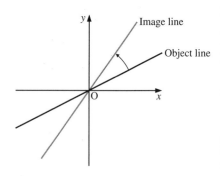

However, in three-dimensional space, a rotation must have an invariant line, namely the line about which the rotation occurs. In three-dimensional space, a plane always maps onto a plane unless the matrix is singular (that is, $\det \mathbf{M} = 0$). When the matrix is singular, a plane sometimes maps onto a line or a point. Similarly, a line always maps onto a line unless the matrix is singular, in which case the line might map onto a point.

Eigenvectors and eigenvalues

An **eigenvector** of a linear transformation T is a vector pointing in the direction of an invariant line under the transformation T.

For example, let T be a reflection in the line $y = x$. Then $(1, -1)$ is on the invariant line $y = -x$, but it maps onto $(-1, 1)$.

The **eigenvalue** for the eigenvector $\begin{pmatrix} 1 \\ -1 \end{pmatrix}$ is -1, since all the points on the line $y = -x$ map onto points whose coordinates are -1 times the original coordinates.

To summarise, if $\mathbf{M}$ is the matrix for a transformation T, then

$$\mathbf{M}\begin{pmatrix} x \\ y \end{pmatrix} = \lambda \begin{pmatrix} x \\ y \end{pmatrix}$$

means that $\begin{pmatrix} x \\ y \end{pmatrix}$ is an eigenvector of T, and λ is the eigenvalue of T associated with $\begin{pmatrix} x \\ y \end{pmatrix}$.

In this case, we have

$$\begin{pmatrix} 0 & -1 \\ 1 & 0 \end{pmatrix}\begin{pmatrix} x \\ y \end{pmatrix} = -1\begin{pmatrix} x \\ y \end{pmatrix}$$

In three-dimensional space,

$$\mathbf{M}\begin{pmatrix} x \\ y \\ z \end{pmatrix} = \lambda \begin{pmatrix} x \\ y \\ z \end{pmatrix}$$

means that $\begin{pmatrix} x \\ y \\ z \end{pmatrix}$ is an eigenvector of T, and that λ is the eigenvalue of T

associated with $\begin{pmatrix} x \\ y \\ z \end{pmatrix}$.

Finding eigenvectors and eigenvalues

To find the eigenvalues of a transformation whose matrix is

$$\mathbf{M} = \begin{pmatrix} a & b & c \\ d & e & f \\ g & h & i \end{pmatrix}$$

we solve the equation $\det(\mathbf{M} - \lambda\mathbf{I}) = 0$ for λ, which we will now prove.

We have

$$\begin{pmatrix} a & b & c \\ d & e & f \\ g & h & i \end{pmatrix} \begin{pmatrix} x \\ y \\ z \end{pmatrix} = \lambda \begin{pmatrix} x \\ y \\ z \end{pmatrix}$$

which gives

$$ax + by + c = \lambda x$$
$$dx + ey + fz = \lambda y$$
$$gx + hy + iz = \lambda z$$

from which we obtain

$$(a - \lambda)x + by + cz = 0$$
$$dx + (e - \lambda)y + fz = 0$$
$$gx + hy + (i - \lambda)z = 0$$

For the eigenvectors to be non-zero, these three equations must have non-unique solutions (see page 87). Hence, we have

$$\begin{vmatrix} a - \lambda & b & c \\ d & e - \lambda & f \\ g & h & i - \lambda \end{vmatrix} = 0$$

which is

$$\det(\mathbf{M} - \lambda\mathbf{I}) = 0$$

To find the eigenvectors, we solve

$$\mathbf{M} \begin{pmatrix} x \\ y \\ z \end{pmatrix} = \lambda \begin{pmatrix} x \\ y \\ z \end{pmatrix}$$

for each value of λ.

Example 9 Find **a)** the eigenvalues and **b)** the eigenvectors of the matrix

$$\begin{pmatrix} 1 & 1 & 2 \\ 0 & 2 & 2 \\ -1 & 1 & 3 \end{pmatrix}$$

SOLUTION

a) We solve the equation

$$\begin{vmatrix} 1-\lambda & 1 & 2 \\ 0 & 2-\lambda & 2 \\ -1 & 1 & 3-\lambda \end{vmatrix} = 0$$

which gives

$$(1-\lambda)[(2-\lambda)(3-\lambda)-2] - 1(2) + 2(2-\lambda) = 0$$
$$\Rightarrow \quad (1-\lambda)(\lambda^2 - 5\lambda + 4) - 2 + 4 - 2\lambda = 0$$
$$\Rightarrow \quad \lambda^3 - 6\lambda^2 + 11\lambda - 6 = 0$$

This equation is known as the **characteristic equation** of the matrix (see page 323).

Factorising the LHS, we obtain

$$(\lambda - 1)(\lambda - 2)(\lambda - 3) = 0$$
$$\Rightarrow \quad \lambda = 1, 2, 3$$

Therefore, the eigenvalues are 1, 2 and 3.

b) The eigenvector for the eigenvalue 1 is given by a solution to the equation

$$\begin{pmatrix} 1 & 1 & 2 \\ 0 & 2 & 2 \\ -1 & 1 & 3 \end{pmatrix}\begin{pmatrix} x \\ y \\ z \end{pmatrix} = \begin{pmatrix} x \\ y \\ z \end{pmatrix}$$
$$\Rightarrow \quad \begin{pmatrix} x + y + 2z \\ 2y + 2z \\ -x + y + 3z \end{pmatrix} = \begin{pmatrix} x \\ y \\ z \end{pmatrix}$$

from which we obtain the simultaneous equations

$$x + y + 2z = x \qquad [1]$$
$$2y + 2z = y \qquad [2]$$
$$-x + y + 3z = z \qquad [3]$$

We note that there are only two different equations from which to solve for three unknowns. Therefore, we cannot obtain a unique solution to such a set of equations (see page 87). Hence, we will let one of the unknowns be t. We also note that subtracting [3] from [1] gives $x = 0$.

So, we let $z = t$ and solve the simultaneous equations for y:

$$x + y + 2t = x \qquad [4]$$
$$2y + 2t = y \qquad [5]$$
$$-x + y + 3t = t \qquad [6]$$

From [4], we obtain $y = -2t$.

Therefore, the direction of the eigenvector is $\begin{pmatrix} 0 \\ -2t \\ t \end{pmatrix}$.

Hence, $\begin{pmatrix} 0 \\ -2 \\ 1 \end{pmatrix}$ is an eigenvector for the eigenvalue 1.

The eigenvector for the eigenvalue 2 is given by a solution to the equation

$$\begin{pmatrix} 1 & 1 & 2 \\ 0 & 2 & 2 \\ -1 & 1 & 3 \end{pmatrix} \begin{pmatrix} x \\ y \\ z \end{pmatrix} = 2 \begin{pmatrix} x \\ y \\ z \end{pmatrix}$$

from which we obtain the simultaneous equations

$$x + y - 2z = 2x \qquad [7]$$
$$2y + 2z = 2y \qquad [8]$$
$$-x + y + 3z = 2z \qquad [9]$$

This time, we do not let $z = t$, since [8] immediately gives $z = 0$.

So, we put $y = t$. Then from [7], we obtain $x = t$.

Therefore, the direction of the eigenvector is $\begin{pmatrix} t \\ t \\ 0 \end{pmatrix}$.

Hence, $\begin{pmatrix} 1 \\ 1 \\ 0 \end{pmatrix}$ is an eigenvector for the eigenvalue 2.

The eigenvector for the eigenvalue 3 is given by a solution to the equation

$$\begin{pmatrix} 1 & 1 & 2 \\ 0 & 2 & 2 \\ -1 & 1 & 3 \end{pmatrix} \begin{pmatrix} x \\ y \\ z \end{pmatrix} = 3 \begin{pmatrix} x \\ y \\ z \end{pmatrix}$$

from which we obtain the simultaneous equations

$$x + y + 2z = 3x$$
$$2y + 2z = 3y$$
$$-x + y + 3z = 3z$$

which give

$$-2x + y + 2z = 0 \qquad [10]$$
$$2z = y \qquad [11]$$
$$-x + y = 0 \qquad [12]$$

We let $x = t$. Then from [12] and [11], we have $y = t$ and $z = \dfrac{t}{2}$.

Therefore, the direction of the eigenvector is $\begin{pmatrix} t \\ t \\ \frac{1}{2}t \end{pmatrix}$.

Hence, $\begin{pmatrix} 1 \\ 1 \\ \frac{1}{2} \end{pmatrix}$ is an eigenvector for the eigenvalue 3.

Since any scalar multiple of an eigenvector is also an eigenvector,

we can write the eigenvector for 3 as $\begin{pmatrix} 2 \\ 2 \\ 1 \end{pmatrix}$.

Example 10 Show that $\begin{pmatrix} 1 \\ -1 \\ -2 \end{pmatrix}$ is an eigenvector of the matrix $\mathbf{A}$, where

$$\mathbf{A} = \begin{pmatrix} 1 & 0 & -1 \\ 1 & 2 & 1 \\ 2 & 0 & 4 \end{pmatrix}$$

Find the associated eigenvalue.

SOLUTION

If $\begin{pmatrix} 1 \\ -1 \\ -2 \end{pmatrix}$ is an eigenvector of $\mathbf{A}$, then we have

$$\mathbf{A} \begin{pmatrix} 1 \\ -1 \\ -2 \end{pmatrix} = \lambda \begin{pmatrix} 1 \\ -1 \\ -2 \end{pmatrix}$$

where λ is the eigenvalue associated with $\begin{pmatrix} 1 \\ -1 \\ -2 \end{pmatrix}$.

Hence, we obtain

$$\mathbf{A} \begin{pmatrix} 1 \\ -1 \\ -2 \end{pmatrix} = \begin{pmatrix} 1 & 0 & -1 \\ 1 & 2 & 1 \\ 2 & 0 & 4 \end{pmatrix} \begin{pmatrix} 1 \\ -1 \\ -2 \end{pmatrix} = \begin{pmatrix} 3 \\ -3 \\ -6 \end{pmatrix}$$

$$\Rightarrow \quad \mathbf{A} \begin{pmatrix} 1 \\ -1 \\ -2 \end{pmatrix} = 3 \begin{pmatrix} 1 \\ -1 \\ -2 \end{pmatrix}$$

We note that

$$\mathbf{A} \begin{pmatrix} 1 \\ -1 \\ -2 \end{pmatrix} = 3 \begin{pmatrix} 1 \\ -1 \\ -2 \end{pmatrix}$$

has the same form as $\mathbf{A}\mathbf{x} = \lambda\mathbf{x}$, therefore $\begin{pmatrix} 1 \\ -1 \\ -2 \end{pmatrix}$ is an eigenvector of $\mathbf{A}$

and its associated eigenvalue is 3.

Diagonalisation

If **M** is a **symmetric matrix**, then $\mathbf{M}^T = \mathbf{M}$. That is, the transpose of the matrix **M** is the same as the original matrix **M**.

For example, $\begin{pmatrix} 3 & 4 & -2 \\ 4 & 1 & 7 \\ -2 & 7 & 4 \end{pmatrix}$ is a symmetric matrix, since we have

$$\begin{pmatrix} 3 & 4 & -2 \\ 4 & 1 & 7 \\ -2 & 7 & 4 \end{pmatrix}^T = \begin{pmatrix} 3 & 4 & -2 \\ 4 & 1 & 7 \\ -2 & 7 & 4 \end{pmatrix}$$

Note that the eigenvectors of a symmetric matrix which have non-equal eigenvalues are **mutually perpendicular**,

A **diagonal matrix** is one in which every element is 0 except those in the leading diagonal.

For example $\begin{pmatrix} 1 & 0 & 0 \\ 0 & 7 & 0 \\ 0 & 0 & -1 \end{pmatrix}$ is a diagonal matrix.

If **P** is the matrix with the eigenvectors of **M** as each of its columns, we have

$$\mathbf{P}^{-1}\mathbf{MP} = \mathbf{D}$$

where **D** is a **diagonal matrix**, the diagonal elements of which are the eigenvalues.

Since $\mathbf{P}^{-1}\mathbf{MP} = \mathbf{D}$, we have

$$\mathbf{PP}^{-1}\mathbf{MP} = \mathbf{PD} \quad \Rightarrow \quad \mathbf{MP} = \mathbf{PD}$$

$$\Rightarrow \quad \mathbf{MPP}^{-1} = \mathbf{PDP}^{-1} \quad \Rightarrow \quad \mathbf{M} = \mathbf{PDP}^{-1}$$

Hence, **M** representing the transformation may be expressed in terms of a diagonal matrix when the eigenvectors are used as axes. (See page 323 for an example of this.)

If **P** is the matrix with the eigenvectors of the **symmetric matrix M** as each of its columns, we have

$$\mathbf{P}^T\mathbf{MP} = \mathbf{D}_1$$

where $\mathbf{D}_1$ is also a diagonal matrix.

$$\nearrow P\left(\frac{v_1}{|v_1|}, \frac{v_2}{|v_2|}, \frac{v_3}{|v_3|} \right)$$

If the eigenvectors used in **P** are **normalised** (that is, converted to unit vectors), then the elements of $\mathbf{D}_1$ are also the eigenvalues.

However, if the eigenvectors of the symmetric matrix **P** are not normalised, then each element in the leading diagonal is the product of an eigenvalue and the square of the modulus of the associated eigenvector.

In Example 9 (pages 316–18), we found that the eigenvectors of $\begin{pmatrix} 1 & 1 & 2 \\ 0 & 2 & 2 \\ -1 & 1 & 3 \end{pmatrix}$

are $\begin{pmatrix} 0 \\ -2 \\ 1 \end{pmatrix}$, $\begin{pmatrix} 1 \\ 1 \\ 0 \end{pmatrix}$ and $\begin{pmatrix} 2 \\ 2 \\ 1 \end{pmatrix}$. Hence, we have

$$\mathbf{P} = \begin{pmatrix} 0 & 1 & 2 \\ -2 & 1 & 2 \\ 1 & 0 & 1 \end{pmatrix} \quad \mathbf{P}^{-1} = \begin{pmatrix} \frac{1}{2} & -\frac{1}{2} & 0 \\ 2 & -1 & -2 \\ -\frac{1}{2} & \frac{1}{2} & 1 \end{pmatrix}$$

which gives the diagonal matrix, $\mathbf{P}^{-1}\mathbf{MP}$, as

$$\begin{pmatrix} \frac{1}{2} & -\frac{1}{2} & 0 \\ 2 & -1 & -2 \\ -\frac{1}{2} & \frac{1}{2} & 1 \end{pmatrix} \begin{pmatrix} 1 & 1 & 2 \\ 0 & 2 & 2 \\ -1 & 1 & 3 \end{pmatrix} \begin{pmatrix} 0 & 1 & 2 \\ -2 & 1 & 2 \\ 1 & 0 & 1 \end{pmatrix} = \begin{pmatrix} \frac{1}{2} & -\frac{1}{2} & 0 \\ 4 & -2 & -4 \\ -\frac{3}{2} & \frac{3}{2} & 3 \end{pmatrix} \begin{pmatrix} 0 & 1 & 2 \\ -2 & 1 & 2 \\ 1 & 0 & 1 \end{pmatrix}$$

$$= \begin{pmatrix} 1 & 0 & 0 \\ 0 & 2 & 0 \\ 0 & 0 & 3 \end{pmatrix}$$

That is, we have

$$\mathbf{P}^{-1}\mathbf{MP} = \mathbf{D}$$

The diagonalisation of a symmetric matrix is given in Example 11.

Example 11 The transformation T is represented by

$$\mathbf{M} = \begin{pmatrix} 3 & 4 & -4 \\ 4 & 5 & 0 \\ -4 & 0 & 1 \end{pmatrix}$$

Find

a) the eigenvalues of $\mathbf{M}$
b) their associated eigenvectors
c) a matrix, $\mathbf{P}$, so that $\mathbf{P}^{\mathrm{T}}\mathbf{MP} = \mathbf{D}$, where $\mathbf{D}$ is a diagonal matrix whose diagonal elements are the eigenvalues.

SOLUTION

a) To find the eigenvalues, we have

$$\mathbf{Mx} = \lambda\mathbf{x}$$
$$\Rightarrow \quad (\mathbf{M} - \lambda\mathbf{I})\,\mathbf{x} = 0$$
$$\Rightarrow \quad |\mathbf{M} - \lambda\mathbf{I}| = 0$$

which gives

$$\begin{vmatrix} 3 - \lambda & 4 & -4 \\ 4 & 5 - \lambda & 0 \\ -4 & 0 & 1 - \lambda \end{vmatrix} = 0$$

$$(3 - \lambda)(5 - \lambda)(1 - \lambda) - 4 \times 4(1 - \lambda) - 4 \times 4(5 - \lambda) = 0$$

$$\lambda^3 - 9\lambda^2 - 9\lambda + 81 = 0$$

Factorising, we obtain

$$(\lambda - 3)(\lambda + 3)(\lambda - 9) = 0$$

$$\Rightarrow \quad \lambda = 3, 9, -3$$

Therefore, the eigenvalues of $\mathbf{M}$ are $3, 9, -3$.

b) When $\lambda = 3$, we find the associated eigenvector from $\mathbf{Mx} = 3\mathbf{x}$, which gives

$$\begin{pmatrix} 3 & 4 & -4 \\ 4 & 5 & 0 \\ -4 & 0 & 1 \end{pmatrix} \begin{pmatrix} x \\ y \\ z \end{pmatrix} = 3 \begin{pmatrix} x \\ y \\ z \end{pmatrix}$$

$$3x + 4y - 4z = 3x \quad \Rightarrow \quad 4y - 4z = 0 \qquad [1]$$

$$4x + 5y = 3y \quad \Rightarrow \quad 4x + 2y = 0 \qquad [2]$$

$$-4x + z = 3z \quad \Rightarrow \quad -4x = 2z \qquad [3]$$

Putting $x = t$, we obtain, from [2] and [3], $y = -2t$ and $z = -2t$.

Therefore, one eigenvector is $\begin{pmatrix} 1 \\ -2 \\ -2 \end{pmatrix}$.

Similarly, we find the other eigenvectors are $\begin{pmatrix} 2 \\ 2 \\ -1 \end{pmatrix}$ and $\begin{pmatrix} 2 \\ -1 \\ 2 \end{pmatrix}$.

c) From part **b**, we have

$$\mathbf{P} = \begin{pmatrix} 1 & 2 & 2 \\ -2 & 2 & -1 \\ -2 & -1 & 2 \end{pmatrix}$$

We find that the magnitude of each of the eigenvectors $\begin{pmatrix} 1 \\ -2 \\ -2 \end{pmatrix}$,

$\begin{pmatrix} 2 \\ 2 \\ -1 \end{pmatrix}$ and $\begin{pmatrix} 2 \\ -1 \\ 2 \end{pmatrix}$ is 3.

Therefore, normalising the eigenvectors, we obtain respectively

$$\begin{pmatrix} \frac{1}{3} \\ -\frac{2}{3} \\ -\frac{2}{3} \end{pmatrix} \quad \begin{pmatrix} \frac{2}{3} \\ \frac{2}{3} \\ -\frac{1}{3} \end{pmatrix} \quad \text{and} \quad \begin{pmatrix} \frac{2}{3} \\ -\frac{1}{3} \\ \frac{2}{3} \end{pmatrix}$$

which give

$$\mathbf{P} = \begin{pmatrix} \frac{1}{3} & \frac{2}{3} & \frac{2}{3} \\ -\frac{2}{3} & \frac{2}{3} & -\frac{1}{3} \\ -\frac{2}{3} & -\frac{1}{3} & \frac{2}{3} \end{pmatrix}$$

Hence, we have

$$\mathbf{P}^{\mathsf{T}}\mathbf{MP} = \begin{pmatrix} \frac{1}{3} & -\frac{2}{3} & -\frac{2}{3} \\ \frac{2}{3} & \frac{2}{3} & -\frac{1}{3} \\ \frac{2}{3} & -\frac{1}{3} & \frac{2}{3} \end{pmatrix} \begin{pmatrix} 3 & 4 & -4 \\ 4 & 5 & 0 \\ -4 & 0 & 1 \end{pmatrix} \begin{pmatrix} \frac{1}{3} & \frac{2}{3} & \frac{2}{3} \\ -\frac{2}{3} & \frac{2}{3} & -\frac{1}{3} \\ -\frac{2}{3} & -\frac{1}{3} & \frac{2}{3} \end{pmatrix}$$

which gives

$$\mathbf{P^T P M} = \begin{pmatrix} 1 & -2 & -2 \\ 6 & 6 & -3 \\ -2 & 1 & -2 \end{pmatrix} \begin{pmatrix} \frac{1}{3} & \frac{2}{3} & \frac{2}{3} \\ -\frac{2}{3} & \frac{2}{3} & -\frac{1}{3} \\ -\frac{2}{3} & -\frac{1}{3} & \frac{2}{3} \end{pmatrix}$$

$$= \begin{pmatrix} 3 & 0 & 0 \\ 0 & 9 & 0 \\ 0 & 0 & -3 \end{pmatrix}$$

which is a diagonal matrix with the eigenvalues of $\mathbf{M}$ as its elements.

We noticed in Example 7 (page 312) that the transformation composed of

i) a one-way stretch in the x-direction, scale factor 3

ii) a one-way stretch in the y-direction, scale factor 9

iii) a one-way stretch in the z-direction, scale factor 3

iv) a reflection in the xy-plane

was represented by

$$\mathbf{N} = \begin{pmatrix} 3 & 0 & 0 \\ 0 & 9 & 0 \\ 0 & 0 & -3 \end{pmatrix}$$

By geometrical consideration of the actual transformation, we can deduce that the eigenvectors of this transformation are the three mutually perpendicular

vectors $\begin{pmatrix} 1 \\ 0 \\ 0 \end{pmatrix}$, $\begin{pmatrix} 0 \\ 1 \\ 0 \end{pmatrix}$ and $\begin{pmatrix} 0 \\ 0 \\ 1 \end{pmatrix}$ with associated eigenvalues $3, 9, -3$.

We have just found that the transformation represented by

$$\mathbf{M} = \begin{pmatrix} 3 & 4 & -4 \\ 4 & 5 & 0 \\ -4 & 0 & 1 \end{pmatrix}$$

also has three mutually perpendicular eigenvectors with associated eigenvalues $3, 9, -3$. Thus, these two transformations (Example 7, page 312, and Example 11, page 320) are the same transformation but about different axes: that represented by $\mathbf{N}$ has its one-way stretches in each of the three mutually perpendicular directions $\mathbf{i}, \mathbf{j}, \mathbf{k}$, whereas that represented by $\mathbf{M}$ has its one-way stretches of the same scale factors in the three mutually perpendicular

directions $\begin{pmatrix} \frac{1}{3} \\ -\frac{2}{3} \\ -\frac{2}{3} \end{pmatrix}$, $\begin{pmatrix} \frac{2}{3} \\ \frac{2}{3} \\ -\frac{1}{3} \end{pmatrix}$, $\begin{pmatrix} \frac{2}{3} \\ -\frac{1}{3} \\ \frac{2}{3} \end{pmatrix}$

Naturally, both matrices have determinant -81, being the scale factor of the volume of the enlargement, which is the volume of the image of the unit cube.

Hence, the transformation $\mathbf{x}' = \mathbf{Mx}$, where

$$\mathbf{M} = \begin{pmatrix} 3 & 4 & -4 \\ 4 & 5 & 0 \\ -4 & 0 & 1 \end{pmatrix}$$

with respect to axes in the direction of the eigenvectors becomes the transformation $\mathbf{X}' = \mathbf{DX}$, where

$$\mathbf{D} = \begin{pmatrix} 3 & 0 & 0 \\ 0 & 9 & 0 \\ 0 & 0 & -3 \end{pmatrix}$$

which is the diagonalised form of $\mathbf{M}$.

The characteristic equation

On page 316, we mentioned that the **characteristic equation** of the matrix

$$\mathbf{M} = \begin{pmatrix} 1 & 1 & 2 \\ 0 & 2 & 2 \\ -1 & 1 & 3 \end{pmatrix}$$

is

$$\lambda^3 - 6\lambda^2 + 11\lambda - 6 = 0$$

where the values of λ are the eigenvalues of $\mathbf{M}$.

$\mathbf{M}$ also satisfies this characteristic equation. Hence, we have

$$\mathbf{M}^3 - 6\mathbf{M}^2 + 11\mathbf{M} - 6\mathbf{I} = 0$$

From this equation, we can find $\mathbf{M}^{-1}$.

Postmultiplying by $\mathbf{M}^{-1}$, we obtain

$$\mathbf{M}^3\mathbf{M}^{-1} - 6\mathbf{M}^2\mathbf{M}^{-1} + 11\mathbf{M}\mathbf{M}^{-1} - 6\mathbf{M}^{-1} = 0$$
$$\Rightarrow \quad \mathbf{M}^2 - 6\mathbf{M} + 11\mathbf{I} - 6\mathbf{M}^{-1} = 0$$

which gives

$$\mathbf{M}^{-1} = \frac{1}{6}\mathbf{M}^2 - \mathbf{M} + \frac{11}{6}\mathbf{I}$$

Exercise 14B

1 The matrix **A** is given by

$$\mathbf{A} = \begin{pmatrix} \frac{1}{2}\sqrt{3} & \frac{1}{2} \\ -\frac{1}{2} & \frac{1}{2}\sqrt{3} \end{pmatrix}$$

Give a full description of the geometrical transformation represented by $\mathbf{A}^4$. (OCR)

2 The matrix **C** is $\begin{pmatrix} -1 & 0 \\ 0 & 2 \end{pmatrix}$. The geometrical transformation represented by **C** may be considered as the result of a reflection followed by a stretch. By considering the effect on the unit square, or otherwise, describe fully the reflection and the stretch.

Find the matrices **A** and **B** which represent the reflection and the stretch respectively. (OCR)

3 The matrix **M** is given by $\mathbf{M} = \begin{pmatrix} 1 & -1 \\ 0 & 1 \end{pmatrix}$.

Describe fully the geometrical transformation represented by **M**.

The matrix **C** is given by

$$\mathbf{C} = \begin{pmatrix} \frac{1}{2} & \frac{1}{2}(\sqrt{3}-1) \\ -\frac{1}{2}\sqrt{3} & \frac{1}{2}(\sqrt{3}+1) \end{pmatrix}$$

C represents the combined effect of the transformation represented by **M** followed by the transformation represented by a matrix **B**.

i) Find the matrix **B**.
ii) Describe fully the geometrical transformation represented by **B**. (OCR)

4 The matrices **A** and **B** are given by

$$\mathbf{A} = \begin{pmatrix} 3 & -4 \\ 4 & 3 \end{pmatrix} \qquad \mathbf{B} = \begin{pmatrix} 1 & 0 \\ 0 & -1 \end{pmatrix}$$

Under the transformation represented by **AB**, a triangle P maps onto the triangle Q whose vertices are $(0, 0)$, $(9, 12)$ and $(22, -4)$.

i) Find the coordinates of the vertices of P.
ii) State the area of P and hence find the area of Q.
iii) Find the area of the image of P under the transformation represented by $\mathbf{ABA}^{-1}$. (OCR)

5 Let $\mathbf{A} = \begin{pmatrix} 1 & -1 & 0 \\ -1 & 0 & -1 \\ -1 & 1 & 0 \end{pmatrix}$. Write down the matrix $\mathbf{A} - \lambda\mathbf{I}$, where $\lambda \in \mathbb{R}$ and **I** is the

3×3 identity matrix.

Find the values of λ for which the determinant of $\mathbf{A} - \lambda\mathbf{I}$ is zero. (SQA/CSYS)

6 The matrix **P** is defined by

$$\mathbf{P} = \begin{pmatrix} 1 & -2 \\ -2 & 1 \end{pmatrix}$$

a) Find the eigenvalues of **P**.

b) Find an eigenvector corresponding to each eigenvalue.

c) Verify that these eigenvectors are orthogonal.　　(NEAB)

7 The matrix **A** is given by $\mathbf{A} = \begin{pmatrix} 1 & 4 \\ 2 & 3 \end{pmatrix}$.

a) i) Find the eigenvalues of **A**.

　　ii) For each eigenvalue find a corresponding eigenvector.

b) Given that $\mathbf{U} = \begin{pmatrix} a & 5 \\ -3 & b \end{pmatrix}$, write down the values of a and b such that

$$\mathbf{U}^{-1}\mathbf{AU} = \begin{pmatrix} -1 & 0 \\ 0 & 5 \end{pmatrix} \qquad \text{(NEAB)}$$

8 The eigenvalues of the matrix $\mathbf{A} = \begin{pmatrix} 2 & 2 & -3 \\ 2 & 2 & 3 \\ -3 & 3 & 3 \end{pmatrix}$ are $\lambda_1, \lambda_2, \lambda_3$.

a) Show that $\lambda_1 = 6$ is an eigenvalue and find the other two eigenvalues λ_2 and λ_3.

b) Verify that $\det \mathbf{A} = \lambda_1 \lambda_2 \lambda_3$.

c) Find an eigenvector corresponding to the eigenvalue $\lambda_1 = 6$.

Given that $\begin{pmatrix} 1 \\ -1 \\ 1 \end{pmatrix}$ and $\begin{pmatrix} 1 \\ 1 \\ 0 \end{pmatrix}$ are eigenvectors of **A** corresponding to λ_2 and λ_3,

d) write down a matrix **P** such that $\mathbf{P}^{\mathrm{T}}\mathbf{AP}$ is a diagonal matrix.　　(EDEXCEL)

9 $$\mathbf{A} = \begin{pmatrix} 3 & 4 & -4 \\ 4 & 5 & 0 \\ -4 & 0 & 1 \end{pmatrix}$$

a) Show that 3 is an eigenvalue of **A** and find the other two eigenvalues.

b) Find an eigenvector corresponding to the eigenvalue 3.

Given that the vectors $\begin{pmatrix} 2 \\ 2 \\ -1 \end{pmatrix}$ and $\begin{pmatrix} 2 \\ -1 \\ 2 \end{pmatrix}$ are eigenvectors corresponding to the other two eigenvalues,

c) write down a matrix **P** such that $\mathbf{P}^{\mathrm{T}}\mathbf{AP}$ is a diagonal matrix.　　(EDEXCEL)

10 The matrix **A** is given by $\begin{bmatrix} 7 & 4 \\ -1 & 3 \end{bmatrix}$. The plane transformation **T** is such that $\mathbf{T}: \begin{bmatrix} x \\ y \end{bmatrix} \mapsto \mathbf{A}\begin{bmatrix} x \\ y \end{bmatrix}$.

a) i) Show that **A** has only one eigenvalue. Find this eigenvalue and a corresponding eigenvector.

　　ii) Hence, or otherwise, determine a cartesian equation of the fixed line of **T**.

b) Under **T**, a square with area $1\,\text{cm}^2$ is transformed into a parallelogram with area $d\,\text{cm}^2$. Find the value of d.　　(AEB 96)

11 The matrix **P** is defined by

$$\mathbf{P} = \begin{pmatrix} 1 & 3 & 0 \\ 2 & 0 & 2 \\ 1 & 1 & 2 \end{pmatrix}$$

a) Show that $v_1 = \begin{pmatrix} 1 \\ -1 \\ 0 \end{pmatrix}$ and $v_2 = \begin{pmatrix} 1 \\ 0 \\ -1 \end{pmatrix}$ are eigenvectors of **P** and find the two

corresponding eigenvalues.

b) Given that the third eigenvalue of **P** is 4, find the corresponding eigenvector, v_3.

c) Show that v_1, v_2 and v_3 are linearly independent.

d) Express the vector $\begin{pmatrix} a \\ b \\ c \end{pmatrix}$ as a linear combination of v_1, v_2 and v_3 with coefficients in terms

of the constants a, b and c.   (NEAB)

12 Let **A** be the matrix $\begin{bmatrix} 3 & 1 \\ 5 & -1 \end{bmatrix}$.

a) Determine the eigenvalues and corresponding eigenvectors of **A**.

b) i) Show that $A^2 - 2A - 8I = Z$, where $I = \begin{bmatrix} 1 & 0 \\ 0 & 1 \end{bmatrix}$ and $Z = \begin{bmatrix} 0 & 0 \\ 0 & 0 \end{bmatrix}$.

    ii) The matrix $B = A^{-1}$. By multiplying the matrix equation $A^2 - 2A - 8I = Z$ by **B**, or
otherwise, find the values of the scalars α and β for which $B = \alpha A + \beta I$.   (AEB 97)

13 a) Determine the eigenvalues of the matrix

$$A = \begin{pmatrix} 3 & -3 & 6 \\ 0 & 2 & -8 \\ 0 & 0 & -2 \end{pmatrix}$$

b) Show that $\begin{pmatrix} 3 \\ 1 \\ 0 \end{pmatrix}$ is an eigenvector of **A**.

$$B = \begin{pmatrix} 7 & -6 & 2 \\ 1 & 2 & 3 \\ 1 & -3 & 2 \end{pmatrix}$$

c) Show that $\begin{pmatrix} 3 \\ 1 \\ 0 \end{pmatrix}$ is an eigenvector of **B** and write down the corresponding eigenvalue.

d) Hence, or otherwise, write down an eigenvector of the matrix **AB**, and state the
corresponding eigenvalue.   (EDEXCEL)

14 The transformation **T** maps points (x, y) of the plane into image points (x', y') such that

$$x' = 4x + 2y + 14$$
$$y' = 2x + 7y + 42$$

a) i) Find the coordinates of the invariant point of **T**.

    ii) Hence express **T** in the form

$$\begin{bmatrix} x' \\ y' + k \end{bmatrix} = A \begin{bmatrix} x \\ y + k \end{bmatrix}$$

where k is a positive integer and **A** is a 2×2 matrix.

b) i) Determine the eigenvalues and corresponding eigenvectors of the matrix $\begin{bmatrix} 4 & 2 \\ 2 & 7 \end{bmatrix}$.

   ii) Deduce the cartesian equations of the invariant lines of **T**, and prove that they are perpendicular.

c) Give a full geometrical description of **T**.   (AEB 98)

15 i) Given that $\mathbf{P} = \begin{pmatrix} 4 & -1 & 0 \\ 1 & 5 & 3 \\ 2 & 1 & 1 \end{pmatrix}$, find $\det \mathbf{P}$ and $\mathbf{P}^{-1}$.

The 3×3 matrix **M** has eigenvalues $-1, 2, 5$ with corresponding eigenvectors

$$\begin{pmatrix} 4 \\ 1 \\ 2 \end{pmatrix} \quad \begin{pmatrix} -1 \\ 5 \\ 1 \end{pmatrix} \quad \begin{pmatrix} 0 \\ 3 \\ 1 \end{pmatrix}$$

respectively.

   ii) By considering **MP**, or otherwise, find the matrix **M**.
   iii) Find the characteristic equation for **M**.
   iv) Find p, q and r such that $\mathbf{M}^{-1} = p\mathbf{M}^2 + q\mathbf{M} + r\mathbf{I}$.   (MEI)

16 A linear transformation of three-dimensional space is defined by $\mathbf{r}' = \mathbf{Mr}$, where

$$\mathbf{r}' = \begin{pmatrix} x' \\ y' \\ z' \end{pmatrix} \quad \mathbf{r} = \begin{pmatrix} x \\ y \\ z \end{pmatrix} \quad \mathbf{M} = \begin{pmatrix} 2 & 1 & -1 \\ -1 & 0 & 3 \\ 2 & k & 4 \end{pmatrix}$$

a) Show that the transformation is singular if and only if $k = 2$.
b) In the case when $k = 2$, show that **M** represents a transformation of three-dimensional space onto a plane and find a cartesian equation of this plane.   (NEAB)

17 The vectors **a**, **b** and **c**, given below, are linearly independent.

$$\mathbf{a} = \begin{pmatrix} 1 \\ 2 \\ -1 \end{pmatrix} \quad \mathbf{b} = \begin{pmatrix} 0 \\ 3 \\ 4 \end{pmatrix} \quad \mathbf{c} = \begin{pmatrix} 1 \\ 2 \\ 0 \end{pmatrix}$$

Find α, β and γ such that the vector

$$\mathbf{d} = \begin{pmatrix} 7 \\ 5 \\ -14 \end{pmatrix}$$

can be expressed as a linear combination of **a**, **b** and **c**, in the form

$$\mathbf{d} = \alpha\mathbf{a} + \beta\mathbf{b} + \gamma\mathbf{c} \qquad \text{(NEAB)}$$

18 The matrix **A** is defined by

$$\mathbf{A} = \begin{pmatrix} 1 & 1 & 1 \\ 1 & k & 1 \\ 1 & 1 & k \end{pmatrix}$$

a) Find the determinant of **A** in terms of k.
b) The matrix **A** corresponds to a linear transformation T in three-dimensional space. When a region in three-dimensional space is transformed by T its volume, V, is increased by a factor of four to $4V$. Find the possible values of k.   (NEAB)

19 A linear transformation T of three-dimensional space is defined by $\mathbf{r}' = \mathbf{Mr}$, where

$$\mathbf{r}' = \begin{pmatrix} x' \\ y' \\ z' \end{pmatrix} \qquad \mathbf{r} = \begin{pmatrix} x \\ y \\ z \end{pmatrix} \qquad \mathbf{M} = \begin{pmatrix} \dfrac{1}{2} & \dfrac{1}{2} & -\dfrac{1}{\sqrt{2}} \\[2mm] \dfrac{1}{2} & \dfrac{1}{2} & \dfrac{1}{\sqrt{2}} \\[2mm] \dfrac{1}{\sqrt{2}} & -\dfrac{1}{\sqrt{2}} & 0 \end{pmatrix}$$

a) Show that every point on the line $x = y$, $z = 0$ is invariant under T.

b) Find $\mathbf{M}^2$ and hence show that $\mathbf{M}^4 = \mathbf{I}$, where $\mathbf{I}$ is the 3×3 unit matrix.

c) Given that T is a rotation, state
   i) the axis of the rotation
   ii) the angle of the rotation.

d) Write down the image under T of the unit vector $\begin{pmatrix} 0 \\ 0 \\ 1 \end{pmatrix}$, and hence indicate by means of a diagram the sense of the rotation.   (NEAB)

20 a) The matrix $\mathbf{A}$ and a non-singular matrix $\mathbf{M}$ are defined by

$$\mathbf{A} = \begin{pmatrix} 5 & -1 & 0 \\ -1 & 10 & 3 \\ 0 & 3 & 1 \end{pmatrix} \qquad \mathbf{M} = \begin{pmatrix} 0 & -1 & 0 \\ 0 & -1 & -2 \\ 2 & 3 & 6 \end{pmatrix}$$

Show that $\mathbf{M}^{\mathrm{T}}\mathbf{AM} = 4\mathbf{I}$, where $\mathbf{M}^{\mathrm{T}}$, the transpose of the matrix $\mathbf{M}$, is given by

$$\mathbf{M}^{\mathrm{T}} = \begin{pmatrix} 0 & 0 & 2 \\ -1 & -1 & 3 \\ 0 & -2 & 6 \end{pmatrix}$$

and $\mathbf{I}$ denotes the 3×3 unit matrix.

b) A closed surface S in three-dimensional space is defined by the equation

$$5x^2 + 10y^2 + z^2 - 2xy + 6yz = 4$$

Verify that this equation can be obtained from the equation

$$\mathbf{r}^{\mathrm{T}}\mathbf{Ar} = 4 \qquad (^{*})$$

where $\mathbf{r} = \begin{pmatrix} x \\ y \\ z \end{pmatrix}$, $\mathbf{r}^{\mathrm{T}} = (x\,y\,z)$ and $\mathbf{A}$ is the matrix defined in part **a**.

c) A linear transformation L is defined by $\mathbf{R} = \mathbf{M}^{-1}\mathbf{r}$, where $\mathbf{R} = \begin{pmatrix} X \\ Y \\ Z \end{pmatrix}$ and $\mathbf{M}$ is the matrix defined in part **a**.

   i) By using the relationships

$$\mathbf{r} = \mathbf{MR} \quad \text{and} \quad \mathbf{r}^{\mathrm{T}} = \mathbf{R}^{\mathrm{T}}\mathbf{M}^{\mathrm{T}}$$

   where $\mathbf{R}^{\mathrm{T}} = (X\,Y\,Z)$, in equation $(^{*})$, or otherwise, show that L maps the surface S on to the surface of a sphere of unit radius centred at the origin which has the equation

$$X^2 + Y^2 + Z^2 = 1$$

   ii) Show that $\det \mathbf{M}^{-1} = \frac{1}{4}$.

iii) Given that the volume enclosed by a sphere of unit radius is $\frac{4}{3}\pi$, find the volume of the region enclosed by S.   (NEAB)

21 A transformation T of three-dimensional space is defined by $\mathbf{r}' = \mathbf{Mr}$, where

$$\mathbf{r}' = \begin{pmatrix} x' \\ y' \\ z' \end{pmatrix} \qquad \mathbf{r} = \begin{pmatrix} x \\ y \\ z \end{pmatrix} \qquad \mathbf{M} = \begin{pmatrix} 0 & 1 & 0 \\ 0 & 0 & 1 \\ 1 & 0 & 0 \end{pmatrix}$$

i) Find the image P' of the point $P(2, -3, 1)$ under T.
ii) Show that there is a line L such that all points on L are invariant under T, and find the cartesian equations of this line.
iii) Obtain the equation of the plane Π through the origin O perpendicular to L and verify that P and P' lie in Π.
iv) Given that T represents a rotation about the line L, find the magnitude of the angle of rotation.

Find $\mathbf{M}^2$ and $\mathbf{M}^3$ and state what transformations are represented by these matrices.   (NEAB)

22 Determine the eigenvalues and corresponding eigenvectors of the matrix $\mathbf{A}$, where

$$\mathbf{A} = \begin{bmatrix} 26 & -5 \\ -5 & 2 \end{bmatrix}$$

The plane transformation $\mathbf{T}$ is defined by $\mathbf{T}: \begin{bmatrix} x \\ y \end{bmatrix} \mapsto \mathbf{A}\begin{bmatrix} x \\ y \end{bmatrix}$.

a) Write down a cartesian equation of the line of invariant points of $\mathbf{T}$.
b) Show that all lines of the form $y = -\frac{1}{5}x + k$ (where k is an arbitrary constant) are invariant lines of $\mathbf{T}$.
c) Evaluate the determinant of $\mathbf{A}$, and explain the geometrical significance of this answer in relation to $\mathbf{T}$.
d) Give a full geometrical description of $\mathbf{T}$.   (AEB 98)

23 A transformation T of three-dimensional space is defined by $\mathbf{r}' = \mathbf{Mr}$, where

$$\mathbf{r}' = \begin{pmatrix} x' \\ y' \\ z' \end{pmatrix} \qquad \mathbf{r} = \begin{pmatrix} x \\ y \\ z \end{pmatrix} \qquad \mathbf{M} = \begin{pmatrix} 1 & 3 & 2 \\ 1 & 1 & 1 \\ -1 & 2 & k \end{pmatrix}$$

where k is real.

i) Find $\mathbf{M}^{-1}$ for $k \neq \frac{1}{2}$
ii) In the case when $k = 1$, find the coordinates of the point whose **image** under T is the point $(2, 1, 2)$.
iii) In the case when $k = \frac{1}{2}$, show that the image under T of every point in space lies in the plane

$$3x - 5y - 2z = 0$$

iv) Show that, for one particular value of k, there is a line L such that every point on L is invariant under T. Find the cartesian equations of L.   (NEAB)

15 Further complex numbers

In his Miscellanea analytica (1730), Abraham de Moivre presented further analytical trigonometric results (some formulated as early as 1707), making use of complex numbers. Although he did not state what is now known as de Moivre's theorem, it is clear that he was making use of it.
ALBERT C. LEWIS

De Moivre's theorem

On page 8, we found that

$$(\cos\theta + i\sin\theta)(\cos\phi + i\sin\phi) \equiv \cos(\theta + \phi) + i\sin(\theta + \phi)$$

Hence, we have

$$(\cos\theta + i\sin\theta)^2 \equiv (\cos\theta + i\sin\theta)(\cos\theta + i\sin\theta)$$

$$\equiv \cos 2\theta + i\sin 2\theta$$

The general case of this result is known as de Moivre's theorem, which states that, for all real values of n,

$$(\cos\theta + i\sin\theta)^n \equiv \cos n\theta + i\sin n\theta$$

When n is not an integer, then $\cos n\theta + i\sin n\theta$ is only one of the possible values.

Proof when n is a positive integer

This proof is an example of **proof by induction** (see page 159).

We assume that the statement is true when $n = k$. Hence, we have

$$(\cos\theta + i\sin\theta)^k \equiv (\cos k\theta + i\sin k\theta)$$

$$\Rightarrow \quad (\cos\theta + i\sin\theta)^{k+1} \equiv (\cos k\theta + i\sin k\theta)(\cos\theta + i\sin\theta)$$

Using $(\cos\theta + i\sin\theta)(\cos\phi + i\sin\phi) \equiv \cos(\theta + \phi) + i\sin(\theta + \phi)$, we obtain

$$(\cos\theta + i\sin\theta)^{k+1} \equiv \cos(k+1)\theta + i\sin(k+1)\theta$$

Therefore, statement is true for $n = k + 1$.

When $n = 1$, we have

$$(\cos\theta + i\sin\theta)^n \equiv \cos\theta + i\sin\theta$$

and

$$\cos n\theta + i\sin n\theta \equiv \cos\theta + i\sin\theta$$

Therefore, the statement is true for $n = 1$.

Therefore, de Moivre's theorem is true for all values of $n \geqslant 1$. That is, for all positive integers.

Proof when n is a negative integer

When n is a negative integer, $n = -p$, where p is a positive integer. Hence, we have

$$(\cos\theta + i\sin\theta)^n \equiv (\cos\theta + i\sin\theta)^{-p}$$

$$\equiv \frac{1}{(\cos\theta + i\sin\theta)^p}$$

Using de Moivre's theorem for the positive integer p, we obtain

$$\frac{1}{(\cos\theta + i\sin\theta)^p} \equiv \frac{1}{(\cos p\theta + i\sin p\theta)}$$

$$\equiv \frac{\cos p\theta - i\sin p\theta}{(\cos p\theta + i\sin p\theta)(\cos p\theta - i\sin p\theta)}$$

which gives

$$\frac{1}{(\cos\theta + i\sin\theta)^p} \equiv \cos p\theta - i\sin p\theta$$

But $n = -p$, hence we have

$$\cos p\theta - i\sin p\theta \equiv \cos(-n\theta) - i\sin(-n\theta)$$

$$\equiv \cos n\theta + i\sin n\theta$$

Therefore, we have

$$(\cos\theta + i\sin\theta)^n \equiv \cos n\theta + i\sin n\theta$$

for all negative integers.

Example 1 Find the value of $(\cos\theta + i\sin\theta)^5$.

SOLUTION

Applying de Moivre's theorem, we have

$$(\cos\theta + i\sin\theta)^5 \equiv \cos 5\theta + i\sin 5\theta$$

Example 2 Find $\left[\cos\left(\dfrac{\pi}{6}\right) + i\sin\left(\dfrac{\pi}{6}\right)\right]^3$.

SOLUTION

Applying de Moivre's theorem, we have

$$\left[\cos\left(\frac{\pi}{6}\right) + i\sin\left(\frac{\pi}{6}\right)\right]^3 \equiv \cos\left(3 \times \frac{\pi}{6}\right) + i\sin\left(3 \times \frac{\pi}{6}\right)$$

$$\equiv \cos\left(\frac{\pi}{2}\right) + i\sin\left(\frac{\pi}{2}\right)$$

which gives

$$\left[\cos\left(\frac{\pi}{6}\right) + i\sin\left(\frac{\pi}{6}\right)\right]^3 = i \quad \left(\text{since}\cos\left(\frac{\pi}{2}\right) = 0 \text{ and} \sin\left(\frac{\pi}{2}\right) = 1\right)$$

Example 3 Find $\left[\sin\left(\dfrac{\pi}{3}\right) + i\cos\left(\dfrac{\pi}{3}\right)\right]^{6}$.

SOLUTION

Using $\cos\left(\dfrac{\pi}{2} - \theta\right) = \sin\theta$, we obtain

$$\left[\sin\left(\frac{\pi}{3}\right) + i\cos\left(\frac{\pi}{3}\right)\right]^{6} \equiv \left[\cos\left(\frac{\pi}{6}\right) + i\sin\left(\frac{\pi}{6}\right)\right]^{6}$$

$$\equiv \cos\pi + i\sin\pi$$

which gives

$$\left[\sin\left(\frac{\pi}{3}\right) + i\cos\left(\frac{\pi}{3}\right)\right]^{6} = -1 \quad (\text{since } \cos\pi = -1 \text{ and } \sin\pi = 0)$$

Alternatively, we can proceed as follows:

$$\left[\sin\left(\frac{\pi}{3}\right) + i\cos\left(\frac{\pi}{3}\right)\right]^{6} \equiv \left\{i\left[\cos\left(\frac{\pi}{3}\right) - i\sin\left(\frac{\pi}{3}\right)\right]\right\}^{6}$$

$$\equiv \left\{i\left[\cos\left(-\frac{\pi}{3}\right) + i\sin\left(-\frac{\pi}{3}\right)\right]\right\}^{6}$$

Applying de Moivre's theorem to the RHS, we obtain

$$\left[\sin\left(\frac{\pi}{3}\right) + i\cos\left(\frac{\pi}{3}\right)\right]^{6} \equiv i^{6}[\cos(-2\pi) + i\sin(-2\pi)] = -1 \times 1 = -1$$

Therefore, we have

$$\left[\sin\left(\frac{\pi}{3}\right) + i\cos\left(\frac{\pi}{3}\right)\right]^{6} = -1$$

as above.

Caution You will have noticed that

$$\left[\cos\left(\frac{\pi}{3}\right) - i\sin\left(\frac{\pi}{3}\right)\right]^{6} \equiv \cos 2\pi - i\sin 2\pi$$

and hence you may have deduced that

$$(\cos\theta - i\sin\theta)^{n} \equiv \cos n\theta - i\sin n\theta$$

However, this **cannot** be used as a correct version of de Moivre's theorem, which is only applicable to $(\cos\theta + i\sin\theta)^{n}$

Thus, if you are asked to use de Moivre's theorem to find the value of, say, $\left[\cos\left(\dfrac{\pi}{3}\right) - i\sin\left(\dfrac{\pi}{3}\right)\right]^{6}$, you **must** change this into $\left[\cos\left(-\dfrac{\pi}{3}\right) + i\sin\left(-\dfrac{\pi}{3}\right)\right]^{6}$, as shown in Example 3.

Example 4 Find the value of $(1 + i)^4$.

SOLUTION

Initially, we convert $(1 + i)^4$ into its (r, θ) form, and then use de Moivre's theorem. Hence, we have

$$(1 + i)^4 = \left\{ \sqrt{2}\left[\cos\left(\frac{\pi}{4}\right) + i\sin\left(\frac{\pi}{4}\right) \right] \right\}^4$$

$$= (\sqrt{2})^4\left[\cos\left(\frac{\pi}{4}\right) + i\sin\left(\frac{\pi}{4}\right) \right]^4$$

$$= 4(\cos\pi + i\sin\pi)$$

which gives

$$(1 + i)^4 = -4$$

Example 5 Find the value of $\dfrac{1}{(4 - 4i)^3}$.

SOLUTION

First, we convert $4 - 4i$ into its (r, θ) form, and then use de Moivre's theorem. Hence, we have.

$$4 - 4i = 4\sqrt{2}\left[\cos\left(-\frac{\pi}{4}\right) + i\sin\left(-\frac{\pi}{4}\right) \right]$$

$$\Rightarrow \quad (4 - 4i)^3 = \left\{ 4\sqrt{2}\left[\cos\left(-\frac{\pi}{4}\right) + i\sin\left(-\frac{\pi}{4}\right) \right] \right\}^3$$

$$\Rightarrow \quad \frac{1}{(4 - 4i)^3} = \frac{1}{\left\{ 4\sqrt{2}\left[\cos\left(-\frac{\pi}{4}\right) + i\sin\left(-\frac{\pi}{4}\right) \right] \right\}^3}$$

$$= \frac{1}{128\sqrt{2}\left[\cos\left(-\frac{\pi}{4}\right) + i\sin\left(-\frac{\pi}{4}\right) \right]^3}$$

$$= \frac{1}{128\sqrt{2}}\left[\cos\left(-\frac{\pi}{4}\right) + i\sin\left(-\frac{\pi}{4}\right) \right]^{-3}$$

Using de Moivre's theorem, we obtain

$$\frac{1}{(4 - 4i)^3} = \frac{1}{128\sqrt{2}}\left[\cos\left(\frac{3\pi}{4}\right) + i\sin\left(\frac{3\pi}{4}\right) \right]$$

$$= \frac{1}{128\sqrt{2}}\left(-\frac{1}{\sqrt{2}} + \frac{1}{\sqrt{2}}i \right)$$

which gives

$$\frac{1}{(4 - 4i)^3} = \frac{1}{256}(-1 + i)$$

Exercise 15A

1 Using de Moivre's theorem, find the value of each of the following.

a) $(\cos\theta + i\sin\theta)^6$

b) $(\cos 2\theta + i\sin 2\theta)^4$

c) $\left[\cos\left(\dfrac{\pi}{3}\right) + i\sin\left(\dfrac{\pi}{3}\right)\right]^9$

d) $\left[\cos\left(\dfrac{\pi}{4}\right) + i\sin\left(\dfrac{\pi}{4}\right)\right]^6$

e) $\dfrac{1}{(\cos 2\theta + i\sin 2\theta)^4}$

f) $\dfrac{1}{\left[\cos\left(\dfrac{\pi}{6}\right) + i\sin\left(\dfrac{\pi}{6}\right)\right]^6}$

g) $\left[\cos\left(\dfrac{2\pi}{5}\right) + i\sin\left(\dfrac{2\pi}{5}\right)\right]^{10}$

h) $\left[\cos\left(-\dfrac{\pi}{18}\right) + i\sin\left(-\dfrac{\pi}{18}\right)\right]^9$

2 Simplify each of the following.

a) $(\cos 3\theta + i\sin 3\theta)(\cos 7\theta + i\sin 7\theta)$

b) $(\cos 5\theta + i\sin 5\theta)(\cos 6\theta - i\sin 6\theta)$

c) $\dfrac{\left[\cos\left(\dfrac{\pi}{3}\right) + i\sin\left(\dfrac{\pi}{3}\right)\right]^5}{\left[\cos\left(\dfrac{\pi}{3}\right) - i\sin\left(\dfrac{\pi}{3}\right)\right]^4}$

d) $(1+i)^4 + (1-i)^4$

3 Simplify each of the following.

a) $(1+i)^8$

b) $(2 - \sqrt{3}i)^6$

c) $(3 - \sqrt{3}i)^6$

d) $(1-i)^4$

e) $(2 + 2\sqrt{3}i)^6$

f) $(2i - \sqrt{3})^9$

4 Simplify each of the following.

a) $(\cos\theta - i\sin\theta)^5$

b) $(\sin\theta - i\cos\theta)^4$

c) $\dfrac{1}{(\sin\theta + i\cos\theta)^6}$

d) $\dfrac{1}{\left[\sin\left(\dfrac{\pi}{5}\right) - i\cos\left(\dfrac{\pi}{5}\right)\right]^{10}}$

5 Show that

$$\frac{\cos 2x + i\sin 2x}{\cos 9x - i\sin 9x}$$

can be expressed in the form $\cos nx + i\sin nx$, where n is an integer to be found. (EDEXCEL)

*n*th roots of unity

When n is not an integer, de Moivre's theorem gives **only one** of the possible values for $(\cos\theta + i\sin\theta)^n$, which is $\cos n\theta + i\sin n\theta$.

However, $(\cos\theta + i\sin\theta)^{\frac{1}{n}}$ can take n different values, as we will now show.

We let

$$(\cos\theta + i\sin\theta)^{\frac{1}{n}} = r(\cos\phi + i\sin\phi)$$

Comparing the moduli of both sides, we have $r = 1$.

Raising both sides to the *n*th power, and using

$$[(\cos\theta + i\sin\theta)^{\frac{1}{n}}]^n = \cos\theta + i\sin\theta$$

we obtain

$$\cos\theta + i\sin\theta = [(\cos\theta + i\sin\theta)^{\frac{1}{n}}]^n = (\cos\phi + i\sin\phi)^n$$

$$\Rightarrow \quad \cos\theta + i\sin\theta = \cos n\phi + i\sin n\phi$$

Therefore, we have

$$\cos\theta = \cos n\phi \quad \text{and} \quad \sin\theta = \sin n\phi$$

which give

$$n\phi = \theta, \ \theta + 2\pi, \ \theta + 4\pi, \ \theta + 6\pi, \ldots$$

since $\cos(\theta + 2\pi) = \cos\theta$, and $\sin(\theta + 2\pi) = \sin\theta$.

That is, we have

$$\phi = \frac{\theta}{n}, \ \frac{\theta + 2\pi}{n}, \ \frac{\theta + 4\pi}{n}, \ \ldots$$

which means that $(\cos\theta + i\sin\theta)^{\frac{1}{n}}$ is identical to

$$\cos\left(\frac{\theta}{n}\right) + i\sin\left(\frac{\theta}{n}\right)$$

or

$$\cos\left(\frac{\theta + 2\pi}{n}\right) + i\sin\left(\frac{\theta + 2\pi}{n}\right)$$

or

$$\cos\left(\frac{\theta + 4\pi}{n}\right) + i\sin\left(\frac{\theta + 4\pi}{n}\right)$$

and so on, adding $\dfrac{2\pi}{n}$ each time until we obtain

$$(\cos\theta + i\sin\theta)^{\frac{1}{n}} \equiv \cos\left[\frac{\theta + (n-1)2\pi}{n}\right] + i\sin\left[\frac{\theta + (n-1)2\pi}{n}\right]$$

All subsequent values are repeats of the *n* different values given above. Therefore, $(\cos\theta + i\sin\theta)^{\frac{1}{n}}$ has *n* different values.

We note that these *n* solutions are symmetrically placed on a circle drawn on an Argand diagram.

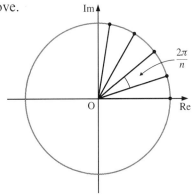

Example 6 Find the value of $(-64)^{\frac{1}{6}}$.

SOLUTION

Expressing -64 in the form $r(\cos\theta + i\sin\theta)$, we have

$$-64 = 64(\cos\pi + i\sin\pi)$$

which gives

$$(-64)^{\frac{1}{6}} = 64^{\frac{1}{6}}(\cos\pi + i\sin\pi)^{\frac{1}{6}}$$

$$= 2\left[\cos\left(\frac{\pi}{6}\right) + i\sin\left(\frac{\pi}{6}\right)\right] \quad \text{(from de Moivre's theorem)}$$

Using symmetry, we find that the other roots are as shown in the diagram below right. That is,

$$2\left[\cos\left(\frac{\pi}{2}\right) + i\sin\left(\frac{\pi}{2}\right)\right]$$

$$2\left[\cos\left(\frac{5\pi}{6}\right) + i\sin\left(\frac{5\pi}{6}\right)\right]$$

$$2\left[\cos\left(-\frac{5\pi}{6}\right) + i\sin\left(-\frac{5\pi}{6}\right)\right]$$

$$2\left[\cos\left(-\frac{\pi}{2}\right) + i\sin\left(-\frac{\pi}{2}\right)\right]$$

$$2\left[\cos\left(-\frac{\pi}{6}\right) + i\sin\left(-\frac{\pi}{6}\right)\right]$$

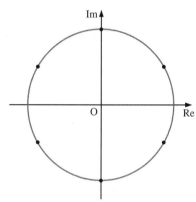

Since all of these values can be expressed simply in the form $a + ib$, it is common to give these answers in the form

$$\pm\left(\frac{\sqrt{3}}{2} \pm \frac{i}{2}\right), \quad \pm i$$

Example 7 Find the values of $(-1 - \sqrt{3}\,i)^{\frac{1}{2}}$.

SOLUTION

Expressing $-1 - \sqrt{3}i$ in the form $\cos\theta + i\sin\theta$, we have

$$-1 - \sqrt{3}i = 2\left[\cos\left(-\frac{2\pi}{3}\right) + i\sin\left(-\frac{2\pi}{3}\right)\right]$$

Therefore, from de Moivre's theorem, one value of $(-1 - \sqrt{3}i)^{\frac{1}{2}}$ is

$$\left\{2\left[\cos\left(-\frac{2\pi}{3}\right) + i\sin\left(-\frac{2\pi}{3}\right)\right]\right\}^{\frac{1}{2}} = 2^{\frac{1}{2}}\left[\cos\left(-\frac{\pi}{3}\right) + i\sin\left(-\frac{\pi}{3}\right)\right]$$

$$= \sqrt{2}\left(+\frac{1}{2} - \frac{\sqrt{3}}{2}i\right)$$

$$= \frac{\sqrt{2}}{2} - \frac{\sqrt{6}}{2}i$$

By symmetry, the other root is as shown in the diagram on the right. That is,

$$-\frac{\sqrt{2}}{2}+\frac{\sqrt{6}}{2}i$$

Therefore, we have

$$\left(-1-\sqrt{3}i\right)^{\frac{1}{2}}=\pm\left(\frac{\sqrt{2}}{2}-\frac{\sqrt{6}}{2}i\right)$$

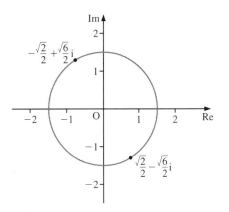

Example 8 Find the solutions of $27z^3 = 8$.

SOLUTION

We take the cube root of both sides, remembering to multiply one side of the resulting equation by each of the three cube roots of unity, taken one at a time. In this case, it is simpler to multiply $\sqrt[3]{8}$ by the three cube roots.

Hence, we have

$$27z^3 = 8$$
$$\Rightarrow \quad 3z = \sqrt[3]{1} \times 2$$

From page 18, we know that $\sqrt[3]{1}$ has the following values:

$$1 \qquad -\frac{1}{2}+\frac{\sqrt{3}}{2}i \qquad -\frac{1}{2}-\frac{\sqrt{3}}{2}i$$

Using $\sqrt[3]{1} = 1$, we obtain

$$3z = 2 \quad \Rightarrow \quad z = \frac{2}{3}$$

Using $\sqrt[3]{1} = -\frac{1}{2}+\frac{\sqrt{3}}{2}i$, we obtain

$$3z = -1+\sqrt{3}i \quad \Rightarrow \quad z = -\frac{1}{3}+\frac{\sqrt{3}}{3}i$$

Using $\sqrt[3]{1} = -\frac{1}{2}-\frac{\sqrt{3}}{2}i$, we obtain

$$3z = -1-\sqrt{3}i \quad \Rightarrow \quad z = -\frac{1}{3}-\frac{\sqrt{3}}{3}i$$

Example 9 Find the solutions of $16z^4 = (z-1)^4$.

SOLUTION

We take the fourth root of both sides, remembering to multiply one side of the resulting equation by each of the four fourth roots of unity, taken one at a time.

Hence, we have

$$16z^4 = (z-1)^4$$

$$\Rightarrow \quad 2z = \sqrt[4]{1}(z-1)$$

We know that $\sqrt[4]{1} = 1, -1, \mathrm{i}, -\mathrm{i}$.

Using $\sqrt[4]{1} = 1$, we obtain

$$2z = z - 1 \quad \Rightarrow \quad z = -1$$

Using $\sqrt[4]{1} = -1$, we obtain

$$2z = -(z-1) \quad \Rightarrow \quad 3z = 1 \quad \Rightarrow \quad z = \tfrac{1}{3}$$

Using $\sqrt[4]{1} = \mathrm{i}$, we obtain

$$2z = \mathrm{i}(z-1)$$

$$\Rightarrow \quad z = -\frac{\mathrm{i}}{2-\mathrm{i}}$$

$$\Rightarrow \quad z = -\frac{\mathrm{i}(2+\mathrm{i})}{(2-\mathrm{i})(2+\mathrm{i})}$$

which gives

$$z = \tfrac{1}{5}(1 - 2\mathrm{i})$$

Using $\sqrt[4]{1} = -\mathrm{i}$, we obtain

$$2z = -\mathrm{i}(z-1)$$

$$\Rightarrow \quad z = \frac{\mathrm{i}}{2+\mathrm{i}}$$

$$\Rightarrow \quad z = \frac{\mathrm{i}(2-\mathrm{i})}{5}$$

which gives

$$z = \tfrac{1}{5}(1 + 2i)$$

Therefore, the four solutions of $16z^4 = (z-1)^4$ are $-1, \tfrac{1}{3}, \tfrac{1}{5}(1 \pm 2\mathrm{i})$.

Exponential form of a complex number

Using the power series expansions studied on pages 177–9, we have

$$\mathrm{e}^{\mathrm{i}\theta} = 1 + \mathrm{i}\theta + \frac{(\mathrm{i}\theta)^2}{2!} + \frac{(\mathrm{i}\theta)^3}{3!} + \ldots$$

$$= 1 + \mathrm{i}\theta - \frac{\theta^2}{2!} - \frac{\mathrm{i}\theta^3}{3!} + \frac{\theta^4}{4!} + \frac{\mathrm{i}\theta^5}{5!} - \ldots$$

$$= \left(1 - \frac{\theta^2}{2!} + \frac{\theta^4}{4!} - \ldots\right) + \mathrm{i}\left(\theta - \frac{\theta^3}{3!} + \frac{\theta^5}{5!} - \ldots\right)$$

$$\Rightarrow \quad \mathrm{e}^{\mathrm{i}\theta} = \cos\theta + \mathrm{i}\sin\theta$$

This is the **exponential form** of a complex number.

Expressed generally, we have

$$z = r(\cos\theta + i\sin\theta) \quad \Rightarrow \quad z = re^{i\theta}$$

We can use the exponential form to simplify many types of problem.

Note Using the exponential form of $(\cos\theta + i\sin\theta)^n$, we have

$$(\cos\theta + i\sin\theta)^n = (e^{i\theta})^n = e^{i(n\theta)} = \cos n\theta + i\sin n\theta$$

which proves de Moivre's theorem.

Example 10 Express $2 + 2i$ in $re^{i\theta}$ form.

SOLUTION

The modulus of $2 + 2i$ is $2\sqrt{2}$ and its argument is $\dfrac{\pi}{4}$. Hence, we have

$$2 + 2i = 2\sqrt{2}e^{\frac{i\pi}{4}}$$

Example 11 Express $1 - i\sqrt{3}$ in $re^{i\theta}$ form.

SOLUTION

The modulus of $1 - i\sqrt{3}$ is 2 and its argument is $-\dfrac{\pi}{3}$. Hence, we have

$$1 - i\sqrt{3} = 2e^{-i\pi/3}$$

Example 12 Find the values of $(-2 + 2i)^{\frac{1}{3}}$ and show their positions on an Argand diagram.

SOLUTION

We proceed as follows:

- First, express $(-2 + 2i)$ in its (r, θ) form.
- Then find one value of $(-2 + 2i)^{\frac{1}{3}}$
- Finally, use symmetry to find the other roots.

Hence, we have

$$(-2 + 2i)^{\frac{1}{3}} = \left\{ 2\sqrt{2}\left[\cos\left(\frac{3\pi}{4}\right) + i\sin\left(\frac{3\pi}{4}\right)\right] \right\}^{\frac{1}{3}}$$

$$= \left\{ 2^{\frac{3}{2}}\left[\cos\left(\frac{3\pi}{4}\right) + i\sin\left(\frac{3\pi}{4}\right)\right] \right\}^{\frac{1}{3}}$$

Therefore, from de Moivre's theorem, one value of $(-2 + 2i)^{\frac{1}{3}}$ is

$$2^{\frac{1}{2}}\left[\cos\left(\frac{\pi}{4}\right) + i\sin\left(\frac{\pi}{4}\right)\right]$$

By symmetry, the other roots are

$$2^{\frac{1}{2}}\left[\cos\left(\frac{\pi}{4}+\frac{2\pi}{3}\right)+i\sin\left(\frac{\pi}{4}+\frac{2\pi}{3}\right)\right]$$

and

$$2^{\frac{1}{2}}\left[\cos\left(\frac{\pi}{4}+\frac{4\pi}{3}\right)+i\sin\left(\frac{\pi}{4}+\frac{4\pi}{3}\right)\right]$$

These three roots (see Argand diagram on the right) may be expressed as

$$2^{\frac{1}{2}}e^{i\pi/4} \qquad 2^{\frac{1}{2}}e^{11i\pi/12} \qquad 2^{\frac{1}{2}}e^{19i\pi/12}$$

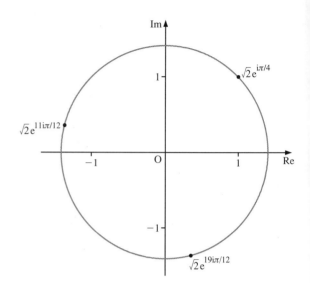

Multiplying one complex number by another

Expressing the two numbers, z_1 and z_2, in their exponential form, we have

$$z_1 z_2 = r_1 e^{i\theta_1} \times r_2 e^{i\theta_2}$$

which gives

$$z_1 z_2 = r_1 r_2 e^{i(\theta_1 + \theta_2)}$$

This is a very simple way of showing that to find the product of two complex numbers, we multiply the moduli and add the arguments. (See page 8.)

Simplifying certain integrals

We can simplify integrals of the type $\int e^{ax}\cos bx\,dx$ using the exponential form, as shown in Examples 13 and 14.

Example 13 Find $\int e^{2x}\sin x\,dx$.

SOLUTION

We have

$$\int e^{2x}\sin x\,dx = \operatorname{Im}\int e^{2x}(\cos x + i\sin x)\,dx$$

where $\operatorname{Im}\int$ is the imaginary part of the given integral.

Using the exponential form of $\cos x + i \sin x$, we obtain

$$\int e^{2x} \sin x \, dx = \text{Im} \int e^{2x} \times e^{ix} \, dx$$

$$= \text{Im} \int e^{(2+i)x} \, dx$$

which gives

$$\int e^{2x} \sin x \, dx = \text{Im} \left(\frac{1}{2+i} e^{(2+i)x} \right)$$

$$= \text{Im} \left[\frac{2-i}{(2+i)(2-i)} e^{2x} (\cos x + i \sin x) + c \right]$$

$$= \text{Im} \left[e^{2x} \frac{(2\cos x + \sin x + 2i \sin x - i \cos x)}{2^2 - (i)^2} \right] + c$$

Hence, we find that

$$\int e^{2x} \sin x \, dx = \frac{e^{2x}}{5} (2 \sin x - \cos x) + c$$

Example 14 Find $\int e^{4x} \cos 3x \, dx$.

SOLUTION

We have

$$\int e^{4x} \cos 3x = \text{Re} \int e^{4x} (\cos 3x + i \sin 3x) \, dx$$

where $\text{Re} \int$ is the real part of the given integral.

Using the exponential form of $\cos 3x + i \sin 3x$, we obtain

$$\int e^{4x} \cos 3x \, dx = \text{Re} \int e^{(4+3i)x} \, dx$$

$$= \text{Re} \left(\frac{1}{4+3i} e^{(4+3i)x} + c \right)$$

$$= \text{Re} \left[\frac{4-3i}{(4+3i)(4-3i)} e^{4x} (\cos 3x + i \sin 3x) + c \right]$$

Hence, we have

$$\int e^{4x} \cos 3x \, dx = \frac{e^{4x}}{25} \left(\frac{4\cos 3x + 3 \sin 3x}{25} \right) + c$$

Exercise 15B

1 For each of the following, find the possible values of z, giving your answers in

   i) $a + ib$ form   **ii)** $re^{i\theta}$ form

   a) $z^4 = -16$

   b) $z^3 = -8 + 8i$

   c) $z^3 = 27i$

   d) $z^2 = 16i$

   e) $z^2 = -25i$

   f) $z^5 = -32$

2 Find the six sixth roots of unity.

3 Solve each of these.

   a) $(z + 2i)^2 = 4$

   b) $(z - 1)^3 = 8$

   c) $z^2 = (z + 1)^2$

   d) $(z + 3i)^2 = (2z - 1)^2$

   e) $(z - i)^4 = 81(z + 2)^4$

4 Find the seven seventh roots of unity in the $e^{i\theta}$ form.

5 Solve $z^5 = 32i$. Give your answers in the $re^{i\theta}$ form, and show them on an Argand diagram.

6 By considering the ninth roots of unity, show that

$$\cos\left(\frac{2\pi}{9}\right) + \cos\left(\frac{4\pi}{9}\right) + \cos\left(\frac{6\pi}{9}\right) + \cos\left(\frac{8\pi}{9}\right) = -\frac{1}{2}$$

7 By considering the seventh roots of unity, show that

$$\cos\left(\frac{\pi}{7}\right) + \cos\left(\frac{3\pi}{7}\right) + \cos\left(\frac{5\pi}{7}\right) = \frac{1}{2}$$

8 When $\cos 4\theta = \cos 3\theta$, prove that $\theta = 0, \dfrac{2\pi}{7}, \dfrac{4\pi}{7}, \dfrac{6\pi}{7}$.

   Hence prove that $\cos\left(\dfrac{2\pi}{7}\right)$, $\cos\left(\dfrac{4\pi}{7}\right)$, $\cos\left(\dfrac{6\pi}{7}\right)$ are the roots of $8x^3 + 4x^2 - 4x - 1 = 0$.

9 Evaluate each of these.

   a) $\displaystyle\int e^{4x} \cos 5x \, dx$

   b) $\displaystyle\int e^{3x} \sin 7x \, dx$

   c) $\displaystyle\int e^{-2x} \sin 4x \, dx$

   d) $\displaystyle\int e^{-4x} \cos 3x \, dx$

10 Find, in polar form, each of the fourth roots of $-8 - 8\sqrt{3}i$.   (WJEC)

11 Verify that $(3 - 2i)^2 = 5 - 12i$, showing your working clearly. Find the two roots of the equation $(z - i)^2 = 5 - 12i$ (OCR)

12 i) Find the exact modulus and argument of the complex number $-4\sqrt{3} - 4i$.
   ii) Hence obtain the roots of the equation

$$z^3 + 4\sqrt{3} + 4i = 0$$

   giving your answers in the form $re^{i\theta}$, where $r > 0$ and $-\pi < \theta \leqslant \pi$.   (OCR)

13 Express $(8\sqrt{2})(1+i)$ in the form $r(\cos\theta + i\sin\theta)$, where $r > 0$ and $-\pi < \theta \leqslant \pi$. Hence, or otherwise, solve the equation $z^4 = (8\sqrt{2})(1+i)$, giving your answers in polar form.   (OCR)

14 Write each of the complex numbers

$$z_1 = 1 - (\sqrt{3})i \qquad z_2 = (\sqrt{3}) + i$$

in the form $re^{i\theta}$, where $r > 0$ and $-\pi < \theta \leqslant \pi$.

Hence show that if $z_1^7 + z_2^7 = x + iy$, where $x, y \in \mathbb{R}$, then

$$\frac{y}{x} = 2 + \sqrt{3} \qquad \text{(OCR)}$$

15 a) State de Moivre's theorem for the expansion of $(\cos\theta + i\sin\theta)^n$, where n is a positive integer or rational number.
   b) Find the modulus and the argument of each of the three cube roots of $1 + i$.
   c) Show that $(1 + i)^{51} = 2^{25}(-1 + i)$.   (WJEC)

16 Write down the modulus and argument of the complex number -64.

Hence solve the equation $z^4 + 64 = 0$, giving your answers in the form $r(\cos\theta + i\sin\theta)$, where $r > 0$ and $-\pi < \theta \leqslant \pi$.

Express each of these four roots in the form $a + ib$ and show, with the aid of a diagram, that the points in the complex plane which represent them form the vertices of a square.

     (AEB 96)

17 a) Solve the equation $z^5 = 4 + 4i$, giving your answers in the form $z = re^{ik\pi}$, where r is the modulus of z and k is a rational number such that $0 \leqslant k \leqslant 2$.
   b) Show on an Argand diagram the points representing your solutions.   (EDEXCEL)

18 i) Show that

$$e^{(3+2i)x} \equiv e^{3x}(\cos 2x + i\sin 2x)$$

   where x is real.
   ii) Find the real and imaginary parts of

$$\frac{e^{3x}(\cos 2x + i\sin 2x)}{(3 + 2i)}$$

   iii) If $C = \displaystyle\int e^{3x}\cos 2x\, dx$ and $S = \displaystyle\int e^{3x}\sin 2x\, dx$, by using parts **i** and **ii** and considering $C + iS$, or otherwise, find C and S. [You may assume the normal rules of integration apply to $\displaystyle\int e^{kx}\, dx$ when k is complex.]   (NICCEA)

19 a) Verify that $z_1 = 1 + e^{\pi i/5}$ is a root of the equation $(z - 1)^5 = -1$.
   b) Find the other four roots of the equation.
   c) Mark on an Argand diagram the points corresponding to the five roots of the equation. Show that these roots lie on a circle, and state the centre and the radius of the circle.
   d) By considering the Argand diagram, or otherwise, find
      i) $\arg z_1$ in terms of π.
      ii) $|z_1|$ in the form $a\cos\dfrac{\pi}{b}$, where a and b are integers to be determined.   (NEAB)

20 i) Find the roots of the equation $(z - 4)^3 = 8i$ in the form $a + ib$, where a and b are real numbers. Indicate, on an Argand diagram, the points A, B and C representing these three roots and find the area of $\triangle$ABC.

ii) The equation $z^3 + pz^2 + 40z + q = 0$, where p and q are real, has a root $3 + i$. Write down another root of the equation.

Hence, or otherwise, find the values of p and q.   (EDEXCEL)

21 Write down the fifth roots of unity in the form $\cos\theta + i\sin\theta$, where $0 \leqslant \theta < 2\pi$.

i) Hence, or otherwise, find the fifth roots of i in a similar form.

ii) By writing the equation $(z - 1)^5 = z^5$ in the form

$$\left(\frac{z - 1}{z}\right)^5 = 1$$

show that its roots are

$$\tfrac{1}{2}(1 + i\cot\tfrac{1}{5}k\pi) \quad k = 1, 2, 3, 4 \qquad \text{(OCR)}$$

22 i) Find the six complex roots of the equation $z^6 + 8i = 0$, expressing each in the form $re^{i\theta}$. Give the exact values of θ in radians.

ii) Show that $(1 + i)$ and $(-1 - i)$ are two of the roots.

iii) Sketch the six roots on an Argand diagram, clearly indicating the significant geometrical features.   (NICCEA)

Trigonometric identities

Expressions for $\cos^n\theta$ and $\sin^n\theta$ in terms of multiples of θ

Let $z \equiv \cos\theta + i\sin\theta$. We then have

$$\frac{1}{z} \equiv (\cos\theta + i\sin\theta)^{-1} \equiv \cos\theta - i\sin\theta$$

which gives

$$z + \frac{1}{z} \equiv 2\cos\theta$$

$$z - \frac{1}{z} \equiv 2i\sin\theta$$

We also have

$$z^n \equiv (\cos\theta + i\sin\theta)^n = \cos n\theta + i\sin n\theta$$

$$\Rightarrow \quad \frac{1}{z^n} \equiv \frac{1}{\cos n\theta + i\sin n\theta}$$

$$\Rightarrow \quad \frac{1}{z^n} \equiv \cos n\theta - i\sin n\theta$$

which gives

$$z^n + \frac{1}{z^n} \equiv 2\cos n\theta$$

$$z^n - \frac{1}{z^n} \equiv 2\mathrm{i}\sin n\theta$$

With the aid of these four identities for $z \pm \dfrac{1}{z}$ and $z^n \pm \dfrac{1}{z^n}$, we can write any

power of $\cos\theta$ or $\sin\theta$ in terms of multiples of θ.

Example 15 If $z \equiv \cos\theta + \mathrm{i}\sin\theta$, express the following in terms of θ.

a) z^4 **b)** z^{-3}

SOLUTION

We know that, when $z \equiv \cos\theta + \mathrm{i}\sin\theta$,

$$z^n \equiv \cos n\theta + \mathrm{i}\sin n\theta$$

for all integer n.

Hence, we have

a) $z^4 \equiv \cos 4\theta + \mathrm{i}\sin 4\theta$

b) $z^{-3} \equiv \cos(-3\theta) + \mathrm{i}\sin(-3\theta)$

$\Rightarrow$ $z^{-3} \equiv \cos 3\theta - \mathrm{i}\sin 3\theta$

Example 16 If $z \equiv \cos\theta + \mathrm{i}\sin\theta$, express the following in terms of z.

a) $\cos 6\theta$ **b)** $\sin 3\theta$

SOLUTION

When $z \equiv \cos\theta + \mathrm{i}\sin\theta$, we know that

$$z^n + \frac{1}{z^n} \equiv 2\cos n\theta$$

and $z^n - \dfrac{1}{z^n} \equiv 2\mathrm{i}\sin n\theta$

Hence, we have

a) $2\cos 6\theta \equiv z^6 + \dfrac{1}{z^6}$

$\Rightarrow$ $\cos 6\theta \equiv \dfrac{1}{2}\left(z^6 + \dfrac{1}{z^6}\right)$

b) $2\mathrm{i}\sin 3\theta \equiv z^3 - \dfrac{1}{z^3}$

$\Rightarrow$ $\sin 3\theta \equiv \dfrac{1}{2\mathrm{i}}\left(z^3 - \dfrac{1}{z^3}\right)$

Example 17 Express $\cos^3\theta$ as the cosines of multiples of θ.

SOLUTION

We proceed as follows:

- Express $\cos\theta$ in terms of z, and hence find $\cos^3\theta$.
- Collect terms of the type $z^n + \dfrac{1}{z^n}$, according to the values of n (as we are required to give the answer as the cosines of multiples of θ).
- Finally, convert these terms into cosines of multiples of θ.

Hence, we have

$$\cos\theta \equiv \frac{1}{2}\left(z + \frac{1}{z}\right)$$

which gives

$$\cos^3\theta \equiv \left[\frac{1}{2}\left(z + \frac{1}{z}\right)\right]^3$$

$$\equiv \frac{1}{2^3}\left(z + \frac{1}{z}\right)^3$$

$$\equiv \frac{1}{8}\left(z^3 + 3z^2 \times \frac{1}{z} + 3z \times \frac{1}{z^2} + \frac{1}{z^3}\right)$$

$$\equiv \frac{1}{8}\left(z^3 + 3z + \frac{3}{z} + \frac{1}{z^3}\right)$$

Rearranging the terms on the RHS, we obtain

$$\cos^3\theta \equiv \frac{1}{8}\left[\left(z^3 + \frac{1}{z^3}\right) + 3\left(z + \frac{1}{z}\right)\right]$$

Converting the RHS, we have

$$\cos^3\theta \equiv \frac{1}{8}(2\cos 3\theta + 3 \times 2\cos\theta)$$

which gives

$$\cos^3\theta \equiv \frac{1}{4}\cos 3\theta + \frac{3}{4}\cos\theta$$

Example 18 Express $\cos^6\theta$ as the cosines of multiples of θ.

SOLUTION

We have

$$\cos^6\theta \equiv \left[\frac{1}{2}\left(z + \frac{1}{z}\right)\right]^6$$

where $z \equiv \cos\theta + \mathrm{i}\sin\theta$.

Using the binomial theorem, we obtain

$$\left(z + \frac{1}{z}\right)^6 = z^6 + 6z^5 \times \frac{1}{z} + 15z^4 \times \frac{1}{z^2} + \ldots + \frac{1}{z^6}$$

which gives

$$\cos^6\theta \equiv \frac{1}{64}\left(z^6 + 6z^4 + 15z^2 + 20 + \frac{15}{z^2} + \frac{6}{z^4} + \frac{1}{z^6}\right)$$

$$\equiv \frac{1}{64}\left[\left(z^6 + \frac{1}{z^6}\right) + 6\left(z^4 + \frac{1}{z^4}\right) + 15\left(z^2 + \frac{1}{z^2}\right) + 20\right]$$

Converting the RHS, we have

$$\cos^6\theta \equiv \frac{1}{64}(2\cos 6\theta + 6 \times 2\cos 4\theta + 15 \times 2\cos 2\theta + 20)$$

$$\Rightarrow \quad \cos^6\theta \equiv \frac{1}{32}\cos 6\theta + \frac{3}{16}\cos 4\theta + \frac{15}{32}\cos 2\theta + \frac{5}{16}$$

Example 19 Express $\sin^5\theta$ as the sines of multiples of θ.

SOLUTION

We have

$$\sin^5\theta \equiv \left[\frac{1}{2i}\left(z - \frac{1}{z}\right)\right]^5$$

where $z = \cos\theta + i\sin\theta$.

Using the binomial theorem, we obtain

$$\sin^5\theta \equiv \frac{1}{32i^5}\left(z^5 - 5z^3 + 10z - \frac{10}{z} + \frac{5}{z^3} - \frac{1}{z^5}\right)$$

$$\equiv \frac{1}{32i}\left[\left(z^5 - \frac{1}{z^5}\right) - 5\left(z^3 - \frac{1}{z^3}\right) + 10\left(z - \frac{1}{z}\right)\right]$$

Converting the RHS, we have

$$\sin^5\theta \equiv \frac{1}{32i}[2i\sin 5\theta - 10i\sin 3\theta + 20i\sin\theta]$$

$$\Rightarrow \quad \sin^5\theta \equiv \frac{1}{16}\sin 5\theta - \frac{5}{16}\sin 3\theta + \frac{5}{8}\sin\theta$$

Expansions of $\cos n\theta$ and $\sin n\theta$ as powers of $\cos\theta$ and $\sin\theta$

To change a function such as $\cos 6\theta$ into powers of $\cos\theta$, we express $\cos 6\theta$ as the **real part** of $\cos 6\theta + i\sin 6\theta$.

By de Moivre's theorem, we have

$$\cos 6\theta + i\sin 6\theta = (\cos\theta + i\sin\theta)^6$$

the RHS of which we expand by the binomial theorem. We then extract the real terms from this expansion.

Similarly, we express, for example, $\sin 7\theta$ as the **imaginary part** of $\cos 7\theta + i\sin 7\theta$.

Example 20 Express $\sin 3\theta$ in terms of $\sin \theta$.

SOLUTION

We put

$$\sin 3\theta = \text{Im} (\cos 3\theta + i \sin 3\theta)$$

where $\text{Im}\,(z)$ is the imaginary part of z.

Hence, we have

$$\sin 3\theta = \text{Im} (\cos \theta + i \sin \theta)^3$$

Expanding the RHS by the binomial theorem, we obtain

$$\sin 3\theta = \text{Im}[\cos^3\theta + 3\cos^2\theta(i \sin \theta) + 3\cos \theta(i \sin \theta)^2 + (i \sin \theta)^3]$$

$$= \text{Im} (\cos^3\theta + 3i \cos^2\theta \sin \theta - 3\cos \theta \sin^2\theta - i \sin^3\theta)$$

$$= 3\cos^2\theta \sin \theta - \sin^3\theta$$

Using $\cos^2\theta = 1 - \sin^2\theta$ (as the answer has to be in terms of $\sin \theta$), we have

$$\sin 3\theta = 3(1 - \sin^2\theta) \sin \theta - \sin^3\theta$$

$$= 3\sin \theta - 3\sin^3\theta - \sin^3\theta$$

which gives

$$\sin 3\theta = 3\sin \theta - 4\sin^3\theta$$

Example 21

a) Express $\cos 6\theta$ in terms of powers of $\cos \theta$.

b) Express $\dfrac{\sin 6\theta}{\sin \theta}$ in terms of powers of $\cos \theta$.

SOLUTION

a) We put

$$\cos 6\theta = \text{Re}(\cos 6\theta + i \sin 6\theta)$$

where $\text{Re}(z)$ means the real part of z.

Hence, we have

$$\cos 6\theta = (\cos \theta + i \sin \theta)^6$$

Expanding the RHS by the binomial theorem, we obtain

$$\cos 6\theta = \text{Re}\left[\cos^6\theta + 6\cos^5\theta\,(i \sin \theta) + \frac{6.5}{2.1}\cos^4\theta\,(i \sin \theta)^2 + \right.$$

$$+ \frac{6.5.4}{3.2.1}\cos^3\theta\,(i \sin \theta)^3 + \frac{6.5.4.3}{4.3.2.1}\cos^2\theta\,(i \sin \theta)^4 +$$

$$\left. + \frac{6.5.4.3.2}{5.4.3.2.1}\cos \theta\,(i \sin \theta)^5 + (i \sin \theta)^6\right]$$

$$\Rightarrow \quad \cos 6\theta = \cos^6\theta - 15\cos^4\theta \sin^2\theta + 15\cos^2\theta \sin^4\theta - \sin^6\theta$$

Using $\sin^2\theta = 1 - \cos^2\theta$, we have

$$\cos 6\theta = \cos^6\theta - 15\cos^4\theta\,(1 - \cos^2\theta) + 15\cos^2\theta(1 - \cos^2\theta)^2 - (1 - \cos^2\theta)^3$$

$$= \cos^6\theta - 15\cos^4\theta + 15\cos^6\theta + 15\cos^2\theta - 30\cos^4\theta + 15\cos^6\theta -$$

$$- 1 + 3\cos^2\theta - 3\cos^4\theta + \cos^6\theta$$

which gives

$$\cos 6\theta = 32\cos^6\theta - 48\cos^4\theta + 18\cos^2\theta - 1$$

b) We put

$$\sin 6\theta = \mathrm{Im}(\cos 6\theta + \mathrm{i}\sin 6\theta)$$

where $\mathrm{Im}(z)$ means the imaginary part of z.

Hence, we have

$$\sin 6\theta = \mathrm{Im}\,(\cos\theta + \mathrm{i}\sin\theta)^6$$

Expanding the RHS by the binomial theorem, we obtain

$$\sin 6\theta = \mathrm{Im}\left[\cos^6\theta + 6\cos^5\theta\,(\mathrm{i}\sin\theta) + \frac{6.5}{2.1}\cos^4\theta\,(\mathrm{i}\sin\theta)^2 + \right.$$

$$+ \frac{6.5.4}{3.2.1}\cos^3\theta\,(\mathrm{i}\sin\theta)^3 + \frac{6.5.4.3}{4.3.2.1}\cos^2\theta\,(\mathrm{i}\sin\theta)^4 +$$

$$\left. + \frac{6.5.4.3.2}{5.4.3.2.1}\cos\theta\,(\mathrm{i}\sin\theta)^5 + (\mathrm{i}\sin\theta)^6\right]$$

$$\Rightarrow \quad \sin 6\theta = 6\cos^5\theta\sin\theta - 20\cos^3\theta\sin^3\theta + 6\cos\theta\sin^5\theta$$

Therefore, we have

$$\frac{\sin 6\theta}{\sin\theta} = 6\cos^5\theta - 20\cos^3\theta(1 - \cos^2\theta) + 6\cos\theta(1 - \cos^2\theta)^2$$

$$= 6\cos^5\theta - 20\cos^3\theta + 20\cos^5\theta + 6\cos\theta - 12\cos^3\theta + 6\cos^5\theta$$

which gives

$$\frac{\sin 6\theta}{\sin\theta} = 32\cos^5\theta - 32\cos^3\theta + 6\cos\theta$$

Example 22

a) Express $\sin 5\theta$ in terms of $\sin\theta$.

b) Hence, prove that $\sin\left(\dfrac{\pi}{5}\right)$, $\sin\left(\dfrac{2\pi}{5}\right)$, $\sin\left(\dfrac{6\pi}{5}\right)$ and $\sin\left(\dfrac{7\pi}{5}\right)$ are the roots of the equation $16x^4 - 20x^2 + 5 = 0$.

c) Deduce that $\sin^2\left(\dfrac{\pi}{5}\right)$ and $\sin^2\left(\dfrac{2\pi}{5}\right)$ are roots of the equation

$16y^2 - 20y + 5 = 0$, and hence find the exact value of

i) $\sin\left(\dfrac{\pi}{5}\right)\sin\left(\dfrac{2\pi}{5}\right)$ **ii)** $\cos\left(\dfrac{2\pi}{5}\right)$

SOLUTION

a) We put

$$\sin 5\theta = \text{Im}\,(\cos 5\theta + i \sin 5\theta)$$

where $\text{Im}\,(z)$ is the imaginary part of z.

Hence, we have

$$\sin 5\theta = \text{Im}\,(\cos\theta + i\sin\theta)^5$$
$$= \text{Im}\,(\cos^5\theta + 5i\cos^4\theta\sin\theta + 10i^2\cos^3\theta\sin^2\theta + 10i^3\cos^2\theta\sin^3\theta +$$
$$+ 5i^4\cos\theta\sin^4\theta + i^5\sin^5\theta)$$

which gives

$$\sin 5\theta = 5\cos^4\theta\sin\theta - 10\cos^2\theta\sin^3\theta + \sin^5\theta$$

Using $\cos^2\theta = 1 - \sin^2\theta$, we obtain

$$\sin 5\theta = 5(1 - \sin^2\theta)^2 \sin\theta - 10(1 - \sin^2\theta)\sin^3\theta + \sin^5\theta$$
$$\Rightarrow \quad \sin 5\theta = 16\sin^5\theta - 20\sin^3\theta + 5\sin\theta$$

b) From part **a**, we have

$$\frac{\sin 5\theta}{\sin\theta} = 16\sin^4\theta - 20\sin^2\theta + 5$$

When $\sin 5\theta = 0$, $16\sin^4\theta - 20\sin^2\theta + 5 = 0$, which gives

$$16x^4 - 20x^2 + 5 = 0$$

on substituting $x = \sin\theta$.

The solutions of $16x^4 - 20x^2 + 5 = 0$ are $x = \sin\theta$, where θ satisfies $\dfrac{\sin 5\theta}{\sin\theta} = 0$. All the x are different, and since $\sin 5\theta$ is divided by $\sin\theta$, we exclude the possible root $\sin\theta = 0$. Hence, we have

$$\sin 5\theta = 0 \quad \Rightarrow \quad \theta = 0 \text{ (excluded)}, \frac{\pi}{5}, \frac{2\pi}{5}, \frac{3\pi}{5}, \dots$$

which give the following values for $\sin\theta$:

$$\sin\left(\frac{\pi}{5}\right), \quad \sin\left(\frac{2\pi}{5}\right), \quad \sin\left(\frac{3\pi}{5}\right) \text{ which is the same as } \sin\left(\frac{2\pi}{5}\right),$$

$$\sin\left(\frac{4\pi}{5}\right) \text{ which is the same as } \sin\left(\frac{\pi}{5}\right),$$

$\sin\pi$ which is zero and hence excluded,

$$\sin\left(\frac{6\pi}{5}\right) \quad \text{and} \quad \sin\left(\frac{7\pi}{5}\right)$$

Therefore, the four **different** non-zero values of x for $16x^4 - 20x^2 + 5 = 0$ are

$$\sin\left(\frac{\pi}{5}\right) \quad \sin\left(\frac{2\pi}{5}\right) \quad \sin\left(\frac{6\pi}{5}\right) \quad \sin\left(\frac{7\pi}{5}\right)$$

c) We substitute $y = x^2$ to obtain the equation $16y^2 - 20y + 5 = 0$, whose roots are the two different values for y given by the substitution.

There are just two values of x^2, $\sin^2\left(\dfrac{\pi}{5}\right)$ and $\sin^2\left(\dfrac{2\pi}{5}\right)$, since

$$\sin\left(\frac{6\pi}{5}\right) = -\sin\left(\frac{\pi}{5}\right) \text{ and } \sin\left(\frac{7\pi}{5}\right) = -\sin\left(\frac{2\pi}{5}\right), \text{ which give}$$

$$\sin^2\left(\frac{6\pi}{5}\right) = \sin^2\left(\frac{\pi}{5}\right) \quad \text{and} \quad \sin^2\left(\frac{7\pi}{5}\right) = \sin^2\left(\frac{2\pi}{5}\right)$$

Therefore, the two different roots of the equation $16y^2 - 20y + 5 = 0$

are $y = \sin^2\left(\dfrac{\pi}{5}\right)$ and $y = \sin^2\left(\dfrac{2\pi}{5}\right)$.

i) Using the product of the roots of a polynomial (see page 147), we have for $16y^2 - 20y + 5 = 0$,

$$\alpha\beta = \frac{5}{16}$$

$$\Rightarrow \quad \sin^2\left(\frac{\pi}{5}\right)\sin^2\left(\frac{2\pi}{5}\right) = \frac{5}{16}$$

$$\Rightarrow \quad \sin\left(\frac{\pi}{5}\right)\sin\left(\frac{2\pi}{5}\right) = \pm\sqrt{\frac{5}{16}}$$

Since both $\sin\left(\dfrac{\pi}{5}\right)$ and $\sin\left(\dfrac{2\pi}{5}\right)$ are positive, we obtain

$$\sin\left(\frac{\pi}{5}\right)\sin\left(\frac{2\pi}{5}\right) = \frac{\sqrt{5}}{4}$$

ii) Since $16y^2 - 20y + 5 = 0$ is a quadratic equation, its roots are

$$y = \frac{20 \pm \sqrt{400 - 320}}{32}$$

$$\Rightarrow \quad y = \frac{20 \pm \sqrt{80}}{32} = \frac{5 \pm \sqrt{5}}{8}$$

Since these two roots are $\sin^2\left(\dfrac{\pi}{5}\right)$ and $\sin^2\left(\dfrac{2\pi}{5}\right)$, and

$\sin\left(\dfrac{2\pi}{5}\right) > \sin\left(\dfrac{\pi}{5}\right) > 0$, we have

$$\sin^2\left(\frac{\pi}{5}\right) = \frac{5 - \sqrt{5}}{8}$$

Using the identity $\cos\theta \equiv 1 - \sin^2\left(\dfrac{\theta}{2}\right)$, we obtain

$$\cos\left(\frac{2\pi}{5}\right) = 1 - 2\sin^2\left(\frac{\pi}{5}\right)$$

$$= 1 - 2 \times \frac{5 - \sqrt{5}}{8}$$

$$\Rightarrow \quad \cos\left(\frac{2\pi}{5}\right) = \frac{\sqrt{5} - 1}{4}$$

Exercise 15C

1 If $z = \cos\theta + i\sin\theta$, find the values of each of the following.

a) $z^2 - \dfrac{1}{z^2}$

b) $z^4 + \dfrac{1}{z^4}$

c) $z^5 + \dfrac{1}{z^5}$

d) $z^2 - \dfrac{2}{z} + \dfrac{2}{z} - \dfrac{1}{z^2}$

2 Express each of the following in terms of z, where $z = \cos\theta + i\sin\theta$.

a) $\cos 6\theta$

b) $\sin 5\theta$

c) $\cos^4\theta$

d) $\sin^3\theta$

e) $\sin^2 5\theta$

f) $\cos^4 3\theta$

3 Express each of the following in terms of $\cos\theta$.

a) $\cos 6\theta$

b) $\cos 4\theta$

c) $\dfrac{\sin 4\theta}{\sin\theta}$

d) $\dfrac{\sin 6\theta}{\sin\theta}$

4 Express each of the following in terms of $\sin\theta$.

a) $\sin 3\theta$

b) $\sin 5\theta$

c) $\dfrac{\cos 7\theta}{\cos\theta}$

d) $\dfrac{\cos 5\theta}{\cos\theta}$

5 Express each of the following in terms of sines or cosines of multiple angles.

a) $\sin^3\theta$

b) $\cos^3\theta$

c) $\cos^5\theta$

d) $\sin^5\theta$

e) $\cos^6\theta$

6 Prove that $\cos^4\theta = \dfrac{1}{8}(\cos 4\theta + 4\cos 2\theta + 3)$.

7 Prove that $\tan 3\theta = \dfrac{3\tan\theta - \tan^3\theta}{1 - 3\tan^2\theta}$. Hence solve $t^3 - 3t^2 - 3t + 1 = 0$.

8 By considering $(\cos\theta + i\sin\theta)^3$, use de Moivre's theorem to establish the identity

$$\cos 3\theta \equiv 4\cos^3\theta - 3\cos\theta$$

Write down the coefficient of θ^4 in the series expansion of $\cos 3\theta$.

Hence, using the identity above, obtain the coefficient of θ^4 in the series expansion of $\cos^3\theta$.

(AEB 96)

9 i) Show that $(2 + i)^4 = -7 + 24i$.

ii) Use de Moivre's theorem to show that

$$\cos 4\theta = \cos^4\theta - 6\cos^2\theta\sin^2\theta + \sin^4\theta$$

and $\quad \sin 4\theta = 4\sin\theta\cos^3\theta - 4\sin^3\theta\cos\theta$

iii) If $t = \tan\theta$, show that

$$\tan 4\theta = \dfrac{4t - 4t^3}{1 - 6t^2 + t^4}$$

iv) By considering the argument of $(2 + i)$, explain why $t = \frac{1}{2}$ is a root of the following equation

$$\dfrac{4t - 4t^3}{1 - 6t^2 + t^4} = -\dfrac{24}{7}$$

v) Using the symmetry properties of the four roots of $z^4 = a^4$, draw an Argand diagram showing the four roots of $z^4 = -7 + 24i$.

vi) Find one other root of the equation in part **iv**.   (NICCEA)

10 Use de Moivre's theorem to prove that

$$\sin 6\theta = 6\cos^5\theta \sin\theta - 20\cos^3\theta \sin^3\theta + 6\cos\theta \sin^5\theta$$

By putting $x = \sin\theta$, deduce that, for $|x| \leqslant 1$,

$$-\tfrac{1}{2} \leqslant x(16x^4 - 16x^2 + 3)\sqrt{(1 - x^2)} \leqslant \tfrac{1}{2}   \text{(OCR)}$$

11 Use de Moivre's theorem to prove that

$$\cos 5\theta = \cos\theta(16\cos^4\theta - 20\cos^2\theta + 5)$$

By considering the equation $\cos 5\theta = 0$, show that the exact value of $\cos^2\left(\tfrac{1}{10}\pi\right)$ is $\dfrac{5 + \sqrt{5}}{8}$.

(OCR)

12 Use de Moivre's theorem to show that

$$\tan 5\theta = \frac{5t - 10t^3 + t^5}{1 - 10t^2 + 5t^4}$$

where $t = \tan\theta$.   (OCR)

13 Let $z = \cos\theta + i\sin\theta$.

a) Use the binomial theorem to show that the real part of z^4 is

$$\cos^4\theta - 6\cos^2\theta \sin^2\theta + \sin^4\theta$$

Obtain a similar expression for the imaginary part of z^4 in terms of θ.

b) Use de Moivre's theorem to write down an expression for z^4 in terms of 4θ.

c) Use your answers to parts **a** and **b** to express $\cos 4\theta$ in terms of $\cos\theta$ and $\sin\theta$.

d) Hence show that $\cos 4\theta$ can be written in the form $k(\cos^m\theta - \cos^n\theta) + p$, where k, m, n, p are integers. State the values of k, m, n, p.   (SQA/CSYS)

14 Use de Moivre's theorem to show that

$$\sin 5\theta = a\cos^4\theta \sin\theta + b\cos^2\theta \sin^3\theta + c\sin^5\theta$$

where a, b and c are integers to be determined.

Hence show that

$$\frac{\sin 5\theta}{\sin\theta} = 16\cos^4\theta - 12\cos^2\theta + 1   (\theta \neq k\pi, \text{ where } k \in \mathbb{Z})$$

By means of the substitution $x = 2\cos\theta$, find, in trigonometric form, the roots of the equation

$$x^4 - 3x^2 + 1 = 0$$

Hence, or otherwise, show that

$$\cos^2\left(\tfrac{1}{5}\pi\right) + \cos^2\left(\tfrac{2}{5}\pi\right) = \tfrac{3}{4}   \text{(NEAB)}$$

15 Find each of the roots of the equation $z^5 - 1 = 0$ in the form $r(\cos\theta + i\sin\theta)$, where $r > 0$ and $-\pi < \theta \leqslant \pi$.

a) Given that α is the complex root of this equation with the smallest positive argument, show that the roots of $z^5 - 1 = 0$ can be written as $1, \alpha, \alpha^2, \alpha^3, \alpha^4$.

b) Show that $\alpha^4 = \alpha^*$ and hence, or otherwise, obtain $z^5 - 1$ as a product of real linear and quadratic factors, giving the coefficients in terms of integers and cosines.

c) Show also that

$$z^5 - 1 = (z - 1)(z^4 + z^3 + z^2 + z + 1)$$

and hence, or otherwise, find $\cos\left(\frac{2}{5}\pi\right)$, giving your answer in terms of surds. (EDEXCEL)

16 a) Use mathematical induction to prove that when n is a positive integer

$$(\cos\theta + i\sin\theta)^n = \cos n\theta + i\sin n\theta$$

b) Hence show that

$$\sin 5\theta = 16\sin^5\theta - 20\sin^3\theta + 5\sin\theta \qquad \text{(EDEXCEL)}$$

17 In the polynomial equation

$$a_n z^n + a_{n-1} z^{n-1} + \ldots + a_0 = 0$$

all the coefficients $a_n, a_{n-1}, \ldots, a_0$ are real. Given that $x + iy$ is a root of the equation, show that the complex conjugate $x - iy$ is also a root.

Show that $e^{\pi i/6}$ is one root of the equation $z^3 = i$. Find the other two roots and mark on an Argand diagram the points representing the three roots. Show that these three roots are also roots of the equation

$$z^6 + 1 = 0$$

and write down the remaining three roots of this equation. Hence, or otherwise, express $z^6 + 1$ as the product of three quadratic factors each with coefficients in integer or surd form.

(NEAB)

Transformations in a complex plane

We need to be able to transform simple loci in a complex plane, such as straight lines and circles, into new loci, which are again usually straight lines and circles.

The method we usually use is to identify the general point on the original locus and find its image.

Example 23 Under the transformation $w = z^2$, find the image of

a) circle, centre O, radius 3, and

b) line $\arg z = \dfrac{\pi}{2}$.

SOLUTION

The original locus is $z \equiv x + iy$, and the new locus is $w \equiv u + iv$.

a) The general point on the original circle is $z = 3e^{i\theta}$, or $z = 3\cos\theta + 3i\sin\theta$. Its image point is $w = z^2 = 9e^{2i\theta}$, or $z = 9(\cos 2\theta + i\sin 2\theta)$. Therefore, the locus of the image is a circle, centre O, radius 9.

b) The general point on the line $\arg z = \dfrac{\pi}{2}$ is

$$z = re^{i\pi/2} = r\left[\cos\left(\frac{\pi}{2}\right) + i\sin\left(\frac{\pi}{2}\right)\right]$$

$$\Rightarrow \quad z = ir$$

Hence, we have for the image point
$$w = z^2 = r^2 e^{i\pi} \quad \text{or} \quad -r^2$$

Therefore, the locus of the image is a line along the real axis in the negative direction from O.

Example 24 Find the image under the transformation $w = \dfrac{2+z}{i-z}$, where z is the circle $|z| = 1$.

SOLUTION

To find the image of $|z| = k$, we usually express z in terms of w and then apply this expression to $|z| = k$.

Hence, we have
$$w(i - z) = 2 + z$$
$$\Rightarrow \quad wi - 2 = (1 + w)z$$
$$\Rightarrow \quad z = \frac{wi - 2}{1 + w}$$

Applying this to $|z| = 1$, we obtain

$$|wi - 2| = |1 + w|$$

Now $|wi - 2| = |i| \, |w + 2i|$, thus we have

$$|w + 2i| = |1 + w| \quad \text{(since } |i| = 1\text{)}$$

Therefore, the locus is the perpendicular bisector
of the line joining $-2i$ to -1 (see page 13).

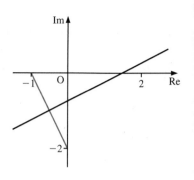

Example 25 Find the image of a circle, centre O, radius 1, under the
transformation $w = \dfrac{1}{1 - z}$.

SOLUTION

The general point on the original circle is $z = e^{i\theta}$ or $z = \cos\theta + i\sin\theta$.
Hence, we have

$$w = \frac{1}{1 - e^{i\theta}} = \frac{1}{1 - \cos\theta - i\sin\theta}$$

Note Do **not** use

$$w = -\frac{1}{1 - e^{i\theta}} = \frac{1 - e^{-i\theta}}{(1 - e^{i\theta})(1 - e^{-i\theta})}$$

as $1 - e^{-i\theta}$ is **not** the complex conjugate of $1 - e^{i\theta}$.

Multiplying both the numerator and the denominator by
$1 - \cos\theta + i\sin\theta$, we obtain

$$w = \frac{1}{1 - \cos\theta - i\sin\theta} = \frac{1 - \cos\theta + i\sin\theta}{(1 - \cos\theta - i\sin\theta)(1 - \cos\theta + i\sin\theta)}$$

$$= \frac{1 - \cos\theta + i\sin\theta}{(1 - \cos\theta)^2 + \sin^2\theta}$$

Using $\cos^2\theta + \sin^2\theta = 1$, we have

$$w = \frac{1 - \cos\theta + i\sin\theta}{2 - 2\cos\theta}$$

$$= \frac{1}{2} + \frac{i\sin\theta}{2 - 2\cos\theta}$$

Using the half-angle identities for $\sin\theta$ and $\cos\theta$, we obtain

$$w = \frac{1}{2} + \frac{i}{2}\cot\left(\frac{\theta}{2}\right)$$

which gives $u = \frac{1}{2}$, since $w = u + iv$.

Therefore, the locus of w is the straight line, $u = \frac{1}{2}$.

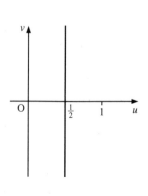

Example 26 Find the image of $|z| = 2$ under the transformation $w = 2z - \dfrac{3}{z}$.

SOLUTION

The general point on the original circle is $z = 2e^{i\theta}$, or $z = 2\cos\theta + 2i\sin\theta$. Hence, we have

$$w = 4\cos\theta + 4i\sin\theta - \frac{3}{2(\cos\theta + i\sin\theta)}$$

$$= 4\cos\theta + 4i\sin\theta - \frac{3}{2}(\cos\theta - i\sin\theta)$$

which gives

$$u + iv = \frac{5}{2}\cos\theta + \frac{11}{2}i\sin\theta$$

$$\Rightarrow \quad u = \frac{5}{2}\cos\theta \quad \text{and} \quad v = \frac{11}{2}\sin\theta$$

Eliminating $\cos\theta$ and $\sin\theta$, we obtain

$$\left(\frac{2u}{5}\right)^2 + \left(\frac{2v}{11}\right)^2 = 1$$

$$\Rightarrow \quad \frac{4u^2}{25} + \frac{4v^2}{121} = 1$$

Therefore, the image is an ellipse with the above equation.

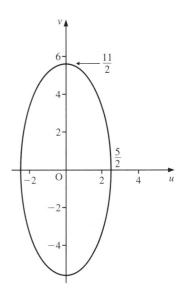

Example 27 Find the image of $|z - 7| = 7$ under the transformation $w = \dfrac{28}{z} \; (z \neq 0)$.

SOLUTION

The general point of $|z - 7| = 7$ is $z = 7 + 7\cos\theta + 7i\sin\theta$. Hence, we have

$$w = \frac{28}{7 + 7\cos\theta + 7i\sin\theta}$$

$$= \frac{4}{1 + \cos\theta + i\sin\theta}$$

$$= \frac{4(1 + \cos\theta - i\sin\theta)}{(1 + \cos\theta + i\sin\theta)(1 + \cos\theta - i\sin\theta)}$$

$$= \frac{4(1 + \cos\theta - i\sin\theta)}{(1 + \cos\theta)^2 + \sin^2\theta}$$

$$= \frac{4(1 + \cos\theta - i\sin\theta)}{2 + 2\cos\theta} = 2 - \frac{4i\sin\theta}{2 + 2\cos\theta}$$

Using the half-angle identities for $\sin\theta$ and $\cos\theta$, we obtain

$$u + iv = 2 - 2i\tan\left(\frac{\theta}{2}\right)$$

which gives $u = 2$.

This line, $u = 2$, is $|w - 4| = |w|$, which is the perpendicular bisector of the line joining 0 and 4.

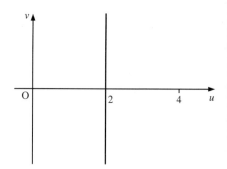

Note We could have found the image from

$$|z - 7| = 7 \quad \Rightarrow \quad \left|\frac{28}{w} - 7\right| = 7$$

$$|28 - 7w| = 7|w|$$

which gives $|w| = |w - 4|$, as required.

Example 28 Find the image of the straight line $3x + 2y = 8$ under the transformation $w = \dfrac{1}{2 - z}$.

SOLUTION

We start by expressing $z \equiv x + iy$ in terms of $w = u + iv$:

$$w = \frac{1}{2 - z} \quad \Rightarrow \quad z = 2 - \frac{1}{w}$$

$$\Rightarrow \quad x + iy = 2 - \frac{1}{u + iv}$$

$$= 2 - \frac{u - iv}{u^2 + v^2}$$

Hence, we have

$$x = 2 - \frac{u}{u^2 + v^2} \qquad y = \frac{v}{u^2 + v^2}$$

Using these values in the equation of the line, $3x + 2y = 8$, we obtain

$$3\left(2 - \frac{u}{u^2 + v^2}\right) + 2\left(\frac{v}{u^2 + v^2}\right) = 8$$

$$\Rightarrow \quad 6(u^2 + v^2) - 3u + 2v = 8(u^2 + v^2)$$

$$\Rightarrow \quad u^2 + v^2 + \frac{3}{2}u - v = 0$$

which gives

$$\left(u + \frac{3}{4}\right)^2 + \left(v - \frac{1}{2}\right)^2 = \frac{13}{16}$$

This is the equation of a circle, centre $(-\frac{3}{4}, \frac{1}{2})$ or $-\frac{3}{4} + \frac{1}{2}i$, radius $\frac{1}{4}\sqrt{13}$.

Exercise 15D

1 For the transformation $w = z^2$, find the locus of w when

   a) z lies on a circle centre O, radius 5
   b) z lies on the real axis
   c) z lies on the imaginary axis.

2 For the transformation $w^2 = z$, find the locus of w when

   a) z lies on a circle centre O, radius 5
   b) z lies on a circle centre O, radius 2
   c) z lies on the imaginary axis.

3 For the transformation $w = z^2$, show that the locus of w, when z moves along a line $y = k$, is a parabola. Find its equation.

4 For the transformation $w = \dfrac{z + i}{iz + 2}$, find

   a) the locus of w when z lies on the real axis
   b) the locus of w when z lies on the imaginary axis
   c) any invariant points.

5 For the transformation $w = 3z + 2i - 5$, find the locus of w for $|z| = 4$.

6 a) For the transformation $w = \dfrac{az + b}{z + c}$, where $a, b, c \in \mathbb{R}$, find a, b and c given that $w = 3i$ when $z = -3i$, and $w = 1 - 4i$ when $z = 1 + 4i$.
   b) Show that the points for which $w = \bar{z}$ lie on a circle. Find its centre and radius.

7 Find the image under the transformation $w = \dfrac{3i + z}{2 - z}$, where z is the circle $|z| = 3$.

8 Find the image of $|z| = 3$ under the transformation $w = 3z + \dfrac{4}{z}$.

9 Find the image of $|z - 5| = 5$ under the transformation $w = \dfrac{30}{z}$ $(z \neq 0)$.

10 The point P in the Argand diagram represents the complex number z.

   a) Given that $|z| = 1$, sketch the locus of P.

   The point Q is the image of P under the transformation

   $$w = \dfrac{1}{z - 1}$$

   b) Given that $z = e^{i\theta}$, $0 < \theta < 2\pi$, show that $w = -\tfrac{1}{2} - \tfrac{1}{2}i \cot \tfrac{1}{2}\theta$
   c) Make a separate sketch of the locus of Q.   (EDEXCEL)

11 i) Solve the equation $z^3 + 8i = 0$, giving your answers in the form $re^{i\theta}$, where $r > 0$ and $-\pi \leqslant \theta < \pi$.

ii) The point P represents the complex number z in an Argand diagram. Given that $|z - 3i| = 2$,

a) sketch the locus of P in an Argand diagram.

Transformations T_1, T_2 and T_3 from the z-plane to the w-plane are given by

$$T_1: \quad w = iz$$
$$T_2: \quad w = 3z$$
$$T_3: \quad w = z^*$$

b) Describe precisely the locus of the image of P under each of these transformations.

(EDEXCEL)

12 A transformation T from the z-plane to the w-plane is given by

$$w = \frac{z + 1}{z - 1} \quad z \neq 1$$

Find the image in the w-plane of the circle $|z| = 1$, $z \neq 1$, under the transformation T.

(EDEXCEL)

13 The transformation, T, from the z-plane to the w-plane is given by

$$w = \frac{1}{z - 2} \quad z \neq 2$$

where $z = x + iy$ and $w = u + iv$.

Show that under T the straight line with equation $2x + y = 5$ is transformed to a circle in the w-plane with centre $(1, -\frac{1}{2})$ and radius $\frac{1}{2}\sqrt{5}$.   (EDEXCEL)

14 The complex numbers z and w are defined by

$$z = e^{(1 + 2i)\phi} \quad \text{and} \quad w = \frac{z}{1 + i}$$

where ϕ is real.

a) i) Show that $|z| = e^{\phi}$ and $\arg z = 2\phi$.

ii) In an Argand diagram, z is represented by the point P. Sketch the locus of P when ϕ varies from 0 to π.

b) i) Show that the imaginary part of w is

$$\frac{1}{2}e^{\phi}(\sin 2\phi - \cos 2\phi)$$

ii) Determine the values of ϕ in the interval $0 \leqslant \phi \leqslant \pi$ for which w is real.   (NEAB)

15 Given that $z = x + iy$ and $w = u + iv$ are complex numbers related by $w = \frac{1}{z} + 1$, obtain expressions for u and v in terms of x and y.

The complex numbers z and w are represented by the points P and Q respectively in the Argand diagram. Given that P moves along the line $y = 2x$, show that Q moves along the line $2u + v - 2 = 0$.   (WJEC)

16 Intrinsic coordinates

It is no paradox to say that in our most theoretical moods we may be nearest to our practical applications.
ALFRED NORTH WHITEHEAD

We have already seen that the position of a point on a curve (and hence the curve's equation) may be given in terms of:

- cartesian coordinates (x, y), or
- polar coordinates (r, θ) (see pages 43–56).

We can also define the position of a point on a curve by means of **intrinsic coordinates** (s, ψ), where s is the length of the arc from a fixed point to the given point, and ψ is the angle which the tangent to the curve at that point makes with the x-axis.

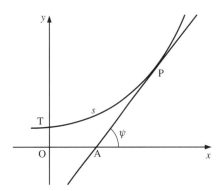

Thus, referring to the figure on the right, intrinsic coordinates would give the position of point P in terms of the arc length PT and the angle which PA makes with Ox.

We must stress, however, that the majority of the equations of curves cannot realistically be given in intrinsic form. Also, only in rare cases is it sensible to try to convert the cartesian, parametric or polar equation of a curve to its intrinsic form.

But two curves in particular are more readily treated in their intrinsic forms. They are the **catenary** (see Example 2, on pages 365–6) and the **cycloid** (see Example 3, on pages 366–7).

Trigonometric functions of ψ

Considering the gradient of a tangent, we have

$$\frac{\mathrm{d}y}{\mathrm{d}x} = \tan \psi$$

When we derived the length of the arc of a curve (see pages 250–3), we found that

$$\frac{\mathrm{d}s}{\mathrm{d}x} = \sqrt{1 + \left(\frac{\mathrm{d}y}{\mathrm{d}x} \right)^2}$$

$$\Rightarrow \quad \frac{\mathrm{d}s}{\mathrm{d}x} = \sqrt{1 + \tan^2 \psi}$$

Using the identity $1 + \tan^2 \psi \equiv \sec^2 \psi$, we obtain

$$\frac{\mathrm{d}s}{\mathrm{d}x} = \sec \psi$$

$$\Rightarrow \quad \cos \psi = \frac{\mathrm{d}x}{\mathrm{d}s}$$

Using $\sin \psi = \tan \psi \cos \psi$, we have

$$\sin \psi = \frac{\mathrm{d}y}{\mathrm{d}x} \frac{\mathrm{d}x}{\mathrm{d}s}$$

$$\Rightarrow \quad \sin \psi = \frac{\mathrm{d}y}{\mathrm{d}s}$$

Radius of curvature

Let P and Q be points on the curve with intrinsic coordinates (s, ψ) and $(s + \delta s, \psi + \delta \psi)$ respectively. Hence, δs is the length of PQ.

If δs is sufficiently small, we may assume that PQ is a segment of a circle.

If C is the centre of the circle passing through P and Q, then the angle PCQ is $\delta \psi$.

Let ρ be the radius of curvature at P. Hence, the length of PQ is $\rho \delta \psi$. That is,

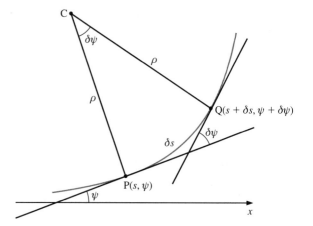

$$\delta s = \rho \delta \psi \quad \Rightarrow \quad \rho = \frac{\delta s}{\delta \psi}$$

As $\delta s \to 0$, this gives

$$\text{Radius of curvature} = \rho = \frac{\mathrm{d}s}{\mathrm{d}\psi}$$

To find the radius of curvature in terms of x and y, we need to differentiate $\frac{\mathrm{d}y}{\mathrm{d}x} = \tan \psi$ with respect to x, which gives

$$\frac{\mathrm{d}^2 y}{\mathrm{d}x^2} = \frac{\mathrm{d}}{\mathrm{d}x}(\tan \psi)$$

$$\Rightarrow \quad \frac{\mathrm{d}^2 y}{\mathrm{d}x^2} = \frac{\mathrm{d}}{\mathrm{d}\psi}(\tan \psi) \frac{\mathrm{d}\psi}{\mathrm{d}x} = (\sec^2 \psi) \frac{\mathrm{d}\psi}{\mathrm{d}x}$$

$$\Rightarrow \quad \frac{\mathrm{d}x}{\mathrm{d}\psi} = \frac{\sec^2 \psi}{\dfrac{\mathrm{d}^2 y}{\mathrm{d}x^2}}$$

Using $\rho = \dfrac{ds}{d\psi} = \dfrac{ds}{dx}\dfrac{dx}{d\psi}$ and substituting for $\dfrac{ds}{dx}$ and $\dfrac{dx}{d\psi}$, we have

$$\rho = \sqrt{1 + \left(\frac{dy}{dx}\right)^2 \left(\frac{\sec^2\psi}{\dfrac{d^2y}{dx^2}}\right)}$$

$$\Rightarrow \quad \rho = \frac{\sqrt{1 + \left(\dfrac{dy}{dx}\right)^2}(1 + \tan^2\psi)}{\dfrac{d^2y}{dx^2}} = \frac{\sqrt{1 + \left(\dfrac{dy}{dx}\right)^2}\left[1 + \left(\dfrac{dy}{dx}\right)^2\right]}{\dfrac{d^2y}{dx^2}}$$

which gives

$$\rho = \frac{\left[1 + \left(\dfrac{dy}{dx}\right)^2\right]^{\frac{3}{2}}}{\dfrac{d^2y}{dx^2}} \qquad [1]$$

When x and y are given in terms of a parameter t, we can find $\dfrac{dx}{dt}$ and $\dfrac{dy}{dt}$, and hence $\dfrac{dy}{dx}$, in terms of t.

We have

$$\frac{d^2y}{dx^2} = \frac{d}{dx}\left(\frac{dy}{dx}\right)$$

$$\Rightarrow \quad \frac{d^2y}{dx^2} = \frac{d}{dt}\left(\frac{dy}{dx}\right)\frac{dt}{dx}$$

$$\Rightarrow \quad \frac{d^2y}{dx^2} = \frac{d}{dt}\left(\frac{dy}{dt}\frac{dt}{dx}\right)\frac{dt}{dx}$$

Using $\dot{y}$ and $\dot{x}$ to indicate that we have differentiated with respect to t (see page 252), we have

$$\dot{y} \equiv \frac{dy}{dt} \quad \text{and} \quad \ddot{y} \equiv \frac{d^2y}{dt^2}$$

$$\dot{x} \equiv \frac{dx}{dt} \quad \text{and} \quad \ddot{x} \equiv \frac{d^2x}{dt^2}$$

which give

$$\frac{d^2y}{dx^2} = \frac{d}{dt}\left(\frac{\dot{y}}{\dot{x}}\right)\frac{1}{\dot{x}}$$

Remembering that $\dfrac{\dot{y}}{\dot{x}}$ is a quotient, we find

$$\frac{d^2y}{dx^2} = \left(\frac{\ddot{y}\dot{x} - \ddot{x}\dot{y}}{\dot{x}^2}\right)\frac{1}{\dot{x}} = \frac{\ddot{y}\dot{x} - \ddot{x}\dot{y}}{\dot{x}^3}$$

Substituting this expression in [1], we obtain

$$\rho = \frac{\left[1 + \left(\dfrac{\dot{y}}{\dot{x}}\right)^2\right]^{\frac{3}{2}}}{\dfrac{\ddot{y}\dot{x} - \ddot{x}\dot{y}}{\dot{x}^3}} \qquad \Rightarrow \qquad \rho = \frac{(\dot{x}^2 + \dot{y}^2)^{\frac{3}{2}}}{\ddot{y}\dot{x} - \ddot{x}\dot{y}}$$

Thus, the radius of curvature is given by

$$\rho = \frac{\left[1 + \left(\dfrac{dy}{dx}\right)^2\right]^{\frac{3}{2}}}{\dfrac{d^2y}{dx^2}} \qquad \text{or} \qquad \rho = \frac{(\dot{x}^2 + \dot{y}^2)^{\frac{3}{2}}}{\ddot{y}\dot{x} - \ddot{x}\dot{y}} \qquad \text{or} \qquad \rho = \frac{ds}{d\psi}$$

Example 1 Find the radius of curvature of the rectangular hyperbola $y = \dfrac{16}{x}$, given that its parametric coordinates are $x = 4t$, $y = \dfrac{4}{t}$.

SOLUTION

Method 1
We use the recommended method of staying in the parametric form throughout. Hence, we have

$$\dot{x} = 4 \quad \Rightarrow \quad \ddot{x} = 0 \qquad \dot{y} = -\frac{4}{t^2} \quad \Rightarrow \quad \ddot{y} = \frac{8}{t^3}$$

Substituting for $\dot{x}$, $\ddot{x}$, $\dot{y}$ and $\ddot{y}$ in

$$\rho = \frac{(\dot{x}^2 + \dot{y}^2)^{\frac{3}{2}}}{\ddot{y}\dot{x} - \ddot{x}\dot{y}}$$

where ρ is the radius of curvature, we obtain

$$\rho = \frac{t^3\left(16 + \dfrac{16}{t^4}\right)^{\frac{3}{2}}}{32}$$

$$\Rightarrow \quad \rho = 2t^3\left(1 + \frac{1}{t^4}\right)^{\frac{3}{2}}$$

Method 2
We could use the cartesian form, which readily gives $\dfrac{dy}{dx}$ but from which $\dfrac{d^2y}{dx^2}$ is rather more difficult to obtain, as the following shows.

We have

$$x = 4t \quad \Rightarrow \quad \frac{dx}{dt} = 4 \qquad y = \frac{4}{t} \quad \Rightarrow \quad \frac{dy}{dt} = -\frac{4}{t^2}$$

which give

$$\frac{dy}{dx} = \frac{dy}{dt}\frac{dt}{dx} = -\frac{4}{t^2} \times \frac{1}{4} \quad \Rightarrow \quad \frac{dy}{dx} = -\frac{1}{t^2}$$

Differentiating again, we obtain

$$\frac{d^2y}{dx^2} = \frac{d}{dx}\left(-\frac{1}{t^2}\right)$$

$$\Rightarrow \quad \frac{d^2y}{dx^2} = \frac{d}{dt}\left(-\frac{1}{t^2}\right)\frac{dt}{dx} = \frac{2}{t^3} \times \frac{1}{4} = \frac{1}{2t^3}$$

Substituting for $\dfrac{dy}{dx}$ and $\dfrac{d^2y}{dx^2}$ in

$$\rho = \frac{\left[1 + \left(\dfrac{dy}{dx}\right)^2\right]^{\frac{3}{2}}}{\dfrac{d^2y}{dx^2}}$$

where ρ is the radius of curvature, we have

$$\rho = 2t^3\left(1 + \frac{1}{t^4}\right)^{\frac{3}{2}}$$

Finding intrinsic equations

Example 2 $y = \cosh x$ passes through the point $(0, 1)$. Find the intrinsic equation of the curve.

SOLUTION

We know that

$$s = \int \sqrt{1 + \left(\frac{dy}{dx}\right)^2}\, dx$$

which gives

$$s = \int \sqrt{1 + \sinh^2 x}\, dx$$

$$\Rightarrow \quad s = \int \cosh x\, dx = \sinh x + c$$

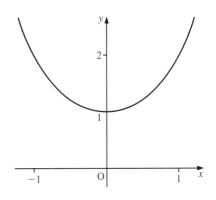

When $x = 0$, $s = 0$, which gives $c = 0$. Hence, we find

$$s = \sinh x$$

Now $\dfrac{dy}{dx} = \sinh x$. So, using $\tan \psi = \dfrac{dy}{dx}$ (see page 361), we obtain

$$\sinh x = \tan \psi$$

Since $s = \sinh x$, we have

$$s = \tan \psi$$

This is an equation with s and ψ as the only variables. Therefore, the intrinsic equation of $y = \cosh x$ is $s = \tan \psi$.

The curve $y = \cosh x$ (which we met on pages 189 and 190) is a **catenary**.

The catenary is the form assumed by a uniform, heavy and flexible cable hanging freely between two points. An example is a slack mooring line between a ship and a quay. In large suspension bridges, where heavy cables are used, the curve assumed by the cables is sometimes close to a catenary.

The standard intrinsic equation of the catenary is $s = a \tan \psi$, where a is the y-intercept, corresponding to the standard cartesian equation $y = a \cosh\left(\dfrac{x}{a}\right)$.

Another curve of practical interest (for example, as the flank profile of the teeth of certain gear wheels) is the **cycloid**. This is the locus of a point fixed on the circumference of a circle which is rolling along a stationary, straight base-line, as shown below.

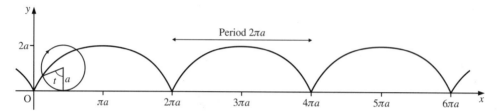

Note that the distance between successive cusps is $2\pi a$, where a is the radius of the rolling circle. Hence, the catenary is **periodic** with period $2\pi a$.

The cartesian equation of the cycloid is difficult to derive, hence we normally work with its parametric equations

$$x = a(t - \sin t) \quad \text{and} \quad y = a(1 - \cos t)$$

where t is the central angle of the circle, as shown in the figure.

Example 3 Find the intrinsic equation of the cycloid.

SOLUTION

We know that

$$s = \int \sqrt{\left(\frac{dx}{dt}\right)^2 + \left(\frac{dy}{dt}\right)^2}\, dt$$

Differentiating the parametric equations for the cycloid and substituting them in the above, we obtain

$$s = \int \sqrt{a^2(1 - \cos t)^2 + a^2 \sin^2 t}\, dt$$

$$\Rightarrow \quad s = a \int \sqrt{2 - 2\cos t}\, dt$$

Using $\cos t \equiv 1 - 2\sin^2\left(\dfrac{t}{2}\right)$, we have

$$s = a \int \sqrt{2 - 2\left[1 - 2\sin^2\left(\frac{t}{2}\right)\right]}\, dt = a \int 2\sin\left(\frac{t}{2}\right) dt$$

which gives

$$s = -4a \cos\left(\frac{t}{2}\right) + c \qquad [1]$$

Using

$$\tan \psi = \frac{dy}{dx} = \frac{dy}{dt} \frac{dt}{dx}$$

we obtain

$$\tan \psi = \frac{\sin t}{1 - \cos t}$$

Using $\sin t \equiv 2 \sin \left(\frac{t}{2}\right) \cos \left(\frac{t}{2}\right)$ and $\cos t \equiv 1 - 2 \sin^2 \left(\frac{t}{2}\right)$, we have

$$\tan \psi = \frac{2 \sin \left(\frac{t}{2}\right) \cos \left(\frac{t}{2}\right)}{1 - \left[1 - 2 \sin^2 \left(\frac{t}{2}\right)\right]} = \frac{2 \sin \left(\frac{t}{2}\right) \cos \left(\frac{t}{2}\right)}{2 \sin^2 \left(\frac{t}{2}\right)}$$

which gives

$$\tan \psi = \cot \left(\frac{t}{2}\right) = \tan \left(\frac{\pi}{2} - \frac{t}{2}\right)$$

$$\Rightarrow \quad \psi = \frac{\pi}{2} - \frac{t}{2} \quad \Rightarrow \quad \frac{t}{2} = \frac{\pi}{2} - \psi$$

Substituting for $\frac{t}{2}$ in [1], we have

$$s = c - 4a \cos \left(\frac{\pi}{2} - \psi\right)$$

Therefore, the intrinsic equation of the cycloid is

$$s = c - 4a \sin \psi$$

The value of c will be different for each arch of the cycloid.

Exercise 16

In Questions **1** to **8**, find the radius of curvature of each curve at the point specified.

1 $y^2 = x^3 + 3$, at $(1, 2)$.

2 $y = e^x$, at $(1, e)$.

3 $y = \sin x$, when $x = \frac{\pi}{3}$.

4 $y = x \ln x$, at $(1, 0)$.

5 $x = t^3$, $y = t^2$, when $t = 1$.

6 $x = ct$, $y = \frac{c}{t}$, when $t = 2$.

7 $x = \cos^2 t$, $y = \sin^2 t$, when $t = \frac{\pi}{4}$.

8 $x = a \cos^3 t$, $y = a \sin^3 t$, when $t = \frac{\pi}{3}$.

9 Find the radius of curvature, in terms of ψ, for

    a) $s = \psi^3 + \cos \psi$ **b)** $s = 3\psi + 4\psi \sin \psi$ **c)** $s = \psi \cos \psi + \psi^2$

10 Find the intrinsic equation of the curve $y = \ln \sec x$, where s is the distance from the origin.

11 A curve has intrinsic equation $s = a \cos \psi$.

   a) Calculate the radius of curvature of the curve in terms of ψ.
   b) Show that the tangent to the curve at the point where $s = 0$ is parallel to the y-axis.

(EDEXCEL)

12 The curve C has equation $y = 3 \cosh\left(\dfrac{x}{3}\right)$.

   a) Show that the radius of curvature, at the point on C where $x = t$, is $3 \cosh^2\left(\dfrac{t}{3}\right)$.

   b) Find the radius of curvature at the point where $t = 1.5$, giving your answer to three significant figures.
   c) Find the area of the surface generated when the arc of C between $x = -3$ and $x = 3$ is rotated through 2π radians about the x-axis, giving your answer in terms of e and π.

(EDEXCEL)

13 A curve has parametric equations $x = 4t - \frac{1}{3}t^3$, $y = 2t^2 - 8$.

   i) Show that the radius of curvature at a general point $(4t - \frac{1}{3}t^3, 2t^2 - 8)$ on the curve is $\frac{1}{4}(4 + t^2)^2$.
   ii) Find the centre of curvature corresponding to the point on the curve given by $t = 3$.

   The arc of the curve given by $0 \leqslant t \leqslant 2\sqrt{3}$ is denoted by C.

   iii) Find the length of the arc C.
   iv) Find the area of the curved surface generated when the arc C is rotated about the y-axis.

(MEI)

14 A curve is given parametrically by $x = e^\theta(2 \sin 2\theta + \cos 2\theta)$, $y = e^\theta(\sin 2\theta - 2 \cos 2\theta)$. P is the point corresponding to $\theta = 0$, and Q is the point corresponding to $\theta = \alpha$ (where $\alpha > 0$).

   i) Show that the gradient of the curve at Q is $\tan 2\alpha$, and find the length of the arc of the curve between P and Q.
   ii) Using intrinsic coordinates (s, ψ), where s is the arc length of the curve measured from P and $\tan \psi = \dfrac{dy}{dx}$, show that $s = 5(e^{\frac{1}{2}\psi} - 1)$.

   iii) Find the radius of curvature at the point Q.
   iv) Show that the centre of curvature corresponding to the point Q is

   $$\left(\tfrac{1}{2}e^\alpha(2 \cos 2\alpha - \sin 2\alpha), \tfrac{1}{2}e^\alpha(2 \sin 2\alpha + \cos 2\alpha)\right) \qquad \text{(MEI)}$$

17 Groups

Before the word 'group' appeared in the mathematical literature, there had been a longer period of development in which mathematicians applied group-theoretical results without the concept of a group being explicitly defined.
WALTER PURKERT AND HANS WUSSING

Binary and unary operations

A **binary operation**, usually denoted by $*$, is a rule which takes an ordered pair of elements, a and b, and gives a uniquely defined third element, c, so that $a * b = c$. (Other symbols used to represent a binary operation include $\bigcirc$, $\otimes$ and $\oplus$.)

For example, multiplication is a binary operation. If we represent $*$ by multiplication, then

$$4 * 3 = 4 \times 3 = 12$$

Addition is also a binary operation. If we represent $*$ by addition, then

$$6 * 3 = 6 + 3 = 9$$

Likewise for division, where we have

$$6 * 3 = \frac{6}{3} = 2$$

But note that in the case of division, the operation is **not commutative**. Hence, we have

$$3 * 6 = \frac{3}{6} = \frac{1}{2}$$

That is,

$$6 * 3 \neq 3 * 6$$

In general, we have

$$a * b \neq b * a$$

Thus, for some binary operations, the order in which we enter the elements does matter.

Unary operations

A **unary operation** is one which uses only one element. For example, $a \rightarrow a^2$ is a unary operation.

Modular arithmetic

We can perform arithmetical operations in different **moduli**. To indicate the use of a particular modulo, say n, we add (mod n) after we have completed the calculation.

Take, for example, the **multiplication** of two integers in modulo 6. We multiply the two integers normally and then subtract 6 repeatedly until the answer is between 0 and 5.

Hence, we have for $3 \times 3 = 9$

$$3 \times 3 = 3 \,(\text{mod } 6) \quad \text{since } 9 - 6 = 3$$

Similarly, for $5 \times 4 = 20$, we have

$$5 \times 4 = 2 \,(\text{mod } 6) \quad \text{since } 20 - 6 - 6 - 6 = 2$$

And for $4 \times 3 = 12$, we have

$$4 \times 3 = 0 \,(\text{mod } 6) \quad \text{since } 12 - 6 - 6 = 0$$

Modular **addition** is similar to multiplication. Suppose we want to add two integers in modulo 4. We add them normally and then subtract 4 repeatedly until the answer is between 0 and 3.

For example, we have

$$2 + 3 = 5 = 1 \,(\text{mod } 4) \qquad 2 + 0 = 2 \,(\text{mod } 4)$$
$$1 + 3 = 4 = 0 \,(\text{mod } 4) \qquad 3 + 3 = 2 \,(\text{mod } 4)$$

Example 1 Express 9×11 in modulo 17.

SOLUTION

We have $9 \times 11 = 99$, which becomes

$$9 \times 11 = 14 \,(\text{mod } 17) \quad \text{since } 99 - 17 - 17 - 17 - 17 - 17 = 14$$

Definition of a group

A **group** comprises

- a set of elements (or members), G, together with
- a binary operation $*$ on this set.

To be a group, G must satisfy the following **four properties** (sometimes referred to as **axioms**).

- **Closure** G must be closed. This means that if a and b are members of G, then $a * b$ must also be a member of G. This is written as

    $a * b \in G$, for all a and $b \in G$

- **Associativity**   Provided their original order is preserved, the result of combining a, b and c does not depend on which two adjacent elements are combined first. This is written as

    $(a * b) * c = a * (b * c)$, for all a, b and $c \in G$

- **Identity**   There is an element e in G for which $a * e = e * a = a$ for every a in G. That is, there is an **identity element** e in G which does not change any other element.

- **Inverses**   For any element a in G, there is an **inverse element** of a in G, denoted by a^{-1}. This is written as

  For any $a \in G$, there exists $a^{-1} \in G$, for which $a * a^{-1} = a^{-1} * a = e$

To confirm that a set of elements, together with an operation on the set, forms a group, we have to verify that the set possesses **every one** of these four properties. This can be difficult, since we need to check **each property** for **every element** or **pairs of elements** of the set.

Note   When you are given a question involving a group, you will also always be given a binary operation (which is usually multiplication or addition). It is essential that you recognise which binary operation is being used.

Example 2 Prove that the set $G = \{1, i, -1, -i\}$ under multiplication is a group (where $i^2 = -1$).

SOLUTION

To prove that this is a group, we need to verify each of the four properties in turn. It is essential to confirm that **all** the properties are satisfied.

Closure   We have to verify that, for any a and $b \in G$, $a * b \in G$.

Therefore, if we take any element in G and multiply it by any other element in G, the result should be an element in G. One way to check this is to take every pair in turn. (This method is only feasible in this case because G is a small group.) Hence, we have

$$1 * 1 = 1 \qquad 1 * i = i \qquad 1 * -1 = -1 \qquad 1 * -i = -i$$
$$i * 1 = i \qquad i * i = -1 \qquad i * -1 = -i \qquad i * -i = 1$$
$$-1 * 1 = -1 \qquad -1 * i = -i \qquad -1 * -1 = 1 \qquad -1 * -i = i$$
$$-i * 1 = -i \qquad -i * i = 1 \qquad -i * -1 = i \qquad -i * -i = -1$$

That is, $a * b \in G$.

Associativity   We have to verify that $(a * b) * c = a * (b * c)$ for all a, b and c in G.

We have, for example,

$$(i * -1) * -i = -i * -i = -1 \quad \text{and} \quad i * (-1 * -i) = i * i = -1$$

This verifies associativity for just this one triple combination. To prove associativity by this method, we would have to check every other triple combination, of which there are 64.

Alternatively, we can simply recall and state the fact that multiplication of complex numbers is associative.

Identity   1 is the identity element of this group. This is because multiplying any number by 1 does not change its value. To confirm that 1 is the identity element, we have to verify that $1 * a = a * 1 = a$ for every a in G.

In this case, it is not too onerous to do all the calculations concerned. Hence, we have

$$1 * 1 = 1 * 1 = 1 \qquad 1 * i = i * 1 = i$$

$$1 * -i = -i * 1 = -i \qquad 1 * -1 = -1 * 1 = -1$$

Alternatively, we can simply state that 1 is the identity, since we know that multiplying any number by 1 does not change its value.

Inverses We need to find the inverse of each of the elements 1, i, −1 and −i to confirm that each inverse is a member of the group. That is, for each element a in G, we need to find an a^{-1}, and verify that $a * a^{-1} = a^{-1} * a = 1$.

Hence, we have

> Inverse of 1 is 1, since $1 * 1 = 1 * 1 = 1$
>
> Inverse of i is $-i$, since $i * -i = -i * i = 1$
>
> Inverse of −1 is −1, since $-1 * -1 = -1 * -1 = 1$
>
> Inverse of −i is i, since $-i * i = i * -i = 1$

Since all of the properties are satisfied for any choice of elements, we have proved that G is a group.

It can take a long time to prove that a set of elements, together with an operation on the set, forms a group, especially if there are many elements. However, there are short-cuts we can take.

- We can use algebraic rules to prove closure. For example, to prove that the set of integers under addition forms a group, we just state that the sum of any two integers is always an integer.

- Associativity is always difficult to prove. However, we recall that the multiplication and the addition of real numbers, the multiplication and the addition of complex numbers, and the multiplication and the addition of square matrices, are all associative.

- To find the identity of a group, we recall that 0 is the identity for addition (since adding zero to a number does not change the number). We recall also that 1 is the identity for multiplication (since multiplying a number by 1 does not change the number). We must be careful, however, because in some unusual cases of multiplication, such as modulo 14, the identity may not be 1 (see page 375).

- To find inverses, we often just need to give a general formula which identifies all the inverses.

Example 3 Prove that the set $G = \{0, 1, 2, 3\}$ under the binary operation addition (mod 4) forms a group.

SOLUTION

As usual, we must verify that all the group properties are satisfied.

Closure   Whenever we add two numbers (mod 4), we always get a number between 0 and 3. Therefore, addition (mod 4) is closed.

Associativity   Addition is associative, and so addition (mod 4) must also be associative.

Identity   Adding 0 to a number (mod 4) does not change the number. So 0 is the identity of addition (mod 4).

Inverses   We have:

Inverse of 0 is 0, since $0 + 0 = 0$ (mod 4)

Inverse of 1 is 3, since $1 + 3 = 3 + 1 = 0$ (mod 4)

Inverse of 2 is 2, since $2 + 2 = 0$ (mod 4)

Inverse of 3 is 1, since $3 + 1 = 1 + 3 = 0$ (mod 4)

Therefore, all four group properties are satisfied.

Hence, the set $G = \{0, 1, 2, 3\}$ under the binary operation addition (mod 4) forms a group.

Group table

A **group table** shows the effect of combining any two elements. (Other descriptions commonly used are Cayley table, composition table, combination table, operation table and multiplication table.) The entry in row a and column b is the composition $a * b$.

The group table for the set $G = \{0, 1, 2, 3\}$ under addition modulo 4 is shown below. As an example, **[3]** identifies the result $1 * 2 = 3$.

+ (mod 4)	0	1	2	3
0	0	1	2	3
1	1	2	[3]	0
2	2	3	0	1
3	3	0	1	2

To complete the table, we need to find each of the 16 results.

We can use the fact that $x * e = x$ and $e * x = x$ to find seven of these results quite simply. All the other entries have to be calculated.

Even though we have to complete all the entries in the table, it is often easier to draw and use a group table to see whether the set under the operation forms a group.

For the group properties, we have:

Closure   This can be seen by noting that all the results in the group table are in the original set.

Associativity   This **cannot** be seen from the group table.

Identity The column under the identity element and the row across from the identity element contain the elements in the same order as the original set.

The row and the column given below show that 0 is the identity element:

$+ \pmod 4$	0	1	2	3
0	0	1	2	3
1	1			
2	2			
3	3			

Note The identity element does not have to be 0 or 1. For example, see the group table on page 375 for the set of integers $\{2, 4, 6, 8, 10, 12\}$ under multiplication (mod 14).

Inverses We can find the position of the identity element in each column and each row. For example, $1 * 3 = 3 * 1 = 0$, which is the identity. Therefore, 3 is the inverse of 1.

In fact, the set $G = \{0, 1, 2, \ldots, m - 1\}$ under the binary operation, addition (mod m), also forms a group. (You can check this for yourself for various values of m.) Notice that, in general, the inverse of k under addition (mod m) is $m - k$.

Example 4 Find whether the set $\{1, 3\}$ under multiplication (mod 11) forms a group.

SOLUTION

We can find the answer by checking each of the group properties in turn, until we find one which does **not** work. We recall that for G to be a group, we need to check that **all** four group properties are satisfied. So, to check that G is **not** a group, we need only to find **one** property which is not satisfied.

In this case, since

$$3 * 3 = 3 \times 3 = 9 \pmod{11}$$

and 9 is **not** a member of the original set, **closure does not hold**.

Since the set $\{1, 3\}$ is not closed under multiplication (mod 11), it does **not** form a group.

Note If in Example 4 we were to consider the other group properties, we would find:

- The group is associative, since multiplication is associative.
- There is an identity element, 1, since 1 is the identity under multiplication.
- There is, however, no element a for which $3 * a \equiv 1 \pmod{11}$, and so the property of possessing an inverse element is not satisfied either.

Example 5 Prove that the set of integers $\{2, 4, 6, 8, 10, 12\}$ under multiplication (mod 14) forms a group.

SOLUTION

Again, we need to check that all the group properties hold. However, the last two are difficult to prove, and so we have to use a group table to work out how all the elements combine.

Closure   If we multiply two even integers together, we obtain an even integer, which is also even (mod 14). Hence, the set is closed.

Associativity   Multiplication is associative.

Identity   There is no obvious identity element. The identity element we would naturally look for, 1, is missing from the group. To overcome this problem, we draw the group table, which shows the effect of combining any two elements.

$\times$ (mod 14)	2	4	6	8	10	12
2	4	8	12	2	6	10
4	8	2	10	4	12	6
6	12	10	8	6	4	2
8	2	4	6	8	10	12
10	6	12	4	10	2	8
12	10	6	2	12	8	4

From this table, we can see the column under 8, or the row across from 8, is 2, 4, 6, 8, 10, 12, which is the same as the original set. Thus, multiplication by 8 changes none of the elements of the group, and so 8 is the identity element of this group.

Inverses   As with most problems involving multiplication (mod n), there is no easy way to prove that every element has an inverse. However, as in Example 4, we can use the group table. To find the inverse of 2, we need an element a for which $2 * a = 8$. (Remember that 8 is the identity element of this group.)

Inverse of 2 is 4, since $2 * 4 = 4 * 2 = 8$ (mod 14)

Inverse of 4 is 2, since $4 * 2 = 2 * 4 = 8$ (mod 14)

Inverse of 6 is 6, since $6 * 6 = 6 * 6 = 8$ (mod 14)

Inverse of 8 is 8, since $8 * 8 = 8 * 8 = 8$ (mod 14)

Inverse of 10 is 12, since $10 * 12 = 12 * 10 = 8$ (mod 14)

Inverse of 12 is 10, since $12 * 10 = 10 * 12 = 8$ (mod 14)

Thus, we have checked that all the group properties hold.

Therefore, the set of integers $\{2, 4, 6, 8, 10, 12\}$ under multiplication (mod 14) forms a group.

Note   We say that the number 6 in Example 5 is **self inverse**, since its inverse is itself. (See also page 393.)

Exercise 17A

In Questions **1** to **4**, prove that each set under the given operation satisfies all the group properties and hence forms a group.

1 The set $\{1, 5\}$ under $(\times, \text{mod } 12)$.

2 The set $\{1, 2, 3, 4\}$ under $(\times, \text{mod } 5)$.

3 The set $\{0, 1, 2, 3, 4, 5\}$ under $(+, \text{mod } 6)$.

4 The set $\{1, 2, 3, 4, 5, 6\}$ under $(\times, \text{mod } 7)$.

5 Show that the set $\{1, 3\}$ under $(\times, \text{mod } 12)$ does not form a group.

6 Show that the set of positive integers under addition is not a group.

Symmetries of a regular *n*-sided polygon

The set of symmetries of a regular polygon forms a group under the composition of symmetries. Hence, this is true of the set of symmetries of, for example, a square, a regular hexagon and a regular heptagon.

Example 6 Prove that the set of symmetries of a regular pentagon under composition forms a group.

SOLUTION

It is easier to specify this group geometrically than to write down all the elements. The symmetries of a pentagon, PQRST, shown on the right, are the five reflections (top row) and the five rotations (bottom row) drawn below.

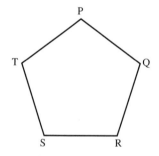

Reflections

    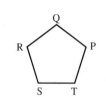

Rotations

The binary operation in this case is the composition of symmetries. For example, the composition of a clockwise rotation through 72° and a clockwise rotation through 216° is a clockwise rotation through 288°.

To prove that the set of symmetries forms a group, we must check each of the four group properties.

Closure The composition of two rotations is another rotation. The composition of a reflection and a rotation is a reflection, as shown in the example below.

Reflection in the perpendicular bisector of RS, which passes through P, . . .

. . . followed by an anticlockwise rotation through 72° . . .

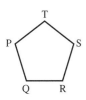

. . . is the reflection in the perpendicular bisector of PT, which passes through R.

The composition of two reflections is a rotation, as shown in the examples below:

Reflection in the perpendicular bisector of RS, which passes through P, . . .

. . . followed by reflection in the perpendicular bisector of PT, which passes through R, . . .

. . . is an anticlockwise rotation through 288°.

So, the set of symmetries of a pentagon is closed.

Associativity   We certainly do **not** want to prove $a * (b * c) = (a * b) * c$ for each of the ten symmetries, giving 1000 possible combinations! Instead, we recall that each symmetry can be represented by a 2×2 matrix (see page 310). Thus, the composition of transformations corresponds to the multiplication of matrices. Since the multiplication of matrices is associative, so is the composition of transformations. Hence, the set of symmetries of a pentagon is associative.

Identity   The identity transformation (rotation of $0°$) is in the set of transformations.

Inverses   Every symmetry has an inverse. In each case, the inverse is a symmetry which returns the pentagon to its original position. For a clockwise transformation of $n°$, the anticlockwise rotation of $-n°$ is the inverse transformation.

Carrying out the same reflection twice, always returns the pentagon to its original position. Thus, the inverse of a given reflection is the same reflection. Hence, all reflections are self-inverse.

Therefore, we have verified that the symmetries of a pentagon form a group.

Dihedral groups

A group of symmetries is called a **dihedral group**. The group of symmetries of a pentagon contains ten (2×5) elements and is denoted by D_{10}. The symmetries of the other regular polygons also form groups. For example, the symmetries of a regular heptagon form a group. Since a heptagon has seven sides, the group contains 14 (2×7) elements. Hence, the group of symmetries of a regular heptagon is denoted by D_{14}.

Non-finite groups

The groups that we have so far considered are all composed of sets which contain a finite number of elements. We will now consider groups whose sets contain an **infinite** number of elements.

When dealing with a non-finite group, we use a similar approach to that which we use with finite groups, except that we **cannot construct** a group table because there is an infinite number of elements. However, this does not make the verification of the group too much harder; it just means that it has to be done **algebraically**.

Example 7 Prove that the set of integers under addition forms a group.

SOLUTION

Since there is an infinite number of integers, we cannot use a group table. We therefore use algebraic methods to verify that all four properties are satisfied.

Closure   If we add together any two integers, we always get an integer. Therefore, if a and b are integers, we know that

$$a * b = a + b = c$$

and hence c is an integer. Therefore, the set of integers under addition is closed.

Associativity We may simply quote the fact that addition is always associative.

Identity As always with addition, 0 is the identity element. For any given integer, a, we have

$$a * 0 = 0 * a = a + 0 = 0 + a = a$$

This proves that 0 is the identity element for the group.

Inverses Given any integer a, its inverse is $-a$. This is because

$$a * -a = a + -a = 0 \quad \text{and} \quad -a * a = -a + a = 0$$

Therefore, we have checked that the four properties are satisfied, and so the set of integers under addition forms a group.

Example 8 Prove that the set of integers under multiplication as the binary operation does not form a group.

SOLUTION

We recall that to prove that a set under an operation does **not** form a group, we just need to check that one of the properties is not satisfied.

In this case, the inverse property does not hold.

The identity element under multiplication would be 1, but the inverse of 2 would be $\frac{1}{2}$, since

$$\tfrac{1}{2} * 2 = \tfrac{1}{2} \times 2 = 1$$

But $\frac{1}{2}$ is not a member of the set of integers, and therefore 2 does not have an inverse in the set.

Since one of the elements does not satisfy one of the properties, the set of integers under multiplication cannot be a group.

Example 9 Prove that the set of real numbers (excluding zero) under the binary operation of multiplication forms a group.

SOLUTION

Again, we need to check that all four properties are satisfied.

Closure   The product of any two real numbers which are not zero is also a real number which is not zero. Therefore, the set is closed.

Associativity   Multiplication is always associative.

Identity 1 is the identity of multiplication, and it is in this group. Hence, there is an identity element.

Inverses   For any real number x, its inverse is $\dfrac{1}{x}$ since

$$\frac{1}{x} \times x = \frac{1}{x} \times x = 1$$

Hence, every element has an inverse which is a member of the set.

Therefore, the set of real numbers (excluding zero) under the binary operation of multiplication forms a group.

Note   The set of real numbers including zero under the binary operation of multiplication is **not** a group. This is because zero does not have an inverse. To find an inverse of zero, would mean finding $\dfrac{1}{0}$, which is impossible.

Example 10   Let G be the set of 3×3 matrices with integer elements and determinant 1, under the multiplication of matrices as the binary operation. Prove that G forms a group.

SOLUTION

To find whether a set of matrices under a particular operation forms a group, we have to apply the rules of matrices. But we also still need to check that G satisfies **every** one of the four group properties.

Closure   If $\mathbf{A}$ and $\mathbf{B}$ are 3×3 matrices, then $\mathbf{AB}$ will also be a 3×3 matrix.

We also need to check that $\mathbf{AB}$ has integer elements.

If $\mathbf{A}$ and $\mathbf{B}$ have integer elements, then consider how we find $\mathbf{AB}$. We multiply the integers in $\mathbf{A}$ by the integers in $\mathbf{B}$, and then add them up. Therefore, the entries in $\mathbf{AB}$ are all integers.

Finally, we also need to check that the value of the determinant $\mathbf{AB}$ is 1.

Using $\det(\mathbf{AB}) = \det \mathbf{A} \times \det \mathbf{B}$, we find that

$$\det(\mathbf{AB}) = 1 \times 1 = 1$$

Hence, $\mathbf{AB}$ is a member of the set and thus the set under multiplication is closed.

Associativity   Multiplication of matrices is associative.

Identity   The identity of matrix multiplication is the identity matrix $\mathbf{I}$. Since $\mathbf{I}$ has integer elements and determinant 1, $\mathbf{I}$ is a member of the set.

Inverse   The inverse of a 3×3 matrix is a 3×3 matrix. However, we need to check that the inverse matrix has integer elements and determinant 1.

To verify that the inverse matrix has integer elements, we consider how we would find the inverse (see pages 304–6). To find the inverse of a 3×3 matrix, we need to find the cofactor of each element. In this case, this means finding the determinants of nine 2×2 matrices, each of which has integer elements. This gives integer results. Then we divide each cofactor by the determinant of the original 3×3 matrix, which in this case is 1.

Therefore, the inverse of any 3×3 matrix with integer elements and determinant 1 is also a 3×3 matrix with integer elements.

Finally, if $\mathbf{A}$ has determinant 1, then $\mathbf{A}^{-1}$ also has determinant 1.

Therefore, all the group properties are satisfied, and so G forms a group.

a^n notation

It is usual to write a^2 for $a * a$. Similarly, $a * a * a$ is written as a^3.

If n is positive, then a^n means $a * a * \ldots * a$. (Here, there are n copies of a.)

a^0 is taken as the identity element, e.

a^{-n} means $a^{-1} * a^{-1} * \ldots * a^{-1}$. (Here, again, there are n copies of a^{-1}.)

Division in a group

In a group G, we **cannot divide** by a. Instead, we **multiply** by its inverse, a^{-1}, which has the same effect as dividing by a.

When multiplying by a^{-1}, we must ensure that we multiply both sides with a^{-1} in the **same position**. For example, if $b = c$, we have

$$a^{-1} * b = a^{-1} * c \quad \text{and} \quad b * a^{-1} = c * a^{-1}$$

We **cannot** have $a^{-1} * b = c * a^{-1}$.

Permutation groups

Suppose that we are given n objects in a particular order. By switching two objects, we can change that order. Switching the objects in positions 1 and 2 is represented by the notation $(1\,2)$ or $\begin{pmatrix} 1 & 2 \\ 2 & 1 \end{pmatrix}$. Similarly, the notation for switching the objects in positions 5 and 8 is $(5\,8)$ or $\begin{pmatrix} 5 & 8 \\ 8 & 5 \end{pmatrix}$. If we want to move the object in position 1 to position 8, the object in position 8 to position 5, and the object in position 5 back to position 1, the notation for this is $(1\,8\,5)$ or $\begin{pmatrix} 1 & 8 & 5 \\ 8 & 5 & 1 \end{pmatrix}$. This means '1 to 8, 8 to 5, and 5 to 1'.

Similarly, the notation $(1\,2)\,(3\,4\,5)$ or $\begin{pmatrix} 1 & 2 \\ 2 & 1 \end{pmatrix} \begin{pmatrix} 3 & 4 & 5 \\ 4 & 5 & 3 \end{pmatrix}$ means '1 to 2 and 2 to 1, then 3 to 4, 4 to 5, and 5 to 3'.

This 'language' of permutations is illustrated below.

$$\begin{pmatrix} A & B & C & D & E & F \\ B & A & D & C & F & E \end{pmatrix} \quad \text{is represented by}$$

$(1\,2)\,(3\,4)\,(5\,6)$

or

$$\begin{pmatrix} 1\,2\,3\,4\,5\,6 \\ 2\,1\,4\,3\,6\,5 \end{pmatrix}$$

The set which contains all possible permutations of n objects forms a group. The binary operation is the composition of the permutations. We can verify all four group properties.

Closure Composing two permutations of n objects gives another permutation of the n objects.

Associativity The composition of permutations is associative, but we will not prove it.

(One way to prove this is to encode the permutations as an $n \times n$ matrix containing 1s and 0s, and then the composition will correspond to the multiplication of matrices, which we know is associative. However, this technique is beyond the scope of this book.)

We can simply state without proof that the composition of permutations is associative.

Identity The identity permutation which does not interchange any objects is a member of the set.

Inverses The inverse of a typical permutation $(a\,b\,c\dots d)$ is the permutation $(d\dots c\,b\,a)$, which is also a member of the set.

Since there are $n!$ ways of arranging n objects, it follows that there are $n!$ different permutations in the permutation group of n elements, which is denoted by S_n.

Generator of a group

If a is a member of a group and $a \neq e$, a^2 is also a member of the group.

If $a^2 \neq e$, $a^2 * a$ or a^3 will also be a member of the group.

If a is a **generator** of a group, then every member of the group may be expressed as a^k for some integer k.

If the group is finite, then $a^r = e$ for some integer r, and the members of the group are

$$a, a^2, a^3, \dots, a^{r-1} \quad \text{and} \quad a^r = e$$

For example, in the group $(\{e, a, a^2, a^3, a^4\}, *)$ with $a^5 = e$, each of a, a^2, a^3 and a^4 is a generator. (See pages 391–2.)

Cyclic groups

Cyclic groups are the simplest type of group. In any cyclic group, there is some element a which **generates** the group. Hence, the elements of the group are

$$\{e, a, a*a, a*a*a, a*a*a*a, \dots\} \quad \text{or} \quad \{e, a, a^2, a^3, \dots\}$$

Examples of cyclic and non-cyclic groups

- The group $\{1, i, -1, -i\}$ is cyclic.

  Since, $i^2 = -1$ and $i^3 = -i$, the group can be written as $\{1, i, i^2, i^3\}$.

- The group of integers under addition (mod 4) is also a cyclic group. The elements of this group are $\{0, 1, 2, 3\}$. The group can be written as $\{e, 1, 1^2, 1^3\}$.

  We have already seen that $1^2 = 1 * 1 = 1 + 1 \pmod{4} = 2 \pmod{4}$.

Notice that these two cyclic groups are very similar, both having four elements. On page 388, we will find that they are **isomorphic**, since they have identical structures.

- The integers under addition (mod n) always forms a cyclic group, which can always be generated by one element.

- The symmetries of a pentagon are **not** cyclic. If we repeat a rotation again and again, we will **never** get a reflection. If we repeat a reflection again and again, we **only ever** get that reflection and its identity. Therefore, there is no way in which we can repeat the same symmetry over and over again and get **all** the symmetries. So, the group of symmetries of a pentagon cannot be cyclic.

Abelian groups

An **abelian group** is a group in which $a * b = b * a$ for **every** pair of elements a and b. In other words, it does not matter which way round we combine the elements. An abelian group is sometimes called a **commutative group**, since every pair of elements commutes.

We note that a group is abelian when the group table has symmetry in the leading diagonal.

(This class of groups is named after the prodigiously gifted Norwegian mathematician Niels Henrik Abel (1802–29).)

Determining which groups are abelian

Consider the binary operation. Addition is always commutative, and the multiplication of numbers is also always commutative. However, the multiplication of matrices is **not** commutative, since, in general, $\mathbf{AB} \neq \mathbf{BA}$, where $\mathbf{A}$ and $\mathbf{B}$ are matrices.

Thus, for example, the group of $\{0, 1, 2, 3\}$ under addition (mod 4) is abelian, since $a + b = b + a$ for any integers a and b.

To show that the group of 2×2 matrices with integer elements and determinant 1 is **not** an abelian group, we need to find **one** pair of matrices $\mathbf{A}$ and $\mathbf{B}$ with $\mathbf{AB} \neq \mathbf{BA}$.

For example, we have

$$\begin{pmatrix} 1 & 1 \\ 0 & 1 \end{pmatrix}\begin{pmatrix} 1 & 0 \\ 1 & 0 \end{pmatrix} = \begin{pmatrix} 2 & 0 \\ 1 & 0 \end{pmatrix}$$

$$\begin{pmatrix} 1 & 0 \\ 1 & 0 \end{pmatrix}\begin{pmatrix} 1 & 1 \\ 0 & 1 \end{pmatrix} = \begin{pmatrix} 1 & 1 \\ 1 & 1 \end{pmatrix}$$

This single example proves that the group of 2×2 matrices with integer elements and determinants is **not** an abelian group.

The set of transformations of a pentagon is **not** abelian, since there is a difference between performing a reflection followed by a rotation, and performing a rotation followed by a reflection.

Before transformation.

Before transformation.

Reflection in the perpendicular bisector of RS, which passes through P.

Rotation anticlockwise through 144°.

Rotation anticlockwise through 144°.

Reflection in the perpendicular bisector of PT, which passes through R.

All cyclic groups are abelian

Proof

We need to prove that $a*b = b*a$, where a and b are any two elements in a cyclic group G. Since G is cyclic, there is some element c which **generates** G. Since c generates G, there are integers n and m for which $c^n = a$ and $c^m = b$.

Hence, we have

$$a*b = c^n * c^m = c^{n+m} = c^m * c^n = b*a$$

$$\Rightarrow \quad a*b = b*a$$

Hence, all cyclic groups are abelian.

Benefit of abelian groups

It is much easier to calculate in abelian groups than in non-abelian groups. When calculating in non-abelian groups, we always have to ensure that the elements are in their correct positions. Given below is an example of calculating in a non-abelian group.

Calculating in a non-abelian group

When a group is not abelian, it is important that we do **not** switch the order of any pair of elements. For example, consider the group of symmetries of a square. This is the dihedral group D_8.

Let a denote rotation by $90°$ anticlockwise. Then the rotations of the square are a, a^2, a^3 and $e\,(a^4 = e)$, as shown below.

Now let b be the reflection shown below.

The other reflections are given by $a * b$, $a^2 * b$, and $a^3 * b$, as shown below.

We note that $b * a = a^3 * b$, which leads to a way of writing down the group.

The group of symmetries of a square is the group whose elements are $\{e, a, a^2, a^3, a * b, a^2 * b, a^3 * b\}$, with the **stipulations** that $a^4 = e$, $b^2 = e$, $b * a = a^3 * b$.

These relations are enough to find the composition of any two elements. But again, we must be careful – the group is **not** abelian. Hence, we **cannot switch the order of two elements**.

For example, consider $(a * b) * (a^3 * b)$. Using $b * a = a^3 * b$, we obtain

$$(a * b) * (a^3 * b) = (a * b) * (b * a) = a * (b * b) * a$$

Using $b^2 = e$, we obtain

$$a * (b * b) * a = a * e * a = a * a = a^2$$

Hence, we have

$$(a * b) * (a^3 * b) = a^2$$

Similarly, we have

$$(a * b) * a^2 = a * (b * a) * a = a * (a^3 * b) * a = (a * a^3) * (b * a) =$$
$$= a^4 * (b * a) = e * (b * a) = (e * b) * a = b * a = a^3 * b$$

Because the group is not abelian, we also need to be careful when writing down inverse elements. The inverse of $a*b$ is $b^{-1}*a^{-1}$, since

$$(a*b)*(b^{-1}*a^{-1}) = a*(b*b^{-1})*a^{-1} = a*e*a^{-1} = a*a^{-1} = e$$

Order of a group

The **order** of a group is the number of elements in the group. Most of the cases we have dealt with so far involve **finite groups**, which are those containing only a finite number of elements.

To find the order of a group, we see how many elements are in the group. Here are four examples:

- The order of the group of integers under addition (mod 4) is 4, since the elements are $\{0, 1, 2, 3\}$.

- The order of the group $\{1, i, -1, -i\}$ is 4, since again there are four elements.

- The order of the group of symmetries of a pentagon is 10 (see page 376), since there are five rotations and five reflections.

- The order of the permutation group S_n is $n!$, since there are $n!$ possible arrangements of n objects (see page 382).

Order of an element

In any finite group, any element combined repeatedly with itself must eventually give the identity element. For example, in the group of addition (mod 4), we have

$$1+1+1+1 = 0 \quad \text{and} \quad 2+2 = 0$$

In the group of symmetries of a pentagon, a rotation through $72°$ repeated four more times, returns the pentagon to its original position. So, five rotations through $72°$ are equivalent to the identity symmetry.

The **order**, or **period**, of an element is the smallest number of times we have to repeat the element before we obtain the identity element. So, in the group G, the order of an element, a, is n, where n is the smallest integer for which $a^n = e$. For example, we have:

- In the group of addition (mod 4), the order of element 1 is 4, since $1+1+1+1 = 0$ (mod 4).

- The order of element 3, under addition (mod 4), is 4, since $3+3+3+3 = 0$ (mod 4).

- The order of element 2 is 2, since we need only combine 2 twice before returning to the identity: $2+2 = 0$ (mod 4). The order of 0, which is also the order of every identity element in every group, is 1.

- In the symmetries of a pentagon, the order of any reflection is 2, since combining the same reflection twice returns us to the identity. The order of any rotation in the symmetries of a pentagon is 5, since we need to repeat a rotation four more times before returning to the identity.

Note

- The order of an element **must not be more** than the order of its group.

- The order of a group and the order of an element are **completely different concepts**.

Subgroups

Consider the group $\{0, 1, 2, 3\}$ of integers under addition (mod 4). If we just take the smaller set $\{0, 2\}$ of integers under addition (mod 4), we have another group. (We can confirm this by verifying all of the group properties: closure, associativity, identity and inverses.) Since $\{0, 2\}$ is a group which is wholly contained within the original group, we say that $\{0, 2\}$ is a **subgroup** of $\{0, 1, 2, 3\}$.

That is, if both G and H are groups under the same binary operation, and every member of H is contained within G, then H is a **subgroup** of G.

It is often much easier to check that H is a subgroup of G than to check from scratch that H is a group. Thus, to confirm that H is a subgroup of G, we need to proceed as follows:

1 Check that H contains the identity of G.
2 Check that if a and b are in H, then $a * b$ is also in H.
3 Ensure that for each a in H, a^{-1} is also in H.

We only need to verify that these three conditions are satisfied, since they will ensure that all four group properties are satisfied.

Closure   Condition **2** checks that H is closed.

Associativity   We know that H must be associative, since it is a subset of G, which is associative.

Identity   Condition **1** checks that H has an identity.

Inverses   Condition **3** checks that every element in H has an inverse element.

Example 11 Prove that the set H of rotations of a pentagon is a subgroup of the set of all symmetries of a pentagon.

SOLUTION

Applying the three subgroup conditions, we find the following:

1 The identity symmetry, a rotation through $0°$, is in H.

2 The composition of two rotations is also a rotation so, if a and b are in H, then $a * b$ is also in H.

3 The inverse of a clockwise rotation through $\alpha°$ is an anticlockwise rotation through $\alpha°$. Therefore, the inverse of any element in H is also in H.

Isomorphic groups

We have already found that the group $G = \{0, 1, 2, 3\}$ under addition (mod 4) and the group $H = \{1, i, -1, -i\}$ under multiplication are similar because they are both cyclic of order 4. By drawing their group tables, we can see that they have identical structures.

+ (mod 4)	0	1	2	3
0	0	1	2	3
1	1	2	3	0
2	2	3	0	1
3	3	0	1	2

$\times$	1	i	-1	$-i$
1	1	i	-1	$-i$
i	i	-1	$-i$	1
-1	-1	$-i$	1	i
$-i$	$-i$	1	i	-1

Two groups which have the same structure are said to be **isomorphic**.

To prove that G and H are isomorphic, we need to identify the way in which we can map elements of G onto elements of H.

In the case above, we can map an integer $n \in G$ onto the complex number $e^{\pi i n/2} \in H$. To confirm that a mapping f from G to H is an isomorphism, we must verify each of the following:

1 Each and every element of G is maps onto a **unique** element of H.
2 Each and every element of H is the image of **exactly one** element of G.
3 The image of the identity of G, f(e), is the identity of H.
4 The composition element $f(a) * f(b)$ in H is the same element as the image $f(a * b)$ of the composition element $(a * b)$ in G.

Example 12 Show that the mapping $f(n) = e^{\pi i n/2}$ from $G = \{0, 1, 2, 3\}$ to $H = \{1, i, -1, -i\}$ is an isomorphism.

SOLUTION

We need to check that f satisfies all four conditions for isomorphism.

1 f identifies the image of each member of G.

2 Each and every member of H is the image of $f(n)$ for some n. This is because

$$1 = e^{0\pi i/2} \quad i = e^{\pi i/2} \quad -1 = e^{2\pi i/2} \quad -i = e^{3\pi i/2}$$

3 The image of the identity of G, 0, is f(0), which is 1. This is the identity of H, which confirms that the identity element in G is mapped onto the identity element in H.

4 We must check that for all integers n and m between 0 and 3

$$f(n * m) = f(n) * f(m)$$

Since the binary operations in G is different from that in H, the equation we have to check becomes

$$f(n + m) = f(n) \times f(m)$$

which gives

$$f(n + m) = e^{\pi i(n+m)/2} = e^{\pi in/2} \times e^{\pi im/2} = f(n) \times f(m)$$

Hence, we have

$$f(n * m) = f(n) * f(m)$$

Since we have proved that all four conditions are satisfied, f is an isomorphism from G to H.

When we have found an isomorphism between two groups, we know that the two groups are essentially the same. The elements are different, the operations are different, but because of condition **4**, they combine in the same way.

These two groups are **isomorphic**.

Lagrange's theorem

Lagrange's theorem states that for a finite group G, of order n, the order m of the subgroup H is a factor of n.

Thus, the subgroups of a group of order 10, have either order 2 or order 5.

There is no necessity for a group of order 10 to have a subgroup of order 2, or a subgroup of order 5, but the **only possible** subgroups **must be** of one of these orders.

Lagrange's theorem helps us understand the structure of groups. The more we can understand how different groups relate to each other, the more we can hope to understand about groups in general.

Examples of Lagrange's theorem

We have found already that $\{0, 2\}$ is a subgroup of the group $\{0, 1, 2, 3\}$ of integers under addition (mod 4). $\{0, 2\}$ has order 2, and $\{0, 1, 2, 3\}$ has order 4. Since we know that 4 is divisible by 2, this example agrees with Lagrange's theorem.

On page 387, we also found that the group of rotations of a pentagon are a subgroup of the symmetries of a pentagon. There are five rotations and ten symmetries of the pentagon. Again, we see that the order of the smaller group (5) is a factor of the order of the larger group (10). This also agrees with Lagrange's theorem.

One consequence of Lagrange's theorem is the following:

The order of an element is a factor of the order of the group.

Take an element a of a group, G, and consider the cyclic group generated by a.

If a has order n, then $a^n = e$. So, the group, H, generated by a is $\{e, a, a^2, a^3, \ldots, a^{n-1}\}$.

Since H is a subgroup of G, we can use Lagrange's theorem, which states that the order of H divides the order of G.

But the order of H is n. Hence, the order of a divides the order of G.

Example 13 A group, G, has subgroups $\{a\}$, $\{a, b, c, d, f\}$, and $\{a, d\}$.

a) What is the identity of G?

b) Could G contain only the five elements $\{a, b, c, d, f\}$? Explain your answer.

c) What is the smallest possible order of G?

SOLUTION

a) The identity of G must be a. This is because every subgroup of G must contain the identity of G, and $\{a\}$ is a subgroup of G.

b) The order of a subgroup divides the order of a group. So, the order of G must be divisible by 1, by 5 and by 2. Therefore, the order of G cannot be 5, and so G cannot contain only the five elements a, b, c, d and f.

c) The smallest order that G could have is the smallest number divisible by 1, 2 and 5. This number is 10.

Groups of order 3

If G is a group of order 3, then the order of every element of G must divide 3. That is, the order of every element of G must be either 1 or 3.

There is only one element of order 1, the identity element e.

Since there are three elements in G, there must be two elements which are not of order 1 and hence must each be of order 3.

Let a be an element of order 3. Then a, a^2 and $a^3 = e$ are three different elements in the group. Since there are only three elements in the group, it follows that a, a^2 and e are **all** the elements in the group, and so G has to be a cyclic group of order 3.

Hence, all groups of order 3 are cyclic. They are also all isomorphic with each other.

Groups of order 4

If G is a group of order 4, then all the elements must have order 1, 2 or 4. One of the elements must have order 1; this is the identity element e. If G has an element a of order 4, then e, a, a^2 and a^3 are the four elements of G, and therefore G must be the cyclic group generated by a.

Thus, if G is **not** the cyclic group generated by a, no element of G has order 4. If G has **no** elements of order 4, then every element apart from e must have order 2. Therefore, the group is $\{e, a, b, c\}$ and $a^2 = b^2 = c^2 = e$.

The only unknown is how different elements combine.

If $a*b = a$, then premultiplying both sides by a^{-1} gives

$$a^{-1}*a = a^{-1}*(a*b), = (a^{-1}*a)*b = e*b = b$$

Since $a^{-1}*a = e$, this gives $b = e$. This cannot be true, since e, a, b, c were assumed to be four different elements of the group. Hence $a*b \neq a$.

If $a*b = b$, then by postmultiplying both sides by b^{-1}, we can prove similarly that $a = e$. A third option would be that $a*b = e$, but then premultiplying both sides by a would give $a*a*b = a*e$. Since $a^2 = e$, this gives $b = a$, which is also impossible. Since the set is closed and $a*b \neq a$, b or e, we have

$$a*b = c$$

Similarly, we can prove that $b*c = a$, and that $a*c = b$. Therefore, there is only one way in which a group of order 4 can be anything other than cyclic, and this happens if the elements combine as described above. The composition table for this group is then:

*	e	a	b	c
e	e	a	b	c
a	a	e	c	b
b	b	c	e	a
c	c	b	a	e

After filling in the row and the column next to e, and inserting e for a^2, b^2 and c^2, we can complete the composition table very simply by using the rule that each member of the group, e, a, b, c, occurs once and only once in every row and column.

Distinguishing between the two groups of order 4

We have proved that there are only two different groups of order 4. Hence, any group of order 4 is either isomorphic to a cyclic group of order 4, or isomorphic to the group whose composition table is given above.

There are two ways to distinguish between groups of order 4:

- If G contains an element of order 4, then it is cyclic.
- If G contains three elements of order 2, then it is not cyclic.

Groups of order 5

If G is a group of order 5, then the order of every element of G must be a factor of 5. Since 5 is prime, the order of each element must be either 1 or 5. There is only one element of order 1, and that is the identity element.

Select any element other than the identity element, and let this be a. Then a has order 5, and so the five elements of the group must be a, a^2, a^3, a^4 and a^5, which is e. Since the group has only five elements, these are all the elements of the group, and the group is cyclic.

We note that every element excluding the identity element is a generator.

Therefore, **any** group of order 5 is **cyclic**, and **all** groups of order 5 are **isomorphic**.

Groups of order 6

There are only two groups of order 6:

- Type 1: the cyclic group.
- Type 2: the group of symmetries of an equilateral triangle.

If we are given a group of order 6, there are several ways to tell whether it is cyclic.

Distinguishing between types 1 and 2

If G is abelian, then it must be cyclic. The group of symmetries of an equilateral triangle is not abelian.

If G has an element of order 6, then it must be a member of the cyclic group of order 6. There is no symmetry that completely generates the group of symmetries of an equilateral triangle.

If G has three elements of order 2, then it is isomorphic to the group of symmetries of an equilateral triangle. These three elements of order 2 correspond to the three reflections of an equilateral triangle. In the cyclic group of order 6, there is only one element of order 2.

Example 14 The symmetries of a square form the dihedral group, D_8. Find

a) any subgroups of D_8 of order 3

b) all the subgroups of D_8 of order 4.

SOLUTION

a) Since D_8 has order 8, there can be no subgroups of order 3, since 3 does not divide 8.

b) Let H be a subgroup of order 4.

If there is a rotation of 90° in H, then H must be the set of **all** rotations of a square. This is because a rotation of 90° generates all four rotations of a square, and because H has only four elements.

If a reflection in a diagonal is in H, then the only other reflection in H is the reflection in the other diagonal.

If we were to include any other reflection, then H must contain **all** the reflections. Thus, H must contain all four reflections **and** the identity element, which is **impossible**, since H has order 4.

Similarly, if a reflection which is not in a diagonal is included, the other reflection which is not a reflection in either diagonal, must be in the subgroup.

Therefore, the only subgroups of D_8 of order 4 are:

- All four rotations (cyclic subgroup).
- One rotation of $180°$, together with the two reflections in the diagonals and the identity.
- One rotation of $180°$, together with the two reflections which are **not** reflections in the diagonals and the identity.

Example 15 G is the group of symmetries of a square. Find all the solutions to the equation $x^3 = x$ in the group G.

SOLUTION

We really want to divide both sides of this equation by x. However, we cannot divide in groups, and so we must first multiply both sides of the equation by x^{-1}. Since the group is not abelian, we must specify whether we are going to pre- or postmultiply by x^{-1}. We will pick postmultiplication (although either way works in this case), which gives

$$x^3 = x$$
$$\Rightarrow \quad x * x * x = x$$
$$\Rightarrow \quad (x * x * x) * x^{-1} = x * x^{-1}$$
$$\Rightarrow \quad (x * x) * (x * x^{-1}) = x * x^{-1}$$
$$\Rightarrow \quad x * x * e = e$$
$$\Rightarrow \quad x * x = e$$
$$\Rightarrow \quad x^2 = e$$

Therefore, the solution is all symmetries which, done twice, give the identity.

Such transformations are called **self inverse**, since they are their own inverses. (See also page 375.)

In the group of symmetries of a square, these transformations are all four reflections, a rotation through $180°$, and the identity transformation.

Exercise 17B

1 Consider the two groups:

G_1: $(\mathbb{R}^+, \times)$, the set of positive real numbers under multiplication

G_2: $(\mathbb{R}, +)$, the set of real numbers under addition

i) What are the identity elements in each of the two groups?
ii) Why must zero be excluded from the set of elements in G_1?

Consider the mapping

$$f: \mathbb{R}^+ \rightarrow \mathbb{R}$$

$$f(x) = \log_e(x)$$

iii) Explain why this mapping defines an isomorphism from G_1 to G_2.   (NICCEA)

2 Consider the following three groups.

$\quad$ G_1: $(\{1, 3, 7, 9\}, \times_{10})$, i.e. the set $\{1, 3, 7, 9\}$ under multiplication mod 10

$\quad$ G_2: $(\{1, 5, 7, 11\}, \times_{12})$

$\quad$ G_3: $(\{1, 3, 5, 7\}, \times_8)$

Draw up the group tables for G_1, G_2 and G_3 and use them to:

i) find which two are isomorphic to each other and write down an isomorphism between them
ii) solve the equation $x^3 = x$ in each of the three groups.   (NICCEA)

3 Show that the set of all matrices of the form $\begin{pmatrix} 1 & n \\ 0 & 1 \end{pmatrix}$, where n is an integer, forms a group under

the operation of matrix multiplication. (You may assume the associativity of matrix multiplication.)
Describe the geometrical transformation represented by the matrix $\begin{pmatrix} 1 & n \\ 0 & 1 \end{pmatrix}$.   (OCR)

4 i) The set of integers $\{1, 3, 5, 7\}$, together with the operation of multiplication modulo 8, forms
a group G. Show the operation table for G.
ii) Identify three proper subgroups of G.
iii) The functions f, g, h, k are defined, for $x \neq 0$, as follows:

$$f : x \mapsto x \quad g : x \mapsto \frac{1}{x} \quad h : x \mapsto -x \quad k : x \mapsto -\frac{1}{x}$$

The set $\{f, g, h, k\}$ under the operation of composition of functions forms a group H. Show
the operation table for H.

iv) State, with a reason, whether or not G and H are isomorphic.   (OCR)

5 a) Let $G = \{1, 3, 5, 7\}$. Construct the Cayley table for G with respect to multiplication (mod 8),
and determine whether or not G is a group with respect to this operation.
b) Explain why the set $\mathbb{Z}_8 - \{0\}$ is **not** a group under multiplication (mod 8).
c) For which values of n does $\mathbb{Z}_n - \{0\}$ form a group under multiplication (mod n)?
$\qquad\qquad\qquad\qquad\qquad\qquad\qquad\qquad\qquad\qquad\qquad\qquad\qquad\qquad$ (SQA/CSYS)

6 Show that the set of all matrices of the form $\begin{pmatrix} 1-n & n \\ -n & 1+n \end{pmatrix}$, where n is an integer (positive,

negative or zero), forms a group G under the operation of matrix multiplication. (You may
assume that matrix multiplication is associative.)

The subset of G which consists of those elements for which n is an even integer (positive,
negative or zero) is denoted by H. Determine whether or not H is a subgroup of G, justifying
your answer.   (OCR)

7 a) It is given that x and y are elements of a multiplicative group G with identity e, and that
$x^2 = e$, $y^2 = e$ and $(xy)^2 = e$. Show that $xy = yx$.
b) The multiplicative group H is commutative. Two elements a and b of H are such that a has
order 2 and b has order 3. Show that ab has order 6.   (OCR)

8 The group G consists of the set of six matrices **I, A, B, C, D, E** defined below, under the operation of matrix multiplication.

$$\mathbf{I} = \begin{pmatrix} 1 & 0 & 0 \\ 0 & 1 & 0 \\ 0 & 0 & 1 \end{pmatrix} \qquad \mathbf{A} = \begin{pmatrix} 0 & 0 & 1 \\ 1 & 0 & 0 \\ 0 & 1 & 0 \end{pmatrix} \qquad \mathbf{B} = \begin{pmatrix} 0 & 1 & 0 \\ 0 & 0 & 1 \\ 1 & 0 & 0 \end{pmatrix}$$

$$\mathbf{C} = \begin{pmatrix} 1 & 0 & 0 \\ 0 & 0 & 1 \\ 0 & 1 & 0 \end{pmatrix} \qquad \mathbf{D} = \begin{pmatrix} 0 & 0 & 1 \\ 0 & 1 & 0 \\ 1 & 0 & 0 \end{pmatrix} \qquad \mathbf{E} = \begin{pmatrix} 0 & 1 & 0 \\ 1 & 0 & 0 \\ 0 & 0 & 1 \end{pmatrix}$$

i) Copy and complete the following group table for G.

	I	A	B	C	D	E
I	I	A	B	C	D	E
A	A	B	I	E	C	
B	B	I	A	D	E	
C	C	D	E			
D	D	E	C			
E						

ii) Show that G is not cyclic.

iii) Find all the proper subgroups of G.

iv) The group H consists of the six elements 1, 2, 3, 4, 5, 6 under multiplication modulo 7. The multiplicative group K consists of the six elements i, a, a^2, b, ab, a^2b, where i is the identity, $a^3 = b^2 = i$ and $ba = a^2b$. Determine whether

  a) H is isomorphic to G

  b) K is isomorphic to G

  c) H is isomorphic to K.

Give reasons for your conclusions.   (OCR)

9 The multiplicative group G has eight elements $e, a, b, c, ab, ac, bc, abc$, where e is the identity. The group is commutative, and the order of each of the elements a, b, c is 2.

i) State the orders of the elements ab and abc.

ii) Find four subgroups of G of order 4.

iii) Give a reason why no group of order 8 can have a subgroup of order 3.

The group H has elements $0, 1, 2, \ldots, 7$ with group operation addition modulo 8.

iv) Find the order of each element of H.

v) Determine whether G and H are isomorphic and justify your conclusion.   (OCR)

10 The multiplication tables for G, a cyclic group of order 6, and H, a non-cyclic group of order 6, are shown below.

G

	e	g	g^2	g^3	g^4	g^5
e	e	g	g^2	g^3	g^4	g^5
g	g	g^2	g^3	g^4	g^5	e
g^2	g^2	g^3	g^4	g^5	e	g
g^3	g^3	g^4	g^5	e	g	g^2
g^4	g^4	g^5	e	g	g^2	g^3
g^5	g^5	e	g	g^2	g^3	g^4

H

	i	h_1	h_2	h_3	h_4	h_5
i	i	h_1	h_2	h_3	h_4	h_5
h_1	h_1	h_2	i	h_5	h_3	h_4
h_2	h_2	i	h_1	h_4	h_5	h_3
h_3	h_3	h_4	h_5	i	h_1	h_2
h_4	h_4	h_5	h_3	h_2	i	h_1
h_5	h_5	h_3	h_4	h_1	h_2	i

i) Give the order of each element of G.

ii) Give the order of each element of H and write down all the proper subgroups of H.

iii) The group M has elements $1, 3, 4, 9, 10, 12$ with operation multiplication modulo 13. State to which of G and H the group M is isomorphic. For the two groups which are isomorphic, write down a correspondence between the elements. (OCR)

11 The group $G = \{e, p_1, p_2, p_3, q_1, q_2, q_3, q_4\}$ has order 8 and its multiplication table is shown below.

	e	p_1	p_2	p_3	q_1	q_2	q_3	q_4
e	e	p_1	p_2	p_3	q_1	q_2	q_3	q_4
p_1	p_1	p_2	p_3	e	q_4	q_3	q_1	q_2
p_2	p_2	p_3	e	p_1	q_2	q_1	q_4	q_3
p_3	p_3	e	p_1	p_2	q_3	q_4	q_2	q_1
q_1	q_1	q_3	q_2	q_4	e	p_2	p_1	p_3
q_2	q_2	q_4	q_1	q_3	p_2	e	p_3	p_1
q_3	q_3	q_2	q_4	q_1	p_3	p_1	e	p_2
q_4	q_4	q_1	q_3	q_2	p_1	p_3	p_2	e

i) Find the orders of p_1 and p_3.

ii) Find two subgroups of order 4.

iii) State whether G has any subgroups of order 6 and justify your answer.

iv) The group H has elements $e^{\frac{1}{4}k\pi i}$, where $k = 0, 1, \ldots, 7$, and the group operation is complex multiplication. Show that H is cyclic.

v) The set $K = \{i, a, a^2, a^3, b, ab, a^2b, a^3b\}$ is a commutative multiplicative group of order 8. The identity element is i and $a^4 = b^2 = i$. Determine whether any two of G, H, K are isomorphic to each other and justify your conclusions. (OCR)

12 a) Explain why $4 \times 14 = 2$ for multiplication modulo 18.

b) Complete the group table shown below for multiplication modulo 18.

	2	4	8	10	14	16
2	4	8	16	2	10	14
4	8	16	14	4	2	10
8	16	14	10	8	4	2
10	2	4				
14	10	2				
16	14	10				

c) State the identity element. Find a subgroup of order 2 and a subgroup of order 3.

d) State, with a reason, whether the group in part **b** is isomorphic to the group of symmetries of an equilateral triangle.   (NEAB/SMP 16–19)

13 Consider the matrices

$$\mathbf{A} = \begin{pmatrix} 3 & 2 \\ 1 & 4 \end{pmatrix} \quad \mathbf{E}_1 = \begin{pmatrix} \frac{1}{3} & 0 \\ 0 & 1 \end{pmatrix} \quad \mathbf{E}_2 = \begin{pmatrix} 1 & 0 \\ -1 & 1 \end{pmatrix} \quad \mathbf{E}_3 = \begin{pmatrix} 1 & 0 \\ 0 & \frac{3}{10} \end{pmatrix}$$

$$\mathbf{E}_4 = \begin{pmatrix} 1 & -\frac{2}{3} \\ 0 & 1 \end{pmatrix}$$

i) Describe the geometrical transformations which correspond to $\mathbf{E}_1$ and $\mathbf{E}_2$.

ii) Calculate the three products: $\mathbf{E}_1\mathbf{A}$, $\mathbf{E}_2(\mathbf{E}_1\mathbf{A})$, $\mathbf{E}_3(\mathbf{E}_2\mathbf{E}_1\mathbf{A})$. Verify that $\mathbf{E}_4(\mathbf{E}_3\mathbf{E}_2\mathbf{E}_1\mathbf{A}) = \mathbf{I}$.

iii) Find the inverse matrices $\mathbf{E}_1^{-1}, \mathbf{E}_2^{-1}, \mathbf{E}_3^{-1}$ and $\mathbf{E}_4^{-1}$.

iv) State how $\mathbf{A}$ can be written as a product of these inverse matrices. Describe fully the geometrical transformation corresponding to $\mathbf{A}$ in terms of a composition of shears and stretches, giving the scale factors and relevant directions in each case.   (NICCEA)

14 i) Form the combination table for the set $\{3, 6, 9, 12\}$ under the operation multiplication modulo 15. Write down any elements which are self-inverse.

ii) A binary operation $*$ is defined on $\mathbb{R}$ by

$$r * s = r + s + rs$$

Given that $S = \{x : x \in \mathbb{R}, x \neq -1\}$, show that S forms a group under the operation $*$.

Solve the equation

$$(x * 2) * x = 3 * (4 * x)   \text{(EDEXCEL)}$$

15 A non-abelian group G consists of eight 2×2 matrices, and the binary operation is matrix multiplication. The eight distinct elements of G can be written as

$$G = \{\mathbf{I}, \mathbf{A}, \mathbf{A}^2, \mathbf{A}^3, \mathbf{B}, \mathbf{AB}, \mathbf{A}^2\mathbf{B}, \mathbf{A}^3\mathbf{B}\}   (*)$$

where $\mathbf{I}$ is the identity matrix, and $\mathbf{A}$, $\mathbf{B}$ are 2×2 matrices such that

$$\mathbf{A}^4 = \mathbf{I} \quad \mathbf{B}^2 = \mathbf{I} \quad \text{and} \quad \mathbf{BA} = \mathbf{A}^3\mathbf{B}$$

i) Show that $(\mathbf{A}^2\mathbf{B})(\mathbf{AB}) = \mathbf{A}$ and $(\mathbf{AB})(\mathbf{A}^2\mathbf{B}) = \mathbf{A}^3$.

ii) Evaluate the following products, giving each one as an element of G as listed in $(*)$,

$$(\mathbf{AB})(\mathbf{A}) \quad (\mathbf{AB})(\mathbf{AB}) \quad (\mathbf{B})(\mathbf{A}^2)$$

iii) Find the order of each element of G.

iv) Show that $\{I, A^2, B, A^2B\}$ is a subgroup of G.

v) Find the other two subgroups of G which have order 4.

vi) For each of the three subgroups of order 4, state whether or not it is a cyclic subgroup.

(MEI)

16 Four of the subgroups of a group, **X**, are $\{A\}$, $\{A, B, C, D\}$, $\{A, C\}$ and $\{A, E\}$.

a) Explain why **X** must contain more than the five elements given above. State the minimum number of extra elements which **X** must have.

b) The subgroup $\{A, B, C, D\}$ is cyclic. State possible geometrical transformations which could correspond to the elements A, B, C and D and construct a table for this subgroup.

(NEAB/SMP 16–19)

17 The matrix $\mathbf{M}(\alpha)$ is defined by

$$\mathbf{M}(\alpha) = \begin{pmatrix} \alpha & \alpha & \alpha \\ \alpha & \alpha & \alpha \\ \alpha & \alpha & \alpha \end{pmatrix}$$

a) Show that the set $G = \{\mathbf{M}(\alpha) : \alpha \in \mathbb{C}, \alpha \neq 0\}$ forms a group under the operation of matrix multiplication, which may be assumed to be associative.

b) Find the order of $\mathbf{M}(\frac{1}{3}i)$ and hence find a subgroup of G of order 4 and a subgroup of G of order 2.

c) Show that the set $H = \{\mathbf{M}(\alpha) : \alpha = 3^k, k \in \mathbb{Z}\}$ is a subgroup of G.

d) Explain why the set $S = \{\mathbf{M}(\alpha) : \alpha = \frac{1}{3}k, k \in \mathbb{Z}, k \neq 0\}$ is not a subgroup of G.

(EDEXCEL)

18 a) Show that if $M = \begin{pmatrix} \cos\theta & \sin\theta \\ \sin\theta & -\cos\theta \end{pmatrix}$ then $M^2 = I$, where I is the 2×2 identity matrix.

By choosing two different values of θ, exhibit two matrices A, B such that $A^2 = I$ and $B^2 = I$ but $(AB)^2 \neq I$.

b) Prove that if C and D are $n \times n$ matrices such that $C^2 = I$, $D^2 = I$ **and C and D commute**, then $(CD)^2 = I$.

c) Let G be an abelian group, and define H by

$$H = \{g \in G : g^2 = e\}$$

where e is the identity element of G. Show that H is a subgroup of G.

d) The following is the multiplication table of the group D_8.

	e	a	b	c	p	q	r	s
e	e	a	b	c	p	q	r	s
a	a	b	c	e	q	r	s	p
b	b	c	e	a	r	s	p	q
c	c	e	a	b	s	p	q	r
p	p	s	r	q	e	c	b	a
q	q	p	s	r	a	e	c	b
r	r	q	p	s	b	a	e	c
s	s	r	q	p	c	b	a	e

i) Determine whether or not D_8 is abelian.

ii) Determine whether or not $\{g \in D_8 : g^2 = e\}$ is a subgroup of D_8. (SQA/CSYS)

19 The six permutations of the set $\{1, 2, 3\}$ are

$$\pi_1 = \begin{pmatrix} 1 & 2 & 3 \\ 1 & 2 & 3 \end{pmatrix} \quad \pi_2 = \begin{pmatrix} 1 & 2 & 3 \\ 2 & 3 & 1 \end{pmatrix} \quad \pi_3 = \begin{pmatrix} 1 & 2 & 3 \\ 3 & 1 & 2 \end{pmatrix}$$

$$\pi_4 = \begin{pmatrix} 1 & 2 & 3 \\ 2 & 1 & 3 \end{pmatrix} \quad \pi_5 = \begin{pmatrix} 1 & 2 & 3 \\ 3 & 2 & 1 \end{pmatrix} \quad \pi_6 = \begin{pmatrix} 1 & 2 & 3 \\ 1 & 3 & 2 \end{pmatrix}$$

i) If $\bigcirc$ denotes the composition of permutations, show that $\pi_5 \bigcirc \pi_3 = \pi_6$.

ii) Show that $\pi_3 \bigcirc (\pi_5 \bigcirc \pi_2) = (\pi_3 \bigcirc \pi_5) \bigcirc \pi_2$.

iii) Draw up a table for the group formed by $\bigcirc$ operating on the set of these six permutations.

iv) State a group which is isomorphic to this group of six permutations. (NICCEA)

20 Consider the binary operation $\otimes$ as defined by

$$a \otimes b = a + b + ab$$

i) Show that

$$a \otimes (b \otimes c) = a + b + c + ab + bc + ac + abc$$

ii) Prove $\otimes$ is associative.

Consider the algebraic system consisting of the set of real numbers, $\mathbb{R}$, and the operation $\otimes$.

iii) Find the identity element for this binary operation.

iv) By considering the inverse of the element a, show that this system is **not** a group.

v) A group can be formed using this operation and a subset of $\mathbb{R}$. State how $\mathbb{R}$ can be amended to form this subset. (NICCEA)

Real vector spaces

Groups have one binary operation. In **vector spaces**, there are two operations: addition and multiplication. The simplest, interesting real vector space is the two-dimensional vector space $\mathbb{R}^2$.

Just as with groups, to check that we have a real vector space, we need to verify that certain properties are satisfied.

A **real vector space** consists of a set of vectors V, which admit of two operations, $+$ and $.$, and have the following six properties:

- $(V, +)$ is an **abelian group**. The identity of this group is the zero vector $\mathbf{0}$.

- If $\mathbf{v}$ is a vector in V and $\lambda \in \mathbb{R}$, then $\lambda . \mathbf{v}$ is a vector in V.

- If $\mathbf{v}$ and $\mathbf{w}$ are vectors in V and $\lambda \in \mathbb{R}$, then $\lambda . (\mathbf{v} + \mathbf{w}) = \lambda . \mathbf{v} + \lambda . \mathbf{v}$.

- If $\mathbf{v} \in V$ and $\lambda, \mu \in \mathbb{R}$, then $(\lambda + \mu) . \mathbf{v} = \lambda . \mathbf{v} + \mu . \mathbf{v}$.

- If $\mathbf{v} \in V$ and $\lambda, \mu \in \mathbb{R}$, then $\lambda(\mu . \mathbf{v}) = (\lambda\mu) . \mathbf{v}$.

- For any $\mathbf{v} \in V$, $1 . \mathbf{v} = \mathbf{v}$.

The last four properties define the operation of multiplication by scalars, which we have already used with geometric vectors (see page 94).

Three-dimensional space

Let V be the three-dimensional vector space $\mathbb{R}^3$. We can define any vector in terms of $\mathbf{i}$, $\mathbf{j}$ and $\mathbf{k}$. For example, $2\mathbf{i} + 3\mathbf{k}$ is a vector in this space. We know how to perform addition in this vector space: we just add the components separately. We also know how to multiply a vector by a real scalar. All of the real vector space properties are satisfied.

Complex numbers

The set of all complex numbers can be treated as a real vector space. We can write any complex number in the form $a + bi$, where a and b are real numbers. Addition is performed in the normal way, and we know how to multiply any complex number by a **real** scalar. Again, we verify that the real vector space properties are satisfied.

Linearly independent sets

Let V be a vector space. We say that $\{\mathbf{v}_1, \mathbf{v}_2, \ldots, \mathbf{v}_n\}$ is a **linearly independent** set if, whenever $\sum_{k=1}^{n} \lambda_k \mathbf{v}_k = \mathbf{0}$, we can deduce that $\lambda_k = 0$ for every k.

Let us consider the vectors $\mathbf{i}$ and $\mathbf{j}$ in normal two-dimensional space. And let us imagine that we are trying to follow paths from the origin back to itself by following the vectors $\mathbf{i}$ and $\mathbf{j}$. Two examples are shown below.

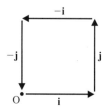 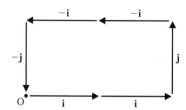

In both cases, the total number of components of vector $\mathbf{i}$ we follow is 0. Similarly, the total number of components of vector $\mathbf{j}$ we follow is 0. Now, the only way that $a\mathbf{i} + b\mathbf{j} = 0$ is if $a = b = 0$. So, $\mathbf{i}$ and $\mathbf{j}$ are linearly independent.

On the other hand, the vectors $\mathbf{i}$, $\mathbf{j}$ and $2\mathbf{i} - 3\mathbf{j}$ are **not** linearly independent. The diagram on the right shows that $-2\mathbf{i} + 3\mathbf{j} + 1(2\mathbf{i} - 3\mathbf{j}) = \mathbf{0}$.

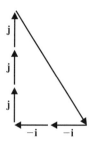

Any three vectors in two-dimensional space are **not** linearly independent.

Spanning sets and basis vectors

Let V be a vector space. The set $\{\mathbf{v}_1, \mathbf{v}_2, \ldots, \mathbf{v}_n\}$ is called a **spanning set** if we can write any element $\mathbf{v}$ of V as a sum,

$$\mathbf{v} = \sum_{k=1}^{n} \lambda_k \mathbf{v}_k$$

A **basis** is a spanning set that is also a linearly independent set. A basis exists in any real vector space. Any two bases of the same vector space have the same number of elements.

The vectors **i**, **j**, and **k** form a basis for three-dimensional space.

The number of elements in a basis of V is called the **dimension** of V.

The vectors 1 and i form a basis for the set of all complex numbers. Thus, the set of all complex numbers have dimension two when regarded as a real vector space.

Linear mappings

A linear mapping $T : V \rightarrow V$ is one which satisfies

$$T(\lambda v) = \lambda T(v) \quad \text{for every } \lambda \in \mathbb{R}, \, v \in V$$

and

$$T(v + w) = T(v) + T(w)$$

Linear mappings are completely determined by their effect on a basis.

For example, consider an anticlockwise rotation of $90°$ in $\mathbb{R}^2$. This moves vector **i** onto **j**, and it also moves **j** onto $-\mathbf{i}$. It follows that the vector $2\mathbf{i} + 3\mathbf{j}$ is carried to $2\mathbf{j} + 3(-\mathbf{i})$.

Now, if we use $\mathbf{e}_1$ to denote **i**, the first basis element, and $\mathbf{e}_2$ to denote **j**, the second basis element, we have

$$T(\mathbf{e}_1) = \mathbf{e}_2 \qquad T(\mathbf{e}_2) = -\mathbf{e}_1$$

For convenience, we represent $\mathbf{e}_1$ by the column vector $\begin{pmatrix} 1 \\ 0 \end{pmatrix}$, and $\mathbf{e}_2$ by the column vector $\begin{pmatrix} 0 \\ 1 \end{pmatrix}$.

We can then represent T by the matrix

$$\mathbf{T} = \begin{pmatrix} 0 & -1 \\ 1 & 0 \end{pmatrix}$$

which gives

$$T(\mathbf{e}_1) = \mathbf{T}\begin{pmatrix} 1 \\ 0 \end{pmatrix} = \begin{pmatrix} 0 \\ 1 \end{pmatrix} = \mathbf{e}_2$$

and

$$T(\mathbf{e}_2) = \mathbf{T}\begin{pmatrix} 0 \\ 1 \end{pmatrix} = \begin{pmatrix} -1 \\ 0 \end{pmatrix} = -\mathbf{e}_1$$

We note that T depends on our choice of basis.

Example 16 illustrates the difference that a change of basis can make to a transformation matrix.

Example 16 Let T be reflection in the *x*-axis. Find the transformation matrix of T with respect to these bases:

a) $e_1 = i$ and $e_2 = j$

b) $e_1 = \dfrac{\sqrt{3}}{2}i + \dfrac{1}{2}j$ and $e_2 = -\dfrac{1}{2}i + \dfrac{\sqrt{3}}{2}j$

SOLUTION

a) When $e_1 = i$ and $e_2 = j$, e_1 is unchanged by reflection in the line $y = 0$, and e_2 is moved to $-e_2$. So, we have

$$T = \begin{pmatrix} 1 & 0 \\ 0 & -1 \end{pmatrix}$$

b) When $e_1 = \dfrac{\sqrt{3}}{2}i + \dfrac{1}{2}j$, e_1 is reflected to $e_1' = \dfrac{\sqrt{3}}{2}i - \dfrac{1}{2}j$. We have to write this in terms of e_1 and e_2, so we solve the equation

$$\frac{\sqrt{3}}{2}i - \frac{1}{2}j = ae_1 + be_2 = a\left(\frac{\sqrt{3}}{2}i + \frac{1}{2}j\right) + b\left(-\frac{1}{2}i + \frac{\sqrt{3}}{2}j\right)$$

Equating components of **i** and **j**, we find that

$$\frac{\sqrt{3}}{2} = \frac{\sqrt{3}}{2}a - \frac{1}{2}b$$

$$-\frac{1}{2} = \frac{1}{2}a + \frac{\sqrt{3}}{2}b$$

Solving these, we obtain $a = \dfrac{1}{2}, b = -\dfrac{\sqrt{3}}{2}$.

Therefore, we have

$$T(e_1) = \frac{1}{2}e_1 - \frac{\sqrt{3}}{2}e_2$$

Similarly, we can find

$$T(e_2) = -\frac{\sqrt{3}}{2}e_1 - \frac{1}{2}e_2$$

Therefore, the transformation matrix is

$$T = \begin{pmatrix} \dfrac{1}{2} & -\dfrac{\sqrt{3}}{2} \\ -\dfrac{\sqrt{3}}{2} & -\dfrac{1}{2} \end{pmatrix}$$

Linear transformations of the set of all complex numbers

The complex conjugate mapping $z \to z^*$ is a linear transformation of the set of all complex numbers regarded as a real vector space. This is because $z \to z^*$ corresponds to a reflection in the real axis of an Argand diagram, which corresponds to a reflection in the line $y = 0$ in $\mathbb{R}^2$.

Example 17 Find the transformation matrix for the mapping

$$f : z \rightarrow (1 + i) z + (2 - i) z^*$$

with respect to the basis $\{1, i\}$.

SOLUTION

We put $e_1 = 1$ and $e_2 = i$. (This reminds us to think of 1 and i as **basis vectors**.)

We now calculate the effect of the transformation on each vector.

$$f(e_1) = f(1) = (1 + i) + (2 - i) = 3 = 3e_1$$
$$f(e_2) = f(i) = (1 + i)i + (2 - i)(-i) = -2 - i = -2e_1 - e_2$$

Therefore, with respect to the basis $\{1, i\}$, the transformation matrix is

$$\begin{pmatrix} 3 & -2 \\ 0 & -1 \end{pmatrix}$$

Exercise 17C

1 The vectors **a**, **b** and **c** are

$$\mathbf{a} = \begin{pmatrix} 1 \\ 2 \\ 3 \end{pmatrix} \quad \mathbf{b} = \begin{pmatrix} 2 \\ 3 \\ 7 \end{pmatrix} \quad \mathbf{c} = \begin{pmatrix} -2 \\ -2 \\ -8 \end{pmatrix}$$

Prove that **a**, **b** and **c** are linearly dependent.

2 What is the span of the three vectors $\begin{pmatrix} 2 \\ 3 \\ 4 \end{pmatrix}, \begin{pmatrix} 1 \\ 0 \\ 1 \end{pmatrix}, \begin{pmatrix} 3 \\ 3 \\ 4 \end{pmatrix}$ in $\mathbb{R}^3$?

Find two vectors with the same span.

3 The vectors **a**, **b** and **c** are

$$\mathbf{a} = \begin{pmatrix} 1 \\ 2 \\ 3 \end{pmatrix} \quad \mathbf{b} = \begin{pmatrix} 0 \\ 1 \\ 4 \end{pmatrix} \quad \mathbf{c} = \begin{pmatrix} 4 \\ 9 \\ 13 \end{pmatrix}$$

Prove that **a**, **b** and **c** are linearly independent.

4 The vectors **a**, **b** and **c** in $\mathbb{R}^3$ are

$$\mathbf{a} = \begin{pmatrix} 1 \\ 2 \\ 3 \end{pmatrix} \quad \mathbf{b} = \begin{pmatrix} 2 \\ 4 \\ 5 \end{pmatrix} \quad \mathbf{c} = \begin{pmatrix} 0 \\ 0 \\ 1 \end{pmatrix}$$

Prove that **a**, **b** and **c** are linearly dependent. Find a vector **d** which, together with **a**, **b** and **c**, spans $\mathbb{R}^3$.

5 Find the transformation matrix (with respect to the standard basis) of the linear transformation which takes $(1, 0)$ to $(2, 1)$ and $(1, 1)$ to $(3, 4)$.

Now find the transformation matrix with respect to the basis $\begin{pmatrix} 2 \\ 0 \end{pmatrix}, \begin{pmatrix} 1 \\ 1 \end{pmatrix}$.

6 The linear transformation $T : \mathbb{R}^2 \to \mathbb{R}^2$ is defined by

$$T\begin{pmatrix} x \\ y \end{pmatrix} = \begin{pmatrix} 2 & 0 \\ 1 & 0 \end{pmatrix} \begin{pmatrix} x \\ y \end{pmatrix}$$

a) Describe all the points in the image of T.

b) Write down the dimension of the image of T.

c) Find a spanning set for the image of T.

7 Certain sets of functions can be viewed as real vector spaces. For example, consider the set

$$T := \left\{ \text{functions} \,|\, f : \mathbb{R} \to \mathbb{R}, f(x) = a_0 + \sum_{n=0}^{\infty} [b_n \sin nx + a_n \cos nx] \right\}$$

'Vectors' in the vector space T are really functions.

a) Prove that T forms a real vector space.

b) Prove that the vectors $\sin x$ and $\sin 2x$ are linearly independent.

8 Let V be a real vector space with operations $+$ and $..$ A set U, with operations $+$ and $.,$ is said to be a subspace of V if the following conditions hold:

- U contains 0.
- U is closed under addition.
- U is closed under scalar multiplication.

a) Prove that if U is a subspace of V, then U is also a vector space.

b) What is the smallest possible subspace of $\mathbb{R}^3$?

9 The set $\mathbb{C}$ of complex numbers can be regarded as a real vector space, where addition of complex numbers, and multiplication of a complex number by a real scalar, are defined in the usual way.

i) Show that $\{1, j\}$ is a basis for this vector space.

Let $u = a + bj$ and $v = c + dj$ (where a, b, c and d are real) be fixed complex numbers. A mapping $T : \mathbb{C} \to \mathbb{C}$ is defined by $T(z) = uz + vz^*$ (where z^* is the complex conjugate of z).

ii) Show that T is a linear mapping.

iii) Find the matrix $\mathbf{M}$ associated with T and the basis $\{1, j\}$.

iv) Given that $\mathbf{M} = \begin{pmatrix} 6 & 1 \\ 11 & 2 \end{pmatrix}$, find the complex number z for which $uz + vz^* = 1 + 4j$.

(MEI)

Answers

Exercise 1A

1 a) $-i$ **b)** 1 **c)** -1 **d)** i **2 a)** $3+2i$ **b)** $6-3i$ **c)** $-4+3i$ **d)** $-2+2\sqrt{2}i$ **e)** $2i$ **3 a)** $3-4i$ **b)** $2+6i$

c) $-4+3i$ **d)** $-8-5i$ **4 a)** $-1\pm\sqrt{3}i$ **b)** $\frac{3}{2}\pm\frac{\sqrt{15}}{2}i$ **c)** $\frac{1}{4}(-1\pm\sqrt{7}i)$ **d)** $1\pm\frac{\sqrt{2}}{2}i$ **5 a)** $10-2i$ **b)** $1-i$ **c)** $-1+2i$

d) $2+55i$ **e)** $1+i$ **f)** $3+8i$ **g)** $10+18i$ **h)** $18+13i$ **6 a)** $3+11i$ **b)** $26+2i$ **c)** $74+7i$ **d)** $42-24i$ **e)** $10+11i$

f) $11-29i$ **7 a)** $\frac{1}{17}(5+14i)$ **b)** $\frac{1}{26}(23+11i)$ **c)** $1-2i$ **d)** $\frac{1}{13}(4-19i)$ **8 a)** $x=4, y=-2$ **b)** $x=-11, y=22$

c) $x=8, y=-1$ **d)** $x=17, y=-17$ **e)** $x=\frac{13}{5}, y=\frac{9}{5}$ **f)** $x=-5, y=-12$ **9** $3.3+0.9i$ **10 a)** $-2+\sqrt{3}i$ **b)** $-1\pm\sqrt{5}i$

c) $\frac{1}{2}(-3\pm3i)$ **d)** $\frac{5}{2}(1\pm\sqrt{3}i)$

Exercise 1B

2 a) $2\sqrt{2}, \pi/4$ **b)** $3\sqrt{2}, 3\pi/4$ **c)** $4, 2\pi/3$ **d)** $\sqrt{2}, -3\pi/4$ **e)** $4, \pi/2$ **f)** $13, \tan^{-1}\left(\frac{12}{5}\right)$ **g)** $4, \pi$ **h)** $7, \tan^{-1}\left(\frac{\sqrt{13}}{6}\right)$

3 a) i) $-7+24i$ **ii)** $-117+44i$ **b) i)** 5 **ii)** 25 **iii)** 125 **c) i)** 0.9273 **ii)** 1.8546 **iii)** 2.7819 **4 a)** $1+\sqrt{3}i$ **b)** $2\sqrt{2}+2\sqrt{2}i$

4 c) $-i$ **d)** $-2\sqrt{2}+2\sqrt{2}i$ **e)** $-\sqrt{3}+i$ **f)** $-3\sqrt{3}-3i$ **5 a)** $-\frac{1}{5}+\frac{2}{5}i$ **b)** $13, \pi-\tan^{-1}\left(\frac{12}{5}\right)$ or 1.9656 **6** $\frac{5}{13}, 2.1033$

7 a) $-\frac{1}{5}+\frac{3}{5}i$ **b)** $\frac{1}{5}\sqrt{10}, \pi-\tan^{-1}3$ or 1.8925 **8 i) b)** $3\pi/4$ **ii)** $\pm(5-3i)$ **9 i)** $\sec\alpha$ **ii)** $4\sec\alpha$ **iii)** $\pi/2-\alpha$ **iv)** $2\pi/5-\alpha$

10 i) $\sin\alpha$ **ii)** $\pi/2-\alpha$ **11 a)** $3-4i$ **c)** 1.95 **d)** $6+14i$ **12 a)** $z=-2\pm\sqrt{3}i$ **c) i)** 2.65 **ii)** ±2.43 **13** $2-3i$ **14** $\frac{6}{5}+\frac{12}{5}i$

15 b) 5 **c)** $\frac{1}{13}(7+17i)$ **d)** 1.18 **e)** $-7, -5$ **16 a)** $-11-2i, 1+2i$ **c)** 2.2 **d)** $-1, -2$ **17 a)** $2\sqrt{2}, 3\pi/4$ **b)** $1/2\sqrt{2}, -3\pi/4$

17 d) $90°$ **18 a)** 2.68 rad **b)** $6, 4$ **d)** 28 **19 i)** $\pm(3+2i)$ **ii)** $-\frac{119}{169}$ **20 a)** $14+2i, 1+i$ **b) i)** $2\sqrt{5}, 0.464$ rad; $\sqrt{10}, -0.3218$ rad

20 b) ii) $\sqrt{10}$

Exercise 1C

6 a) -1 **b)** 2 **c)** 0 **7 a)** 0 or 3 **b)** 2 **c)** w^2 **d)** -1 or 2 **8** $3, \frac{3}{2}+\frac{1}{2}\sqrt{3}i, \frac{3}{2}-\frac{1}{2}\sqrt{3}i$ **10 b)** $5i, 3\sqrt{2}$ **11 i)** 2 **ii)** $5\pi/6$

14 2 **15 i) a)** $\frac{1}{2}(1+i)$ **b)** $\frac{1}{2}(1-i)$ **ii)** $(1-w)/w$ **16 b) i)** $3/2, \sqrt{3}/2$ **ii)** $\pi/6$ **17 a)** $\pi/4, -4(1+a)^4$ **b)** $y=x$

Exercise 2A

1 $n\pi+(-1)^n\frac{\pi}{4}, 180n°+(-1)^n45°$ **2** $2n\pi\pm\frac{2\pi}{3}, 360n°\pm120°$ **3** $n\frac{\pi}{2}+(-1)^n\frac{\pi}{12}, 90n°+(-1)^n15°$ **4** $n\frac{\pi}{3}+\frac{\pi}{12}, 60n°+15°$

5 $n\pi+\frac{\pi}{8}$ **6** $\frac{2}{3}n\pi+\frac{2\pi}{9}, \frac{2}{3}n\pi$ **7** $n\frac{\pi}{2}-\frac{\pi}{24}$ **8** $n\frac{\pi}{4}+\frac{\pi}{16}, 45n°+11.25°$ **9** $\frac{2}{3}n\pi+\frac{\pi}{3}, 120n°+60°$

10 $n\pi+\frac{\pi}{2}, 2n\pi\pm\frac{\pi}{3}; 180n°+90°, 360n°\pm60°$ **11** $n\pi+\frac{\pi}{2}, 180n°+90°$ **12** $n\frac{\pi}{2}, \frac{2}{3}n\pi\pm\frac{\pi}{9}; 90n°, 120n°\pm20°$ **13** $n\pi, 180n°$

14 $2n\pi+\frac{3\pi}{2}, n\pi+(-1)^n0.848; 360n°+270°, 180n°+(-1)^n48.59°$ **15** $n\frac{\pi}{2}-\frac{\pi}{8}, n\frac{\pi}{5}-\frac{\pi}{20}; 90n°-22.5°, 36n°-9°$ **16** $n\pi, n\pi\pm\frac{\pi}{6}$

17 $n\pi+\frac{\pi}{2}, 2n\pi+(-1)^n\frac{\pi}{6}$ **18** $\frac{\pi}{3}+2n\pi, \frac{2}{3}n\pi-\frac{\pi}{9}$ **19** $\frac{2t}{1-t^2}; n\pi, n\pi\pm\frac{\pi}{3}$

Exercise 2B

1 a) $13, 67.3°$ **b)** $5, 53.13°$ **c)** $5, 53.13°$ **d)** $\sqrt{2}, 45°$ **e)** $10, 53.13°$ **2 a) i)** $15, -15$ **ii)** $306.87°, 143.13°$

2 b) i) $10, -10$ **ii)** $18.43°, 108.43°$ **c) i)** $\frac{4}{3}, \frac{4}{13}$ **ii)** $53.13°, 233, 13°$ **d) i)** $\frac{3}{4-\sqrt{5}}, \frac{3}{4+\sqrt{5}}$ **ii)** $296.56°, 116.56°$

405

ANSWERS

2 e) i) $-\frac{1}{2}$ (max), $\frac{1}{3}$ (min) **ii)** $233.13°$, $53.13°$ **f) i)** $-\frac{3}{2}$ (max), $\frac{1}{6}$ (min) **ii)** $216.87°$, $36.87°$ **3 a)** $360n° + 53.13° \pm 60°$

3 b) $360n° - 22.62° \pm 60°$ **c)** $180n° - 7.5°, 180n° + 52.5°$ **d)** $60n° + 15° + (-1)^n 10°$ **e)** $60n°, 60n° - 17.7°$

Exercise 2C

1 a) $-\pi/6$ **b)** $\pi/6$ **c)** $5\pi/6$ **d)** $\pi/4$ **e)** $\pi/4$ **f)** $18.4°$ or 0.3218 rad **3** $\pi/10$ **5** $360n° - 66.8° \pm 142.0°$ **6 a)** 13 **b)** $67.4°$

6 c) $360n° - 67.4° \pm 72.1°$ **7 i)** $\sqrt{58}\cos(\theta + 23.2°)$ **ii)** $\sqrt{58}, -\sqrt{58}$ **iii)** $360n° - 23.2° \pm 82.5°$ **8 a)** $126.9°, 270°$

8 b) $180n° + 90°, 60n° + 15°$ **9 a)** $25, 73.7°$ **b)** $360n° + 20.6°$ or $360n° + 126.8°$ **c)** $\frac{1}{630} \leqslant f(x) \leqslant \frac{1}{5}$ **10** $60°; 195°, 345°$

11 $2\cos(\theta - 60°), 360n° + 60° \pm 40°$ **12 a)** $85, 0.154$ rad **b)** $2\pi n - 0.154 \pm 1.369$ **13 a)** 25 **b)** -25 **c)** 1.85 rad

13 d) 3.84 rad, 6.16 rad **14 i) a)** $15, 53.13°$ **b)** $156.9°, 276.9°$ **ii)** $n\pi + \pi/3$

Exercise 2D

1 (a) $\dfrac{5}{\sqrt{1-25x^2}}$ **b)** $\dfrac{3}{1+9x^2}$ **c)** $\dfrac{\sqrt{2}}{\sqrt{1-2x^2}}$ **d)** $\dfrac{12}{16+9x^2}$ **e)** $\dfrac{2x}{\sqrt{1-x^4}}$ **f)** $\dfrac{1-x^2}{1+3x^2+x^4}$ **g)** $\dfrac{6(\sin^{-1}2x)^2}{\sqrt{1-4x^2}}$ **h)** $\dfrac{1620(\tan^{-1}5x)^3}{1+25x^2}$

1 i) $\dfrac{1}{x\sqrt{x^2-1}}$ **j)** $-\dfrac{1}{x^2+1}$ **2 a)** $\sin^{-1}\left(\dfrac{x}{2}\right)+c$ **b)** $\sin^{-1}\left(\dfrac{x}{3}\right)+c$ **c)** $\dfrac{1}{2}\sin^{-1}\left(\dfrac{2x}{5}\right)+c$ **d)** $\dfrac{1}{9}\sin^{-1}\left(\dfrac{3x}{4}\right)+c$ **e)** $\dfrac{1}{3}\tan^{-1}\left(\dfrac{x}{3}\right)+c$

2 f) $\dfrac{1}{4}\tan^{-1}\left(\dfrac{x}{4}\right)+c$ **g)** $\dfrac{1}{20}\tan^{-1}\left(\dfrac{4x}{5}\right)+c$ **h)** $\dfrac{1}{15}\tan^{-1}\left(\dfrac{5x}{3}\right)+c$ **3 a)** $\pi/2$ **b)** $\pi/8$ **c)** $\pi/2$ **d)** $\pi/6\sqrt{3}$ **e)** $\pi/5$

4 a) 0.0505 **b)** 0.0444 **c)** 0.615 **d)** 0.0741 **e)** 0.841 **f)** 0.0207 **5** $\pi/24$ **6** $9-(x-2)^2, \pi/3$ **7** $2+\dfrac{1}{1+x}+\dfrac{2x+1}{x^2+4}$

8 $\dfrac{\sqrt{3}}{3}-\dfrac{\pi}{12}$ **9** $\dfrac{y+x\dfrac{dy}{dx}}{1+x^2y^2}$ **10 ii)** $-\dfrac{x}{\sqrt{1-x^2}}$ **iii)** $\dfrac{\pi}{2}-1$ **11** $-\sin^{-1}\left(\dfrac{1}{x}\right)+c$ **12 ii)** $\dfrac{1}{2}t\sqrt{1-t^2}$

Exercise 3B

1 a) $x^2+y^2=16$ **b)** $x=3$ **c)** $y=7$ **d)** $x^2+y^2=ax+a\sqrt{x^2+y^2}$ **e)** $x^2+y^2+ax=a\sqrt{x^2+y^2}$ **f)** $y^2=4-4x$

2 a) $r=3$ **b)** $r^2\sin 2\theta=32$ **c)** $\dfrac{r^2\cos^2\theta}{9}+\dfrac{r^2\sin^2\theta}{16}=1$ **d)** $r=6\cos\theta$ **e)** $r^2+8r\sin\theta=16$ **f)** $r^2=\cos 2\theta$

Exercise 3D

1 $\dfrac{7\pi^3 a^2}{48}$ **2 a)** $\dfrac{\pi a^2}{8}$ **b)** $\dfrac{\pi a^2}{8}$ **c)** $\dfrac{\pi a^2}{16}$ **3** $\dfrac{\pi a^2}{4}$ **4** $5\sqrt{5}+\dfrac{17}{2}\cos^{-1}\left(-\dfrac{2}{3}\right)$ **5 a)** $r^2=\sin^4\theta$ **b) ii)** $\dfrac{3\pi}{8}$

6 $\left(1,\dfrac{\pi}{6}\right), \left(1,\dfrac{5\pi}{6}\right); \dfrac{7\pi}{3}-4\sqrt{3}$ **7 b)** $\dfrac{\pi a^2}{12}$ **8 b)** $\dfrac{9\pi a^2}{2}$

Exercise 3E

1 $y=0.185a, y=\pm 0.88a$ **2 a)** $y=\dfrac{e^{-\pi/4}}{\sqrt{2}}, y=\dfrac{e^{3\pi/4}}{\sqrt{2}}, \dots$ **b)** $x=\dfrac{e^{-3\pi/4}}{\sqrt{2}}, x=\dfrac{e^{\pi/4}}{\sqrt{2}}, \dots$ **3 a)** $y=\pm\dfrac{2}{3\sqrt{6}}$ **b)** $x=\pm a$

4 b) $\dfrac{\pi}{16}$ **c) iii)** $\dfrac{16}{27}$ **5** $\sqrt{3}\cos\theta-\cos^2\theta; 0, \dfrac{\pi}{6}, -\dfrac{\pi}{6}, \pi$ **6 b)** $(0.667, 0.421), (0.667, -0.421)$ **7 b)** $a^2\left(2-\dfrac{\pi}{4}\right)$

8 b) $(0, 0), \left(3, \dfrac{\pi}{3}\right)$ **c)** 1.33 **9 a)** $r=2$ **b)** $r\cos\theta=3$ **c)** $r\sin\left(\theta+\dfrac{\pi}{3}\right)=2\sqrt{3}$

Exercise 4A

1 a) x^2 **b)** $\sqrt{x^2+1}$ **c)** $\dfrac{1}{x^3}$ **d)** $\sec x$ **e)** $\sqrt{x^2-1}$ **f)** 2^{3x} **2** $y=\dfrac{1}{3}x-\dfrac{1}{9}+ce^{-3x}$ **3** $y=-\dfrac{1}{3}e^{2x}+ce^{5x}$ **4** $y=\dfrac{x^2}{3}+\dfrac{c}{x}$

5 $y=x^3+cx^2$ **6** $y=5(x-1)^4\ln(x-1)+c(x-1)^4$ **7** $y\sin x=\dfrac{2}{5}e^{2x}\sin x-\dfrac{1}{5}e^{2x}\cos x+c$ **8 a)** $ye^x=x+c$

8 b) i) $y=(x-1)e^{-x}$ **ii)** $y=0$ **9** $y=(c+\dfrac{1}{2}x^2)e^{x^3}, y=(1+\dfrac{1}{2}x^2)e^{x^3}$ **10** $y=x\cos x+c\cos x$

11 $y=\dfrac{1}{5}x+cx^{-4}, y=\dfrac{1}{5}(x+4x^{-4})$ **12** $y=x^2-x+cx^{-4}$ **13 i)** $v=\dfrac{1}{\alpha+\beta}e^{\beta t}+ce^{\alpha t}$ **14** $v=200\tan 0.2+\sec 0.2$

15 i) $y \sec x = x + c$ **ii)** $y = (x+2)\cos x$ **16** $y = (3x^4 + 5)e^{-x^2 - x}$ **17** $N = \dfrac{\mu}{\lambda} t + \dfrac{\mu}{\lambda^2} + \left(N_0 - \dfrac{\mu}{\lambda^2}\right)e^{\lambda t}$ **18 i)** $y = ce^{kx} - x - \dfrac{1}{k}$

18 ii) a) $y = -x - \frac{1}{4} - \frac{3}{4}e^{4(x-1)}$ **19 ii)** $y \sin^2 x = -\frac{1}{2}\cos 2x + c$ **20** $s = \frac{1}{2}\tan^{-1}\left(\dfrac{t}{2}\right) + c,\ s = \frac{1}{2}\tan^{-1}\left(\dfrac{t}{2}\right) - \dfrac{\pi}{8}$

21 a) $y = cx - xe^{-x}$ **b) i)** $y = -xe^{-x}$ **ii)** 0

Exercise 4B

1 $y = e^{(3+\sqrt{17})x} + Be^{(3-\sqrt{17})x}$ **2** $y = Ae^{-x} + Be^{-2x}$ **3** $y = Ae^{2x} + Be^{-3/2x}$ **4** $y = Ae^x + Be^{-7x/3}$ **5** $x = Ae^{8t} + Be^{-t}$

6 $x = Ae^{7t} + Be^{4t}$ **7** $y = (A + Bx)e^{2x}$ **8** $y = (A + Bx)e^{-3x}$ **9** $y = e^{-x/2}\left[A\cos\left(\dfrac{\sqrt{3}}{2}x\right) + B\sin\left(\dfrac{\sqrt{3}}{2}x\right)\right]$

10 $y = e^{-2x}(A\cos 2x + B\sin 2x)$ **11** $x = Ae^{(3+\sqrt{2})t} + Be^{(3-\sqrt{2})t}$ **12** $x = e^{-t}(A\cos 2\sqrt{3}t + B\sin 2\sqrt{3}t)$

Exercise 4C

1 $y = Ae^{-8x} + Be^x - 2x - \frac{7}{4}$ **2** $y = Ae^{-x} + Be^{-3x} - 4e^{-2x}$ **3** $y = Ae^{-x} + Be^{5x/2} - 2x^2 + \frac{12}{5}x - \frac{81}{25}$

4 $y = Ae^{-x} + Be^{x/3} - \frac{304}{5876}\sin 5x - \frac{40}{5876}\cos 5x$ **5** $x = Ae^{5t} + Be^{-t} - \frac{3}{8}e^{3t}$

6 $s = Ae^{5t} + Be^{3t} + \frac{55}{377}\cos 2t - \frac{80}{377}\sin 2t$ **7** $y = Ae^{-4x} + Be^{-x} + \frac{2}{3}xe^{-x}$ **8** $y = (A + Bx + \frac{5}{2}x^2)e^{3x}$

9 $y = e^x(A\cos\sqrt{2}x + B\sin\sqrt{2}x) + 2e^{4x}$ **10** $y = e^{-3x}(A\cos x + B\sin x) + \frac{3}{2}e^{-4x}$ **11** $x = (A + Bt + 2t^2)e^t$

12 $x = A\cos 4t + B\sin 4t + \frac{3}{8}t\sin 4t$ **13** $x = (2\sin 2t - 3\cos 2t)e^t$ **14 a)** $y = Ae^{2x} + Be^{-2x} + 2e^{3x}$ **b)** $y = e^{-2x} - 5e^{2x} + 2e^{3x}$

15 $y = A + Bx + \frac{1}{4}e^{2x} - 4\cos\frac{1}{2}x$; $y(0)$ and $y'(0)$, or y for two values of x **16 i)** $y = e^{-2x}(4\cos 3x + 3\sin 3x)$ **ii)** $\dfrac{3}{2} - \dfrac{\pi}{24}$

17 $y = e^{2x}(A\cos x + B\sin x) + \frac{1}{65}(\sin 2x + 8\cos 2x)$ **18 i)** $x = A\cos 4t + B\sin 4t$ **iii)** 1.424 **19** $x = 2t + 3 + 4e^{-t/10}\sin\left(\dfrac{t}{5}\right)$

20 $x = Ae^{-4t} + Be^{-t} + \sin 3t$ **21** $y = Ae^{4x} + Be^{-x} + 3\cos 2x - 4\sin 2x,\ y = 3\cos 2x - 4\sin 2x - 3e^{-x}$

22 i) $x = e^{2t}(A\cos 5t + B\sin 5t) + 2\sin 2t$ **ii)** $x = 3e^{2t}\cos 5t + 2\sin 2t$ **23 a)** $\ln(1 + y) = c + \frac{1}{2}x^2$

23 b) $y = Ce^{3x};\ y = -\frac{2}{9} - \frac{2}{3}x + e^{4x},\ y = Ce^{3x} - \frac{2}{9} - \frac{2}{3}x + e^{4x}$ **24 a)** $y = e^{-2x}(A\cos 3x + B\sin 3x)$ **b)** $\frac{3}{4},\ \frac{1}{4}$

24 d) $y = \frac{3}{4}\cos 3x + \frac{1}{4}\sin 3x - \frac{3}{4}e^{-2x}(\cos 3x + \sin 3x)$ **25 a)** $y = Ae^{4x} + Be^{-x/2} + \sin x + 2\cos x$ **b)** $y = 2e^{4x} - 4e^{-x/2} + \sin x + 2\cos x$

26 i) $x = e^{-t}\left[A\cos\left(\dfrac{t}{2}\right) + B\sin\left(\dfrac{t}{2}\right)\right] + 2\cos t$ **ii)** \$9 550 000 **27** $p = 0,\ q = -\frac{1}{4},\ y = A\sin 2x + B\cos 2x - \frac{1}{4}x\cos 2x,\ y \approx \frac{1}{4}(n + \frac{1}{2})\pi$

28 a) $\frac{1}{2}$ **b)** $y = (1 + t)e^{-t} + \frac{1}{2}t^2e^{-t}$

Exercise 4D

1 a) $x^3(x - 4y) = c$ **b)** $y^2 = 2x^2\ln x + cx^2$ **c)** $\ln\left[\dfrac{x^4(x - y)}{x + y}\right] = \dfrac{2x}{x + y} + c$ **d)** $\dfrac{3}{8}\ln\left[\dfrac{(y - 2x)(y + 2x)}{y^2}\right] = \ln x + c$

2 $x + y - \frac{1}{4}\ln(4x + 4y + 5) = 4x + c$ **3** $2x + 3y + \frac{21}{8}\ln(16x + 24y - 13) = 8x + c$ **4 a)** $y = Ax + \dfrac{B}{x^2}$ **b)** $y = Ax^{3+\sqrt{15}} + Bx^{3-\sqrt{15}}$

4 c) $y = Ax^2 + Bx^2\ln x$ **d)** $y = x^{-\frac{1}{2}}\left[A\cos\left(\dfrac{\sqrt{3}}{2}\ln x\right) + B\sin\left(\dfrac{\sqrt{3}}{2}\ln x\right)\right]$ **5** $\dfrac{dy}{dx} = 2\sqrt{t}\dfrac{dy}{dt},\ y = Ae^{x^2} + Be^{-4x^2} + \frac{1}{6}e^{2x^2}$

6 a) $z = \frac{1}{2}e^x + ce^{-x}$ **b)** $y = \frac{1}{2}xe^x + cxe^{-x}$ **7 b)** 2 **c)** $xy = 2e^4(1 - x)e^{-2x} + 2e^{2x}$ **d)** Infinite

8 $x = (A + Bt)e^{-t},\ y = (A + B + Bt)e^{-t};\ x = (1 - t)e^{-t},\ y = -te^{-t}$ **9 a)** $y = \dfrac{(\cosh^{-1}x + c)}{\sqrt{x^2 - 1}}$ **b) ii)** $y = \dfrac{1}{x}(A\sin 5x + B\cos 5x)$

Exercise 5A

1 a) 177 **b)** 15 **c)** 0 **d)** -255 **2 a)** $(a + b + c)(a - b)(b - c)(c - a)$ **b)** $-24pqr$ **c)** $(ab + bc + ca)(b - c)(c - a)$

d) 0 **3** $(a - b)(b - c)(c - a)(a + b + c + 1),\ -1$

Exercise 5B

1 $(a - b)(b - c)(c - a)(a + b + c);\ 1, 2, -3;\ (t, 0, t)$ **2 ii)** $(5 - t, t - 1, t)$ **3 i)** $\left(\dfrac{7t - 19}{5}, \dfrac{37 - 11t}{5}, t\right)$ **4 a)** $a = b = c$

4 b) $x = a + b,\ y = b - a + t,\ z = t$ **i)** Planes intersect in a line **ii)** Two planes parallel **5** $\frac{1}{2}, -2$; planes form triangular prism

6 10 **7** $1 - k^2;\ k \neq 1, -1$ **i)** $2, 1, 0$

ANSWERS

7 iii) $2 + 3t, -1 - 2t, t$ Three planes meet at point; planes form triangular prism; two planes coincident

8 a) $(a-b)(b-c)(c-a)(a+b+c)$ **b)** -5 **9 ii)** $x = -17, y = \frac{17}{2}, z = -1$ **iv)** Two parallel planes, intersecting third plane

10 i) $(p+1)(q-2)$ **iii)** 7 **iv)** Sheaf of planes **11** $(-\frac{1}{3}t, \frac{2}{3}t, t)$; three planes intersect in line $\dfrac{x}{1} = \dfrac{y}{-2} = \dfrac{z}{-3}$ **12 i)** $q \neq 2$

12 ii) $p = -4, q = 2$ **iii)** $q = 2, p \neq -4$

Exercise 6A

1 a) $\mathbf{r} = \begin{pmatrix} 2 \\ -7 \\ 5 \end{pmatrix} + t \begin{pmatrix} 3 \\ 4 \\ -7 \end{pmatrix}$ **b)** $\mathbf{r} = \begin{pmatrix} 4 \\ 8 \\ -6 \end{pmatrix} + t \begin{pmatrix} -2 \\ 3 \\ 6 \end{pmatrix}$ **c)** $\mathbf{r} = \begin{pmatrix} 7 \\ 4 \\ -1 \end{pmatrix} + t \begin{pmatrix} 2 \\ -1 \\ -3 \end{pmatrix}$ **d)** $\mathbf{r} = \begin{pmatrix} -8 \\ 1 \\ -3 \end{pmatrix} + t \begin{pmatrix} 1 \\ 3 \\ -7 \end{pmatrix}$

2 a) $\mathbf{r} = \begin{pmatrix} 4 \\ 8 \\ -2 \end{pmatrix} + t \begin{pmatrix} 3 \\ 11 \\ -6 \end{pmatrix}$ **b)** $\mathbf{r} = \begin{pmatrix} -1 \\ 8 \\ 3 \end{pmatrix} + t \begin{pmatrix} 3 \\ -11 \\ 6 \end{pmatrix}$ **c)** $\mathbf{r} = \begin{pmatrix} 1 \\ 7 \\ -2 \end{pmatrix} + t \begin{pmatrix} 4 \\ 3 \\ -10 \end{pmatrix}$ **d)** $\mathbf{r} = \begin{pmatrix} 3 \\ -5 \\ -9 \end{pmatrix} + t \begin{pmatrix} -5 \\ 2 \\ 16 \end{pmatrix}$

3 a) $\dfrac{x-2}{3} = \dfrac{y+7}{4} = \dfrac{z-5}{-7}$ **b)** $\dfrac{x-4}{-2} = \dfrac{y-8}{3} = \dfrac{z+6}{6}$ **c)** $\dfrac{x-7}{2} = \dfrac{y-4}{-1} = \dfrac{z+1}{-3}$ **d)** $\dfrac{x+8}{1} = \dfrac{y-1}{3} = \dfrac{z+3}{-7}$

4 a) $\mathbf{r} = \begin{pmatrix} 3 \\ -2 \\ 4 \end{pmatrix} + t \begin{pmatrix} 4 \\ 3 \\ -5 \end{pmatrix}$ **b)** $\mathbf{r} = \begin{pmatrix} -2 \\ 1 \\ -3 \end{pmatrix} + t \begin{pmatrix} 5 \\ -7 \\ -2 \end{pmatrix}$ **c)** $\mathbf{r} = \begin{pmatrix} -5 \\ 2 \\ -4 \end{pmatrix} + t \begin{pmatrix} 1 \\ -3 \\ 2 \end{pmatrix}$ **d)** $\mathbf{r} = \begin{pmatrix} \frac{3}{2} \\ 5 \\ 2 \end{pmatrix} + t \begin{pmatrix} 2 \\ 3 \\ -1 \end{pmatrix}$

4 e) $\mathbf{r} = \begin{pmatrix} \frac{5}{3} \\ -2 \\ 2 \end{pmatrix} + t \begin{pmatrix} 2 \\ 4 \\ -3 \end{pmatrix}$ **5 a)** $109°$ **b)** $93.3°$ **6 a)** $\mathbf{r} = \begin{pmatrix} 2 \\ 1 \\ 4 \end{pmatrix} + t \begin{pmatrix} 2 \\ 6 \\ 1 \end{pmatrix}$ **b)** $\mathbf{r} = \begin{pmatrix} -1 \\ -4 \\ 3 \end{pmatrix} + t \begin{pmatrix} 3 \\ 12 \\ 1 \end{pmatrix}$

6 c) $\mathbf{r} = \begin{pmatrix} 4 \\ 1 \\ -5 \end{pmatrix} + t \begin{pmatrix} -1 \\ 1 \\ -1 \end{pmatrix}$ **7** $2\sqrt{2}$ **8** $19/5\sqrt{2}$ **9 a) i)** $1 : 2 : -2$ **ii)** $\frac{1}{3}, \frac{2}{3}, -\frac{2}{3}$ **b) i)** $3 : -4 : -5$ **ii)** $3/5\sqrt{2}, -4/5\sqrt{2}, -1/\sqrt{2}$

9 c) i) $3 : 2 : -5$ **ii)** $3/\sqrt{38}, 2/\sqrt{38}, -5/\sqrt{38}$ **d) i)** $1 : -2 : -3$ **ii)** $1/\sqrt{14}, -2/\sqrt{14}, -3/\sqrt{14}$

10 a) $\mathbf{r} = 2\mathbf{i} + \mathbf{j} + \mathbf{k} + t(-2\mathbf{i} + 4\mathbf{j} + 2\mathbf{k})$ **d)** $18.7°$ **11 b)** $\mathbf{r} = -9\mathbf{j} + 13\mathbf{k} + t(\mathbf{i} + 2\mathbf{j} - 3\mathbf{k})$ **c)** $(5, 1, -2)$ **e)** $43°$ **f)** $(4.5, 0, -0.5)$

12 a) $(4, -1, -3)$ **b)** $71.4°$ **13 a)** $\mathbf{r} = \begin{pmatrix} 1 \\ 2 \\ 3 \end{pmatrix} + t \begin{pmatrix} 0 \\ 4 \\ -3 \end{pmatrix}$ **b)** $21°$ **14** $47\mathbf{i} - 9\mathbf{j} + 62\mathbf{k}, a = -5$ **15 a)** $\mathbf{i} - 3\mathbf{j} - \mathbf{k}$ **c)** $95.2°$

16 i) $(2, 3, 5)$ **ii)** $40.9°$ **17 a)** 0.148 rad **18** $(1, 1, -12)$ **19 i)** $\mathbf{r} = \begin{pmatrix} 7 \\ -8 \\ 7 \end{pmatrix} + t \begin{pmatrix} 1 \\ -5 \\ 1 \end{pmatrix}$ **ii)** $5\mathbf{i} + 2\mathbf{j} + 5\mathbf{k}$

Exercise 6B

1 a) $-5\mathbf{i} + 7\mathbf{j} + 11\mathbf{k}$ **b)** $31\mathbf{i} + 22\mathbf{j} + \mathbf{k}$ **c)** $22\mathbf{i} + 14\mathbf{j} + 16\mathbf{k}$ **d)** $-32\mathbf{i} + 23\mathbf{j} - 10\mathbf{k}$ **2 a)** $\mathbf{r} . \begin{pmatrix} 3 \\ -5 \\ 4 \end{pmatrix} = -13$ **b)** $\mathbf{r} . \begin{pmatrix} 9 \\ 7 \\ -2 \end{pmatrix} = 47$

2 c) $\mathbf{r} . \begin{pmatrix} 28 \\ -17 \\ 18 \end{pmatrix} = 41$ **3 a)** $3x + y + 7z = 4$ **b)** $2x + 4y + 3z = 8$ **c)** $-x + 5y + 3z + 7 = 0$ **4 a)** $68.5°$ **b)** $34.1°$ **c)** $28.1°$

4 d) $48.5°$ **5** $29.1°$ **6** $0°$ **7** $\mathbf{r} . \begin{pmatrix} 3/5\sqrt{2} \\ 4/5\sqrt{2} \\ -1/\sqrt{2} \end{pmatrix} = 2\sqrt{2}, 2\sqrt{2}$ **8 a) i)** -13 **ii)** $-12\mathbf{i} + 8\mathbf{j} + 8\mathbf{k}$ **b) i)** $+\frac{13}{21}$ **ii)** $2\sqrt{17}$

8 b) iii) $\mathbf{r} . \begin{pmatrix} -3 \\ 2 \\ 2 \end{pmatrix} = 3$ **9 a)** $-\mathbf{i} + 8\mathbf{j} - 4\mathbf{k}$ **b)** $3\mathbf{i} + \mathbf{j} - \mathbf{k}$ **d)** $\mathbf{r} = \begin{pmatrix} 1 \\ 1 \\ 1 \end{pmatrix} + t \begin{pmatrix} 4 \\ 13 \\ 25 \end{pmatrix}$ **10 a)** $\mathbf{r} = 24\mathbf{i} + 6\mathbf{j} + t(\mathbf{i} + \mathbf{j} + 2\mathbf{k})$

10 c) $(35, 17, 32)$ **d)** $18\sqrt{101}$ **11 a)** $36\mathbf{i} + 12\mathbf{j} + 9\mathbf{k}$ **b)** $\mathbf{r} . \begin{pmatrix} 36 \\ 12 \\ 9 \end{pmatrix} = 9$ **c)** $(1, -3, 1)$ **12 i)** $-\mathbf{i} + \mathbf{j} + \mathbf{k}$ **13** $\mathbf{r} . \begin{pmatrix} 11 \\ 5 \\ -7 \end{pmatrix} = 14$

14 $\mathbf{r} = \begin{pmatrix} 1 \\ -8 \\ 7 \end{pmatrix} + t \begin{pmatrix} 1 \\ 2 \\ -1 \end{pmatrix}$ **i)** $-x + 5y + 3z = 46$ **iii)** $12\mathbf{i} + 14\mathbf{j} - 4\mathbf{k}$ **15 i)** $-\mathbf{i} + 2\mathbf{j} + \mathbf{k}$ **iv)** $31.8°$ **16** $3\mathbf{i} + 8\mathbf{k} - 15\mathbf{j}, 14.4°$

17 ii) $3x + y + 2z = 15$ **iii)** $5 - x = y = z$ **18 a)** $2\mathbf{i} - 3\mathbf{j} - 2\mathbf{k}$ **b)** $\frac{1}{2}\sqrt{17}$ **c)** $\mathbf{r} . (2\mathbf{i} - 3\mathbf{j} - 2\mathbf{k}) = -7$ **d)** $2x - 3y - 2z = -7$

18 e) $\dfrac{7}{\sqrt{17}}$ **f)** $3.2°$ **19** $\mathbf{i}+3\mathbf{j}-3\mathbf{k}$ **20 i)** $(2,1,1)$ **ii)** $3\mathbf{i}+2\mathbf{j}+4\mathbf{k}$ **iii)** $8\mathbf{i}-2\mathbf{j}-5\mathbf{k}$ **iv)** $\mathbf{r}=\begin{pmatrix}2\\1\\1\end{pmatrix}+t\begin{pmatrix}2\\-47\\22\end{pmatrix}$

21 $-2\mathbf{a}\times\mathbf{b};\,0,\,\pi$ **22 b)** $7\mathbf{i}+3\mathbf{j}-4\mathbf{k}$ **c)** $72.8°$ **d)** $(-3,\tfrac12,4\tfrac12)$ **e)** $\tfrac12\sqrt{26}$ **23 a)** $\dfrac{x}{1}=\dfrac{y-2}{-1}=\dfrac{z+3}{-1}$ **b)** $x+z=0$

24 a) $(1,-2,3)$ **b)** $2x+y-3z=-9$ **c)** $70.9°$ **25 ii)** $7+\tfrac23 k$ **iii)** $-\tfrac{21}{2}$ **iv)** $\mathbf{r}=\begin{pmatrix}4\\12\\5\end{pmatrix}+t\begin{pmatrix}2\\10\\11\end{pmatrix}$ **26 i)** $2\mathbf{i}-2\mathbf{j}-2\mathbf{k}$

26 ii) $x-y-z=12$ **27 a)** $(-3,0,1)$ **b)** $-3,-1$ **c)** $x-3y-z+4=0$ **28** $1:-2:4,\,2x-y-z=0$ **29 a) i)** $2\mathbf{i}-3\mathbf{j}+6\mathbf{k}$

29 a) iii) $\mathbf{r}\cdot\begin{pmatrix}2\\-3\\6\end{pmatrix}=14$ **b) i)** $\mathbf{r}=\begin{pmatrix}3\\-1\\2\end{pmatrix}+t\begin{pmatrix}2\\1\\1\end{pmatrix}$ **iii)** Opposite side of plane to origin; distance 1 from Π

30 a) ii) $\mathbf{r}\cdot\begin{pmatrix}1\\0\\-1\end{pmatrix}=1$ **b) ii)** $3x+5y+3z+1=0$ **31 a)** $\mathbf{r}=\mathbf{i}+2\mathbf{j}+\mathbf{k}+t(2\mathbf{i}+\mathbf{j}+3\mathbf{k})$ **b)** $(3,3,4)$ **c)** $5\mathbf{i}-\mathbf{j}-3\mathbf{k}$ **d)** $\sqrt{\dfrac{35}{34}}$

31 e) $(5,4,7)$ **32 i)** $\mathbf{r}=\begin{pmatrix}1\\2\\0\end{pmatrix}+t\begin{pmatrix}2\\2\\-1\end{pmatrix},\dfrac{\sqrt{65}}{3}$ **ii)** $\mathbf{r}=\begin{pmatrix}1\\2\\0\end{pmatrix}+t\begin{pmatrix}-1\\0\\3\end{pmatrix},50.8°$ **iii)** $\dfrac{3}{\sqrt{10}}$ **33 i)** 3 **ii)** $\begin{pmatrix}-17\\-10\\14\end{pmatrix}$ **iii)** $\dfrac{21}{\sqrt{65}}$

34 $3\mathbf{i}-5\mathbf{j}-7\mathbf{k},\,3x-5y-7z=1$ **35** $3\mathbf{i}+4\mathbf{j}+\mathbf{k},\,3x-2y-2z=-1$ **36 b) i)** $(4,0,0)$ **iv)** $(2,1,4)$ **v)** 8

Exercise 6C

1 -73 **2** 177 **3** 21 **4** 8 **5 i)** 0 **ii)** $\overrightarrow{EF}$ **iii)** $\sqrt{\dfrac{2}{3}}$ **6 a)** $-30\mathbf{i}-15\mathbf{j}+45\mathbf{k}$ **b)** $\mathbf{r}=\begin{pmatrix}3\\1\\2\end{pmatrix}+t\begin{pmatrix}2\\1\\-3\end{pmatrix}$ **d)** 35

7 a) $5\mathbf{i}-3\mathbf{j}-4\mathbf{k}$ **b)** 100 **c)** 50 **8 a)** $-6\mathbf{i}-2\mathbf{j}+5\mathbf{k}$ **b)** $\tfrac12\sqrt{65},\,\tfrac16$ **c)** $\mathbf{r}\cdot(6\mathbf{i}+2\mathbf{j}-5\mathbf{k})=1$ **d)** $(\tfrac12,\tfrac14,\tfrac12)$ **e)** $9.52°$

Exercise 7A

7 a) $y\leqslant1,y\geqslant\tfrac{49}{25}$ **b)** $y\geqslant1,y\leqslant\tfrac{5}{13}$ **8** $x=-\tfrac23,x=\tfrac57,y=\tfrac{4}{21}$ **9** $x=-1,y=x-2$ **10 i)** $x=-2$ **ii)** 7

11 i) $x=2,x=-2,y=10$ **ii)** $\dfrac{16}{(x-2)^3}-\dfrac{54}{(x+2)^3}$ **iii)** $(10,8\tfrac34)$ **12 a)** $y=1,x=2\sqrt3-1,x=-1-2\sqrt3$ **b)** $(5,\tfrac56),(1,\tfrac12)$

13 a) i) $x=-\tfrac12,y=\dfrac{x}{2}$ **ii)** $(0,0),(-1,-1)$ **14 a)** $\tfrac54$ **b) i)** $x=2$ **ii)** $y=2x+1$ **c)** $(3,8)\min$ **15 i)** $y=2+\dfrac{3}{x-1}+\dfrac{1}{x+2}$

15 iii) $y=2,x=1,x=-2$ **iv)** $(-\tfrac54,2)$ **16 a)** $\dfrac{2t}{(1+t)^2}$ **c)** $y=\dfrac{x}{2},x=0$ **d) ii)** $(1,1),(-1,-1)$

Exercise 7B

1 a) $x>-1,x<-2$ **b)** $x>3$ **c)** $-10<x<-3$ **d)** $x>5,x<-14$ **e)** $-\tfrac{3}{10}>x>-\tfrac12$ **f)** $\tfrac15<x<\tfrac{6}{11}$

2 a) $x<-2,\,-1<x<0$ **b)** $x>8,2<x<3$ **c)** $5<x<\tfrac12(3+\sqrt{65}),\tfrac12(3-\sqrt{65})<x<-1$ **d)** $-7<x,3<x<\tfrac{44}{13}$

2 e) $-2<x<-1\tfrac12$ **3 a)** $-\tfrac52<x\,(x\neq-2)$ **b)** $-5<x<-1\,(x\neq-2)$ **c)** $\tfrac53<x<11\,(x\neq4)$ **d)** $x>6,-\tfrac43>x\,(x\neq-5)$

3 e) $x>5,x<-\tfrac35\,(x\neq-2)$ **f)** $x>-2\tfrac12\,(x\neq-3)$ **4 a)** $1>x>-2$ **b)** $x>1,x<-\tfrac32$ **c)** $x<-2\,(x\neq-1)$

5 $x>4,-3<x<0$ **6** $x>1,x<-1$ **7** $x>2,-1<x<1$ **8** $-2<x<-1,1<x<4$ **9** $2<x<5,x<-1$

11 a) $\mathrm{f}(x)\leqslant\tfrac16,\mathrm{f}(x)\geqslant\tfrac32;(-4,\tfrac32)(6,\tfrac16)$ **b)** One   **12** $x<-\tfrac32,x>\tfrac74$ **13** $x>-1,x<-\tfrac52$ **14** $1,2,4;y=x+2$

15 a) $(0,3),(-\tfrac32,0)$ **b)** $x>-\tfrac{13}{6}$

Exercise 8A

1 a) $-3,-7$ **b)** $11,5$ **c)** $-5,-4$ **d)** $-\tfrac{11}{3},\tfrac23$ **e)** $-2,-5$ **f)** $-2,-\tfrac72$ **2 a)** $x^2-7x+15=0$ **b)** $x^2+3x+5=0$

2 c) $x^2+2x-4=0$ **d)** $x^2+5x-11=0$ **3 a)** 0 **b)** -10 **c)** -9 **4 i)** $9+\mathrm{i}$ **ii)** $\tfrac52-2\mathrm{i}$ **5 i)** $-10-37\mathrm{i}$ **ii)** $-\tfrac83-\tfrac73\mathrm{i}$

6 $x^3-4x^2-4x-25=0$ **7** $6,-6$ **8** $3x^2-11x+10=0$ **9 a)** 15

ANSWERS

Exercise 8B

1 $x^2 + 14x + 44 = 0$ **2** $x^2 - 45x + 63 = 0$ **3** $3x^3 - 8x^2 + 32x - 56 = 0$ **4** $8x^3 - 12x^2 - 22x + 5 = 0$ **5** $4x^2 + 59x + 289 = 0$

6 $9x^2 + 41x + 225 = 0$ **7 a)** $x^2 + 7x + 6 = 0$ **b)** $6x^2 + 7x + 1 = 0$ **c)** $4x^2 - 37x + 9 = 0$ **d)** $2x^2 - x - 3 = 0$

8 a) $3x^2 + 36x - 32 = 0$ **b)** $6x^2 + 9x - 1 = 0$ **c)** $9x^2 - 93x + 4 = 0$ **d)** $3x^2 + 27x + 52 = 0$ **9 a)** $x^3 + 9x^2 + 45x + 189 = 0$

9 b) $x^3 + x^2 - 17x - 49 = 0$ **c)** $x^3 - 6x^2 + 14x - 8 = 0$ **10** $x^4 + 9x^3 + 63x^2 - 297x + 81 = 0$ **11** $x^2 + 10x + 75 = 0$

12 $4x^2 - 3x + 1 = 0$

Exercise 8C

1 $i, -i, \dfrac{5}{2} \pm \sqrt{\dfrac{21}{2}}$ **2** $2i, -2i, 2, -\dfrac{5}{3}$ **3** 1 **4** 3 **5** $0 < k < 4$ **6** $1 - i, \dfrac{3}{2} \pm \dfrac{i\sqrt{11}}{2}$ **7 b)** $3 - i, -11$ **8 b) i)** $2 + 3i, -1$ (twice)

8 b) ii) $(z^2 - 4z + 13)(z^2 + 2z + 1)$ **9 b)** $-i, 1 + i, 1 - i$ **10** $2 - i, \dfrac{2}{3}$ **11 i)** $1 + 3i$ **ii)** $5, -\dfrac{1}{2}$ **12** $-3i, \dfrac{5}{3}$

13 b) $(z - 2)(z^2 - 6z + 10)$; $2, 3 + i, 3 - i$ **14 i)** $1 - 3i$ **ii)** $z^3 - 4z^2 + 14z - 20 = 0$ **15** $1 - i; 2, 18$ **16 a)** $\dfrac{8}{7}$ **b)** $1 - 2i, -\dfrac{6}{7}$

17 i) $3p^2$ **iii)** $z^3 - 3pz^2 + 3p^2z - (p^3 + q^3) = 0$ **18** $5 + 4i, 5 - 4i, \sqrt{2}(1 + i), \sqrt{2}(1 - i), -\sqrt{2}(1 + i), -\sqrt{2}(1 - i)$

Exercise 9A

6 4 **11** $\dfrac{n}{3}(4n^2 - 1)$ **14 a)** $\dfrac{n^2}{(2n - 1)(2n + 1)}$ **16 a)** $\dfrac{n}{4}(n + 1)(n + 2)(n - 1)$

18 Counter-example: $\begin{pmatrix} 1 \\ 0 \\ 0 \end{pmatrix} \times \begin{pmatrix} 1 \\ 0 \\ 0 \end{pmatrix} = \begin{pmatrix} 0 \\ 0 \\ 0 \end{pmatrix}, \dfrac{1}{3}\begin{pmatrix} 2 \\ -2 \\ 1 \end{pmatrix}$ **23** $\dfrac{n}{6}(n + 1)(2n + 7)\ln 2$ **25 b)** Not convergent. $(-2)^n$ does not tend to 0

27 Yes **28** $\dfrac{1}{6}$ **29 i)** Not true: for example, $u = 2, v = 3, w = 6$ **ii)** True

29 iii) True. Converse of **ii** is 'If u divides $v + w$, then u divides both u and w'. This is not true.

Exercise 9B

1 $\dfrac{2n}{3}(n + 1)(n + 2)$ **2** $\dfrac{n}{2}(n + 1)(n^2 + n + 1)$ **3** $\dfrac{n}{3}(n^2 - 7)$ **4** $\dfrac{n}{6}(4n^2 + 33n - 1)$ **5 b)** $61\,907$ **6** $n^2(2n^2 - 1)$ **7** $18\,760$

8 $N^2(2N + 1)^2, -N^2(4N + 3)$

Exercise 9C

1 $\dfrac{3}{4}$ **2** $\dfrac{n^2 + 3n}{4(n + 1)(n + 2)}$ **3 a)** $\dfrac{1}{2}\left(\dfrac{1}{2r - 1} - \dfrac{1}{2r + 1}\right)$ **b)** $a = b = 1$ **c)** 0 **4** 8 **5** $\dfrac{1}{2}\left(\dfrac{1}{2r + 1} - \dfrac{1}{2r + 3}\right), \dfrac{1}{6} - \dfrac{1}{2(2n + 3)}, \dfrac{1}{6}$

6 2 **7 i)** $1 - e^{Nx}$ **ii)** $x > 0, 1$ **8** $\dfrac{1}{7} - \dfrac{1}{\sqrt{2N + 1}}, \dfrac{1}{7}$ **9 i)** $1 - \dfrac{1}{(n + 1)!}$ **ii)** $2e - 3$ **10 b)** $\dfrac{n}{2(n + 2)}$

Exercise 9D

1 b) $\dfrac{25}{8}$ **c)** 0.110 **2 i)** $x - \dfrac{x^2}{2} + \dfrac{x^3}{3}$ **ii)** $\dfrac{5}{6}, -\dfrac{3}{5}$

3 a) $2(1 + x)\cos x - (1 + x)^2\sin x, 2\cos x - 4(1 + x)\sin x - (1 + x)^2\cos x; -6\sin x - 6(1 + x)\cos x + (1 + x)^2\sin x$

3 b) $1 + 2x + \dfrac{x^2}{2} - x^3$ **4 i)** $x - \dfrac{x^3}{3!} + \dfrac{x^5}{5!}$ **5 b)** $0.029\,565$ **6** $(1 + \sin x + \cos x)e^x; \dfrac{1}{2}, 4$ **7 i)** $A = 0, B = 1, C = 0, D = -\dfrac{1}{3}$

7 ii) $1 - u^2 + u^4$ **iii)** $x - \dfrac{x^3}{2} + \dfrac{x^5}{5}$ **iv)** $\tan^{-1} x$ **8** $1, \dfrac{1}{2}, \dfrac{1}{8}$ **9 a)** $-\dfrac{6}{(1 + x)^4}$ **c)** $\dfrac{(-1)^r x^r}{r(r - 1)}$ **10 a) i)** $\dfrac{2x}{2 + x^2}$ **ii)** $\ln 2 + \dfrac{x^2}{2}$

10 b) $-\dfrac{x^4}{8}$ **11 i)** $1 + rx + \dfrac{r(r - 1)}{2!}x^2 + \dfrac{r(r - 1)(r - 2)}{3!}x^3 + \dfrac{r(r - 1)(r - 2)(r - 3)}{4!}x^4$ **ii)** $\sqrt[3]{\dfrac{3}{2}}$ **12 i)** $1 + x + \dfrac{x^2}{2!} + \dfrac{x^3}{3!} + \dfrac{x^4}{4!}$

12 ii) $\dfrac{(3n - 2)}{n!}$ **iii)** $e + 2$

Exercise 9E

1 a) $2x - \frac{4x^3}{3} + \frac{4x^5}{15} - \ldots$ **b)** $5x - \frac{125}{6}x^3 + \frac{625}{24}x^5$ **c)** $1 + 8x + 32x^2 + \frac{256}{3}x^3 + \ldots$ **d)** $x^2 - \frac{x^4}{2} + \frac{x^6}{3} - \frac{x^8}{4} + \ldots$

1 e) $-\left(2x + 2x^2 + \frac{8}{3}x^3 + \ldots\right)$ **2 a)** $x^2 - \frac{x^6}{3!} + \frac{x^{10}}{5!} - \ldots + (-1)^n\frac{x^{(4n+2)}}{(2n+1)!} + \ldots$ **b)** $1 + 4x + \frac{15}{2}x^2 + 9x^3 + \frac{63}{8}x^4$

2 c) $2 - 8x^2 + \frac{9}{4}x^4$ **d)** $e\left(1 - \frac{x^2}{2} + \frac{x^4}{6}\right)$ **e)** $\ln 2 - \frac{x^2}{4} - \frac{x^4}{96}$ **3 a)** Converge   **b)** Converge   **c)** Converge

4 $1 - \frac{x^6}{2!} + \frac{x^{12}}{4!} - \ldots + (-1)^n\frac{x^{6n}}{(2n)!} + \ldots$, all values of x **5** $1 + 2x^2 + 2x^4 + \frac{4}{3}x^6 + \ldots$ **6** $|x| < 3$ **7 a)** $2 - \frac{x}{4} - \frac{x^2}{64} - \frac{x^3}{512}$

7 b) $6x - \frac{3}{4}x^2 - \frac{579}{64}x^3$ **8 a)** $\frac{1}{2} - \frac{3}{4}x + \frac{9}{8}x^2 - \frac{27}{16}x^3$ **b)** $x - \frac{3}{2}x^2 + \frac{19}{12}x^3 - \frac{19}{8}x^4$ **9 a)** $\frac{1}{2}, -\frac{1}{2}\sqrt{3}$

10 a) $a - \frac{1}{3}b, \frac{1}{2}a^2 + \frac{1}{9}b^2, \frac{1}{6}a^3 - \frac{5}{81}b^3$ **b) i)** 1, 3

Exercise 10A

1 a) i) $\frac{1}{2}(e^2 + e^{-2})$ **ii)** 3.76   **b) i)** $\frac{1}{2}(e^3 - e^{-3})$ **ii)** 10.0   **c) i)** $\frac{e^8 - 1}{e^8 + 1}$ **ii)** 0.999   **3 a)** $2\sinh 2x$ **b)** $5\cosh 5x$ **c)** $3\operatorname{sech}^2 3x$

3 d) $8\sinh 4x - 15\cosh 3x$ **e)** $6\sinh 2x + 30\cosh 5x$ **f)** $-\operatorname{cosech}^2 x$ **g)** $-\operatorname{sech} x \tanh x$ **h)** $45\sinh 3x \cosh^4 3x$

3 i) $64\cosh 8x \sinh^3 8x$ **j)** $\tanh x$ **k)** $2\cosh 2x\, e^{\sinh 2x}$ **l)** $5\operatorname{cosech} 5x \operatorname{sech} 5x$ **4 a)** $\frac{1}{3}\cosh 3x$ **b)** $\frac{1}{4}\sinh 4x$ **c)** $3\cosh\left(\frac{x}{3}\right)$

4 d) $10\sinh\left(\frac{x}{5}\right)$ **e)** $\frac{3}{5}\sinh 5x - 4\cosh\left(\frac{x}{2}\right)$ **f)** $\frac{1}{4}\ln\cosh 4x$ **5 a)** 0.540   **b)** 0.693   **c)** 0.457   **d)** 0.191   **e)** 1.10, −0.625

5 f) 0.514   **6** $\ln\frac{1}{2}, \ln\frac{2}{3}$ **7 a) i)** $\frac{e^x - e^{-x}}{e^x + e^{-x}}$ **b)** $\ln 2$ **8** $(-\ln 2, 4)$, min   **10 i)** $x\sinh x + 4\cosh x$ **ii), iii)** $x\sinh x + 2n\cosh x$

11 $-\ln 3, 0$ **16 b) i)** $(1.32, 2.15)$ **ii)** 3.3000   **c)** $-1, 4$

Exercise 10B

1 a) $\frac{5}{\sqrt{1+25x^2}}$ **b)** $\frac{3}{\sqrt{9x^2-1}}$ **c)** $\frac{\sqrt{2}}{\sqrt{1+2x^2}}$ **d)** $\frac{3}{\sqrt{9x^2-16}}$ **e)** $\frac{2x}{\sqrt{1+x^4}}$ **f)** $-\frac{1}{x\sqrt{1-x^2}}$ **g)** $\frac{1}{1-x^2}$ **2 a)** $\cosh^{-1}\left(\frac{x}{2}\right) + c$

2 b) $\cosh^{-1}\left(\frac{x}{3}\right) + c$ **c)** $\frac{1}{2}\cosh^{-1}\left(\frac{2x}{5}\right) + c$ **d)** $\frac{1}{3}\cosh^{-1}\left(\frac{3x}{4}\right) + c$ **e)** $\sinh^{-1}\left(\frac{x}{3}\right) + c$ **f)** $\sinh^{-1}\left(\frac{x}{4}\right) + c$ **g)** $\frac{1}{4}\sinh^{-1}\left(\frac{4x}{5}\right) + c$

2 h) $\frac{1}{5}\sinh^{-1}\left(\frac{5x}{3}\right) + c$ **3 a)** $\ln(1+\sqrt{2})$ **b)** $\ln(1+\sqrt{2})$ **c)** $\ln(2+\sqrt{3})$ **d)** $\frac{1}{\sqrt{3}}\ln(2+\sqrt{3})$ **e)** $\frac{1}{5}\ln(5+\sqrt{24})$

4 a) $\frac{1}{5}\ln\left(\frac{10+4\sqrt{6}}{5+\sqrt{21}}\right)$ **b)** $\frac{1}{3}\ln\left(\frac{6+2\sqrt{10}}{3+\sqrt{13}}\right)$ **c)** $\ln\left(1+\sqrt{\frac{2}{3}}\right)$ **d)** $\frac{1}{2}\ln\left(\frac{4+\sqrt{21}}{5}\right)$ **e)** $\ln\left(1+\sqrt{\frac{2}{3}}\right)$ **f)** $\frac{1}{4}\ln\left(\frac{13+2\sqrt{71}}{5+2\sqrt{35}}\right)$

5 a) $\cosh^{-1}\left(\frac{x+2}{4}\right) + c$ **b)** $\ln\left(\frac{3+\sqrt{8}}{2+\sqrt{3}}\right)$ **6 b)** $\ln(1+\sqrt{2})$ **7 a)** $x\tanh x - \ln\cosh x + c$

7 b) $y = x\sinh x - \cosh x \ln\cosh x + c\cosh x$ **8 a)** 2, 1, 4   **b)** $\frac{1}{4}\tanh^{-1}\left(\frac{2x+1}{2}\right) + c$ **9 b)** 0.2763   **10 a)** $(2x+1)^2 + 25$

10 b) $\frac{1}{2}\sinh\left(\frac{2x+1}{5}\right) + c$ **11 i)** 3, 4, −25   **12** $(x-3)^2 - 1^2, \ln(2+\sqrt{3})$ **13 c)** $x\sinh^{-1}x - \sqrt{1+x^2} + c$

17 iii) $\frac{1}{3}\ln(6+\sqrt{37})$ **18 a)** $x \geqslant 1$ **b) ii)** $\sqrt{2} - \ln(1+\sqrt{2}); \sqrt{2}, 1+\sqrt{2}$ **19 a)** $2\tan^{-1}e^x + c$ **c)** 2.604   **20** $\pi\ln(2+\sqrt{5})$

21 c) $\frac{3}{x^2 - 9}$ **d)** $-\frac{3}{x} - \frac{9}{x^3} - \frac{243}{5x^5} - \frac{2187}{7x^7}, -\frac{3^{2n+1}}{2n+1}$ **e)** $\frac{3}{x^2} + \frac{27}{x^4} + \frac{243}{x^6} + \frac{2187}{x^8}, 3^{2n-1}$ **22 a)** $x + \frac{x^3}{3} + \frac{x^5}{5}, \frac{1}{2n+1}$

22 b) $\frac{1}{2}\ln\left(\frac{1}{5}\right)$ **c)** $\frac{1}{4}\ln\left(\frac{27}{2}\right)$ **23 b ii)** $2\ln 2 - 1$

Exercise 10C

1 a) $1 + 2x + \frac{2}{3}x^4$ **b)** $3x + \frac{9}{32}x^3$ **c)** $1 + x + \frac{25}{2}x^2 + \frac{25}{2}x^3 + \frac{625}{24}x^4$ **d)** $6x + 12x^2 + 36x^3 + 72x^4$

2 $\frac{1}{2}x\sqrt{x^2-9} - \frac{9}{2}\cosh^{-1}\left(\frac{x}{3}\right) + c$ **3** $\frac{1}{2}x\sqrt{x^2+16} + 8\sinh^{-1}\left(\frac{x}{4}\right) + c$ **4** $\frac{1}{2}x\sqrt{x^2+25} + \frac{25}{2}\sinh^{-1}\left(\frac{x}{5}\right) + c$

ANSWERS

5 $\frac{1}{2}x\sqrt{x^2-25} - \frac{25}{2}\cosh^{-1}\left(\frac{x}{5}\right) + c$ **6** $\frac{1}{2}x\sqrt{x^2-4} + 2\cosh^{-1}\left(\frac{x}{2}\right) + c$ **7** $\frac{1}{8}x\sqrt{x^2+9} - \frac{5}{2}\sinh^{-1}\left(\frac{x}{3}\right) + c$

8 $\frac{x}{2}\sqrt{x^2+4} + 2\sinh^{-1}\left(\frac{x}{2}\right) + c$ **9 i)** 0; $\frac{1}{2}\left[\dfrac{1-(n+1)\cosh nx + n\cosh(n+1)x}{\cosh x - 1}\right]$ **12** $y = \frac{1}{2}\coth x - \frac{1}{2}x\,\mathrm{cosech}\,x + c\,\mathrm{cosech}^2 x$

14 $\dfrac{x}{\sqrt{x^2-1}}$; $\frac{5}{4}, -\frac{3}{4}$ **15 c) ii)** 1.76 **17 d)** $\frac{1}{2}\ln(2+\sqrt{3}) - \frac{1}{4}\sqrt{3}$; $\frac{1}{2}, -\frac{1}{4}$ **18 b)** $1, -\frac{1}{2}, \frac{13}{120}$ **c)** $x^3 + \frac{1}{6}x^5$

Exercise 11A

1 a) $(4, 0), x = -4$ **b)** $(7, 0), x = -7$ **c)** $(0, 2), y = -2$ **d)** $(0, -4), y = 4$ **e)** $(-3, 0), x = 3$ **f)** $(8, -1), x = -8$ **g)** $(5, 2), x = 1$

2 a) $y^2 = 12x$ **b)** $y^2 = 16x$ **c)** $x^2 = 8y$ **d)** $x^2 = -20y$ **3 a)** $ty = x + 5t^2$ **b)** $py = x + 5p^2$ **c)** $y = 5 + x$ **d)** $2y = x + 20$

4 a) $x + y = 6$ **b)** $(18, -12)$

Exercise 11B

1 a) $\frac{\sqrt{7}}{4}, (\pm\sqrt{7}, 0), x = \pm\frac{16}{\sqrt{7}}$ **b)** $\frac{\sqrt{33}}{7}, (\pm\sqrt{33}, 0), x = \pm\frac{49}{\sqrt{33}}$ **c)** $\frac{3}{5}; (\pm 3, 0), x = \pm\frac{25}{3}$ **d)** $\frac{\sqrt{5}}{3}, (0, \pm 2\sqrt{5}), y = \pm\frac{18}{5}$

1 e) $\frac{4}{5}, (5, -2), x = \frac{29}{4}, x = -\frac{21}{4}$ **2 a)** $\frac{x^2}{36} + \frac{y^2}{27} = 1$ **b)** $\frac{x^2}{36} + \frac{y^2}{32} = 1$ **c)** $\frac{x^2}{16} + \frac{y^2}{32} = 1$ **d)** $\frac{x^2}{36} + \frac{y^2}{45} = 1$

3 a) $4x\cos\theta + 5y\sin\theta = 20$ **b)** $4y\cos\theta = 5x\sin\theta - 9\sin\theta\cos\theta$ **4** $(-\frac{12}{5}a, 0), (-\frac{6}{5}a, 0)$

Exercise 11C

1 a) $\frac{5}{4}, (\pm 5, 0), x = \pm\frac{16}{5}$ **b)** $\frac{\sqrt{65}}{7}, (\pm\sqrt{65}, 0), x = \pm\frac{49}{\sqrt{65}}$ **c)** $\frac{\sqrt{41}}{5}, (\pm\sqrt{41}, 0), x = \pm\frac{25}{\sqrt{41}}$ **d)** $\frac{\sqrt{13}}{2}, (\pm\sqrt{13}, 0), x = \pm\frac{4}{\sqrt{13}}$

1 e) $\frac{\sqrt{34}}{5}, (1 \pm \sqrt{34}, -2), x = 1 \pm\frac{25}{\sqrt{34}}$ **2 a)** $\frac{x^2}{36} - \frac{y^2}{108} = 1$ **b)** $\frac{x^2}{36} - \frac{y^2}{288} = 1$ **c)** $\frac{y^2}{32} - \frac{x^2}{32} = 1$ **d)** $\frac{y^2}{45} - \frac{x^2}{180} = 1$

3 a) $4x\sec\theta - 5y\tan\theta = 20$ **b)** $5x\sin\theta + 4y = 41\tan\theta$

Exercise 11D

2 a) $(0, 2t)$ **b)** $y^2 = 9x$ **3 b)** $(apq, a(p+q))$ **c)** $pq = -1$ **4 b)** $3x + 7y = 37, y + 3x = 13$ **5 a)** $\frac{x}{a}\cos t + \frac{y}{b}\sin t = 1$

5 b) $ax\sin t - by\cos t = (a^2 - b^2)\sin t\cos t$ **c)** $\left(\dfrac{a^2 - b^2}{2a}\right)\cos t, \dfrac{b}{2\sin t}$ **6 a)** $3x\cos\theta + 2y\sin\theta = 6$

6 c) $\theta = 0°, 233.1°, 360°; (-1.2, -2.4)$ **8 c)** 1 **9 a)** $\dfrac{(x-2)^2}{9} + \dfrac{y^2}{4} = 1, \dfrac{\sqrt{5}}{3}$ **b)** $(\pm\sqrt{5}, 0)$ **d)** 33.84

10 c) $C_1: (\pm 3a, 0), y = \pm\frac{x}{3}; C_2: \left(\pm\frac{10a}{3}, 0\right), y = \pm 3x$ **d)** $C_1: \dfrac{\sqrt{10}}{3}, (\pm\sqrt{10}a, 0), x = \pm\frac{9}{10}; C_2: \sqrt{10}, \left(\pm\frac{100a}{3}, 0\right), x = \pm\frac{a}{3}$

12 iv) $(-1, 2t)$

Exercise 12A

The constant of integration is omitted from these answers.

1 a) $\frac{1}{6}(x^2 + 1)^6$ **b)** $\frac{1}{10}(x^2 - 1)^5$ **c)** $\frac{1}{32}(x^4 - 1)^8$ **d)** $-\frac{1}{15}(1 - x^3)^5$ **e)** $-\frac{1}{6}\cos^6 x$ **f)** $\frac{1}{5}\sinh^5 x$ **g)** $\frac{1}{15}\cosh^5 3x$ **h)** $\frac{1}{12}\sin^6 2x$

2 a) $\frac{1}{2}e^x(\cos x + \sin x)$ **b)** $\frac{1}{5}e^x(\cos 2x + 2\sin 2x)$ **c)** $\frac{1}{5}e^{2x}(\sin x + 2\cos x)$ **d)** $\frac{1}{34}e^{3x}(5\sin 5x + 3\cos 5x)$

2 e) $\frac{1}{10}e^{4x}(2\cosh 2x - \sinh 2x)$ **f)** $-\frac{1}{40}e^{-7x}(3\cosh 3x + 7\sinh 3x)$ **3 a)** $x - \tan^{-1} x$ **b)** $x - 5\tan^{-1}\left(\frac{x}{4}\right)$ **c)** $\frac{1}{4}x - \frac{23}{32}\ln(8x + 3)$

3 d) $-\frac{7}{4}x - \frac{47}{16}\ln(5 - 4x)$ **e)** $\ln(x^2 + 2x + 3) - \frac{3}{\sqrt{2}}\tan^{-1}\left(\frac{x+1}{\sqrt{2}}\right)$ **f)** $\frac{1}{2}\ln(x^2 + x + 1) - \frac{1}{\sqrt{3}}\tan^{-1}\left(\frac{2x+1}{\sqrt{3}}\right)$

3 g) $\sqrt{x^2 + x - 1} - \frac{3}{2}\cosh^{-1}\left(\frac{2x+1}{\sqrt{5}}\right)$ **h)** $\sqrt{2x^2 - 4x + 5} - \frac{5}{\sqrt{2}}\sinh^{-1}\left(\sqrt{\frac{2}{3}}(x - 1)\right)$ **i)** $-2\sqrt{1 - 4x - x^2} + \sin^{-1}\left(\frac{x+2}{\sqrt{5}}\right)$

3 j) $-\sqrt{2 - 5x - 3x^2} - \frac{19}{2\sqrt{3}}\sin^{-1}\left(\frac{6x+5}{7}\right)$ **4 a)** $-\sqrt{1 - x^2} + \sin^{-1} x$ **b)** $1 - \frac{\sqrt{3}}{2} + \frac{\pi}{6}$ **5** $\sin^{-1}\left(\frac{x+3}{5}\right)$

6 i) $1 + \frac{2}{1+x} - \frac{1}{9+x^2}$ **7** $\frac{1}{3x+4} + \frac{2x+1}{x^2+9}$

Exercise 12B

1 a) $5\pi/32$ **b)** $\frac{16}{35}$ **2** $I_n = x^n e^n - n I_{n-1}$ **3** $120 - 326/e$ **4 a)** $\frac{1}{4}\pi - \frac{2}{3}$ **b)** $\frac{5}{12} - \frac{1}{2}\ln 2$

6 $\frac{1}{5}\cosh^4 1 \sinh 1 + \frac{4}{15}\cosh^2 1 \sinh 1 + \frac{8}{15}\sinh 1$ **11 b)** $\frac{316}{81}$ **12** $\frac{1}{16}\pi^4 - 3\pi^2 + 24$ **13** $\frac{8}{15}$ **14 b)** $2 - \dfrac{2}{\sqrt{e}}$ **c)** $16 - \dfrac{26}{\sqrt{e}}$

15 ii) $\frac{26}{15}$ **16** $\frac{1}{2}\ln 2 - \frac{1}{4}$ **17 b)** $\sqrt{3} + \frac{1}{2}\ln(2+\sqrt{3})$ **18** $\frac{1}{3}, -\frac{1}{3}, \frac{2}{3}; \frac{1}{3}\ln 2 + \dfrac{\pi}{3\sqrt{3}}; \frac{1}{6} + \dfrac{2\pi}{9\sqrt{3}} + \frac{2}{9}\ln 2$ **21** $\dfrac{4\pi}{3} - 2\sqrt{3}$ **24** $\frac{1}{8}e^2 + \frac{3}{8}$

26 b) $-\frac{17}{15}\sqrt{2} + \frac{4}{5}\sqrt{3}$ **27 a)** $\frac{5}{4}, \frac{3}{4}$ **b) ii)** $\frac{57}{64}$ **28 a)** π **c)** $\frac{5}{16}\pi$ **d)** $\dfrac{5a^2\pi}{32}$ **29 b) i)** $\ln 2$ **iv)** $\frac{3}{5}\ln 2$ **30 i)** $-\dfrac{1}{n+1}\cos^{n+1}\theta$

31 b) $\pi\sqrt{5} - \dfrac{\pi}{2}\ln(2+\sqrt{5})$ **c)** 1.641%

Exercise 12C

1 $\frac{1}{27}(31\sqrt{31} - 8)$ **2** $\dfrac{(14\sqrt{14} - 11\sqrt{11})}{9\sqrt{2}}$ **3** $\frac{1}{2}t\sqrt{1+t^2} + \frac{1}{2}\ln(t + \sqrt{1+t^2})$ **4** $4a$ **5** $c\sinh 1$ **6 a)** $\dfrac{12\pi}{5}(2 + 782\sqrt{17})$

6 b) $\dfrac{8\pi}{3}(5\sqrt{5} - 1)$ **c)** $\dfrac{12\pi a^2}{5}$ **d)** $\dfrac{5\pi}{6}(61^{3/2} - 41^{3/2})$ **e)** $\pi(\frac{1}{2}\sinh 2 + 1)$

6 f) $\dfrac{\pi}{3}e^{12}\sqrt{1+9e^{24}} + \dfrac{\pi}{9}\sinh^{-1}(3e^{12}) - \dfrac{\pi}{3}e^3\sqrt{1+9e^6} - \dfrac{\pi}{9}\sinh^{-1}(3e^3)$ **7** $8a$ **8 a)** $-1, 2$ **b) i)** $\dfrac{2x}{1-x^2}$ **10** $\frac{1}{4}(e^{2\pi} - 1)$ **11 i)** 3

12 $\dfrac{64\pi a^2}{3}$ **13** $2 - 2e^{-\pi/2}, 2\displaystyle\int_0^{\pi/2}[4 + e^{-t}(\cos t - \sin t)]2e^{-t}\,dt$ **14** 6 **16 i)** $3\sqrt{2}$ **ii)** $\dfrac{24\sqrt{2}}{5}\pi$ **17 b) i)** 12 **ii)** $\dfrac{576}{5}\pi$

19 $\dfrac{\pi}{27}(10\sqrt{10} - 1)$ **20 b)** $(\frac{1}{5}e^\pi - \frac{2}{5})2\sqrt{2}\pi$ **22 a)** $\frac{1}{3}\cosh^3 t + c$ **c)** 438

Exercise 12D

1 $\frac{3}{2}$ **2** Does not exist **3** Does not exist **4** Does not exist **5** Does not exist **6** Does not exist **7** Does not exist **8** $\pi/2a$

9 Does not exist **10** Does not exist **11 a)** $\dfrac{x^2}{2}\ln x - \dfrac{x^2}{4} + c$ **b)** $-\frac{1}{4}$ **12 a)** $\frac{1}{2}$ **b)** $\ln\left(\dfrac{3}{2}\right)$ **13 a)** $-\dfrac{k\pi}{2}$ **b) ii)** $\frac{1}{2}\ln 2$

14 i) $\ln 2$ **iii)** $\ln 2$ **iv)** $\pi/4$ **15** $\frac{1}{4}$ **16 iii)** -4

Exercise 13A

1 1.12 **2** 0.87 **3** 2.46 **4** 0.95 **5** 0.93 **6** 0.60 **7** 1.41 **8 a)** $-3.8125, -3.7936, -3.7915, -3.7913, -3.7912, -3.7912, -3.791$

8 b) -3.791 **c)** Five **9** $[0.375, 0.5]$, fourteen **10 b)** 1.432 **d)** 0.669 **11 b)** 1.373

12 Starting with $x_0 = 0$: $x_1 = 1$, $x_2 = 0.806\,824\,2641$, $x_3 = 0.792\,134\,9597$, $x_4 = 0.792\,059\,9704$, $x_5 = 0.792\,059\,9684$, $x_6 = 0.792\,059\,9684$

13 a) $0.109, -0.402$ **b)** 2.11 **c)** 2.13 **14** 0.53 Tangent parallel to x-axis **15** 4.026 **16 b)** 6.135 **17 b)** 1.54 **c)** -0.54

18 a) i) 1.4973 **ii)** $1.497\,3043$ **19 c)** 1.07 **20 a)** $0.2443, 0.2553, 0.2582, 0.2589$ **b)** 2.544 **21 a)** $0.337\,609$ **22 a)** $\frac{1}{2}, \frac{3}{2}$

22 b) i) $4x^3 - 12x^2 + 9x + 3 = 0$ **ii)** -0.2460 **23** $-0.670, 0.78, [1.2, 1.206\,25]$

Exercise 13B

1 a) i) 42.0000 **ii)** 41.3333 **b) i)** 328.0000 **ii)** 320.0000 **c) i)** 3.6734 **ii)** 3.7175 **d) i)** 5.0898 **ii)** 5.1795 **2 i)** 1255.81

2 ii) 1403.734 **3 b)** 3.6281 **c)** $\sinh 2$ **d)** 0.03% **4 b)** $2 - \dfrac{2}{\sqrt{e}}$ **c)** $16 - \dfrac{26}{\sqrt{e}}$ **5 a) i)** 0.4285 **ii)** 0.4217 **b)** 0.4207

6 b) $\dfrac{\pi}{4}\left[2\sqrt{5} - \ln(2+\sqrt{5})\right]$ **c)** 1.641% **7** 0.579 **8** $2.53, 1.26$ **9** $0.52, 0.52$ **10** $0.749\,88$ **11** 0.1026

12 0.82 **i)** Underestimate **ii)** Larger estimate and better approximation

Exercise 13C

1 $y = 1 + x + \dfrac{3x^2}{2} + \dfrac{5x^3}{2} + \dfrac{35x^4}{8}$ **2** $y = 2 + 2x^2 + \dfrac{2x^3}{3} + 2x^4$ **3** $y = x - \dfrac{5x^3}{6}$ **4** $y = 1 - \dfrac{x^4}{6}$, $0.999\,999\,998$

5 $y = (x-1) + \frac{1}{2}(x-1)^2 + \frac{1}{2}(x-1)^3 + \ldots$ **a)** 0.1055 **b)** -0.0955 **6** $y = 2x + \dfrac{8x^3}{3}$

ANSWERS

7 $\sin x = \dfrac{1}{\sqrt{2}} + \dfrac{1}{\sqrt{2}}\left(x - \dfrac{\pi}{4}\right) - \dfrac{1}{2\sqrt{2}}\left(x - \dfrac{\pi}{4}\right)^2,\ 0.719\,448$

8 $f(0.5 + h) = -2 - 28h - 784h^2$ Sensitive because derivative large near $x = 0.5$ **9** $y = 1 + 2x - \dfrac{3x^2}{2} - \dfrac{4x^3}{3}$

10 a) $3\left(1 + \dfrac{dy}{dx}\right)(x + y)^2,\ 3\dfrac{d^2y}{dx^2}(x + y)^2 + 6\left(1 + \dfrac{dy}{dx}\right)^2(x + y)$ **b)** $y = 1 + x + 3x^2 + 7x^3$ **c)** 0.9 **11** $y = 3 + 2x + 3x^2 - \dfrac{7x^3}{3}$

12 a) 0.9 **b)** 0.8362 **c)** 0.7607 **13 a)** 1.1 **b)** 1.221 **14 a)** $x = 2.27,\ y = -0.33$ **b)** $\dfrac{d^2x}{dt^2} + 16x = 20$ **15 a)** 1.628

15 b) $y = x(2\ln x + 1)$ **c)** 0.6% **16** 0.0025, 0.0051, 0.0078 **17 a) i)** $\dfrac{dv}{dt} = 4 - \dfrac{3x}{v}$ **ii)** 0.22 **b) i)** $x = Ae^{3t} + Be^{t}$

17 b) ii) $x = e^{3t} - e^{t},\ 0.245$ **18** 2.21, 0.64 **19** 2.0766, 2.0743 **20** 0.049, $\frac{1}{5}$

Exercise 14A

1 $\begin{pmatrix} 14 & 0 \\ 8 & 5 \end{pmatrix}$, $\begin{pmatrix} 2 & -2 \\ 18 & 17 \end{pmatrix}$, $\mathbf{PQ} \neq \mathbf{QP}$ **2 a)** $\begin{pmatrix} -5 & 4 \\ 4 & -3 \end{pmatrix}$ **b)** $\begin{pmatrix} 4 & -7 \\ -1 & 2 \end{pmatrix}$ **c)** $\begin{pmatrix} -20 & 4 & -9 \\ -5 & 1 & -2 \\ 11 & -2 & 5 \end{pmatrix}$ **d)** $\begin{pmatrix} 0 & -1 & 2 \\ 1 & -1 & -3 \\ -2 & 3 & 5 \end{pmatrix}$

2 e) $-\dfrac{1}{141}\begin{pmatrix} -21 & -12 & 18 \\ -17 & -3 & -19 \\ 5 & -24 & -11 \end{pmatrix}$ **3** $\begin{pmatrix} -\frac{1}{3} & \frac{1}{3} & 0 \\ -\frac{k}{3} & \frac{k}{3} & -1 \\ \frac{2}{3} & \frac{1}{3} & 0 \end{pmatrix}$ **5 a)** $\dfrac{1}{2x - 5}\begin{pmatrix} -2 & -1 & 2x \\ 2 & 1 & -5 \\ 3 & x-1 & -3x \end{pmatrix}$ **b)** $19, -14, -27$

6 $\dfrac{1}{20}\begin{pmatrix} -1 & 11 & -4 \\ -4 & 4 & 4 \\ 11 & -21 & 4 \end{pmatrix}$ **7** $\begin{pmatrix} -1 & 0 & 1 \\ 6\frac{1}{2} & -2\frac{1}{2} & -\frac{1}{2} \\ 1 & 0 & 0 \end{pmatrix}$ **9 i)** $\dfrac{1}{1+a}\begin{pmatrix} 1 & -a & 0 \\ 1 & 1 & 0 \\ -5-a & a^2-5 & 1+a \end{pmatrix}$ **ii)** $(-1, 1, 0)$

10 $\dfrac{1}{6a-6}\begin{pmatrix} 3a-1 & a+1 & -4 \\ 1 & 2a-1 & -2 \\ -3 & -3 & 6 \end{pmatrix}$, $\left(\dfrac{2}{1-a}, \dfrac{1}{1-a}, \dfrac{3}{1-a}\right)$ **11 i)** $\begin{pmatrix} 1 & 0 & 0 \\ 1 & -1 & 0 \\ -\frac{1}{a} & 0 & \frac{1}{a} \end{pmatrix}$ **ii)** $\begin{pmatrix} \frac{1}{a} & 1 & -\frac{1}{a} \\ 1-\frac{1}{2a} & -1 & \frac{1}{2a} \\ -\frac{1}{2a} & 0 & \frac{1}{2a} \end{pmatrix}$

12 i) $\dfrac{1}{10-a}\begin{pmatrix} 5 & 5 & -5 \\ 3a-5 & 5-2a & -1 \\ -a & -a & 2 \end{pmatrix}$ **ii)** $\dfrac{2}{5}y_1 - \dfrac{3}{5}y_2 + \left(\dfrac{a}{10-5a}\right)y_3$ **13 c)** $-\mathbf{A}, \mathbf{A}, -\mathbf{A}; (-1)^n\mathbf{A}$

13 d) $x - y - z = 1,\ -x + y - z = 0,\ -x - y + z = 0;\ x = 0,\ y = z = -\frac{1}{2}$ **e)** $\frac{1}{2}, -\frac{1}{2}$ **14 a)** $\begin{pmatrix} 3 & 3 & 7 \\ 1 & 4 & 4 \\ 3 & 1 & 6 \end{pmatrix}$ **d)** $\begin{pmatrix} 4 & -2 & -3 \\ 1 & 0 & -1 \\ -2 & 1 & 2 \end{pmatrix}$

15 a) $\begin{pmatrix} 1 & 0 & 0 \\ 0 & 4 & 3 \\ 0 & 0 & 1 \end{pmatrix}$ **b)** $\begin{pmatrix} 1 & 0 & 0 \\ 0 & 8 & 7 \\ 0 & 0 & 1 \end{pmatrix}$ **d)** $\dfrac{1}{2^n}\begin{pmatrix} 2^n & 0 & 0 \\ 0 & 1 & 1-2^n \\ 0 & 0 & 2^n \end{pmatrix}$

Exercise 14B

1 Rotation clockwise about O through $2\pi/3$

2 Reflection in y-axis, followed by one-way stretch in y-direction, scale factor 2 $\begin{pmatrix} -1 & 0 \\ 0 & 1 \end{pmatrix}$, $\begin{pmatrix} 1 & 0 \\ 0 & 2 \end{pmatrix}$

3 Shear in x-direction moving $(0, 1)$ to $(-1, 1)$ **i)** $\begin{pmatrix} \frac{1}{2} & \frac{\sqrt{3}}{2} \\ -\frac{\sqrt{3}}{2} & \frac{1}{2} \end{pmatrix}$ **ii)** Rotation clockwise about O through $\pi/3$

4 i) $(0, 0), (3, 0), (2, 4)$ **ii)** 6, 150 **iii)** 6 **5** 0, 1 **6 a)** $3, -1$ **b)** $\begin{pmatrix} 1 \\ -1 \end{pmatrix}, \begin{pmatrix} 1 \\ 1 \end{pmatrix}$ **7 a) i)** $5, -1$ **ii)** $\begin{pmatrix} 1 \\ 1 \end{pmatrix}, \begin{pmatrix} 2 \\ -1 \end{pmatrix}$ **b)** 6, 5

8 a) $4, -3$ **c)** $\begin{pmatrix} 1 \\ -1 \\ -2 \end{pmatrix}$ **d)** $\begin{pmatrix} 1 & 1 & 1 \\ -1 & -1 & 1 \\ -2 & 1 & 0 \end{pmatrix}$ **9 a)** $9, -3$ **b)** $\begin{pmatrix} -1 \\ 2 \\ 2 \end{pmatrix}$ **c)** $\begin{pmatrix} 2 & 2 & -1 \\ 2 & -1 & 2 \\ -1 & 2 & 2 \end{pmatrix}$ **10 a) i)** $5, \begin{pmatrix} 2 \\ -1 \end{pmatrix}$ **ii)** $x + 2y = 0$

10 b) 25 **11 a)** $-2, 1$ **b)** $\begin{pmatrix} 1 \\ 1 \\ 1 \end{pmatrix}$ **d)** $\frac{1}{3}(a - 2b + c)\mathbf{v}_1 + \frac{1}{3}(a + b - 2c)\mathbf{v}_2 + \frac{1}{3}(a + b + c)\mathbf{v}_3$ **12 a)** $4, -2; \begin{pmatrix} 1 \\ 1 \end{pmatrix}, \begin{pmatrix} 1 \\ -5 \end{pmatrix}$

12 b) ii) $\frac{1}{8}, -\frac{1}{4}$ **13 a)** $2, -2, 3$ **c)** 5   **d)** $\begin{pmatrix} 3 \\ 1 \\ 0 \end{pmatrix}, 10$ **14 a) i)** $(0, -7)$ **ii)** $k = 7,\ \mathbf{A} = \begin{pmatrix} 4 & 2 \\ 2 & 7 \end{pmatrix}$ **b) i)** $3, 8;\ \begin{pmatrix} 2 \\ -1 \end{pmatrix}, \begin{pmatrix} 1 \\ 2 \end{pmatrix}$

14 b) ii) $x + 2y = 0,\ y = 2x$ **14 c)** One-way stretch centred at $(0, -7)$, scale factor 3, in direction of $x + 2y = 0$, followed by

one-way stretch centred at $(0, -7)$, scale factor 8, in direction of $y = 2x$ **15 i)** $3, \begin{pmatrix} \frac{2}{3} & \frac{1}{3} & -1 \\ \frac{5}{3} & \frac{4}{3} & -4 \\ -3 & -2 & 7 \end{pmatrix}$ **ii)** $\begin{pmatrix} -6 & -4 & 12 \\ -29 & -17 & 66 \\ -13 & -8 & 29 \end{pmatrix}$

15 iii) $\lambda^3 - 6\lambda^2 + 3\lambda + 10 = 0$ **iv)** $-\frac{1}{10}, \frac{3}{5}, -\frac{3}{10}$ **16 b)** $2x + 2y - z = 0$ **17** $2, -3, 5$ **18 a)** $k^2 - 2k + 1$ **b)** $3, -1$

19 b) $\begin{pmatrix} 0 & 1 & 0 \\ 1 & 0 & 0 \\ 0 & 0 & -1 \end{pmatrix}$ **c) i)** $x = y, z = 0$ **ii)** $90°$ **d)** $\begin{pmatrix} -1/\sqrt{2} \\ 1/\sqrt{2} \\ 0 \end{pmatrix}$ **20 c) iii)** $\frac{16\pi}{3}$ **21 i)** $(-3, 1, 2)$ **ii)** $x = y = z$

21 iii) $x + y + z = 0$ **iv)** $\frac{2\pi}{3}$ $\begin{pmatrix} 0 & 0 & 1 \\ 1 & 0 & 0 \\ 0 & 1 & 0 \end{pmatrix}, \begin{pmatrix} 1 & 0 & 0 \\ 0 & 1 & 0 \\ 0 & 0 & 1 \end{pmatrix}$ Rotation about L through $\frac{4\pi}{3}$; identity   **22** $1, 27;\ \begin{pmatrix} 1 \\ 5 \end{pmatrix}, \begin{pmatrix} 5 \\ -1 \end{pmatrix}$

22 a) $5x = y$ **c)** 27   Scale factor of area enlargement   **d)** One-way stretch in $\begin{pmatrix} 5 \\ -1 \end{pmatrix}$ direction, scale factor 27

23 i) $\frac{1}{1-2k} \begin{pmatrix} k-2 & 4-3k & 1 \\ -1-k & k+2 & 1 \\ 3 & -5 & -2 \end{pmatrix}$ **ii)** $(-1, -1, 3)$ **iv)** $\frac{x}{3} = \frac{y}{2} = -\frac{z}{3}$

Exercise 15A

1 a) $\cos 6\theta + i \sin 6\theta$ **b)** $\cos 8\theta + i \sin 8\theta$ **c)** -1 **d)** $-i$ **e)** $\cos 8\theta - i \sin 8\theta$ **f)** -1 **g)** 1 **h)** $-i$ **2 a)** $\cos 10\theta + i \sin 10\theta$

2 b) $\cos \theta - i \sin \theta$ **c)** -1 **d)** -8 **3 a)** 16 **b)** $512i$ **c)** $24\sqrt{3}$ **d)** -4 **e)** 2^{12} **f)** $-512i$ **4 a)** $\cos 5\theta - i \sin 5\theta$

4 b) $\cos 4\theta + i \sin 4\theta$ **c)** $-\cos 6\theta + i \sin 6\theta$ **d)** -1 **5** $\cos 11x + i \sin 11x$

Exercise 15B

1 a) i) $\pm\sqrt{2}(1 \pm i)$ **ii)** $2e^{i\pi/4}, 2e^{-i\pi/4}, 2e^{3i\pi/4}, 2e^{-3i\pi/4}$ **b) i)** $2^{\frac{2}{3}}(1 + i), 2^{\frac{7}{6}}\left[\cos\left(\frac{11\pi}{12}\right) + i \sin\left(\frac{11\pi}{12}\right)\right], 2^{\frac{7}{6}}\left[\cos\left(\frac{5\pi}{12}\right) - i \sin\left(\frac{5\pi}{12}\right)\right]$

1 b) ii) $2^{\frac{7}{6}} e^{i\pi/4}, 2^{\frac{7}{6}} e^{11i\pi/12}, 2^{\frac{7}{6}} e^{-5i\pi/12}$ **c) i)** $\frac{3}{2}(\sqrt{3} + i), \frac{3}{2}(-\sqrt{3} + i), -3i$ **ii)** $3e^{i\pi/6}, 3e^{5i\pi/6}, 3e^{-i\pi/2}$ **d) i)** $\pm 2\sqrt{2}(1 + i)$

1 d) ii) $4e^{i\pi/4}, 4e^{-3i\pi/4}$ **e) i)** $\frac{5}{\sqrt{2}}(-1 + i), \frac{5}{\sqrt{2}}(1 - i)$ **ii)** $5e^{3i\pi/4}, 5e^{-i\pi/4}$ **f) i)** $-2, 2\left[\cos\left(\frac{\pi}{5}\right) \pm i \sin\left(\frac{\pi}{5}\right)\right], 2\left[\cos\left(\frac{3\pi}{5}\right) \pm i \sin\left(\frac{3\pi}{5}\right)\right]$

1 f) ii) $2e^{i\pi/5}, 2e^{3i\pi/5}, 2e^{i\pi}, 2e^{-i\pi/5}, 2e^{-3i\pi/5}$ **2** $e^{\pm i\pi/3}, e^{\pm 2i\pi/3}, e^{i\pi}, e^{i0}$ or $1, -1, \pm\left(\frac{1}{2} \pm \frac{\sqrt{3}}{2}i\right)$ **3 a)** $2 - 2i, -2 - 2i$ **b)** $3, \pm\sqrt{3}i$

3 c) $-\frac{1}{2}$ **d)** $1 + 3i, \frac{1}{3} - i$ **e)** $-3 - \frac{1}{2}i, -\frac{3}{2} + \frac{1}{4}i, -\frac{21}{10} + \frac{7}{10}i, -\frac{3}{2} - \frac{1}{2}i$ **4** $e^{i0}, e^{2i\pi/7}, e^{4i\pi/7}, e^{6i\pi/7}, e^{-2i\pi/7}, e^{-4i\pi/7}, e^{-6i\pi/7}$

5 $2e^{i\pi/10}, 2e^{i\pi/2}, 2e^{9i\pi/10}, 2e^{-3i\pi/10}, 2e^{-7i\pi/10}$ **9 a)** $\frac{1}{41}(4\cos 5x + 5\sin 5x)e^{4x} + c$ **b)** $\frac{1}{58}(3\sin 7x - 7\cos 7x)e^{3x} + c$

9 c) $-\frac{1}{20}(2\sin 4x + 4\cos 4x)e^{-2x} + c$ **d)** $\frac{1}{25}(3\sin 3x - 4\cos 3x) + c$ **10** $2e^{-i\pi/6}, 2e^{-2i\pi/3}, 2e^{5i\pi/6}, 2e^{i\pi/3}$ **11** $3 - i, -3 + 3i$

12 i) $8, -5\pi/6$ **ii)** $2e^{-5i\pi/18}, 2e^{7i\pi/18}, 2e^{-17i\pi/18}$ **13** $16\left[\cos\left(\frac{\pi}{4}\right) + i \sin\left(\frac{\pi}{4}\right)\right], 2\left[\cos\left(\frac{\pi}{16}\right) + i \sin\left(\frac{\pi}{16}\right)\right],$

$2\left[\cos\left(\frac{9\pi}{16}\right) + i \sin\left(\frac{9\pi}{16}\right)\right], 2\left[\cos\left(-\frac{7\pi}{16}\right) + i \sin\left(-\frac{7\pi}{16}\right)\right], 2\left[\cos\left(-\frac{15\pi}{16}\right) + i \sin\left(-\frac{15\pi}{16}\right)\right]$ **14** $2e^{-i\pi/3}, 2e^{i\pi/6}$

15 b) $2^{\frac{1}{6}}, \pi/12; 2^{\frac{1}{6}}, 3\pi/4; 2^{\frac{1}{6}}, -7\pi/12$

16 $64, \pi; 2\sqrt{2}\left[\cos\left(\frac{\pi}{4}\right) + i \sin\left(\frac{\pi}{4}\right)\right], 2\sqrt{2}\left[\cos\left(\frac{3\pi}{4}\right) + i \sin\left(\frac{3\pi}{4}\right)\right], 2\sqrt{2}\left[\cos\left(-\frac{3\pi}{4}\right) + i \sin\left(-\frac{3\pi}{4}\right)\right], 2\sqrt{2}\left[\cos\left(-\frac{\pi}{4}\right) + i \sin\left(-\frac{\pi}{4}\right)\right];$

$2 + 2i, -2 + 2i, -2 - 2i, 2 - 2i$ **17 a)** $\sqrt{2}e^{i\pi/20}, \sqrt{2}e^{9i\pi/20}, \sqrt{2}e^{17i\pi/20}$ **18 ii)** $\frac{1}{13}e^{3x}(3\cos 2x + 2\sin 2x), \frac{1}{13}e^{3x}(3\sin 2x - 2\cos 2x)$

18 iii) $\frac{1}{13}e^{3x}(3\cos 2x + 2\sin 2x) + c, \frac{1}{13}e^{3x}(3\sin 2x - 2\cos 2x) + c$ **19 b)** $1 + e^{3i\pi/5}, 1 + e^{i\pi}$ (or 0), $1 + e^{-3i\pi/5}, 1 + e^{-i\pi/5}, 1 + e^{i\pi/5}$

19 c) $(1, 0), 1$ **d) i)** $\frac{\pi}{10}$ **ii)** $2\cos\left(\frac{\pi}{10}\right)$ **20 i)** $4 - 2i, 4 + \sqrt{3} + i, 4 - \sqrt{3} + i; 3\sqrt{3}$ **ii)** $-i; -11, -50$

21 $\cos\left(\frac{2r\pi}{5}\right) + i \sin\left(\frac{2r\pi}{5}\right)$ $(r = 0, 1, 2, 3, 4)$ **i)** $\cos\left(\frac{\pi}{2}\right) + i \sin\left(\frac{\pi}{10}\right), \cos\left(\frac{\pi}{2}\right) + i \sin\left(\frac{\pi}{2}\right), \cos\left(\frac{9\pi}{10}\right) + i \sin\left(\frac{9\pi}{10}\right),$

$\cos\left(\frac{13\pi}{10}\right) + i \sin\left(\frac{13\pi}{10}\right), \cos\left(\frac{17\pi}{10}\right) + i \sin\left(\frac{17\pi}{10}\right)$ **22 i)** $\sqrt{2}e^{i\pi/4}, \sqrt{2}e^{7i\pi/12}, \sqrt{2}e^{11i\pi/12}, \sqrt{2}e^{-3i\pi/4}, \sqrt{2}e^{-5i\pi/12}, \sqrt{2}e^{-i\pi/12}$

ANSWERS

Exercise 15C

1 a) $2i \sin \theta$ **b)** $2 \cos 4\theta$ **c)** $2 \cos 5\theta$ **d)** $2i \sin 2\theta - 2i \sin \theta$ **2 a)** $\frac{1}{2}\left(z^6 + \frac{1}{z^6}\right)$ **b)** $\frac{1}{2i}\left(z^5 - \frac{1}{z^5}\right)$ **c)** $\frac{1}{2}\left(z^4 + \frac{1}{z^4}\right)$

2 d) $\frac{1}{2i}\left(z^3 - \frac{1}{z^3}\right)$ **e)** $-\frac{1}{4}\left(z^5 - \frac{1}{z^5}\right)^2$ **f)** $\frac{1}{16}\left(z^3 + \frac{1}{z^3}\right)^4$ **3 a)** $32 \cos^6\theta - 48 \cos^4\theta + 18 \cos^2\theta - 1$ **b)** $8 \cos^4\theta - 8 \cos^2\theta + 1$

3 c) $8 \cos^3\theta - 4 \cos \theta$ **d)** $32 \cos^5\theta - 32 \cos^3\theta + 6 \cos \theta$ **4 a)** $3 \sin \theta - 4 \sin^3\theta$ **b)** $16 \sin^5\theta - 20 \sin^3\theta + 5 \sin \theta$

4 c) $-64 \sin^6\theta + 80 \sin^4\theta - 24 \sin^2\theta + 1$ **d)** $16 \sin^4\theta - 12 \sin^2\theta + 1$ **5 a)** $\frac{3}{4} \sin \theta - \frac{1}{4} \sin 3\theta$ **b)** $\frac{1}{4} \cos 3\theta + \frac{3}{4} \cos \theta$

5 c) $\frac{1}{16} \cos 5\theta + \frac{5}{16} \cos 3\theta + \frac{5}{8} \cos \theta$ **d)** $\frac{1}{16} \sin 5\theta - \frac{5}{16} \sin 3\theta + \frac{5}{8} \sin \theta$ **e)** $\frac{1}{32} \cos 6\theta + \frac{3}{16} \cos 4\theta + \frac{15}{32} \cos \theta + \frac{5}{16}$

7 $\tan\left(\frac{\pi}{12}\right)$, $\tan\left(\frac{5\pi}{12}\right)$, $\tan\left(\frac{3\pi}{4}\right)$ **8** $\frac{27}{8}$, $\frac{7}{8}$ **9 vi)** $-1 + 2i$ **13 a)** $4 \cos^3\theta \sin \theta - 4 \cos \theta \sin^3\theta$ **b)** $z^4 = \cos 4\theta + i \sin 4\theta$

13 c) $\cos 4\theta = \cos^4\theta - 6 \cos^2\theta \sin^2\theta + \sin^4\theta$ **d)** $8, 4, 2, 1$ **14** $5, -10, 1; \pm 2 \cos\left(\frac{\pi}{5}\right), \pm 2 \cos\left(\frac{2\pi}{5}\right)$

15 $\cos \theta + i \sin \theta$, $\cos\left(\frac{2\pi}{5}\right) + i \sin\left(\frac{2\pi}{5}\right)$, $\cos\left(\frac{4\pi}{5}\right) + i \sin\left(\frac{4\pi}{5}\right)$, $\cos\left(-\frac{2\pi}{5}\right) + i \sin\left(-\frac{2\pi}{5}\right)$, $\cos\left(\frac{4\pi}{5}\right) - i \sin\left(\frac{4\pi}{5}\right)$

15 b) $(z - 1)\left[z^2 - 2z \cos\left(\frac{2\pi}{5}\right) + 1\right]\left[z^2 - 2z \cos\left(\frac{4\pi}{5}\right) + 1\right]$ **c)** $\frac{1}{4}(\sqrt{5} - 1)$

17 $e^{5i\pi/6}$, $e^{-i\pi/2}$, $e^{-i\pi/6}$, $e^{-5i\pi/6}$, $e^{i\pi/2}$; $(z^2 + \sqrt{3}z + 1)(z^2 - \sqrt{3}z + 1)(z^2 + 1)$

Exercise 15D

1 a) w lies on circle, centre O, radius 25 **b)** w lies on that part of real axis with argument 0

1 c) w lies on that part of real axis with argument π **2 a)** w lies on circle, centre O, radius $\sqrt{5}$

2 b) w lies on circle, centre O, radius $\sqrt{2}$ **c)** w lies on line $u = v$ **3** $v^2 = 4k^2(u + k^2)$ **4 a)** Circle, centre $-\frac{1}{4}i$, radius $\frac{3}{4}$

4 b) Imaginary axis **c)** $\pm\frac{1}{2}\sqrt{3} + \frac{1}{2}i$ **5** $|w + 5 - 2i| = 12$ or circle, centre $(-5, 2i)$, radius 12 **6 a)** $4, 9, -4$ **b)** $(4, 0), 5$

7 Circle $|w - \frac{9}{5} - \frac{21}{20}i| = \frac{3}{4}\sqrt{17}$ **8** Ellipse: $\left(\frac{4u}{31}\right)^2 + \left(\frac{3v}{23}\right)^2 = 1$ **9** Straight line: $w = 3$ **11 i)** $2e^{i\pi/2}$, $2e^{-i\pi/6}$, $2e^{-5i\pi/6}$

11 ii) b) Circle, centre -3, radius 2; circle, centre $9i$, radius 6; circle, centre $-3i$, radius 2 **12** $w = 0$ **14 b) ii)** $\pi/8, 5\pi/8$

15 $1 + \dfrac{x}{x^2 + y^2}$, $-\dfrac{y}{x^2 + y^2}$

Exercise 16

1 $\frac{125}{78}$ **2** $\dfrac{(1 + e^2)^{3/2}}{e}$ **3** $\dfrac{5\sqrt{5}}{4\sqrt{3}}$ **4** $2\sqrt{2}$ **5** $\dfrac{13\sqrt{13}}{6}$ **6** $\dfrac{17\sqrt{17}c}{16}$ **7** ∞ **8** $\dfrac{\sqrt{3}}{2}a$ **9 a)** $3\psi^2 - \sin \psi$ **b)** $3 + 4 \sin \psi + 4\psi \cos \psi$

9 c) $\cos \psi - \psi \sin \psi + 2\psi$ **10** $s = \ln |\sec \psi + \tan \psi|$ **11 a)** $-a \sin \psi$ **12 b)** 3.81 **c)** $\dfrac{9\pi}{2}(e^2 + 4 - e^{-2})$ **13 ii)** $(42, 26\frac{1}{4})$

13 iii) $16\sqrt{3}$ **iv)** $\left(384 - \dfrac{192\sqrt{3}}{5}\right)\pi$ **14 i)** $5e^\alpha - 5$ **iii)** $\frac{5}{2}e^\alpha$

Exercise 17A

5 $3 \times 3 = 9$, set is not closed **6** No inverse of -1

Exercise 17B

1 i) G_1:1, G_2:0 **ii)** Because 0 does not have an inverse under $\times$ **iii)** $f(xy) = \ln(xy) = \ln x + \ln y = f(x) + f(y)$

2 i) G_1

$\times$	1	3	7	9
1	1	3	7	9
3	3	9	1	7
7	7	1	9	3
9	9	7	3	1

G_2

$\times$	1	5	7	11
1	1	5	7	11
5	5	1	11	7
7	7	11	1	5
11	11	7	5	1

G_3

$\times$	1	3	5	7
1	1	3	5	7
3	3	1	7	5
5	5	7	1	3
7	7	5	3	1

G_2 and G_3 isomorphic. Isomorphism G_2 to G_3: $1 \to 1, 5 \to 3, 7 \to 5, 11 \to 7$

2 ii) G_1: $x = 1, 9$; G_2: $x = 1, 5, 7, 11$; G_3: $x = 1, 3, 5, 7$

3 Shear in x-direction moving $(0, 1)$ to $(n, 1)$ **4 i)**

	1	3	5	7
1	1	3	5	7
3	3	1	7	5
5	5	7	1	3
7	7	5	3	1

ii) $\{1\}, \{1, 3\}, \{1, 5\}, \{1, 7\}$

4 iii)

	f	g	h	k
f	f	g	h	k
g	g	f	k	h
h	h	k	f	g
k	k	h	g	f

iv) Yes. Composition table of $\{f, g, h, k\}$ obtained from composition table of $\{1, 3, 5, 7\}$ by replacing 1 with f, 3 with g, 5 with h, and 7 with k.

5 a)

	1	3	5	7
1	1	3	5	7
3	3	1	7	5
5	5	7	1	3
7	7	5	3	1

b) Not closed, $2 \times 4 = 0$ (mod 8)   **c)** Whenever n is not a prime

6 H subgroup of G **8 i)**

	I	A	B	C	D	E
I	I	A	B	C	D	E
A	A	B	I	E	C	D
B	B	I	A	D	E	C
C	C	D	E	I	A	B
D	D	E	C	B	I	A
E	E	C	D	A	B	I

iii) $\{I\}, \{I, C\}, \{I, D\}, \{I, E\}, \{I, A, B\}$

iv) a) No   **b)** Yes   **c)** No

9 i) 2, 2   **ii)** $\{e, a, b, c\}, \{e, a, bc, abc\}, \{e, b, ac, abc\}$ and $\{e, c, ab, abc\}$ **iii)** Lagrange's theorem, 8 not divisible by 3

9 iv)

Element	0	1	2	3	4	5	6	7
Order	1	8	4	8	2	8	4	8

v) No. Three elements of order 2 in G, only one in H.

10 i)

Element	e	g	g^2	g^3	g^4	g^5
Order	1	6	3	2	3	6

ii)

Element	i	h_1	h_2	h_3	h_4	h_5
Order	1	3	3	2	2	2

Proper subgroups of H: $\{i\}, \{i, h_1, h_2\}, \{i, h_3\}, \{i, h_4\}, \{i, h_5\}$

10 iii) M isomorphic to G. Correspondence is: $\begin{matrix} e & g & g^2 & g^3 & g^4 & g^5 \\ 1 & 4 & 3 & 12 & 9 & 10 \end{matrix}$ or $\begin{matrix} e & g & g^2 & g^3 & g^4 & g^5 \\ 1 & 10 & 9 & 12 & 3 & 4 \end{matrix}$ **11 i)** 4, 4

11 ii) $\{e, p_1, p_2, p_3\}$ and $\{e, p_2, q_1, q_2\}$ **iii)** No, 8 not divisible by 6   **v)** G isomorphic to K **12 b)**

8	10	14	16
4	14	16	8
2	16	8	4

12 c) 10; $\{10, 8\}, \{10, 4, 16\}$ **d)** Yes   **13 i)** E_1 one-way stretch in x-direction, scale factor $\frac{1}{3}$. E_2 shear

13 ii) $E_1A = \begin{pmatrix} 1 & \frac{2}{3} \\ 1 & 4 \end{pmatrix}$, $E_2(E_1A) = \begin{pmatrix} 1 & \frac{2}{3} \\ 0 & \frac{10}{3} \end{pmatrix}$, $E_3(E_2E_1A) = \begin{pmatrix} 1 & \frac{2}{3} \\ 0 & 1 \end{pmatrix}$

13 iii) $\mathbf{E}_1^{-1} = \begin{pmatrix} 3 & 0 \\ 0 & 1 \end{pmatrix}$, $\mathbf{E}_2^{-1} = \begin{pmatrix} 1 & 0 \\ 1 & 1 \end{pmatrix}$, $\mathbf{E}_3^{-1} = \begin{pmatrix} 1 & 0 \\ 0 & \frac{10}{3} \end{pmatrix}$, $\mathbf{E}_4^{-1} = \begin{pmatrix} 1 & \frac{2}{3} \\ 0 & 1 \end{pmatrix}$ **iv)** $\mathbf{A} = \mathbf{E}_1^{-1}\mathbf{E}_2^{-1}\mathbf{E}_3^{-1}\mathbf{E}_4^{-1}$

14 i)

	3	6	9	12
3	9	3	12	6
6	3	6	9	12
9	12	9	6	3
12	6	12	3	9

6 self-inverse **ii)** $-1, \frac{17}{3}$ **15 ii)** $\mathbf{B}, \mathbf{I}, \mathbf{A}^2\mathbf{B}$

15 iii) $\mathbf{I}$, 1; $\mathbf{A}$, 4; $\mathbf{A}^2$, 2; $\mathbf{A}^3$, 4; $\mathbf{B}$, 2; $\mathbf{AB}$, 2; $\mathbf{A}^2\mathbf{B}$, 2; $\mathbf{A}^3\mathbf{B}$, 2 **v)** $\{\mathbf{I}, \mathbf{A}, \mathbf{A}^2, \mathbf{A}^3\}$ and $\{\mathbf{I}, \mathbf{AB}, \mathbf{A}^3\mathbf{B}, \mathbf{A}^2\}$ **vi)** Only $\{\mathbf{I}, \mathbf{A}, \mathbf{A}^2, \mathbf{A}^3\}$ cyclic

16 a) By Lagrange's theorem: if **X** has n elements, then 2 divides n and so does 4. Minimum 8

17 b) $\mathbf{M}(\frac{1}{3}\mathrm{i})$ has order 4. Required groups are: group of order 4: $\{\mathbf{M}(\frac{1}{3}\mathrm{i}), \mathbf{M}(-\frac{1}{3}), \mathbf{M}(-\frac{1}{3}\mathrm{i}), \mathbf{M}(\frac{1}{3})\}$; group of order 2: $\mathbf{M}(-\frac{1}{3}), \mathbf{M}(\frac{1}{3})\}$

17 d) Inverse of $\mathbf{M}(\frac{2}{3})$ is $\mathbf{M}(\frac{1}{6})$, which is not in S. Hence, S is not a group, and so cannot be a subgroup of G.

18 a) For A, let $\theta = 0$; for B, let $\theta = \frac{\pi}{2}$. Then $A^2 = B^2 = I$, but $(AB)^2 = -I$ **d) i)** D_8 not abelian since $q * a \neq a * q$

18 d) ii) Set has 6 elements. Because 8 not divisible by 6, there is no subgroup of D_8 with 6 elements.

19 iii)

Second ○ First	π_1	π_2	π_3	π_4	π_5	π_6
π_1	π_1	π_2	π_3	π_4	π_5	π_6
π_2	π_2	π_3	π_1	π_6	π_4	π_5
π_3	π_3	π_1	π_2	π_5	π_6	π_4
π_4	π_4	π_6	π_5	π_1	π_2	π_3
π_5	π_5	π_4	π_6	π_3	π_1	π_2
π_6	π_6	π_5	π_4	π_2	π_3	π_1

iv) Symmetries of equilateral triangle

20 iii) 0 **iv)** No inverse of -1 **v)** Delete -1

Exercise 17C

2 $\mathbf{r}\cdot\begin{pmatrix} 3 \\ 2 \\ -3 \end{pmatrix} = 0$, $\begin{pmatrix} 2 \\ 3 \\ 4 \end{pmatrix}$ and $\begin{pmatrix} 1 \\ 0 \\ 1 \end{pmatrix}$ **4** For example: $\begin{pmatrix} 1 \\ 1 \\ 0 \end{pmatrix}$ **5** $\begin{pmatrix} 2 & 1 \\ 1 & 3 \end{pmatrix}, \begin{pmatrix} 1 & 3 \\ 5 & 4 \end{pmatrix}$

6 a) Points on line $2y = x$ **b)** 1 **c)** $\begin{pmatrix} 2 \\ 1 \end{pmatrix}$ **8 b)** Point O **9 iii)** $\mathbf{M} = \begin{pmatrix} a+c & d-b \\ b+d & a-c \end{pmatrix}$ **iv)** $z = -2 + 13\mathrm{j}$

Index